CASSELL'S

CARPENTRY AND JOINERY

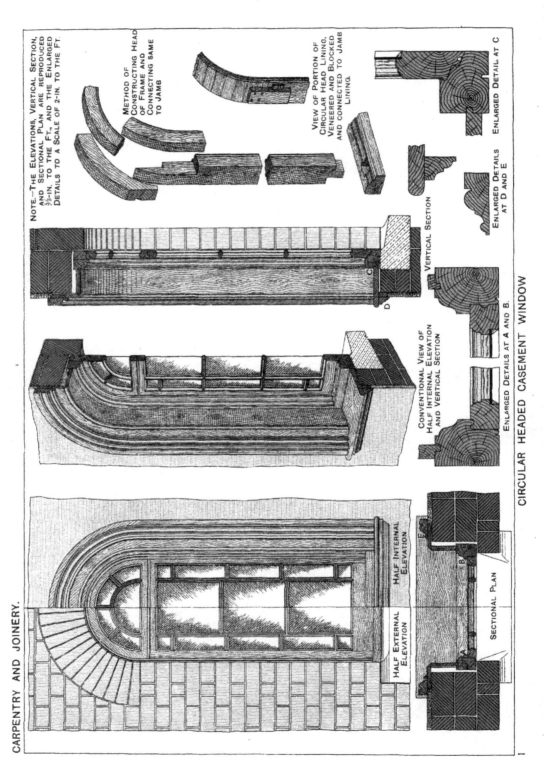

CARPENTRY AND JOINERY.

NOTE—THE ELEVATIONS, VERTICAL SECTION, AND SECTIONAL PLAN ARE REPRODUCED ⅜-IN. TO THE FT., AND THE ENLARGED DETAILS TO A SCALE OF 2-IN. TO THE FT.

METHOD OF CONSTRUCTING HEAD OF FRAME AND CONNECTING SAME TO JAMB

VIEW OF PORTION OF CIRCULAR HEAD LINING, VENEERED AND BLOCKED AND CONNECTED TO JAMB LINING

ENLARGED DETAIL AT C

ENLARGED DETAILS AT D AND E

VERTICAL SECTION

CONVENTIONAL VIEW OF HALF INTERNAL ELEVATION AND VERTICAL SECTION

ENLARGED DETAILS AT A AND B.

HALF EXTERNAL ELEVATION

HALF INTERNAL ELEVATION

SECTIONAL PLAN

CIRCULAR HEADED CASEMENT WINDOW

CASSELL'S

CARPENTRY AND JOINERY

COMPRISING NOTES ON MATERIALS, PROCESSES, PRINCIPLES,
AND PRACTICE, INCLUDING ABOUT 1,800 ENGRAVINGS
AND TWELVE PLATES

EDITED BY

PAUL N. HASLUCK

Editor of " Building World," etc

WITH A NEW INTRODUCTION BY

ROY UNDERHILL

CASSELL AND COMPANY, LIMITED
LONDON, PARIS, NEW YORK & MELBOURNE   MCMVII

REPRINTED BY BT&C BOOKS 2010
www.brooklyntoolandcraft.com

ISBN: 978-0-9828632-0-6

Published by: BT&C Books
32 33rd street 5th floor
Brooklyn NY 11232

Please direct all enquiries to: support@brooklyntoolandcraft.com

The original 1907 edition of this book contained twelve color plates which, due to the cost of color reproduction, are reproduced here, in their appropriate locations, in black and white. To download a PDF of the original plates in color, please visit http://brooklyntoolandcraft.com/hasluckplates.

Safety is your responsibility. This book was originally published over a century ago and the instructions in this book do not follow current safety procedures. The publisher and author cannot assume any responsibility for injuries suffered, damages, or other losses incurred that may result from the material presented in this book.

INTRODUCTION

Paul Hasluck, editor of this volume (and countless others) states in his preface that "Students preparing for examinations in which Carpentry & Joinery are involved will welcome this as a textbook of the utmost value and importance." By 1907 when this work was published, indeed you might learn carpentry and joinery from books and formal instruction. Still hanging on, though, particularly in England, was the old way of learning and of working—through apprenticeship. These two worlds existed side by side, and so, by way of introduction to this fine text, I have chosen to interweave my appreciation of Hasluck's *Carpentry & Joinery* with the experiences of a young apprentice of that same time and place.

I have a copy of an unpublished memoir of Robert Simms as recorded by Angus Murdoch which gives us an unvarnished view of the working conditions of a hardscrabble English joiner in the early twentieth century. The things he learned in the school of hard knocks seem far removed from the ideal world of the textbook. For example, one job that young Simms despised was repair work at pubs. "The oak sills on the doors of the pubs, the busiest business in England, used to wear down about two inches in the middle. You had to get out there at 6 o'clock in the morning, make the sill and measure it and tear the old one out before they opened at ten. Boy, I botched up a lot of those. I used to bang a wedge underneath and get out of there. You learned to do things like that."

Wedging up a sill rather than replacing it is plainly a cheat as well as a quick fix, but in *Carpentry and Joinery* we find plenty of ethical work-saving nuance. When you saw out the waste between the double tenons of door rails, for example, saw it out in the shape of the wedges that you'll use later. After you have glued up the door, rub olive oil (not linseed!) on your tenon saw to keep it from gumming up. In this book, you'll learn the fast teamwork required to clamp up a door with hot hide glue, along with the methods to lay out both the stiffest and the strongest beam from a round log, and how to line out a log for pit sawing. You'll learn the arcana of the sticking board for hand-planed moldings as well as the techniques for faux half-timber construction. All told with the voice of experience.

The voices that instructed young Robert Simms were not always helpful. The experienced joiners Simms learned from "drank beer all day at work. The older ones had a two-gallon copper water can which the boy had to take down to the pub to get filled up. There was a tin cup on each end of the bench, and if he hollered, you had to go and fill the cup. It used to make me shudder, at 6 o'clock in the morning they'd come in and drink an ice-cold beer."

Between refills, the old joiners kept Simms in his place. "The standard rule was, and I was taught this from the start, that unless you could make a four panel door in ten hours, you were no damned good at all. Now that means ripping your stock out, planing it, mortices and tenons, setting it up, making your panels and putting it together. Well, I made hundreds of doors in my time and when I heard these stories I tried to beat them. I never got within miles." Only years later, when he confessed his inadequacy to a sympathetic older joiner, was he told that this was just the old guys way of razzing him. "Don't believe everything people tell you." Again, the voice of experience.

That's where Hasluck's *Carpentry and Joinery* strikes just the right note. Unlike Joseph Moxon's outsider stance (albeit spiced with poetry and earwax jokes) and Peter Nicholson's academic obtuseness, Paul Hasluck turned to experienced workmen who were also experienced writers. The sub-section "Preparing Joiners' Work principally by Hand" is the best step-by-step guide to making a panel door that has ever seen print. Starting with advice to first cut the "black end" off of any timber to protect your plane irons from being notched by "dirt brought, perhaps, from the timber perch," the text and illustrations then carry us through making the frame and preparing the panels. "Cut them out in the rough, jack them over, and bend each one on the edge of the bench. If there should be a shake it will betray itself." (The instruction to "jack them over" assumes a complementary level of knowledge in the reader.) The author continues; "… for a first-class job the panel would be rejected; for commoner work, however, these shakes should have a little whiting and glue rubbed in with a hammer and allowed to dry."

"Hand manufacture has now given place largely to machine work," says our author, but he then goes on to show us how to block in cornice moldings with plow planes and complete the work with hollows and rounds. This is not a holdover, but the practical reality of custom work for fine homes. As Simms recalled "When you got a big job come in, the big contracts like a super-duper new house, the architect would bring the plans and then we would make it all in the shop. We not only made the joinery, we also made all the moldings for these houses. It wouldn't pay them to set out a machine to do it unless they were going to make thousands of feet of it. So we made all the moldings, all the doors - everything. I never even saw the houses. I worked with the architects' plans, where they gave you the measurements to a sixteenth of an inch, and if you copied their drawings everything was going to fit in nice and right."

That's the trick, isn't it? Joinery has to fit, not only within itself, but within the work of the masons and carpenters as well. Simms never mentions joiner's rods in his memoir, but "Carpentry and Joinery" devotes a full chapter to them. These two boards, one showing all the vertical dimensions and one showing the horizontal, encoded all the dimensions for doors, windows and casework. Made from the architect's drawings and models, they guided the joiners in the shop, and in moments of doubt, could be taken to the worksite for a trial fitting.

Building alterations also required visits to the worksite as well as skilled handwork. After the First World War, much of the English upper crust could no longer afford to live in big Georgian mansions, and these were being broken up into separate apartments. Simms was on the job. "The interior rooms had these massive big cornices, 18 inches deep. When they altered the rooms they had to have lengths of all these different moldings to make up where they put partitions in. Most of the shops either couldn't or wouldn't make the moldings by hand, but we did. They worked my butt off, you know."

It's still hard work today, and that's why every little hint that saves you from error and wasted labor is so valuable. It's also part of the pleasure of this undertaking. As Simms put it, "There was so much nuance in the business that if you didn't have a little brain power you'd kill yourself. You had to know every angle in order to save yourself a little bit more of hard work. If your attitude was right and you enjoyed your work, it was a pleasure to be alive. Smelling and feeling the wood and every piece you plane up has a different character. You can spot all the differences and it keeps your interest alive. Maybe I was crazy. I don't know."

We're crazy too, perhaps, but with this fine book to guide us, we will at least have first-class method to our madness. So I'm going to build myself a timber belfry. I'm going to build it atop a half-timbered house with great hammer beam trusses, lots of gables and round windows with tilting sash. I'll make all the six paneled doors and wainscot and then add on a conservatory with skylights and folding partitions. Yes, and some double-margined doors with circular frames and splayed linings to boot. Or, perhaps, I'll just keep reading.

Invest just a few minutes looking through this wonderful volume and your timber castles will surely grow with every page. We again have the voices of first rate masters at our sides.

And now—back to work.

By Roy Underhill May 2010

Former master housewright at Colonial Williamsburg,
host of the "Woodwright's Shop" on Public Television and
author of numerous books on traditional woodworking.

PREFACE.

CASSELL'S CARPENTRY AND JOINERY is a practical work on practical handicrafts, and it is published in the confident belief that it is by far the most exhaustive book on these subjects hitherto produced. Throughout this book actual practice is recorded, mere discussion of theory has been excluded, except where it is essential in explaining the principles underlying a method, a process, or the action of a tool. The tools and processes described are those commonly found in daily use in the workshop. The expert and well-informed reader will of course make due allowance for the great diversities of trade practice in different localities.

Much of the matter appearing in these pages has been written and illustrated by eminent authorities such as Mr C W D Boxall, Prof Henry Adams, Mr F W Loasby, and several other well-known practical contributors to "Building World." The names of these experts are a guarantee of competency and thoroughness.

Students preparing for examinations in which Carpentry and Joinery are involved will welcome this as a text book of the utmost value and importance, and its intensely practical character—in every possible instance the information is imparted by investigating, describing, and illustrating cases that have occurred in actual experience—renders the work extremely useful as a guide to everyday practice, as the volume includes virtually everything that relates to the materials, processes, principles, and practice of Carpentry and Joinery.

The comprehensive scope of the work is evident from a glance at the list of contents. Each of the various sections is dealt with in exhaustive detail, some of the sections extending to nearly 100 pages, and the studiously plain language used throughout the book is further assisted by the use of skilfully drawn diagrams, which are supplemented by twelve full-page coloured plates.

P N HASLUCK

CONTENTS.

LIST OF COLOURED PLATES.

CARPENTRY AND JOINERY.

HAND TOOLS AND APPLIANCES.

Introduction.

The reader of this book is assumed to have some acquaintance with woodworking, and not to stand in need of detail instruction as to the shape, action, care and use of each and all of a woodworker's tools. This information is given in comprehensive style in a companion volume, entitled "Woodworking," produced by the Editor of this present book, and sold by the same

Fig. 1.—Two-foot Rule with Slide Rule.

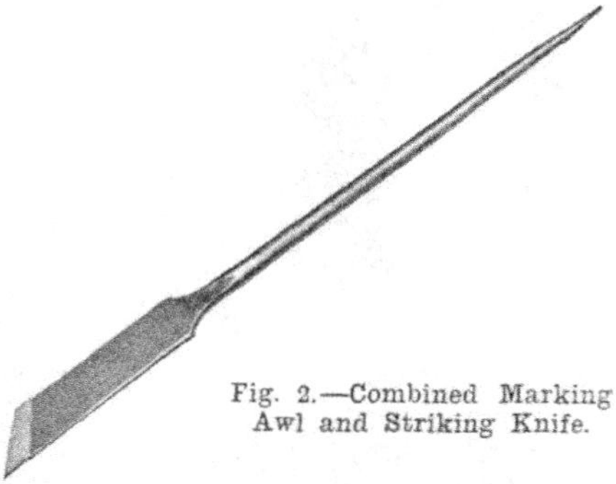

publishers at 9s. Should any reader of this chapter desire further particulars of the tools and appliances here briefly mentioned, he is recommended to consult that work, which undoubtedly contains the most complete description of woodworking tools yet published.

Classification of Tools.

Tools may be classed according to their functions and modes of action, as follows : (1) Geometrical tools for laying off and testing work—such tools are rules, straight-edges, gauges, etc. (2) Tools for supporting and holding work ; such tools are benches, vices, stools, etc. (3) Paring or shaving tools, such as chisels, spokeshaves, planes, etc. (4) Saws. (5) Percussion or impelling tools, such as hammers, mallets,

Fig. 2.—Combined Marking Awl and Striking Knife.

screw-drivers, and (combined with cutting) hatchets, axes, adzes, etc. (6) Boring tools, such as gimlets, brace-bits, etc. (7) Abrading and scraping tools, such as rasps, scrapers, glasspaper, and implements such as whetstones, etc., for sharpening edged tools.

Geometrical Tools.

Rules.—For all-round purposes a 2-ft. four-fold boxwood rule, with or without a slide rule (Fig. 1). is best. Rules are made in great variety, but the average worker's requirements will be best met by a simple one.

1

Fig. 3.—Straight-Edge.

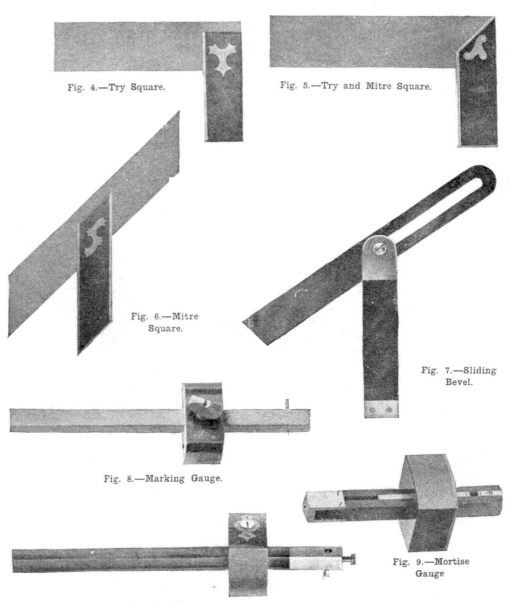

Fig. 4.—Try Square.

Fig. 5.—Try and Mitre Square.

Fig. 6.—Mitre Square.

Fig. 7.—Sliding Bevel.

Fig. 8.—Marking Gauge.

Fig. 9.—Mortise Gauge.

Fig. 10.—Cutting Gauge.

Marking and Scribing.—The carpenter's lead pencil is of a flat, oval section, sharpened to a chisel edge, which has a greater body of lead than a point, and lasts a reasonable time before requiring to be re-sharpened. The marking awl and the striking knife, shown in Fig. 2 as a combined tool, is used by joiners principally for scribing, or cutting-in, the shoulders of framing, etc. Greater accuracy can be attained and a sharper arris is left when sawing than when working to a pencil line.

Straight-edge.—Mechanics in the building trades use a straight-edge, usually made to the shape shown by Fig. 3, and not larger than 15 ft. long, 6 in. wide, and 1¼ in. thick, made from a pine board cut from a straight-

Fig. 11.—Panel Gauge.

grown tree. All straight-edges should be tested occasionally.

Squares and Bevels.—These are used for setting out and testing work. The joiner's steel square is a mere right angle of steel, sometimes nickel-plated, graduated in inches, ¼ in. and ¹⁄₁₆ in., or otherwise as required. The try square (Fig. 4) has a rosewood or ebony stock. The tool shown by Fig. 5 is also of use in setting out and testing mitres, but the proper mitre square is shown by Fig. 6. Try squares are also made with iron frames which are channelled and perforated to reduce weight. Adjustable squares with graduated blades are useful in putting fittings on doors and windows. By means of the sliding bevel (Fig. 7) angles are set off in duplicate; the set screw secures the blade at any desired angle with the stock. A crenellated square has a blade which is notched at every principal graduation, and is used chiefly for setting out mortises and tenons.

Marking Work for Sawing.—The chalk line is used for long pieces of timber, the pencil and rule for ordinary applications, and the scribe for particular work. The "chalk line" is a piece of fine cord rubbed with chalk or black pigment, and strained taut. To mark the work the chalk line is lifted vertically and near the centre, and when released makes a fine and perfectly straight line upon the work. Coloured chalks and pigments are also used.

Marking and Cutting Gauges.—Ordinarily the carpenter draws a line close to and parallel to the edge of a board by means of a rule held in one hand, with the fore-

Fig. 12.—Wing Compasses.   Fig. 13.—Spring Dividers.

finger against the edge of the work and the pencil held close against the end of the rule; but the marking gauge (Fig. 8) gives more accurate results. The gauge may have a pencil point instead of the steel point shown. Developments are the mortise gauge (Fig. 9) and the cutting gauge (Fig. 10), having either a square or oval sliding stock or head. The panel gauge (Fig. 11) is used to mark a line parallel to the true edge of a panel or of any piece of wood too wide for the ordinary gauge to take in.

Compasses, Dividers, and Callipers.—Ordinary wing compasses (Fig. 12) are generally used, but for particular work instruments with fine or sensitive adjustments are obtainable. Spring dividers (Fig. 13) are used for stepping off a number of equal distances, for transferring measurements and for scribing. Callipers (Figs. 14 and 15), obtainable in many styles, are used for

Fig. 14.—Outside Callipers.   Fig. 15.—Inside Callipers.

measuring diameters of cylindrical solids and recesses.

Shooting Boards.—The shooting board (Fig. 16) is used for trueing up with a plane the edges of square stuff. That shown is the simplest possible, but other and improved shapes are obtainable.

Appliances for Mitreing.—The simplest appliance used in cutting mitres is the ordinary mitre block, the work being laid

for shooting or planing the mitred ends of stuff previously sawn in the mitre block or box; in the illustration the rebate or bed for the work is cut out of the solid, but it is general to build up the block with three thicknesses of stuff, and so avoid cutting a rebate. The donkey's-ear shooting block (Fig. 19) is used for mitreing or bevelling the edges of wide but thin material with the cut at right angles to that adopted for stouter mouldings; another form of

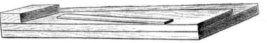

Fig. 16.—Shooting Board.

this block (Fig. 20) has a rest A for the material, a bed B for the shooting plane, a guide C for the plane, and a frame D which is fixed in the bench screw or to the tail of the bench. The mitre template (Fig. 21) is another aid to cutting mitres. Its use will be explained on a later page.

Spirit Levels.—The spirit level is used for determining the planes of the horizon—that is, the plane forming a right angle to the vertical plane. A frame firmly holds a closed glass tube nearly filled with anhydrous ether, or with a mixture of ether and alcohol (see Fig. 22). Good spirit levels have a graduated scale engraved on the glass tube or on a metal rule fastened to the frame beside it. There are many varieties of spirit levels, but all are made on the same principle.

Fig. 17.—Mitre Box with Dovetail Saw.

Fig. 18.—Mitre Shooting Block.

upon a rebate, and saw kerfs in the upper block serving as a guide for the tenon saw. Inclined and other varieties of mitre blocks are in use. The mitre box (Fig. 17) is generally used for broader mouldings. The mitre shooting block (Fig. 18) is used

Plumb Rule and Square.—The plumb rule (Fig. 23) is used by the carpenter and fixer for testing the vertical position of pieces of timber, framing, doorposts, sash frames, etc., which should be fixed upright. The plumb square (Fig. 24) is useful for testing the squareness of work and at the same time the levelness of a head, it being for this purpose sometimes more useful than a spirit level.

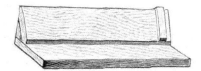

Fig. 19.—Donkey's-ear Shooting Block.

worker and by the kind of work to be done. A joiner's bench of the usual pattern is shown by Fig. 25. It is 12 ft. long, by 2 ft. 6 in. wide, and 3 ft. high. The legs are 4 in. by 4 in. ; bearers and rails, 4 in. by

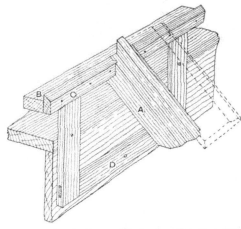

Fig. 20.—Donkey's-ear Block for Shooting Wide Surfaces.

Fig. 22.—Spirit Level.

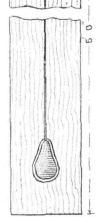

Fig. 23.—Plumb Rule.

Fig. 21.—Mitre Template.

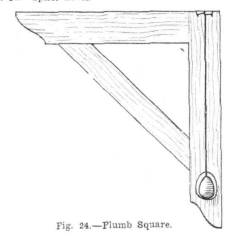

Fig. 24.—Plumb Square.

Tools for Supporting and Holding Work.

Benches. — The ordinary joiner's bench should not be less than about 8 ft. long, 2 ft. 6 in. to 3 ft. high, and 2 ft. 6 in. wide, and should be fitted with wood or iron bench screws so as to accommodate one or two workers. Of course, the height of the bench will be influenced by the stature of the

3 in. ; sides, 1¼ in. by 9 in. ; top, 1½ in. by 9 in. The bench top is mortised at A to

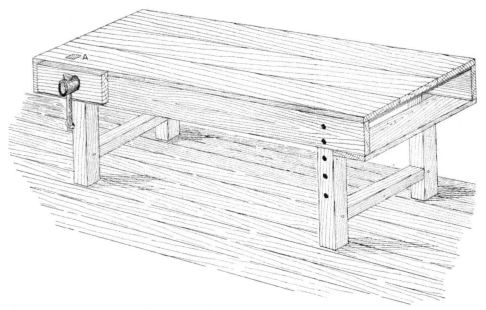

Fig. 25.—Ordinary Pattern Joiner's Bench.

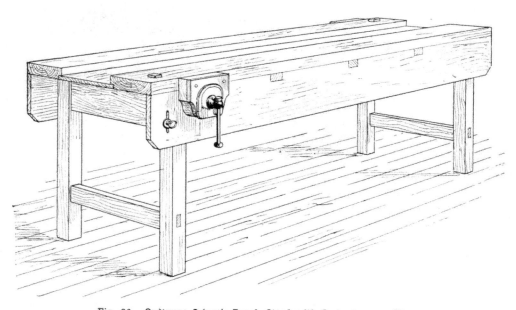

Fig. 26.—Ordinary Joiner's Bench fitted with Instantaneous Vice.

receive the stop, which is of the pattern shown by Fig. 32, so that it fits tightly against the leg of the bench. Fig. 26 shows a joiner's bench of another pattern, fitted with a good form of instantaneous grip vice ; and another variety of this useful

fits the runner shown inside it, is fixed to the top rail connecting the legs, and to the top and side of the bench. Care is taken to keep the runner at right angles

Fig. 28.—Wooden Bench Screw Vice.

Fig. 27.—Bench with Side and Tail Vices.

class of vice is shown in section at Fig. 31. A bench with side and tail vices is illustrated at Fig. 27, and, although not much used by joiners, is a very useful form for small work or as a portable bench. The top and tail vice cheeks contain holes for the reception of bench stops of iron or wood, against which, or between which, work is held for framing, etc.

to the vice cheeks. To fasten the vice outer cheek and screw together, so that upon turning the latter the former will follow it, a groove E is cut. Then from the under edge of the cheek a mortise is made, and a hardwood key is driven to fit fairly tight into the mortise, its end

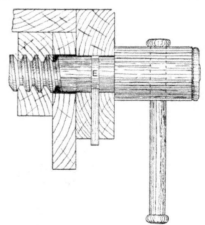

Fig. 30.—Section through Screw Vice.

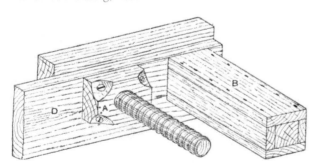

Fig. 29.—Inside View of Screw Vice.

Bench Screws.—A common form of joiner's bench screw is shown by Fig. 28, inside and sectional views being shown respectively by Figs. 29 and 30. D is the side or cheek of the bench to which a wooden nut (A) is screwed. The box B, which accurately

entering E. The screw cheek is usually about 1 ft. 9 in. long, 9 in. wide, and 2 in. to 3 in. thick. The runner is about 3 in. by 3 in. and 2 ft. long. The wooden screws and nuts can be bought ready made. Bench screws are known in great variety, and include an instantaneous grip vice (Fig. 31), a most useful appliance.

Bench Stops.—There are many varieties of iron bench stops on the market, but the ordinary "knock up" stop, which is a piece of hard wood about 2 in. to 2½ in. square, and 9 in. to 18 in. long, fitting tightly into a mortise in the top of the

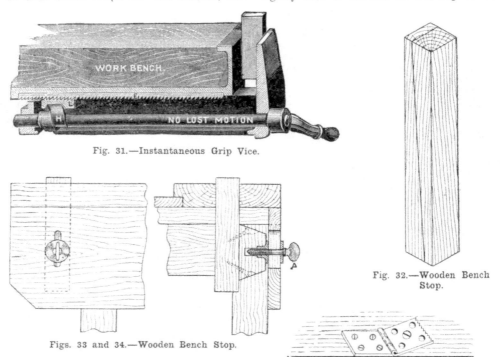

Fig. 31.—Instantaneous Grip Vice.

Figs. 33 and 34.—Wooden Bench Stop.

Fig. 32.—Wooden Bench Stop.

Fig. 35.—Hinge used as Bench Stop.

bench, is still largely used. This is the best form of stop for all ordinary purposes. It is cut wedge shape, as shown by Fig. 32. This stop is raised and lowered by knocking

Fig. 37.—Adjustable Iron Bench Stop.

Fig. 36.—Iron Bench Stop.

Fig. 38.—Morrill's Adjustable Bench Stop.

with a hammer at top or bottom. The plane is not injured if it comes into contact with the stop, which also has greater strength than temporary stops screwed to the face of the bench top. An improved form of this is shown at Figs. 33 and 34. A block is screwed to the stop, and to this the nut of an ordinary shutter bolt is fixed. A slot is cut in the cheek of the bench, as shown. The shoulder of the bolt works against a large washer. This stop can readily be raised or lowered. Two or three steel nails driven in near the top of the stop and filed to form teeth can be used to hold the work. A very useful stop may be contrived, as shown by Fig. 35, by filing one end of a back flap hinge so as to form teeth, the other flap being screwed down to the bench. A long screw through the middle hole in the loose flap affords means of adjustment. By loosening this long screw, the front edge of the stop may be raised,

Fig. 39.—Sawing Stool.

but to retain it in its position it should be packed up with a piece of wood, and the screw tightened down again. A plain iron stop with a side spring to keep it at any desired height is shown by Fig. 36. This form of stop fits into holes mortised through the bench top. Figs. 37 and 38 show good forms of adjustable bench stops that are obtainable from tool-dealers; their principle is fairly obvious on reference to the illustrations.

Sawing Stools or Trestles.—The three-leg sawing stool is of but little service and almost useless for supporting work in course of sawing. Probably one of the best forms of this useful appliance is the four-legged stool shown by Fig. 39. This needs to be built substantially.

Cramps.—A hold-fast for temporarily securing work to the bench is shown by Fig. 40. The old-fashioned hand-screw cramp

1 *

(Fig. 41) is made of wood entirely. It is a very useful tool in the joiner's shop, and is used for holding together pieces of wood when glued for thicknessing up. It is indispensable when glueing up face veneers for shop fittings, etc.; these screws are made in different sizes suitable for heavy and light work. Iron G-cramps are a very useful form, the smaller sizes being made with a thumbscrew (Fig. 42) and being used for light purposes. The stronger and larger kinds will take in work up to 12 in.; greater

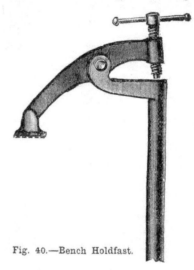

Fig. 40.—Bench Holdfast.

force being required, the screws are rotated by means of the usual lever. The many varieties of iron cramps include the Hammer instantaneous adjustment cramp and a sliding pattern G-cramp, both of which have advantages in many cases over the common G-cramp, a form of which, with thumbscrew, is illustrated by Fig. 42 (p. 10). Sash cramps and joiner's cramps (also shown on p. 10—see Fig. 43) are in common use, a number of patent cramps with special advantages also being known. Figs. 44 and 45 show a useful cramp for thin work. The wedge cramp (Fig. 46), known as a cleat, is also very useful for holding boards together after they have been jointed and glued. The cleats are kept on till the glue in the joint is dry. The wedges prevent the board from casting. Iron dogs (Figs. 47 and 48) are

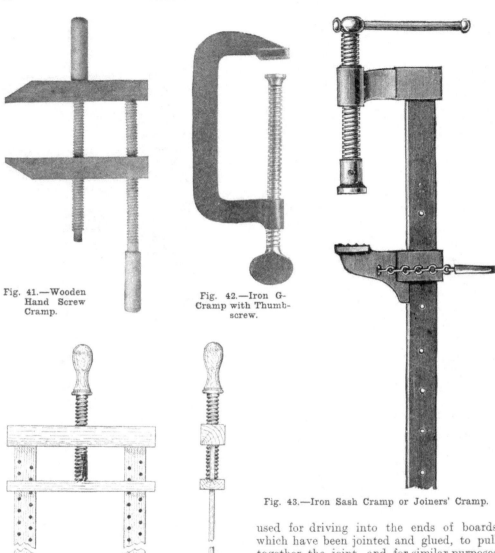

Fig. 41.—Wooden Hand Screw Cramp.

Fig. 42.—Iron G-Cramp with Thumb-screw.

Fig. 43.—Iron Sash Cramp or Joiners' Cramp.

Fig. 44.

Fig. 45.

Figs. 44 and 45.—Wooden Cramp for Thin Work.

used for driving into the ends of boards which have been jointed and glued, to pull together the joint, and for similar purposes by the joiner. They are also used for drawing together face joints when glued, but only in cases where the holes made by the dogs are to be covered afterwards by another piece of lining. A stronger form is also sometimes used by the carpenter for common flooring. The dog is driven into the joist firmly, there being enough space between the dog and the edge of the floorboard to admit a pair of folding wedges, which are then driven tight home, and the floorboards nailed down before removing the dog. More suitable cramps for this purpose are those usually known as floor cramps or dogs, illustrations and particulars of which

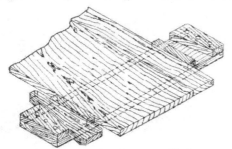

Fig. 46.—Wedge Cramp or Cleat.

will be found in the section on floors. Cramps and similar appliances in less general use, but of importance in special cases, will be illustrated and their use explained, in each of the particular sections to which they belong. For cramping circular work there are many special devices, the flexible steel cramp (Fig. 49) being typical of them. The flexible cramp is shown in use, tightening up

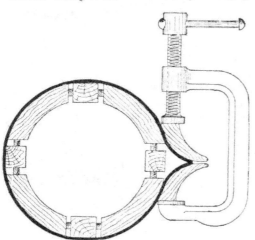

Fig. 49.—Circular Seat with Flexible Cramp.

the four joints used in the construction of a circular seat or other piece of work.

Pincers.—These are used for extracting and beheading nails, and in cases where a form of hand vice is wanted for momentary

use. Two chief patterns are available, Fig. 50 showing the Lancashire pattern. Tower

Fig. 47.—Joiners' Dog.

pincers have a round knob in place of the cone on the end of the handle.

Paring and Shaving Tools.

Chisels.—For full information as to the action of these tools readers are referred to " Woodworking," the companion volume already alluded to. Firmer chisels shown by

Fig. 48.—Dog made from Sheet Steel.

Fig. 51 range from $\frac{1}{8}$ in. to $1\frac{1}{2}$ in. in width, and their use is to cut away superfluous wood in thin chips. The ordinary kind is strong and is made of solid steel, and is used with the aid of a mallet. A lighter form made

Fig. 50.—Lancashire-Pattern Pincers.

with bevelled edges (Fig. 52) is used, generally without a mallet, for fine work and for cutting dovetailed mortises. For paring, a longer chisel is generally employed (Fig. 53). Mortise chisels (Fig. 54) have various shapes, according to their particular uses, and require to be strongly made.

Fig. 51.—Ordinary Firmer Chisel.

Fig. 52.—Firmer Chisel with Bevelled Edges.

Fig. 53.—Long Paring Chisel.

Fig. 54.—Mortise Chisel.

Fig. 55.—Firmer Gouge.

Fig. 56.—Draw Knife.

Fig. 57.—Wooden Spokeshave.

Fig. 58.—Iron Spokeshave.

Gouges.—These have the same action as that of a chisel, but instead of being

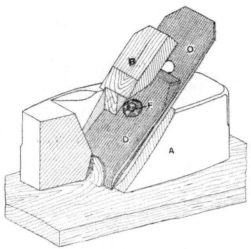

Fig. 59.—Sectional View of Plane.

flat their sections form arcs of circles (see Fig. 55).

Draw Knife.—The draw knife (Fig. 56) is used for roughing stuff to shape preparatory to working with finer tools.

Spokeshaves.—An ordinary spokeshave is merely a knife edge in a suitable holder (Fig. 57); it may jump if the iron is loose, or if the back part of the iron touches the work before the cutting edge.

Planes.—These are the tools chiefly used for smoothing work which has been sawn to approximate size. The simplest plane is a chisel firmly fixed into a wooden block. The construction of an ordinary plane is shown in the sectional view (Fig. 59), in which A shows the stock; B, the wedge; O, cutting iron; D, back iron; F, screw and nut for fastening the cutting and back irons together; the mouth through which the shavings pass upwards is shown. The jack plane (Fig. 60) is the first plane applied to the sawn wood; its parts are: the stock, 17 in. long; the toat, or handle; the wedge; the cutting iron, or cutter, about $2\frac{1}{8}$ in. wide; and the back iron. The trying or trueing plane (Fig. 61) is of similar construction, but is much longer, so as to produce truer surfaces. A still longer trying plane called the jointer is used for jointing boards in long lengths; since the introduction of machinery it is seldom used. The smoothing plane (Fig. 62) smooths the work to form a finished surface; for pine or other soft woods it is 9 in. long, and its iron is $2\frac{1}{4}$ in. wide on the cutting face. Some

Fig. 60.—Jack Plane.

Fig. 61.—Trying Plane.

Spokeshaves are best made with iron stocks and with screws to regulate the cutting iron (Fig. 58).

smoothing planes have iron fronts, as shown in the sectional view, Fig. 63; these can be adjusted for the finest shaving desired.

A good form of iron smoothing plane is shown by Fig. 64; this is intended for superior work. The rebate plane (Fig. 65) is without a back iron, and its cutting iron extends the full width

Fig. 62.—Smoothing Plane.

of the tool, thus enabling the angles of rebates to be cleaned up. Other varieties of planes include the bead plane (Figs. 66 to 68), used for working single and return beads and round rods. Hollows, rounds, etc. (Figs. 69 to 73), are used for working

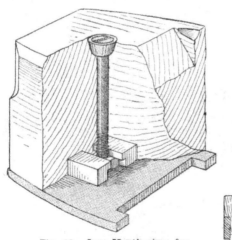

Fig. 63.—Iron Mouth-piece for Smoothing Plane.

straight mouldings of all kinds, but machinery has of late years been increasingly used for such work. Small planes of varying shapes are used for forming mouldings on circular work. The compass plane, used for forming the face of concave ribs, etc., was formerly made of beech wood. The one generally used at the present time

is made of steel entirely. The sole of the plane is about 10 in. long, $2\frac{1}{4}$ in. wide,

Fig. 64.—Iron Smoothing Plane.

and $\frac{1}{32}$ in. thick. It is adjusted by means of a screw, and with it both concave and convex surfaces may be worked perfectly

Fig. 65.—Rebate Plane.

true and even. There are also employed ovolo lamb's-tongue planes for forming the mouldings on sash stiles and

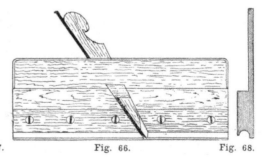

Fig. 67.   Fig. 66.   Fig. 68.

Figs. 66, 67, and 68.—Bead Plane.

rails. The sash fillister (Fig. 74) is generally used for making rebates adjacent to the back side of the stuff, its fence working against the face side. When rebates have to be made next to the face side of the work a side fillister (Fig. 75) is most useful;

its fence is adjustable to the face, allowing a rebate to be made of any width within the breadth of the plane iron. These planes,

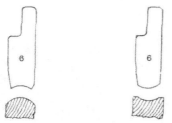

Fig. 69.—Hollow Plane.   Fig. 70.—Round Plane.

and also the plough (Fig. 76), are principally used for grooving with the grain. They are not used so much as formerly, owing to the introduction of machinery in large

Fig. 71.—Sash Plane.   Fig. 72.—Sash Plane.   Fig. 73.—Ogee Moulding Plane.

shops, but they are still indispensable to most joiners. For the working of hard woods, to obtain perfect joints, gun-metal or iron planes known as the shoulder plane and bullnose plane are considered indispensable, as is also the steel smoothing

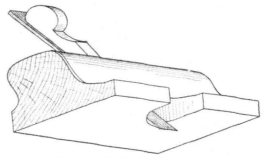

Fig. 74.—Sash Fillister.

plane which is used for cleaning up face work. The router, or " old woman's tooth " (Fig. 77), is used for working out

the bottoms of rectangular cavities ; the chariot plane (almost obsolete), is used for the small parts of work which the smooth-

Fig. 75.—Side Fillister.

ing plane cannot get at, and for planing end grain and cross-grain work ; chamfer planes are used for taking off sharp edges to form chamfers ; mitre shooting planes

Fig. 76.—Plough.

are sufficiently described by their name ; and the plough or plough plane (Fig. 76), used for cutting or " ploughing " grooves. There are many other varieties of planes ; the names and uses of the more important

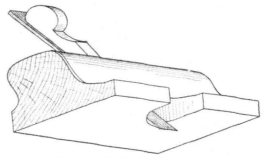

Fig. 77.—Ordinary Router.

will be treated upon in some of the following sections. Particulars of these may be found readily by reference to the index.

Hand Saws.

The saw cannot be classified with any other tool. It is essentially a tool for use across the fibre of the wood, and the separation is a cutting, not a tearing action, as fully explained in the work already alluded to. The carpenter and joiner has some six or

tremes it would be impossible to substitute the ripping and panel saws one for the other. The hand saw, however, which is a kind of compromise between extremes, is used indiscriminately for all purposes,

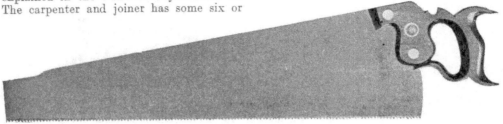

Fig. 78.—Hand Saw.

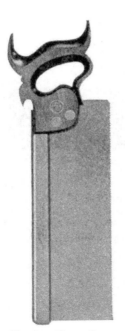

Fig. 79.—Tenon Saw.

Fig. 80.—Bow or Frame Saw.

eight saws, comprising the rip, cross-cut, hand, panel, tenon, dovetail, bow or turning, and keyhole. The hand-saw type includes the hand saw proper, the ripping, half-ripping, and panel saws, all of similar outline, but differing in dimensions and in form and size of teeth. There is no sharp distinction between these tools, as they merge one into the other; yet at the ex-

especially by the carpenter. Fig. 78 is a saw with nibbed back. Straight back and skew back or round back saws are made, and the teeth of the latter do not require to be set. The typical hand saw has a blade which is from 24 in. to 28 in. long. Its blade is as thin as possible, consistent with sufficient strength to prevent the saw buckling under thrust; the taper of the blade is calculated to withstand the thrusting stress without unduly increasing the mass of metal. The teeth are bent to right and left alternately—this being known as the

its teeth, three to the inch, are sharpened square across the blade and set very much forward; this saw is used for cutting along the grain, known as ripping. The tenon saw (Fig. 79) is used for cutting shoulders and in all cases where a clean cut is essential: it obtains this by means of its fine teeth. The dovetail saw is a small tenon saw, it being 6 in. or 8 in. long, whereas the ordinary tenon saw is 12 in. or 14 in.

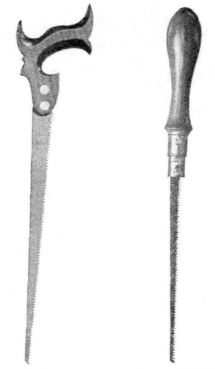

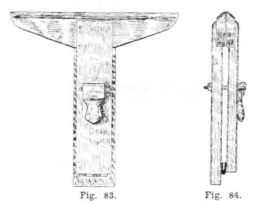

Fig. 83. Fig. 84.

Figs. 83 and 84.—Improved Saw Vice.

long. The bow saw, known also as the turning saw or frame saw (Fig. 80), cuts out curved work with or across the grain, the compass, turning, or lock saw (Fig. 81) being used for a similar purpose, and in cases where a large saw could not be employed. A key-hole or pad saw (Fig. 82) is used for small internal curved work.

Appliances for Sharpening Saws.—For holding a hand saw during the process of sharpening, a saw vice (Figs. 83 and 84) is used, there being many designs to choose from. For sharpening English hand saws, the triangular file (Fig. 85) is com-

Fig. 81.—Compass Saw. Fig. 82.—Pad Saw.

set—and their outline is angular. The teeth are so sharpened that their outer points enter the wood first, the fibre being divided by a gradually incisive kind of action. Six teeth to the inch are suitable for a hand saw used for cutting rough stuff, trimming joists, cutting rafters, etc. For

Fig. 85.—Triangular Saw File.

joiners' work the panel saw, 2 in. or 3 in. shorter and much narrower, thinner, and lighter than the hand saw, is preferable. The rip saw has a blade about 28 in. long, and

monly used; its size varies with that of the saw for which it is required. Special shapes of files are necessary for sharpening American cross-cut and rip saws. Saw files are made in three degrees of fineness. For levelling down or topping saw teeth preparatory to sharpening, a flat file is necessary. The angles of saw teeth are set off with a protractor or hinged rule. For setting the teeth after they have been sharpened—that is, to bend each alternate tooth to one side—saw sets (Fig. 86) are used, or instead, patent contrivances are brought into requisition, these being so arranged that all the teeth can readily be set to one line. A useful form of plier saw set is shown by Fig. 87, and the method of

Fig. 86.—Saw Set with Gauge.

using it by Fig. 88. The amount of set can be regulated by the adjusting screw A. For hammer setting, however, a setting iron with bevelled edges is secured in the vice, the saw laid flat upon it, and the teeth struck one at a time with the pene of a small hammer (Fig. 89). This is the most satisfactory method of setting saw teeth when the operator has the necessary skill.

hammer head. There is the Exeter or London pattern (Fig. 90), the Warrington

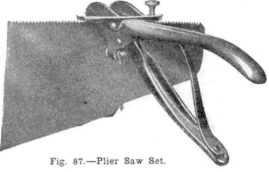

Fig. 87.—Plier Saw Set.

(Fig. 91), and the adze-eye claw pattern (Fig. 92), the last named being less used

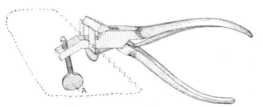

Fig. 88.—Method of using Plier Saw Set.

than the others in the workshop, but being very convenient for many kinds of handi-

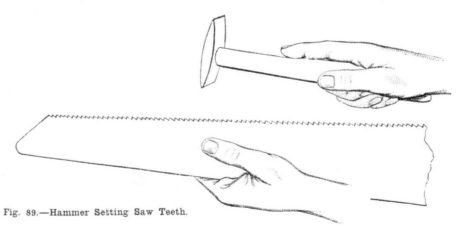

Fig. 89.—Hammer Setting Saw Teeth.

Tools of Percussion and Impulsion.

Hammers.—The carpenter and joiner has the choice between two or three shapes of

work. The hammer heads are of iron, with steel faces and penes. Two hammers, one weighing from 1 lb. to 1½ lbs., and the other from ½ lb. to ¾ lb., will be found use-ful, and it should be remembered that a heavy hammer applied lightly and skil-fully leaves fewer marks and does less

Axes, Hatchets, and Adzes.—These are both percussion and cutting tools, as they combine the offices of the hammer and chisel. Axes have long handles, and may be slung as sledge-hammers, and they have heads more or less of the shape shown by Fig. 94, which illustrates the Kent pat-

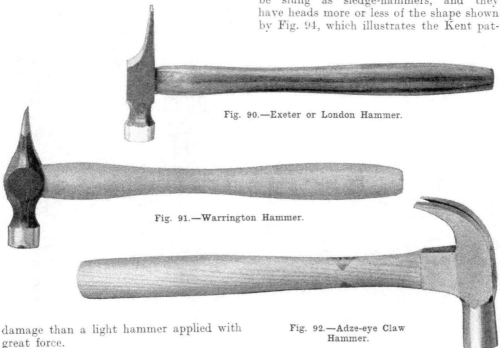

Fig. 90.—Exeter or London Hammer.

Fig. 91.—Warrington Hammer.

Fig. 92.—Adze-eye Claw Hammer.

damage than a light hammer applied with great force.

Mallets.—These are used for driving wood chisels, for knocking light framing together, and in cases where a hammer would probably damage both tools and

tern, many other patterns, however, being in use. Hatchets have short handles, and

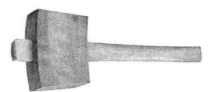

Fig. 93.—English Mallet.

material. An English beech mallet is shown by Fig. 93, but the American hickory or lignum vitæ pattern is perhaps more con-venient, it having all the sharp edges cham-fered off and the handle being round and easier to grasp. In some American mallets the handle screws into the head.

Fig. 94.—Kent Axe Head.

Fig. 95.—American Axe Head.

are used with one hand. The Kent pat-tern already illustrated is common, as is also the Canadian or American pattern shown by Fig. 95. The adze has a long, curved handle, and the operator stands with one foot upon the wood in the line of the fibre, and thus assists in steadying the work. The variety in the shape of the adze heads is very great, but it is sufficient here to show the Scotch pattern (Fig. 96).

Screwdrivers.—These are tools of impulsion, and at least two or three will be required—long and short, and with wide and narrow blade. For general work, a *tool of medium length* should be obtained, although there are, on the one hand, enthusiastic advocates of a short tool, and on the other hand of a long tool for each and every purpose. Any advantage gained by a short over a long tool, or the reverse, is one of advantage in special circumstances only, and not one of saved energy ; theoretic-

Fig. 96.—Scotch Adze Head.

ally, the length does not enter into consideration at all, except when, in starting to extract a difficult screw, the driver is tilted from the upright ; but this is at the risk of a broken tool edge and defaced screw-head. The worker then must decide for himself as to which sizes will best suit his purposes. London screwdrivers have a plain handle (Fig. 97) or oval handle ; cabinet screwdrivers are lighter tools, and there is, indeed, a great variety of patterns from which the worker can choose the tools that suit him. The gimlet-handle screwdriver has certain proved advantages ; and the brace screwdriver—a screwdriver bit used in an ordinary brace—is useful for driving good-sized screws easily and quickly. Short screwdrivers are used in screwing on drawer locks, there being a much heavier though just as short a tool used for screwing up plane irons. Automatic screwdrivers (Fig. 98) were introduced from

America, and by their means the screw is driven home merely by pressure on the top of the handle.

Boring Tools.

Bradawls.—These have round stems and chisel edges (Fig. 99) ; thus the edge cuts the fibres of the wood and the wedge-like form of the tool pushes them aside. Its special use is for making comparatively

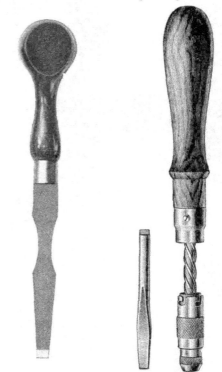

Fig. 97.—Plain Handle London Screwdriver.   Fig. 98.—Miller's Falls Automatic Screwdriver.

small holes in soft wood, and the principal limitation of the tool is that there is no provision for the waste material.

Gimlets.—These are made in many forms, the best known being the twist (Fig. 100) and the shell (Fig. 101), lesser known shapes being the twist-nose (Fig. 102) and the auger (Fig. 103). Gimlets will bore end grain as well as across the fibres, but there is a risk, in boring a narrow strip, of the pointed screw splitting the wood.

Brace and Bits.—There are various kinds of braces on the market, but the more generally used are the two following : A wooden brace with brass mountings is shown at Fig. 104. It is better to buy the bits fitted to this brace, as they are more truly centred than those bought separate are likely to be, and the centering of the bits is essential to their proper action. With the American pattern brace (Fig. 105) this is not necessary, as by turning the socket

of the shape shown for the twist-nose gimlet by Fig. 102 ; it screws itself into the wood. and the chips tend to rise out of the hole, It is found to split narrow strips of wood, but it answers well for all other purposes. All the above bits can be obtained in a great variety of sizes ; but exact size is not guaranteed by the dealers, and the best plan is to bore a hole and measure, rather

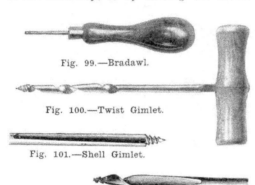

Fig. 99.—Bradawl.

Fig. 100.—Twist Gimlet.

Fig. 101.—Shell Gimlet.

Fig. 102.—Twist-nose Gimlet.

Fig. 103.—Auger Gimlet.

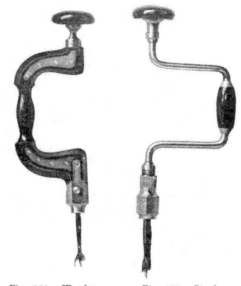

Fig. 104.—Wooden Brace.   Fig. 105.—Steel Brace.

the jaws are expanded or contracted so as to grip the shank of any size bit. This kind of brace may be had with a ratchet movement, which is very useful for boring or turning screws in corner positions where a complete revolution is not possible. Bits are known in very numerous forms. The pin bit is like a gouge sharpened both inside and outside ; when its corners are removed, it becomes a shell bit suited for boring at right angles to the fibre of the wood. The spoon bit resembles the shell bit, but is pointed ; it is found to work easily, freely, and well. The nose bit is of similar shape, but its cutting edge is a part of the steel bent nearly to a right-angle and sharpened to form a kind of chisel ; this tool is efficient for boring the end way of the grain, but not across the grain. The twist-nose bit or half-twist, or Norwegian bit, is exactly

than measure the bit. Holes are enlarged by means of a hollow taper bit. Patent twist bits (Fig. 106) having a screw centre are known as screw bits, and these bore well in any wood and in any direction, relieving themselves of the chips and cutting

Fig. 106.—Gedge's Twist Screw Bit.

true to dimensions. Centre bits are perhaps the most commonly used bits (see Figs. 107 and 108) ; they are useful for boring large holes, and are much superior to shell type bits in the important point of boring exactly where the hole is required. Expanding centre bits (Fig. 109) are known, and are a great convenience if of good quality. The Forstner auger bit (Fig. 110) is guided by its periphery instead of its centre, and consequently it will bore any arc of a circle, and can be guided in any

Fig. 107.—Centre Bit with Pin. Fig. 108.—Centre Bit with Screw.

Fig. 109.—Anderson's Expanding Bit.

Fig. 110.—Forstner Auger Bit.

Fig. 111.—Auger.

direction regardless of grain or knots, leaving a true, polished, cylindrical hollow.

Augers.—The auger (Fig. 111) bores well in the direction of the grain of the wood, and is complete in itself. It is a steel rod, having a round eye at one end, through which a round wooden handle passes. At the other end is a spiral twist terminating in a conical screw with a sharp point. The

edge of the spiral is a nicker which cuts the grain of the wood around the edge of the hollow.

Fig. 112.—Handled Steel Scraper.

Abrading and Scraping Tools.

Steel Scrapers.—The scraper is a thin and very hard steel plate, approximately 5 in. by 3 in., with or without a handle (see Fig. 112). Its action is really that of a cutting tool. It is used on a surface pre-

Fig. 113.—Action of Steel Scraper.

viously made as smooth and level as the plane can make it. The scraper is so sharpened that a burr or feather is formed along its edges (see the diagram, Fig. 113). The thickness of the scraper blade, which is about $\frac{1}{32}$ in., is shown exaggerated. The edge is filed straight and flat, it is then rubbed along the oilstone to remove file marks, and finally the edge is turned and sharpened by two heavy rubs with a round steel burnisher held at an angle of about 30° with the plate.

Glasspaper.—This is the chief abrading material used in woodworking, and consists of strong paper coated with powdered glass secured to the paper with glue. The dif-

Fig. 114.—Glasspaper Rubber.

ferent grades of glasspaper are numbered from 3 to 0, and even finer. For properly

Washita, and Arkansas. The Charnley Forest is of a greenish-slate colour, and sometimes has small red or brown spots— the lighter the colour the better; it may take a little more rubbing than other stones to get an edge on the tool, but that edge will be keen and fine. Some Turkey oilstones are of a dark slate colour when oiled, with white veining and sometimes white spots; they give a keen edge, but wear unevenly, and also are very brittle; they are

Fig. 115.—Half-round Wood Rasp.

Fig. 116.—Half-round Wood File.

using glasspaper a rubber (Fig. 114) is required, this being a shaped wooden block faced with cork. In use the glasspaper is folded round it. Glasspaper has entirely replaced the old-fashioned sandpaper, which was a similar material, except that fine, sharp sand was used instead of powdered glass.

Rasps and Files.—Woodworkers' rasps are generally half-round, though sometimes flat. The wood rasp (Fig. 115) is coarser than the cabinet rasp. The ordinary half-round wood file (bastard cut) is shown at Fig. 116. Both range from 4 in. to 14 in. in length. The usual files used for keeping saws in order chiefly are known as triangular taper (Fig. 85), and have already been alluded to.

Grindstones.—The grindstone (Fig. 117), many varieties of which are obtainable, is an appliance for removing a superfluous thickness of metal, not for producing a good edge. It should be of a light grey colour, even throughout.

Oilstones.—On an oilstone the joiner sharpens his tools, which have been previously ground to shape on the grindstone. The oilstones in most general use are four in number—the Charnley Forest, Turkey,

notoriously slow-cutting, and are expensive. The cheapest oilstone at first cost is the Nova

Fig. 117.—Treadle Grindstone on Iron Stand

Scotia, or Canada stone, which is brownish yellow in colour when new, changing to a yellowish grey by use, and wearing away rather quickly. The Washita stone cuts more quickly than a Turkey stone, and also more regularly. Some kinds are of a whitish grey or light buff colour when oiled. The Arkansas stone is compact and white, and finer in grain than the Washita. It wears well and cuts slowly, producing fine edges. Oilstones generally are about 8 in. long, 2 in. wide, and 1 in. thick, a very convenient width being $1\frac{1}{2}$ in. A small oil-stone of 4 in. by $1\frac{1}{4}$ in. is useful for sharpening spokeshaves, and pieces or slips of

Fig. 118.—Oilstone in Plain Case.

stone of various sizes and shapes are required for gouges, router cutters, etc. It is usual to keep an oilstone in a box or case (Fig. 118). Neat's-foot oil or sperm oil commonly is considered best for oilstone use; lard oil containing sufficient paraffin to prevent it going thick in cold weather is also recommended. Many other oils are used for the purpose, but all tend to harden the surface of the stone much more quickly than neat's-foot or sperm. The oil can be kept in a bench oil-can, which will come in generally useful.

Emery Oilstones and Oilstone Substitutes.—Emery oilstones are an American introduction, and are made of Turkish emery, one face being of fine and the other of medium coarse material. They have the advantage over any natural oilstone of being uniform in texture, and of not being brittle. Oilstone substitutes are strips of zinc upon

which is sprinkled a little flour emery and oil, this working more quickly than a proper stone, but not giving so finished an edge.

Nails, Screws, and Glue.

Nails.—Nails may be of iron, steel, etc., wrought, cast, cut, or made of wire. Formerly nails were said to be 6-lb., 8-lb., etc., according as 1,000 of the variety weighed that amount—hence now such meaningless terms as sixpenny, eightpenny, and tenpenny nails, in which "penny" is a corruption of "pound." Of the nails commonly used in carpentry and joinery, the cut clasp nail, machine-made from sheet "iron" (probably steel), may be used for almost any purpose, and is not liable to split the work. Rose-head nails have a shank parallel in width, but tapered to a chisel point in thickness; these are made of tough wrought iron, and are used chiefly for field-gates and fencing. Wrought clasp nails resemble the cut clasp, but have sharper points, and are used chiefly in common ledged doors, as they will readily clinch. Oval steel nails are nicely shaped, very tough, and are less likely to split the material than any other kind of nail; slight shallow grooves in the shank increase the holding power. Brads are known in more than one variety. The cut-steel large brad is used in flooring, and does not make such a large hole as a cut nail. The cut-steel small brad is used for general purposes. French nails are of round wire, pointed, and have round, flat heads; they are strong, but their unsightly heads cause their use to be confined to rough work.

Fig. 119.—Square Nail Set or Punch.

The double-pointed nail is intended for dowelling and other purposes.

Nail Sets or Punches.—For punching nail heads below the surface of the work a steel set (Fig. 119) of square or round section is used.

Screws.—The screw nail commonly used for uniting woodwork is known as the wood screw, and, although it has been in use a long time, the present pointed screw was not made prior to the year 1841. The screw replaces nails in all fixing where the hammer cannot conveniently be used or where jarring must be avoided. The screw

Fig. 120.—Flat Head Wood Screw.

Fig. 121.—Round Head Wood Screw.

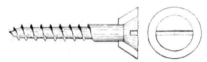

Fig. 122.—Cup Wood Screw.

possesses ten times the compression and attractive strength of ordinary nails, and, besides, is convenient for use in putting work together which is soon to be taken down. Screws are made in almost endless variety, but the best known three are: flat-head screw (Fig. 120), made of iron, steel, or brass; round-head screw (Fig. 121), which is generally japanned and used for fixing bolts, locks, etc.; cup screw (Fig. 122), the head of which fits into a cup (as illustrated) which is let into the work flush with the surface.

Glue, Glue-pots, and Glue-brushes.—Glue, size, and gelatine are varieties of the same substance, differing only in the quantity of moisture and of impurities which they contain. Gelatine-yielding substances employed in glue manufacture include skins of all animals, tendons, intestines, bladders, bones, hoofs, and horns. Glue is manufactured by boiling the animal matter and straining the product into coolers, where it thickens into a jelly, which is cut into sheets and dried in the open air on frames of wire netting. Glue should be of a bright brown or amber colour, free from specks or blotches,

2

nearly transparent, and with but little taste or smell. It should be hard and moderately brittle, not readily affected by moisture in the atmosphere, and should break sharply, but if it shivers as easily as a piece of glass it is much too brittle, though at the same time it must not be tough and leathery. Roughly speaking, a glue which will absorb more water than another is preferable. Good glue does not give off an unpleasant smell after being prepared a few days. In the workshop, different kinds of glue-pots are used, according to the quantity required. The usual glue-pot has an outer and an inner vessel and is shown in section at Fig. 123. When glue is used in large quantities, and steam pipes are laid on for heating purposes, the glue is kept hot on a water bath heated by steam pipes. The joiner prepares glue by breaking it into small pieces, soaking these

Fig. 123.—Section through a Glue-pot.

in clean, cold water for several hours, and then boiling the resulting lumps of jelly—the superfluous water having been poured off—in a double-vessel glue-pot for an hour or two, or until the glue runs easily from the brush without breaking into drops. A glue-brush can be bought for a few pence, and its bristles should be comparatively short. A cane brush is preferred by many workers, this being made with a piece of rattan cane about 8 in. long, the flinty skin for an inch or so at one end being cut away, the end soaked in boiling water for a minute or two, and then hammered till the fibres are loosened; this brush lasts as long as there is any cane left from which to hammer out a fresh end.

Other Tools and Appliances.—Many other tools and appliances not in such general use will be illustrated and described in connection with the matter treated in some of the other sections (see index).

TIMBER.

Growth of Timber Trees.

Structure of Tree Trunk.—Trees which produce timber are known botanically as exogens, or outward growers, because the new wood is added underneath the bark outside that already formed. The whole section (Fig. 124) consists of (*a*) pith in the centre, which dries up and disappears as the tree matures ; (*b*) woody fibre or long

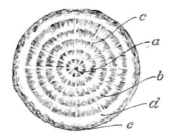

Fig. 124.—Cross Section of Stem of Timber Tree.

tapering bundles of vascular tissue forming the duramen or heartwood, arranged in rings, each of which is considered to represent a year's growth, and interspersed with (*c*) medullary rays or transverse septa consisting of flat, hard plates of cellular tissue known to carpenters as " silver-grain," or " felt," or " flower," and showing most strongly in oak and beech : the heartwood is comparatively dry and hard, from the compression produced by the newer layers ; (*d*) alburnum, or sapwood, which is the immature woody fibre recently deposited. In coniferous trees the sapwood is only distinguishable by a slight greenish tinge when dry, but when wet it holds the moisture much longer than the heartwood, and can often be detected in that way ; (*e*) the bark, which is a protecting coat on the outside

of the tender sapwood ; it receives additions on the inside during the autumn, which cause it to crack and become very irregular in old trees. The mode of growth is as follows : In the spring moisture from the earth is absorbed by the roots, and rises through the stem as sap to form the leaves. The leaves give off moisture and absorb carbon (in the form of carbonic acid gas), which thickens the sap. In the autumn the sap descends inside the bark and adds a new layer of wood to the tree. The actual growth is less regular than appears in Fig. 124, and more resembles Fig. 125.

Formation of Wood.—Fig. 125 further illustrates the manner in which the stem of a timber tree grows by the deposit of successive layers of wood on the outside under the bark, while at the same time the bark becomes thicker by the deposit of layers on its under side. Upon examining the cross section of an oak log as Fig. 125, it is found that the wood is made up of several concentric layers or rings, each ring consisting in general of two parts, the outer part being usually darker in colour, denser, and more solid than the inner part, the difference between the parts varying in different kinds of trees. These layers are called annual rings, because one of them is, as a rule, deposited every year in a manner which will be presently explained. In the centre of the first layer is a column of pith, from which planes, seen in section as thin lines (in many woods not discernible) radiate towards the bark, and in some cases similar lines from the bark converge towards the centre, but do not reach the pith (see Figs. 125 and 126). These radiating lines are known as medullary rays or transverse septa. When they are of large size and strongly marked, as in some kinds of oak, they present the beautiful figured appearance called silver-grain or felt, as illustrated by the longitudinal section (Fig. 126). To produce this effect, the timber must be sawn in the radial planes of the medullary rays, or slightly oblique to them. As already mentioned, the wood is composed of bundles of cellular tubes, which serve to convey

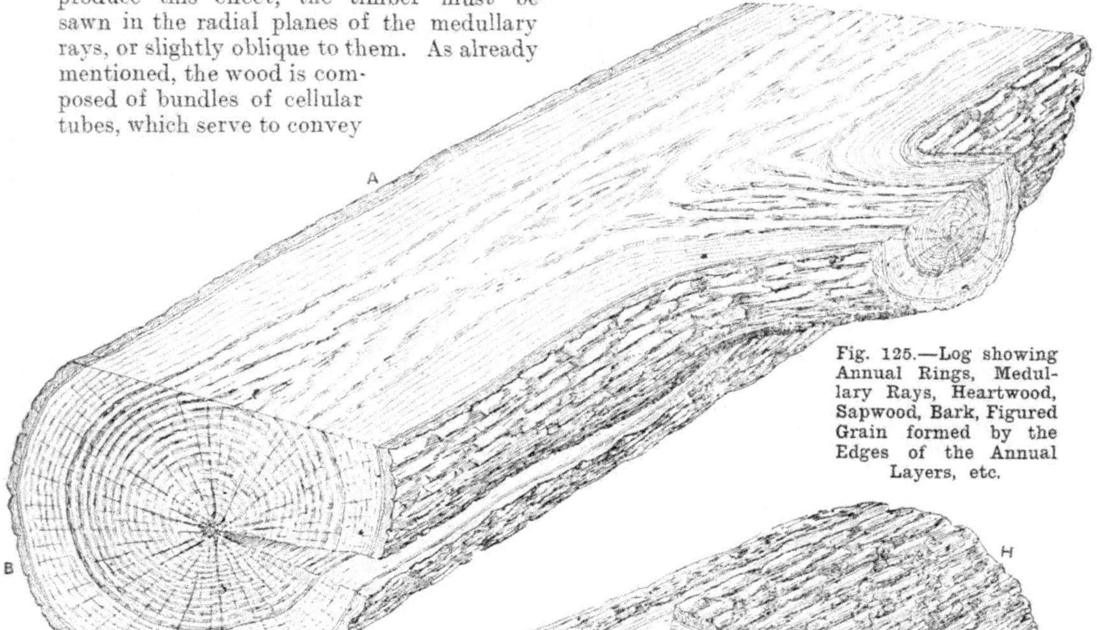

medullary rays, the sapwood being on the outside and the remainder heartwood. F, Fig. 126, shows the longitudinal section

Fig. 125.—Log showing Annual Rings, Medullary Rays, Heartwood, Sapwood, Bark, Figured Grain formed by the Edges of the Annual Layers, etc.

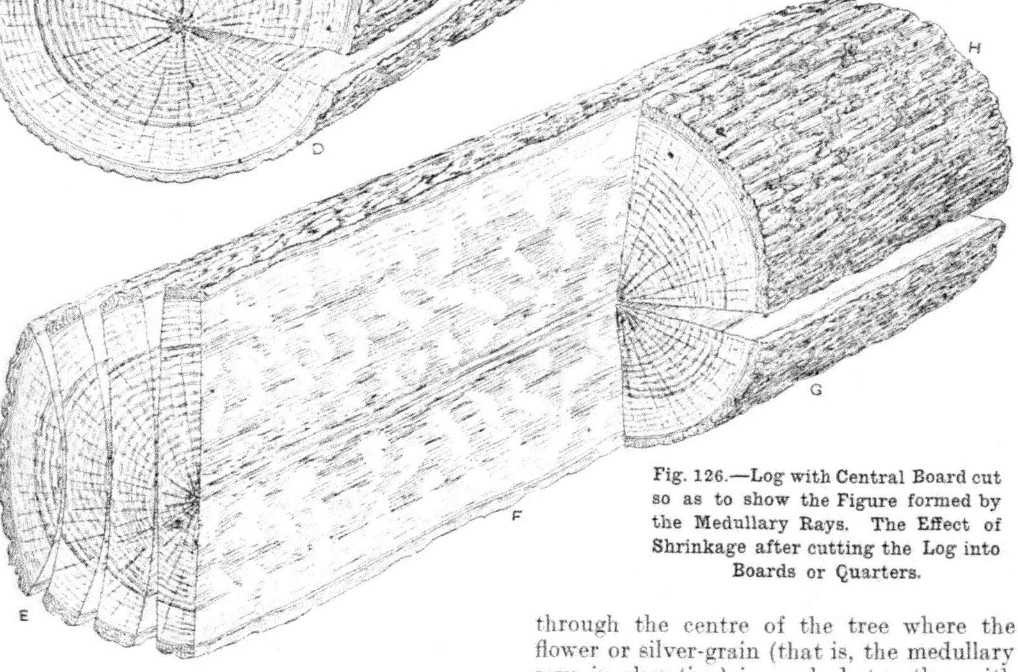

Fig. 126.—Log with Central Board cut so as to show the Figure formed by the Medullary Rays. The Effect of Shrinkage after cutting the Log into Boards or Quarters.

the required nourishment from the earth to the leaves. Fig. 125 shows the cross section with the annual rings and the

through the centre of the tree where the flower or silver-grain (that is, the medullary rays in elevation) is marked, together with the edges of the annual rings. A, Fig. 125, shows a longitudinal section nearer to the bark where the graining is formed by the section of the annual rings, owing to the straight cut through the bent tree. The medullary rays are seen edgeways as fine lines in this section, whilst the annual layers form beautiful wavy and hearty grain. A plank cut so as to contain part of the centre pith of the tree as shown at F. in Fig. 126, would be least affected in breadth by shrinking.

Difference Between Exogenous and Endogenous Timber.—Exogens and endogens are very different in internal structure and in outward appearance. The exogens, as has been explained, increase in size by the addition of new material at the outside of the stem—just under the bark. They continue to increase in diameter as well as in height throughout their whole lifetime. This growth may be carried on continuously, as in the cactuses, or intermittently, by abrupt periodical advances and cessations, as in the forest trees. The hardest portion of the stem is towards the centre. The fibro-vascular bundles are " open "—that is, capable of further development. There is a distinct and separable bark, and usually a number of branches. The trunk and branches are frequently crooked. The leaves are articulated, and drop off neat or clean from the tree. The veins in the leaves ramify, forming an irregular network. The flowers, when present, have, as a rule, four or five sepals and petals, etc., or multiples of these. The seeds (except in conifers) split in two. The oak, apple, laburnum, and the wallflower are examples of exogens. Some exogens live to be more than a thousand years old. Endogens mainly increase in size by end growth. There is lateral distension for a time, but this soon ceases, and then the tree remains of nearly uniform diameter throughout its life. There are no annual rings—the growth being mostly continuous. The hardest portion of the stem is at the outside, where a false rind made up of broken leaf-ends, etc., is formed, but no bark. The fibro-vascular bundles become " limited," or " closed," after a certain period, after which they serve only to strengthen the stem. The trunk is straight, or nearly so, and seldom has any branches. If it does have any branches, as in bamboo, then these are straight too.

At the top end, where the growing is taking place, the new leaves arise inside the old ones, and press them outwards and downwards as they grow. The old leaves eventually die, and hang like a ragged sheath around the stem. The leaves are parallel-veined. The flowers are mostly on the plan of three. The seed is entire : hence Monocotyledons. Few endogens live to be 300 years old. Nearly all the principal kinds belong to tropical or sub-tropical climates—examples are the palms, bamboos, grasses, and lilies. There are no endogenous trees indigenous to England, and it is believed that the only British endogenous shrub is the butcher's broom—*Ruscus aculeatum.*

Function of Sap.—The action of the sap may now be described in fuller detail. In the spring the roots absorb from the soil moisture, which, converted into sap, ascends through the cellular tubes to form the leaves. At the upper surface of the leaves the sap gives off moisture, absorbs carbon from the air, and becomes denser ; after the leaves are full-grown, vegetation is suspended until the autumn, when the sap in its altered state descends, by the under side of the leaves, chiefly between the wood and the bark, where it deposits a layer of new wood (the annual ring for that year), a portion at the same time being absorbed by the bark. During this time the leaves drop off, the flow of sap then almost stops, and vegetation is at a standstill for the winter. With the next spring the operation recommences, so that after a year a distinct layer of wood is added to the tree. The above description refers to temperate climates, in which the circulation of sap stops during the winter ; in tropical climates it stops during the dry season. Thus, as a rule, the age of the tree can be ascertained from the number of annual rings ; but this is not always the case. Sometimes a recurrence of exceptionally warm or moist weather will produce a second ring in the same year.

Heartwood and Sapwood.—A young tree is almost all sapwood, but as it matures this is gradually changed into heartwood more rapidly than sapwood is added, and as the tree increases in age, the inner layers are filled up and hardened, becoming duramen or heartwood, the remainder being alburnum or sapwood. The sapwood is softer and lighter in colour than the heartwood, and can generally be easily distinguished from it. In addition to the strengthening of the wood caused by the drying up of the sap, and consequent hardening of the rings, there is another means by which it is strengthened —that is, by the compressive action of the bark. Each layer, as it solidifies, expands, exerting a force on the bark, which eventually yields, but in the meantime offers a slight resistance, compressing the tree throughout its bulk. The sapwood is generally distinctly bounded by one of the

this makes it drier, lighter, and more resilient or springy. It is less liable to twist, warp, or split. The advantages of using seasoned timber are that it works more easily under the saw and plane, and retains its size and shape after it leaves the hands of the carpenter or joiner. Unseasoned stuff warps and shrinks, and, besides being unsightly, is liable to cause failures in structures of which it may form a part ; it is also very liable to decay from putrefaction of its sap.

Natural Processes of Seasoning Timber.—Timber produced from a newly felled tree is full of moisture, and this must be extracted by drying or seasoning. Timber cut down in the autumn, after the sap has

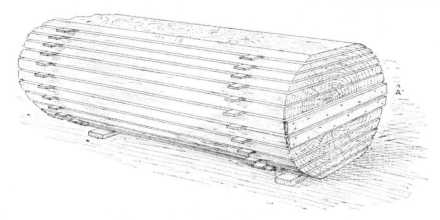

Fig. 127.—Hardwood Stacked for Seasoning.

annual rings, and can thus be sometimes distinguished from stains of a similar colour, which are caused by dirty water soaking into the timber while it is lying in the ponds. These stains do not generally stop abruptly upon a ring, but penetrate to different depths, colouring portions of the various rings. The heartwood is stronger and more lasting than the sapwood, and should alone be used in good work. The annual rings are generally thicker on the side of the tree that has had most sun and air, and the heart is, therefore, seldom in the centre.

Seasoning Timber.

Advantages of Seasoned Timber.—Seasoned timber differs from unseasoned principally in having the sap and moisture removed ;

formed the new layers of wood, is best seasoned by cutting it into planks and stacking them horizontally in open order under cover, exposed to a free current of air, and protected from ground moisture. Hard woods are generally stacked with thin strips between them placed transversely every 2 ft. or so (Fig. 127), and soft woods by laying them on edge with spaces between, the direction being crossed in adjacent courses. The time occupied is, say, two years. Balk timber is best seasoned by putting it under water in a running stream for a few weeks, then stacking it loosely with some protection from sun and rain. These are termed natural processes. For protecting the stacked timber from the action of the sun and high winds, a shed with open ends, or with louvred sides—that is, with sides after the fashion of Venetian blinds—proves satisfactory. In stacking timber horizontally, it should be laid perfectly flat and level in breadth and straight in length. The usual plan is to lay "sleepers" or cross-bearers on the ground, and then stack upon these. The ground on which the timber is to be seasoned should be properly drained so as to carry off driving rain. It should also be protected from vegetable growth; therefore

by 1 in. between each layer, about 2 ft. or 4 ft. apart (Fig. 127), or arranged in some similar manner, the object to be kept in view being to allow free circulation of air round nearly the whole of each piece, gradually carrying off a greater part of the sap and moisture from the timber. To prevent planks and boards splitting from the ends up the centre, they are clamped by nailing strips of wood to the ends as indicated at A (Fig. 127). Timber seasoned as above is said

Fig. 128.—Single "Sturtevant" Apartment Drying Kiln, Section.

it is a good plan to have the ground covered with asphalted paving, or with a layer of smith's or furnace ashes to prevent vegetable growth contaminating the stacked timber and bringing about wet rot, or in some cases from becoming the source of the development of dry rot after the timber has been inserted in a building. The lowest layer of timber should rest upon bearers which should be arranged all in one plane —that is, out of winding, otherwise the timber stacked upon them would become permanently twisted. This is very important. The timber should be stacked in layers, with a space between each piece in the same row, and strips of wood about ½ in.

to retain properties that render it stronger, heavier, more elastic and flexible, and much more durable than timber seasoned by artificial processes.

Artificially Seasoning Timber.—There are various artificial processes of seasoning in use which expedite the work and shorten the time necessary between felling and using, but the strength and toughness of the timber are reduced. The methods are desiccating, or using hot-air chambers, smoking, steaming, and boiling. To reduce the risk of splitting the ends in the drying process, they are clamped—that is, thin pieces are nailed over the end grain so that the ends may dry uniformly with the other parts. McNeile's process is said to be very good : the wood to be seasoned is exposed to a moderate heat in a moist atmosphere charged with the products of combustion, say CO_2, which is supposed to convert the sap to woody fibre and drive out the moisture. Smoke-drying over an open wood

Modern Method of Artificially Seasoning Timber.

Nature seasoning takes so long that it keeps idle a vast amount of capital. By artificial means timber can be dried in fewer days than it takes months by the natural

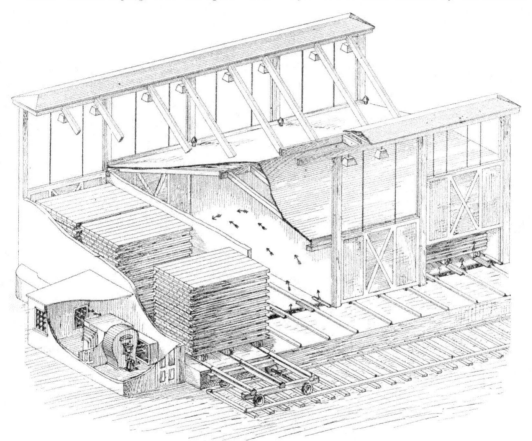

Fig. 129.—Four-chamber "Sturtevant" Drying Kiln, Section.

fire drives out the sap and moisture and renders the wood more durable and less liable to attack by worms. Burying logs in sand is a method of artificial seasoning. The disadvantage of artificial seasoning is that the method of drying is too rapid, and seems to take away the stability of the material, leaving it less firm, more brittle, and duller in appearance.

process, consequently improvements in the methods of seasoning are constantly being sought for. A large quantity of deals, battens, planks, etc., receive a first seasoning before being placed on the market. The most effective artificial methods of seasoning are probably of American origin. The following two systems are largely in use.

The "Sturtevant" System of Drying Timber.

Rapid and efficient drying is effected by subjecting the timber to a continual passage of warm dry air in a kiln constructed of wood or brick into which hot air is introduced by a fan. Fig. 128 shows a sectional view. The air is first heated by a Sturtevant heater E to the desired temperature by either live or exhaust within the kiln, and thus prevents the exterior of the stack drying too quickly and becoming simply skin dried. Perfectly green coniferous timber one inch thick can be dried within six days, other thicknesses in proportion. It is claimed that by this process the outside of the wood is kept open, which allows the moisture from the heart to escape without splitting, warping, or discolouring taking place. Fig. 129 is a sectional view of a large kiln having four compartments. Timber is erected in stacks, on trucks running on rails,

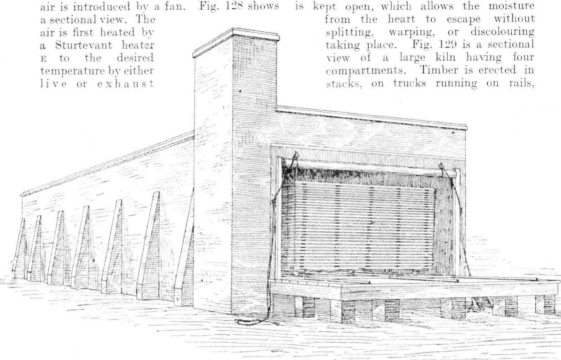

Fig. 130.—Erith's Patent Automatic Drying Kiln, General View.

steam, which ensures that the temperature never exceeds 212° F. Then, by means of the fan F. it is forced through the outlets of the supply duct B into the kiln, circulating completely round the timber. Owing to the high temperature of the air it rapidly absorbs moisture and is then passed into the atmosphere, or it may be returned to the apparatus to be reheated and the absorbing process repeated. The return ducts A and C serve a double purpose by utilising the remaining air which comes gradually laden with moisture; the process of reheating serves, by regulation, to maintain any desired degree of humidity

and is thus easily conveyed in and out of the kiln.

Erith's Patent Automatic Timber Drier.

These kilns may be of wood or brick. For carrying out this system of drying timber one form of kiln is shown at Fig. 130. The timber is conveyed into the kiln by being stacked upon trucks running on rails, and as the timber is dried, it is passed out at the opposite end. A canvas roller door is provided at each end which works on the roller-blind principle, but fitting almost air-tight. This system dries wood by the circulation of warm but very moist air. Its operation is automatic, no machinery or power being required. The apparatus consists of specially arranged steam radiator coils, in which exhaust or live steam may be used ; they are placed beneath the rails near the discharging end of the building. Air flows under the radiator coils, and rises, at the same time travelling through the stacks of wood, thus gradually drawing

Artificially Seasoning Timber Small Stuff.

A method sometimes adopted for seasoning small pieces of timber, especially for tool making, and other purposes, is possible wherever a supply of steam—from the boiler or exhaust of a steam engine—is available. The pieces of wood are stacked in a steam chest (see Fig. 131) or a barrel (Fig. 132) and allowed to become thoroughly

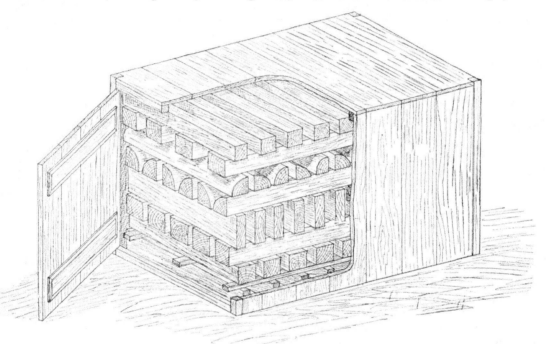

Fig. 131.—Steam Chest for Small Pieces of Timber.

moisture from it. As the air becomes more laden with moisture it sinks between the rails and flows towards the loading end, where it is allowed to escape. This circulation may be regulated by a few simple dampers. The timber is dried from the centre outwards, the surfaces finishing last ; therefore case-hardening, splitting, warping and other injuries are prevented. By this system it is claimed that timber which would require a year to dry in the open can be dried in a week ; this, of course, is a great advantage.

2*

saturated with steam: This will take from two to twelve hours, according to the kind and thickness of the wood. No pressure is required, but the door of the chest or top of the barrel should be closed with a lid ; the fitting is not close, allowing the steam which has circulated round the wood to escape. For this reason the apparatus is kept outside a building. The material being treated is kept from the bottom proper to allow the steam to become evenly distributed. The use of this method is very limited, because by it the natural colours of many woods are more or less changed, especially in the case of beech, the colour of which is changed from a dull white to the familiar reddish tint. After it is taken out the wood is piled under cover in the ordinary manner and allowed to dry; this, in small thin material, usually takes three weeks or a month. The drying time might be considerably shortened by utilising the space above the boiler as a drying loft. A temperature of 120° F. to 180° F. (obtainable above most

Fig. 132.—Barrel for Seasoning Small Timber.

boilers) would get the drying over in a day or two, but the material should not be transferred to such a position direct from the steam-box; let it have a few days' ordinary drying first. The apparatus illustrated by Fig. 132 is also suitable for steam-bending purposes.

Shrinkage During Seasoning.—During seasoning a large proportion of the moisture evaporates, causing the fibres to shrink and the timber to become less in bulk and weight. Timber is considered fit for carpenters' work when it has lost one-fifth of its weight, and for joiners' work when it has lost one-third. It also becomes lighter in colour and more easily worked. The shrinkage is scarcely perceptible in the length, but is very considerable in the width, measuring circumferentially on the annual rings (see E and G, Fig. 126). Radially, or in the

direction of the medullary rays, the shrinkage is only slight, as shown by the board F, Fig. 126. If the log is whole, the shrinkage causes shakes and wind-cracks; if cut up into planks or quartering, the shrinkage is determined by the position of the annual rings, and, with care, shakes are not caused. The wood curls or bends breadthwise, with the edges turning on the side which is farthest away from the heart. This is illustrated at E, G, and H, Fig. 126. This circumstance must always be considered in fixing timber in position.

Preserving Timber.

Bethell's Process.—There are a number of preservative processes other than seasoning which are of value in increasing the durability of timber. Bethell's process, also known as creosoting, consists in placing pieces of seasoned timber in closed wrought-iron cylinders, from which, and also from the pores of the wood, the air is extracted. Oil-of-tar, known as creosote, is then forced into the cylinders and pores of the wood, at a temperature of about 120°, and under a pressure of 60 lb. to 170 lb. per square inch, according to the porosity of the wood and the purpose for which it is required. The quantity forced into the wood varies from 3 lb. per cubic foot in some hard woods to 12 lb. in soft woods.

Bouchere's Process.—This consists in placing a reservoir, containing 100 parts in weight of water to 1 part of sulphate of copper, in a position about 40 ft. or 50 ft. above the timber, and connecting it by a flexible tube to a cap which is fixed tight to one end of the piece of timber under treatment. The pressure is sufficient for the fluid to force out the sap at the other end and take its place in the pores of the timber.

Burnett's System.—By this system a fluid is prepared in the proportion of 1 lb. of chloride of zinc to 4 gal. of water. The timber is sometimes laid in a bath of this fluid until it has absorbed sufficient; or the solution is forced under pressure into the timber. The value of the above processes lies in the preservation of the timber from dry and wet rot, and, in the case of the latter two systems, from most insects, so long as the salts remain in the timber; but by some authorities the salts are said to be gradually removed by the action of water, and thus in time the timber becomes a prey to insects and decay. When, however, timber is treated thoroughly by Bethell's process, its durability is greatly increased, and it is rendered proof against the attacks of every insect, including the white ant.

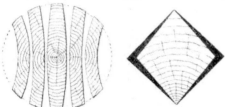

Fig. 133.   Fig. 134.

Fig. 133.—Planks Warped according to position in Tree. Fig. 134.—Shrinkage of Quartering in Seasoning.

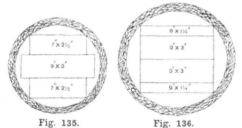

Fig. 135.   Fig. 136.

Fig. 135.—Old Method of Converting Logs into Deals. Fig. 136.—Modern Method of Converting Logs into Deals.

Converted Timber.

A log is the trunk of a tree after the branches are lopped off. A balk is a log which has been squared by means either of sawing or adzing; the latter would be known as a hewn balk. Planks are pieces of sawn timber from 2 in. to 6 in. thick, 11 in. to 18 in. wide, and from 8 ft. and upward in length. Deals are from 2 in. to 4 in. thick and 9 in. wide. Battens are from 4½ in. to 7 in. wide, and from 2 in. to 4 in. thick. Boards are pieces of sawn timber of any length and breadth, but not exceeding 2 in. in thickness. Scantlings are pieces of timber which have been sawn to 4 in. by 4 in., 4 in. by 3 in., 4 in. by 2 in., 3 in. by 3 in., 3 in. by 2 in., etc. The smallest pieces are frequently called quarterings.

Converting Timber.

In converting timber into planks or boards the shrinkage and warping to be expected in use depend upon what part of the tree the piece is cut from. Practically, the stuff will only shrink along the curved lines of the annual rings, and not from the outside towards the centre; so that, a tree being cut into planks, the alteration produced by seasoning is shown in Figs. 126 and 133. A piece of quartering would, in the same way, if originally die-square, become obtuse angled on two opposite edges, and acute

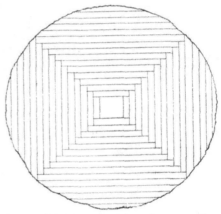

Fig. 137.—Converting Pitchpine Logs into Boards to show the Grain.

angled on the other two, as in Fig. 134. In the conversion of fir, the old system is shown at Fig. 135, which is objectionable on account of the centre deal containing the pith enclosed, and being therefore more subject to dry rot. Fig. 136 shows the modern method of conversion, where the 9 × 3 deals go to the English market, and the 9 × 1¼ to the French market. Of the remainder in each case, some is cut up into battens and fillets for slating and tiling, and similar purposes, and the rest used as fuel. The method of converting a pitchpine log so as to show the best possible grain is indicated by Fig. 137 (see also p. 48). In converting oak, the method will depend upon the purpose for which it is required. For thin stuff, where the silver grain or "flower" is desired to appear, the method shown at A (Fig. 138) is best, and that at B second best, the object being to get the greatest number of pieces with the face nearly parallel to the medullary rays. The method shown at C makes less waste, but does not show up the grain so well; while

and seasoning. It will be noticed that Fig. 140 undergoes the least change. At Fig. 143 two planks are represented occupying adjacent positions in the same log. Fig. 144 indicates the change in shape of each after conversion and seasoning. The centre

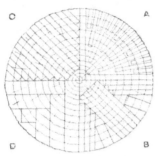

Fig. 138.—Converting Oak into Boards.

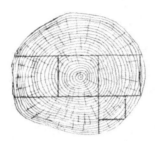

Fig. 139.—Square Scantlings.

Fig. 140.   Fig. 141.

Fig. 140.—Scantling from Centre.
Fig. 141.—Scantling from Side.

Fig. 142.—Scantling from Edge.

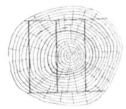

Fig. 143.—Planks.

Fig. 144.—Alteration of Form in Planks.

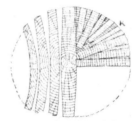

Fig. 145.—Warping of Planks.

Fig. 146.—Plank Cut to show Figure.

Fig. 147.—Oak Plank showing Figure.

Fig. 148.—Well Jointed Planks.

Fig. 149.—Badly Jointed Planks.

D is the most economical when larger scantlings are required.

How the Cutting of Timber Affects its Use.

The method of cutting timber has a big effect upon its use. Fig. 139 shows square scantlings occupying three different positions in the same log; Figs. 140 to 142 show the alteration of form in each piece after sawing

plank and those to the left in Fig. 145 indicate how boards cut from the log tend to shrink and warp if unrestrained. If the boards are cut as shown at K, there would be the least alteration in form. Timber should be cut as represented at Fig. 146 in order to show the figure formed by the annual rings. When it is required to obtain oak panels, etc., showing the beautiful markings of the medullary rays, the timber should be cut as shown at Fig. 147. By arranging boards as in Fig. 148 a better joint is made than that shown at Fig. 149. When mouldings are prepared from wood which has been cut so that the annual rings are nearly parallel to the breadth (see Fig. 150), there is almost sure to be more or less shrinkage, which will, of course, take place in the breadth and thus produce an

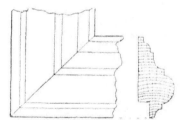

Fig. 150.—Shrinkage of Moulding.

open mitre as shown, although the workmanship may be first rate. Fig. 151 shows the best arrangement, the annual rings being at right angles to the breadth.

Cutting Strongest Beam from Round Log.

By mathematical investigation Fig. 152 shows the graphic method of finding the strongest rectangular beam that can be cut

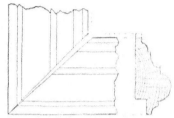

Fig. 151.—Best Arrangement of Grain in Mouldings.

out of a round log of timber. The diameter is divided into three equal parts, and perpendiculars are raised on opposite sides on the inner ends of the outer divisions. The four points in which the circumference is touched are then joined to give the beam A E B F. The proportion is $\dfrac{FB}{AF} = \dfrac{1}{\sqrt{2}}$, because by Euclid II. 14, $\sqrt{AD \times DB} = FD$, and by Euclid I. 47, $FB =$

$\sqrt{(FD)^2 + DB^2}$ also $AF = \sqrt{(AB)^2 - (FB)^2}$.

Let $AB = 1$, then $FD = \sqrt{\frac{2}{3} \times \frac{1}{3}} = \frac{\sqrt{2}}{3}$,

$FB = \sqrt{\left(\frac{\sqrt{2}}{3}\right)^2 + \left(\frac{1}{3}\right)^2} = \sqrt{\frac{2}{9} + \frac{1}{9}} = \sqrt{\frac{1}{3}} = \frac{1}{\sqrt{3}}$, $AF = \sqrt{1^2 - \frac{1}{3}} = \sqrt{\frac{2}{3}} = \frac{\sqrt{2}}{\sqrt{3}}$ and $\frac{FB}{AF} =$

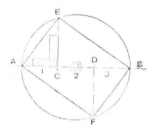

Fig. 152.—Strongest Beam from a Round Log.

$\frac{1}{\sqrt{3}} \div \frac{\sqrt{2}}{\sqrt{3}} = \frac{1}{\sqrt{2}}$. When the diameter of log $AB = d$, the depth $AF = \frac{\sqrt{2}}{\sqrt{3}} = \frac{1\cdot414}{1\cdot732} = \cdot816d$, and the breadth $FB = \frac{1}{\sqrt{3}} = \frac{D}{1\cdot732} = \cdot577d$.

It must be fully understood, of course, that the above shows only the mathematical calculation corresponding to the graphic

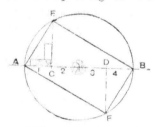

Fig. 153.—Stiffest Beam from a Round Log.

diagram, and does not in any way prove the statement that this beam will be the strongest that can be cut out of a round log. The calculations necessary to prove that statement would probably be a laborious matter. But given such a beam, its strength could be calculated by ordinary formula, and then another beam slightly narrower, and a beam slightly broader, both inscribed in the circle, could be tested by the same formula.

Cutting Stiffest Beam from Round Log.

The stiffest rectangular beam that can be cut out of a round log of timber is shown in Fig. 153, where the diameter is divided into four equal parts, but otherwise the construction and calculation will be on

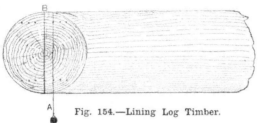

Fig. 154.—Lining Log Timber.

similar lines to the above, resulting in $\dfrac{FB}{AF}$ $= \dfrac{1}{\sqrt{3}}$; and with a log of diameter $AB =$ D the depth of the stiffest beam will be ·866D and the breadth ·5D.

Selecting Timber.

See that timber is free from sap, large or loose knots, flaws, shakes, stains or blemishes of any kind. A light portion near one edge indicates sap, and an absence of grain will be observed on it. This portion decays first and gets soft. The darker the natural

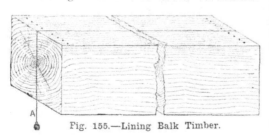

Fig. 155.—Lining Balk Timber.

wood, the lighter is the sappy portion usually when dry. Good timber should be uniform in substance, straight in fibre, and not twisted, warped, or waney. Diagonal knots are particularly objectionable in timber for piles. Good timber should smell sweet when fresh cut, and it has a firm, bright surface, and does not clog the saw. The annular rings should be fairly regular and approximately circular; the closer and nar-

rower the rings the stronger the timber. The colour should be uniform throughout, and not become suddenly lighter towards the edges. Good timber is sonorous when struck; a dull sound indicates decay. In specimens of the same class of timber the heavier is generally the stronger.

Marking Out Timber for Pit Sawing.

Pit sawyers employ various methods for lining timber that is to be sawn; one reli-

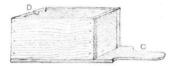

Fig. 156.—Ochre Box.

able method is shown by Fig. 154. The wedge boy (as he is termed) holds the string centrally at one end of the log, and the sawyer holds it at the other end. The string is then pulled tight, and one of the sawyers raises it at the centre (the string should not be raised exactly vertical, but pulled slightly at one side), then lets it go, so as to strike a line the whole length of the log. A plumb-line A is then hung over the end of the log by one of the sawyers, and when perfectly plumb with the centre line, it is pulled tightly against the bottom edge of the log. The other sawyer pulls the string and strikes a central vertical line, as shown at B. A similar line is struck at the other end of the log, after which the thick-

Fig. 157.—Wood for Pulling String.

ness of the planks that are to be sawn is pricked off with a pair of compasses, as indicated by the dots on the end of the log. The plumb-line is again hung over the end of the log in perfect line with the compass marks. Vertical lines, as before, are then struck, then corresponding longitudinal lines. When lining a square balk, a centre line is first struck, then the thickness of the planks is pricked off, as shown in Fig. 155, and the lines are struck. The plumb-line is then hung over the end, as shown, and the vertical lines are struck. The log or balk is now turned over, and longitudinal lines corresponding with the vertical lines are struck. To make an impression that may be clearly seen, the top and end lines are struck with a string that has been passed through a mixture of red ochre and water of the consistency of thin paste. The string used for lining the under side of the timber is passed through whiting. A red or dark line can be better followed by the top sawyer, while from underneath a white line can be best seen. The ochre is placed in a little box (see Fig. 156) and water added. There is a handle at c, and a notch at D. The string is placed in the box and drawn through the notch. A thin piece of wood, as Fig. 157, is placed on the string while it is being pulled through the notch, otherwise it would be necessary for the finger and thumb to guide the string, and to remove the surplus ochre that may be on it.

Weight and Strength of Timber.

These particulars are given in the accompanying table.

(1) Name. Selected Quality.	(2) Weight cub. ft.	(3) Ultimate Tensile Strength.	(4) Ultimate Compression.	(5) Coefficient of Transverse Strength.	(6) Ultimate Bearing Pressure across Grain.
	lbs.	tons per sq. in.	tons per sq. in.		tons per sq. in.
American red pine ..	37	—	2·2	4·0	—
Ash	45	2·0	3·5	5·0	—
Baltic oak	48	3·0	3·2	4·3	—
Beech..	47	1·9	3·8	4·5	—
Elm	37	2·0	3·0	3·0	—
English oak	50	3·0	3·2	5·0	·90
Greenheart	60	—	5·8	8·0	—
Honduras mahogany ..	35	1·5	2·8	4·9	·58
Kauri pine	38	—	2·8	4·8	—
Larch..	35	1·5	2·5	3·5	—
Northern pine	37	1·5	2·9	4·0	·60
Pitchpine	50	—	2·9	5·0	·76
Spanish mahogany..	53	1·8	3·0	5·0	1·9
Spruce fir	31	1·5	2·5	3·6	·22
Teak	50	3·0	3·8	5·0	—
White pine	28	—	1·8	3·8	·27

The safe load in tension and compression (columns 3 and 4) would be from one-tenth to one-fifteenth of the amounts given. The safe bearing pressure across the grain of timber as at the ends of a beam will be about one-fifth of the amounts given in column 6. Column 5 gives the coefficient c in the formula $w = c\,b\,d^2 \div L$, and the safe load would be about one-sixth of w for temporary work, or one-tenth for permanent loads.

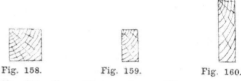

Fig. 158.　　　　Fig. 159.　　　　Fig. 160.

Fig. 158.—Beam 6 in. × 6 in.　Fig. 159.—Beam 6 in. × 3 in.　Fig. 160.—Beam 12 in. × 3 in.

Calculating Strength of Timber Beams.

The strength of solid timber beams varies as the square of the depth, directly as the width, and inversely as the span. Thus, in a beam 6 in. square (Fig. 158), multiplying the width by the square of the depth gives $6 \times 6^2 = 6 \times 36 = 216$; and if this beam was sawn down the middle, there would be $3 \times 6^2 = 3 \times 36 = 108$ (Fig. 159). Another case is that of a beam 12 in. deep and 3 in. wide (Fig. 160), and the corresponding figure then is $3 \times 12^2 = 3 \times 144 = 432$. The ordinary formula for the strength of a beam lying loose on the bearings at each end, and with central load (Fig. 161), is as

Fig. 161.—Loose Beam with Central Load.

follows, when B = breadth in inches, D^2 = square of depth in inches, L = length of bearing in feet, c = constant, for which Barlow and Tredgold give a value for Riga of $c = 4\frac{3}{4}$ cwt. This constant is obtained from the results of trials, but it must be noted that such tests vary considerably. The strength of timber will vary in the same cargo, and allowance must be made for the difference in the growth and fibres of the various pieces, and for the effect of shakes, knots, etc. For example, the case of a balk 13½ in. square and 10 ft. 6 in. span gave 1,114 cwt. as the result according to Tredgold, and 1,120 cwt. according to Barlow. The same size and quality of timber tested by the Mersey Dock Engineers

Fig. 162.—Beam with Wrought Iron Strap.

gave a result of 610 cwt. only, against the preceding figures. It appears from the latter example that the constant should be reduced to, say, 2·6 or 2·3, or say 2·5 to 3 for Memel timber. The formula then is

$$\frac{B\ D^2\ C}{L} = W$$

where w is the breaking weight in cwt. in the centre of the span. We then have for

An experiment was made some time ago by Kirkaldy on the strength added to a beam by the fixing on the top of the beam of a flat iron bar. The span of the beam was 24 ft., and the depth and width were 14 in. and 12 in. respectively. According to the above formula, with a constant of 3, the central breaking load should be

$$\frac{14^2 \times 12 \times 3}{24} = 294 \text{ cwt.} = 14\tfrac{3}{4} \text{ tons, or}$$

with constant of 2·5

$$\frac{14^2 \times 12 \times 2.5}{24} = 245$$

cwt. = 12¼ tons. When the experiment was made, however, the beam snapped suddenly with a central load of 10 tons, showing that the above constants were too high for this case. A similar beam (Fig. 162) was then provided with a wrought-iron strap fixed on the top, and it was then found that the beam failed slowly and gradually under a load of 13 tons—an experiment which showed that added strength was given to

Fig. 163.—Cup-shake in Log.

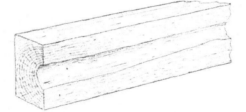

Fig. 164.—Cup-shake in Balk.

Fig. 165.—Heart-shake in Log.

a beam 12 in. wide, 11 in. deep, and 24 ft. span, a strength of

$$\frac{12 \times 11^2 \times 3}{24} = 181.5 \text{ cwt.} = 9 \text{ tons } 1\tfrac{1}{2} \text{ cwt.}$$

Kirkaldy's experiments with a beam of this span and size showed the strength to be 10 tons. It is considered that timber has a set with only one-fifth of its breaking load, and is really safe when loaded to only one-sixth of the breaking load. If the beam is fixed at both ends, it is stronger than when only supported at the ends as 3 is to 2. Some qualities of timber are stronger in tension than in compression, whilst others have just the opposite qualities. Experiments show that Dantzig fir is crushed before it is torn asunder; or, in other words, that its ultimate compressive stress is less than its ultimate tensile stress as 4 is to 5.

the beam by the addition of the iron bar. In other kinds of timber possibly the iron strengthening bar should be on the other side of the beam.

Fig. 166.—Heart-shake in Balk.

Defects in Balk Timber.

Cup-shakes.—These are cracks extending circumferentially at one or more places, caused by the separation of the annual rings, as in Figs. 163 and 164.

Doatiness.—This is a speckled stain found in beech, American oak, and other timber, due to incipient decay. It is produced by imperfect seasoning or by exposure for a long period to a stagnant atmosphere.

Fig. 167.—Sapwood in Balk.   Fig. 168.—Star-shake in Log.

Dry Rot.—If the balks have been stacked on land with insufficient ventilation, the growth of a fungus over them, like white or brown roots, may indicate that dry rot has already begun, although it is chiefly found under kitchen floors.

Foxiness.—A reddish or yellowish brown tint in the grain, caused by incipient decay.

Fig. 169.—Star-shake in Balk.

Heart-shakes.—These are splits or clefts occurring in the centre of the tree, as in Figs. 165 and 166. They are common in nearly every variety of timber, and are very serious when they twist in the length, as they interfere with the conversion of the tree into boards or scantlings. They some-

Fig. 170.—Twisted Fibres.

times divide the log in two for a few feet from the end.

Knots.—Large, or dead and loose knots are objectionable, as they weaken the timber, and are unsightly. A timber pile, in which knots occur diagonally, is liable to be sheared through the knots or severely damaged by the blows of the ram. Dantzic timber has the largest knots, spruce the hardest.

Rind-galls.—These are curved swellings caused by the growth of new layers over a part damaged by insects, or by tearing

Fig. 171.—Upset or Crushed Fibres.

off or imperfect lopping of a branch. These are shown by the grain being irregular and vacuous.

Sapwood.—This occurs more in some trees than in others—say, Dantzic much, pitch-pine little. It may be known by its greenish tinge, and holding the water longer than the sound parts after having been wet. If creosoted, the sapwood is as lasting, but not so strong as the heartwood. It generally occurs at the corners only of the balks, which arises from the desire to save as much timber as possible. (See Fig. 167.)

Fig. 172.—Waney Edge in Balk.

Star-shakes.—When several heart-shakes occur in one tree (see Figs. 168 and 169) they are called star-shakes from the appearance produced by their radiation from the centre.

Thunder-shakes.—These are irregular fractures across the grain, occurring chiefly in Honduras mahogany.

Twisted Fibres.—These are caused by the tree being twisted in its growth, from the action of the wind upon the head. Timber so affected is not suitable for cutting up into joists or planks, owing to the fibres running diagonally in any longitudinal cut, as in Fig. 170. Oak with twisted fibres will not retain its shape when squared, but is very suitable for splitting up into wall plugs.

Upsets.—These are portions of the timber where the fibres have been injured by crushing, as in Fig. 171.

Waney Edges.—These occur when the top end of the tree is not large enough to hold up to the full size to which the lower end is squared, as shown by Fig. 172. These balks may be used for piling without detriment if the top end be driven downwards.

Wide Annual Rings.—These generally indicate soft and weak timber.

Wind-cracks.—Shakes or splits on the sides of a balk of timber, caused by shrinkage of the exterior surface, as in Fig. 172, are called wind-cracks.

Wet Rot.—Timber that has been lying long in the timber ponds, and subjected to alternations of wet and dry, may be so soft and sodden as to have reached the stage of wet rot. The term " wet rot " implies chemical decomposition of the wood ; whereas dry rot is the result of a fungous growth.

Dry Rot.

Cause, Cure, and Prevention.—Dry rot is a special form of decay in timber, caused by the growth of a fungus, *Merulius lachrymans*, which spreads over the surface as a close network of threads, white, yellow, or brown, and causes the inside to perish and crumble. Various causes may combine to render the timber favourable to the growth of this fungus—namely, large proportion of sapwood ; felled at wrong season when full of sap ; if cut down in the spring or the fall of the year instead of in midwinter or midsummer, when the sap is at rest ; stacked for seasoning without sufficient air spaces being left ; fixed before thoroughly seasoned ; painted or varnished while containing moisture ; built into wall without air space ; covered with linoleum ; exposed to warm, stagnant air, as under kitchen floors. There is no cure when the fungus has obtained a good hold. The worst must be cut out and remainder painted with blue vitriol (cupric sulphate). The best preventive is to use only well-seasoned timber and to keep it well ventilated.

Detection and Treatment of Dry Rot.—When dry rot is suspected in a floor the floor-boards should be lifted at the corners of the room, or at dead ends of passages, or wherever signs of weakness show themselves, and the surfaces of the joists, wall-plates, and under side of the floor-boards should be closely examined for fungus, mildew, or any unhealthy sign, such as a brown semi-charred appearance. If any is found, the worst parts should be cut out and renewed, the remainder well scraped over, including the walls, and well washed with a solution of blue copperas (sulphate of copper). If the earth below is found to be damp, a layer of cement concrete should be spread over it, not less than 4 in. thick. Air bricks and ducts should be placed in the walls on opposite sides, to get a through current, as moist, warm, stagnant air is the most potent aid to dry rot ; and every endeavour should be made to obtain thorough ventilation. The means of prevention are : Thorough seasoning, free ventilation, creosoting or charring if necessarily exposed to damp earth, and painting with vitriol or cupric sulphate.

Preservation of Wood Underground.

The best way to preserve from decay wood that is to be buried in the ground is to creosote the wood ; this does not mean painting the wood over with tar, but proper creosoting by the regular process. The butt-end of a post to be placed in the ground may be charred over a wood fire, quenching with water when the wood is charred, say, $\frac{1}{4}$ in. to $\frac{1}{2}$ in. deep. This will prevent rotting and the attacks of worms, but it is necessary that the wood should be previously well seasoned, or the confined moisture will cause decay. Chloride of zinc and water, about 1 to 4, in which wood is steeped under Sir Wm. Burnett's system (see p. 34), preserves the timber from decay and renders it incombustible. A method sometimes adopted is to bed the posts in cement concrete, but this is not quite so good as creosoting.

Soft Woods and Hard Woods.

Timber trees are usually divided into two great classes :—(a) Soft woods or coniferous woods ; (b) hardwoods or leaf woods.

So far as the texture and hardness of each is concerned, some of the former are really more difficult to work than the latter. For example, pitchpine, owing to its resinous nature, is usually more difficult to work than basswood, or the softer kinds of mahogany. The general distinguishing features of soft woods are :—All the trees bear cones and never have broad flat leaves. The timber usually has distinct annual rings formed of two layers, the inner one (known as springwood) being soft, porous, and pale in colour ; the outer (called autumn wood) is harder, more compact, and filled with resinous matter. The whole annual ring is formed of long tapering tubes, interlaced, and breaking joint with each other, and having a small portion of cellular tissue at intervals, and resinous matter in the interstices. Hardwood trees bear broad flat leaves, the timber is never resinous. The annual rings, owing to slower growth, are often much closer together than in softer wood. They are of more uniformity in colour and hardness, but have more or less distinct radial lines, consisting of thin, hard vertical plates, formed entirely of cellular tissue, the medullary rays or silvergrain or flower already mentioned.

Varieties of Timber.

Fir Timber, Converted.—Fir timber, when sawn into convenient sizes suitable for joiner's work, is called deal. It is brought into the market sawn into different widths, which are often classed as deals, or distinguished as battens, deals, and planks. They vary from 2 in. to 4 in. thick, but are mostly 3 in. thick, and from 8 ft. to over 20 ft. in length. All that are under 8 ft. in length are classed as ends, and are sold at a cheaper rate. When 7 in. wide and under they are termed battens ; deals, from 8 in. to 9 in. ; and planks when above 9 in.

Baltic Yellow or Red Deal.—The best yellow deal for building purposes is shipped from the Russian ports of Petersburg, Onega, and Archangel, and the Swedish ports of Söderhamn, Gefle, Stockholm, and Holmsund. Onega, Archangel, and Gefle supply deals of the best quality. The greater portion of the Swedish timber is coarse, but at the same time some of the very best deals, both yellow and white, come from Gefle and Söderhamn. The best Swedish deals run more sound and even in quality than the Russian, owing to the different way in which the timber is converted. A balk of Russian timber is cut into deals, etc., of one quality, and thus they show very many hearts or centres. In Swedish timber the inner and outer wood in the same balk are converted into different quality deals, the centre being put into the lower class ; hence the high price put upon first-class Swedish deals. Deals cut from the centre of the log should not be cut into boards ; 4-in. deals are in nearly all cases cut from the centre of the balk, and consequently are subject to shakes, and unsuitable for boards. Swedish 2-in. and $2\frac{1}{2}$-in. deals of good quality are preferred to 3-in., as they are cut from the sound outer wood. Their value not being generally known, they do not fetch such high prices as the 3-in. deals. The export of deals from the Prussian ports of Dantzic, Memel, Stettin, etc., is almost entirely confined to yellow planks and deck deals (also called red deals), from 2 in. to 4 in. These are suitable for scantlings, framing of roofs, and many purposes connected with housebuilding, engineering, etc. The reason of the timber from the above ports being shipped in an unconverted state is that the wood, being grown in a warmer climate, is coarse in the grain, and could not compete in a converted state with the closergrained exports from the more northerly ports of the Baltic. Baltic yellow deal or red deal is from the *Pinus sylvestris*, or Northern pine. The colour of the wood is generally of a reddish yellow or of a honey yellow of various degrees of brightness, annual rings about $\frac{1}{16}$ in. wide, the outer part being of a bright and reddish colour. When knots occur they are from 1 in. upwards in diameter, and not very hard ; they are of a rich red brown colour, and thin shavings of them are semi-transparent. This timber is stronger and more durable than white deal (*Abies excelsa*, described below).

Baltic White Deal or Spruce Fir.—This is *Abies excelsa*, and is used in the common qualities for the roughest work—scaffold poles, scaffold boards, centering, packing cases, etc.—and in the better qualities for dressers and table tops, bedroom floor-boards cupboard shelves, etc. The wood is of yellowish white, or sometimes of a brownish red colour, becoming of a bluish tint when exposed to the weather. The annual rings are generally clearly defined, the surface when planed has a silky lustre, and the timber contains a large number of very hard, glassy knots. The sapwood is not distinguishable from the heart. Baltic white deal is recognised chiefly by its small hard and dark knots, by its woolliness on leaving the saw, and by its weathering to a greyish tint. When fresh cut, the grain may be more or less pronounced than that of yellow deal. It is subject to streaks of resin in long cavities, and to loose dead knots.

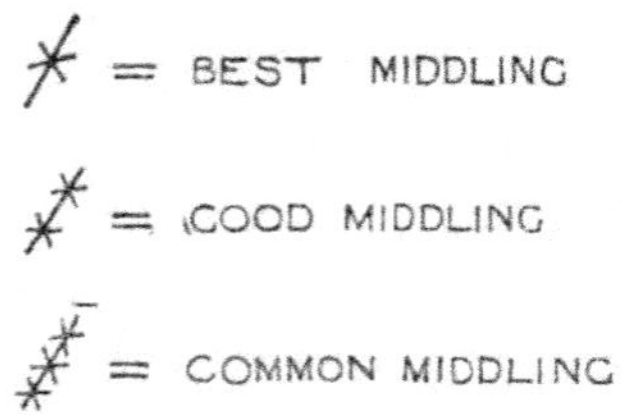

$$\divideontimes = \text{BEST MIDDLING}$$

$$\divideontimes = \text{GOOD MIDDLING}$$

$$\divideontimes = \text{COMMON MIDDLING}$$

Fig. 173.—Dantzic Timber Quality Marks.

In white deal or spruce fir the knots are small, darker, more brittle, and opaque.

Scotch Fir.—This is the wood of *Pinus sylvestris*, and is called also the Northern pine and red or yellow pine. From this the timber known as yellow or red deal is obtained ; it is tough and strong for its weight, durable and easily worked, cheap and plentiful. Comes principally from the north of Europe, and is shipped at Baltic ports. Characteristics : Colour varies according to soil and habitat ; generally of a honey yellow, with distinct annual rings darker and harder on the outside of each, some specimens changing to a reddish cast in seasoning, and others brownish. There are no medullary rays visible. The best has close grain and a medium amount of resin in it. The wood is silky when planed, and when well seasoned crisp and dry to the touch. Its tenacity is 5 tons per square inch, and weight 36 lb. per cubic foot. It requires periodical painting when

exposed to the weather. It is used for all kinds of carpentry and joinery. Its source of supply is chiefly the Baltic ports, whence it comes as deals and logs.

Fir Timber, Unconverted.—All Baltic fir is akin to the Scotch fir (*Pinus sylvestris*) or the spruce fir (*Abies excelsa*), the wood of the former being known as red fir, Baltic fir, Memel fir, etc., in the unconverted state, whilst the wood of the spruce is known as spruce fir, or white fir if unconverted ; but as white planks, deals, or battens if converted. At the outset this peculiarity of calling the same wood red or yellow under different circumstances should be noticed, since the terms applied have led to the very prevalent and mistaken notion that red, yellow, and white denote three, instead of only two, kinds of Baltic fir.

Riga Fir comes from the Russian port of that name, north of Memel, and is inferior in strength to Dantzic and Memel fir of best quality, and does not average so large. It runs about 12 in. square and 40 ft. long, but it is often preferred for cutting into scantlings, being of straighter grain and freer from knots. It is, however, subject to heart-shakes.

White Fir.—But little Baltic white fir comes into the market as square timber. When it does, it is termed white timber or spruce fir ; but spruce poles, or the young trees felled and stripped of their branches, are imported from Sweden and Norway for scaffold poles, the very best being selected as ladder poles. They run in lengths of from 18 ft. to 50 ft.

Prussian Fir Timber.—Sources : Memel, Dantzic, Stettin, Königsberg. The use of the balks is almost entirely confined to heavy timber work, as they are too coarse and open in the grain for being wrought for joiners' work. They are used for outdoor carpentry and heavy woodwork, such as piles, girders, roofs, and joists. Dantzic —Size : 14 in. to 16 in. square, 20 ft. to 50 ft. long. Appearance : Subject to cupand star-shakes and wind-cracks. Knots large and numerous, often dead and loose ; they are very objectionable when grouped near the centre of a beam, or for piles when diagonal. Annual rings wide, large proportion of sapwood (frequently the whole of the four corners of the circumscribing square), 20 ft. to 45 ft. long, heart sometimes loose and "cuppy." Marks: Scribed near centre, as in Fig. 173. It is used for heavy *outdoor carpentry, where large scantlings are required.* Memel fir is tolerably free from knots, but when they occur the grain near them is irregular, and is apt to tear up with the plane.

Norwegian Deals and Balks.—Sources: Christiania, Friedrichstadt, Drontheim, Dram. *Size: Average 8 in. to 9 in.* square, generally tapered; scarcely called balk timber; is known as "under-sized." Appearance: Much sap. Marks: on balks, others by letters, stencilled in blue on ends. Uses: Staging, scaffolding, and coarse carpentry, the best converted into deals, *flooring, and imported joinery. Norwegian* timber is clean and carefully converted, but is imported chiefly in the shape of prepared flooring and matchboarding. Scarce in form of yellow deals, but of high quality. Christiania best, but often contains sap. *Christiania white deal used for best joinery. Christiania and Dram used* for upper floors on account of white colour. Friedrichstadt has small black knots. Some Drammen deals warp and split in drying.

Swedish Deals. — Sources: Stockholm, Gefle, Söderhamn, Gothenburg, Sundsvall, Holmsund, Hernosand. *The greater portion of this is coarse and bad, but some of* the very best Baltic deal comes from Gefle and Söderhamn. First qualities have a high character for freedom from sap, heart-shakes, etc. The lower qualities have the usual defects, being sappy and containing large, coarse knots. *In the best qualities the knots are small, and larger in the lower* qualities. The yellow deal is generally small, coarse, and bad, with large loose knots, sappy, liable to warp and twist, but variable, the best being equal to Norwegian, owing to care in conversion and sorting out into different qualities. *The cheap imported joinery is made from these deals.* They are suitable for floors where warping can be prevented. Gefle and Söderhamn deals are sometimes very good. White deals from Gothenburg, Hernosand and Sundsvall are used for packing-cases. Gefle and

Söderhamn deals are good for upper flooring, dressers, shelves, etc., and backing to veneers. There are also said to be red deals from the Baltic ports and from Canada, from the *Pinus rubra*, used for mouldings and best joinery, very like Memel. Swedish woods are never hammer-marked, but invariably branded with letters or devices stencilled on the ends in red paint, which makes it difficult to judge of their quality by inspection, as they are stacked in the timber yards with their ends only showing. Some of the common fourth- and fifth-quality Swedish goods are left unmarked, but they may generally be distinguished from Russian shipments by the bluer colour of the sapwood. The first and second qualities in Swedish deals are classed together as "mixed," being scarcely ever

Fig. 174.—Riga Timber Quality Marks.

sorted separately, after which come third-down to fifth-quality goods. Deals of lower quality than third are nearly always shaky, *or very full of defects of some kind.*

Russian Timber.—Sources: Petersburg, Archangel, Onega, Riga, Wyborg, Narva. These yellow deals are the best for general building work, more free than other sorts from knots, shakes, sap, etc., clean hard grain and good wearing surface, but do *not stand damp well.* First three used for best floors—all of them for warehouse floors and staircases. Wyborg—very good, but inclined to sap. Riga—best balk timber. Size: Up to 12 in. square, and 40 ft. long. Appearance: Knots few and small, very little sap, annual rings close, wood close *and straight-grained, more colour than Dantzic.* Marks: Scribed at centre, as in Fig. 174. Uses: For masts and best carpentry when large enough, also for flooring and internal joinery. Petersburg—inclined to be shaky. Archangel and Onega—knots often surrounded by dead bark, and drop out when timber is worked. The Russian white deals shrink and swell with the weather, even after painting. Best from Onega. Russian deals generally come unmarked into the market, or only dry stamped or marked at their ends with the blow of a branding hammer, such marks being also termed hard brands. In some cases where the goods are not branded, the second quality have a red mark across the ends, third- being easily distinguished from first-quality goods. The well-known Gromoff Petersburg deals are, however, marked with " C. and Co.," the initials of the shippers (Clarke & Company).

First Quality Marks

Gromoff   or   1

Second Quality Marks

Gromoff   or   0

Another good Petersburg brand is " P. B." (Peter Belaieff) for best, and " P. B. 2 " for second quality.

St. Petersburg Brands :—

Belaieff's Shipment.

First Quality Marks.
P B S & Co.

Second Quality Marks.
P B S & Co.
2

Third Quality Marks.
P B S & Co.
3

Russanoff & Co.
First
N R¹ & S
Second.
N R² & S
Third.
N R³ & S
N. Pavloff.
First.
W O D A
L O G
Second.
W O L O G D A

Russian Goods :—

Imperial Appanage Shipment (Czar's Stock).

1st.	2nd.
Double Eagle 7ʳ	Double Eagle Ƶ

3rd.	4th.
Double Eagle T	Double Eagle Ч

All hammered on butt with the Imperial Arms (double eagle).

Best Archangel Stock :—

Maimax Shipment.

Yeo.	1	△	3	4
Bds.	1	△	∩	0
White	①	②	③	

Onega Wood Co.'s Shipment.

| Deals } Battens } | 1 | 2 | 3 | 4 |
| Boards | X Red | V Red | V Black | V Red |

Amossoff, Gernet Shipment.

A G & Co.   A G & Co.
A   B

A G & Co.   A G & Co.
C   D

Archangel :—

E. H. Brandt & Co. Shipment.

1st.   2nd.   3rd.   4th.

Marks, Archangel :—

Olsen & Stampe Shipment.

O S   O S   O S   O S
 I   II   III   IV

Surkow & Shergold Shipment.

S & S   S & S   S & S   S & S
 I   II   III   IV

Surkow & Shergold K. E. M. Shipment.

KEM. I   KEM. 2   KEM. 3   KEM. 4

Russanoff's Mesane Shipment.

R   R   R   R
N & S   N & S   N & S   N & S
 I   2   3   4

Timbers from Russian and Finland ports are dry-stamped on the ends without colour.

American Yellow Pine.—This is the wood of *Pinus strobus*, and is known also as American yellow deal, Weymouth pine, American white pine, pattern-maker's pine, etc. It is used chiefly for panels on account of its great width, for moulding on account of its uniform grain and freedom from knots, and for patterns for casting from on account of its softness and easy working. It is very uniform in texture, of a very pale honey-yellow or straw colour, turning brown with age, usually free from knots, and specially recognised by short, dark, hair-like markings in the grain when planed, and its light weight. It is subject to cup-shakes and to incipient decay, going brown and "mothery." It takes glue well, but splits in nailing. American woods are not branded, as a rule, though some houses use brands in imitation of the Baltic marks, though without following any definite rules. The qualities may, however, very often be known by red marks "I.," "II.," "III.," upon the sides or ends, but the qualities of American yellow deals are easily told by inspection, the custom in the London Docks being to stack them on their sides, so as to expose their faces to view, and allow of free ventilation. Woods from Canadian ports have black letters and white letters on the ends, and red marks on the edges. American yellow pine may be purchased in balks over 60 ft. in length and 24 in. square. It is not so strong as the American red pine, but is much lighter, and so is distinguished when floating by the height it stands above the water. First-quality pine costs more than any other soft wood used for joinery.

American Red Pine.—This is the wood of the *Pinus mitis*, which is called in America the yellow pine, and is very like the wood of the Scotch fir, though it does not equal it in strength or durability, neither does it grow so large as the Dantzic and Memel timber. Being very straight-grained and free from knots, it is valuable for joiners' work, when a stronger wood than yellow pine (described in the previous paragraph) is required. It is more expensive than Baltic fir, consequently is not so largely used in England. Red pine has of late years been used rather extensively owing to the scarcity of good yellow deal

and the high price of yellow pine. The cost is about the same as Gromoff.

American White Spruce.—This is very like Baltic white timber, but, not being equal to it in durability or strength, it does not command such a large sale as the Baltic white timber. It is the produce of two different trees, the *Abies alba*, or white spruce, and the *Abies nigra*, or black spruce, so named from the colour of their bark; the colour of the wood is white in both cases. The black spruce timber is far better than the white, is more plentiful, and grows to a greater size.

Elm.—Common English elm (*Úlmus campestris*) is of a reddish brown colour with light sapwood, the grain being very irregular and there being numerous small knots. It warps and twists freely, but is very durable if kept constantly under water or constantly dry, but it will not bear alternations of wet and dry. One peculiarity characteristic of elm is that the sap turns white and becomes foxey, and decays quickly. It is used for coffins, piles under foundations, pulley blocks, stable fittings, etc. It is chiefly home-grown.

American Elm.—The wood generally known as American elm is one of the United States timbers (*Úlmus Americána*, L.) locally known as white elm, or water elm. The wood is highly valued, has many properties similar to those of American rock elm *Úlmus racemósa* Thomas)—though not, perhaps, quite so tough as that timber—and is very extensively used in cooperage, saddlery, axe-helves, etc., and wagon- and boat-building. The tree which furnishes the wood grows to large dimensions, and is widely distributed over all the States east of the Mississippi River.

Pitchpine.—This is *Pinus Australis* or *Pinus resinosa*, and is recognised by its weight and strong reddish yellow grain, with distinct and regular annual rings. It must be well seasoned and free from sap and shakes. Pitchpine is very free from knots, but when they occur they are large and transparent, and give variety to the grain. It is used chiefly for treads of stairs and flooring, on account of its hardness and wear-resisting qualities; for doors, staircases, strings, handrails, and balusters on account of its strongly marked and handsome grain ; for open timber roofs on account of its strength and appearance ; and for outdoor carpentry, such as jetties, on account of its length and size. The ornamental grain of pitchpine is due to the annual rings, not the medullary rays as in oak, and the method of sawing oak will therefore not suit at all. The object should be to cut as many boards as possible tangent to the annual rings. About one log in a hundred will show more or less waviness of the grain owing to an ir-regular growth in the tree, and about one in a thousand will be worth very careful conversion. To avoid turning the log so often in cutting the boards, as in Fig. 137 (p. 35), they might be all cut parallel, so obtaining a greater number of wide boards but not of such good figure, the grain showing straight lines towards the edges instead of a fair pattern throughout. Careful and complete seasoning would be required, on account of the great shrinkage occurring.

Oak.—English oak (*Quercus*) is of a light brown or brownish yellow, close-grained, tough, more irregular in its growth than other varieties, and heavier. Its tenacity is, say, $6\frac{1}{2}$ tons per square inch, and its weight, 55 lb. per cubic foot. Baltic oak from Dantzic or Riga is rather darker in colour, close-grained, and compact, and its weight is 49 lb. per cubic foot. Riga oak has more flower than Dantzic. American or Quebec oak is a reddish brown, with a coarser grain, not so strong or durable as English oak, but straighter in the grain. Its tenacity is 4 tons per square inch, and weight 53 lb. per cubic foot. African oak is not a true oak. Exposed to the weather, oak changes from a light brown or reddish grey to an ashen grey, and becomes striated from the softer parts decaying before the harder. In presence of iron it is blackened by moisture owing to the formation of tannate of iron, or ordinary black ink.

Wainscot Oak.—This, known also as " Dutch wainscot," is a variety of oak. It has a straight grain free from knots, is easily worked, and not liable to warp. In conversion it is cut to show the flower or sectional plates of medullary rays. It is used for partitions, dados, and wall panel-ling generally ; also for doors and windows in high-class joinery. Its sources are Hol-land and Riga, being imported in semi-circular logs. Wainscot oak obtained from Riga is spoken of as Riga wainscot. The term " wainscot " describes the method that is adopted, when converting the log into boards, in order to show a large amount of silver grain ; such ornamental boards are specially suitable for wainscoting. The oaks that grow around the Baltic are closely related to those that grow in England. *Quercus robur* affords the best timber as regards strength and durability under exposure, though some of the other varieties (as, for instance, the *Quercus sessiliflora*, or cluster-fruited oak) have an equally pretty figure. Riga oak is not held in such high esteem as English oak for outdoor work and for purposes in which great ten-sile or compressional strength is necessary ; but as the medullary rays are very promin-ent, Riga oak affords a very pretty silver grain. Riga oak may be the wood of either of the two varieties of *Quercus* mentioned above.

Chestnut.—The chestnut timber used for building is the sweet or Spanish chestnut (*Castanea edibilis*), not the common horse chestnut, which is a whitish wood of but little use. The Spanish chestnut is grown only to a small extent in Great Britain at the present time ; it may be known by the leaves being smoother, more parallel, and not radiating so decidedly from one stalk. Spanish chestnut closely resembles coarse-grained oak in colour and in texture, and the wood in all its stages of manufac-ture is frequently mistaken for oak. The bark of the log is like oak bark. The planks are of practically identical appearance, and even after the wood is dressed up the like-ness is still very close. However, when the chestnut is old it has rather more of a cin-namon cast of colour, has less sapwood, and generally a closer grain, although softer and not so heavy as oak. The chief dis-tinguishing characteristic of the chestnut is the absence of the distinct medullary rays which produce the flower in oak ; and old roof-timbers, benches, and church-fittings may be discriminated in this way, also by the chestnut being more liable to split in nailing, while the nails never blacken the

CARPENTRY AND JOINERY.

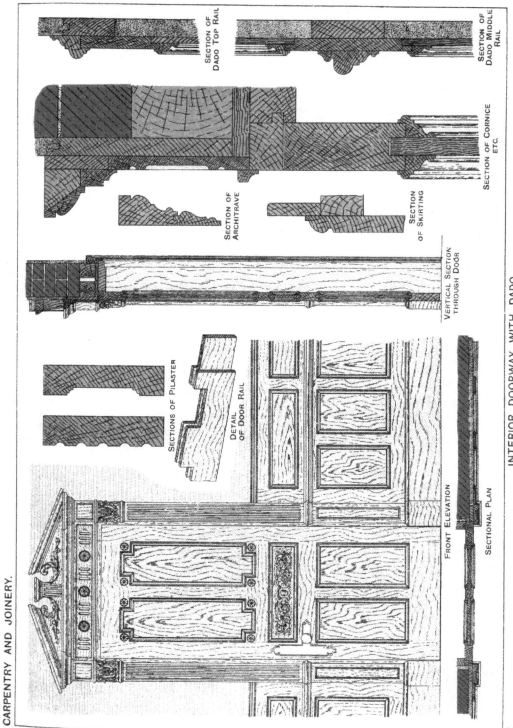

SECTION OF DADO TOP RAIL

SECTION OF DADO MIDDLE RAIL

SECTION OF CORNICE ETC.

SECTION OF ARCHITRAVE

SECTION OF SKIRTING

VERTICAL SECTION THROUGH DOOR

SECTIONS OF PILASTER

DETAIL OF DOOR RAIL

FRONT ELEVATION

SECTIONAL PLAN

INTERIOR DOORWAY WITH DADO

2

timber as they do in oak (see p. 48). Reports vary, but it seems to be decided that the roof of Westminster Hall is of oak, and that of the circular part of the Temple Church of chestnut. Chestnut is largely used in cabinet work for the interior fittings of bureaux, writing cases, etc. Horse chestnut is a white, even, close-grained wood, which could not possibly be confounded with Spanish chestnut when the difference has been once pointed out. Horse chestnut is, however, often substituted for other white woods of close texture, such as lime, holly, and sycamore, and is the inferior material of these three woods. Horse chestnut warps and twists badly, and is liable to turn yellow. At the present time both woods are largely used in fretwork. Spanish chestnut was for-

Fig. 175.—Owner's Mark on Teak.

merly used in conjunction with, and also as a substitute for oak.

Teak.—This wood (*Tectona grandis*) is generally used in situations where red deal and other similar woods are liable to decay or to be destroyed by worms, but its cost prevents it from being used extensively for constructional work. In the construction of hothouses it would be a decided advantage to use teak for all the sills, and for any parts of the staging that are likely to be alternately wet and dry; or for any timbers that, through being in contact with the earth, are always wet; but it is doubtful whether the extra cost of teak would be repaid if it were used for all the woodwork of the house, such as bars, rails, mullions, etc. Teak is used principally for staircases and doors in public buildings. Being considered more fireproof than any other wood, its use is enforced by district surveyors under the Building Act. In special positions, such as those enumerated above, where it is not practicable to paint the woodwork, or where, as in the case of sills, the paint becomes worn off, the greasy nature of teak and the poisonous oil that it contains preserve the wood from decay, and enable it to withstand the attacks

of the spores of dry-rot fungus, without any necessity for covering the surface with any protective coating of paint, varnish, etc. As the life of a piece of teak under such conditions would in all probability be three or four times that of northern pine, it may be inferred that the extra initial cost is compensated by the saving in repairs. Teak logs vary generally from 10 in. to 24 in. square, and 15 ft. to 40 ft. long. When first removed from the ship they are of a good cinnamon brown colour, but soon bleach in the sun, and might at first sight

$$\text{\Large ⚹ |||} = 63 \text{ CUB. FT.}$$

$$\text{\Large ✕} = 21 \text{ CUB. FT.}$$

$$\text{\Large ✕//} = 26 \text{ CUB. FT.}$$

$$\text{\Large ✕/} = 26 \text{ CUB. FT.}$$

$$\text{\Large ✕/} = 16 \text{ CUB. FT.}$$

Fig. 176.—Chalk Marks on Logs Showing Cubic Contents.

be mistaken for oak. They are stacked in piles according to the ownership, with the butt ends flush and the other ends irregular. A few business cards of the timber-broker's firm are generally nailed here and there. The balks are squared up fairly straight and true, but sometimes waney at top end, with heart out of centre owing to the tree having been bent during growth. The ends are stamped with the mark of the firm, often in two or three places, initial letters in a heart, as in Fig. 175, standing for Messrs. ———. Also with the number of the log, and alongside it a mark, thus *, or the word No. preceding the figures to show in which direction they should be read. The dimensions of the log are stamped in 1-in. figures, thus 193 × 22½ × 21 meaning 19 ft. 3 in. long by 22½ in. wide by 21 in. thick. The cubic contents are marked in red chalk, as in the strokes of Fig. 176. These are similar in composition to the quantity marks on Baltic timber; their value is shown added in

Fig. 176. After the logs are all stacked, the invoice mark, as $\frac{21}{743}$, and number of the log are painted on the end of each with white paint to identify them more rapidly; and on the end of a log showing on the outside of the stack the name of ship and number of pile are also painted. The number of pile and name of ship are also painted at the side of each pile, on one of the logs, as in Fig. 177. The principal teak yard in London is at the South-West India Dock.

Mahogany.—The true mahogany (*Swietenia mahogani*) is a dense, hard, strong wood, of straight growth and close texture, and is a rich brownish red in colour, with dark wavy markings; the pores are small and are filled with a chalk-like substance. The weight of the wood, when dry, should average about 65 lb. per cubic foot. The commoner substitutes for true

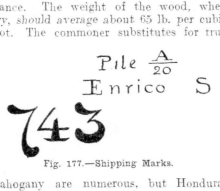

Pile $\frac{A}{20}$

Enrico S

743

Fig. 177.—Shipping Marks.

mahogany are numerous, but Honduras mahogany (baywood) and Panama mahogany may be taken as examples. The weights of these kinds of mahogany vary from 29 lb. to 35 lb. per cubic foot; hence, weight is evidently one of the surest tests of the quality of mahogany. Generally speaking, these commoner varieties are much lighter in colour than true mahogany, and are without the fine black lines running through the grain that form one of the distinguishing characteristics of true mahogany. The hardness of true mahogany is about twice as great as that of the commoner kinds; that is, the best mahogany may be taken as being equal in wearing power to hornbeam, while the inferior kinds would not be harder than Weymouth pine. A large quantity of the light-coloured inferior mahogany is used for furniture, etc., and after having been treated with a specially

prepared darkening oil, or darkened by some other method and polished, in no way differs in appearance (as far as the uninitiated can judge) from the best kinds of mahogany. Light weight and lack of resistance to indentation should, however, make one suspect the quality of any wood that claims to be true mahogany. However, it may be said that "Panama mahogany" includes several useful varieties of woods, to this class belonging St. Domingo, City St. Domingo, and Cuba mahoganies. The grain of these varieties is very fine and even, and the surface of the wood is lustrous, and often has a watered or satiny appearance. These varieties of mahogany are not so dark coloured nor so dense as the true mahogany (*Swietenia mahogani*). Some of the most prominent timber experts state that the characteristics of the various mahoganies are so confusing that great difficulty occurs at times in distinguishing one variety from another; in fact, some go so far as to say that the wood that is sold and known commercially as *Swietenia mahogani*, or true mahogany, is almost entirely different from the specimens of that wood that are exhibited at Kew. This does not prove that these woods are in any way inferior to true mahogany, but that they are obtained from another and an entirely different kind of tree. Hence, therefore, density and colour are two important factors that should be considered when comparing one variety of mahogany with another. Other more minute points of difference are only apparent when the wood is examined through a microscope. The term "Spanish mahogany" is used principally in connection with Cuban mahogany, but sometimes St. Domingo mahogany is termed Spanish mahogany. The term is at best but a vague one, and is rarely used in connection with shipments of timber that arrive in this country from abroad. In such cases the cargo is usually spoken of as so many logs of Cuban (St. Domingo, Mexican, or other) mahogany. In a general sense the term "Spanish mahogany" is used in contradistinction to baywood or Honduras mahogany. No rule or regulation states precisely that Spanish mahogany must fulfil certain specified conditions, and a timber merchant may fairly claim that he acts justly by his customer if he supplies him with a moderately dense, sound quantity of Cuban or St. Domingo mahogany, and has not supplied him with baywood (Honduras mahogany) nor any of the many other so-called mahoganies, the marked characteristics of which differ widely from Cuban mahogany. Such spurious mahoganies are, for example, the so-called African mahogany (*Khaya Senegalensis*) or the wood that is known as Australian mahogany (*Dysoxylum Fraserianum*).

Timbers for Various Purposes.

In the following list the timbers are stated in order of superiority for the purposes named. All the timber should be specified according to the precise quality required, and not merely as "the best."

Dock Gates.—Greenheart, oak, creosoted Memel. The specification of the 60-ft. entrance lock gates at the Victoria Dock, Hull, provided for ribs, heads, and heels of single squared timbers, either of English oak of the very best and quickest grown timber, or of African oak, but no mixture of the two. The planking was specified to be of greenheart.

Doors External for Public Buildings.—Oak used most frequently, next in order mahogany, teak, and pitchpine.

Doors Internal for more Important Buildings.—Oak, mahogany, teak, walnut, pitchpine. Other hardwoods are also used, according to, or in keeping with, other internal fittings. For ordinary buildings, yellow deal for framing and yellow pine for panels.

Floor Boards.—Oak, pitchpine, Stockholm or Gefle yellow deal; and for upper floors, Dram or Christiania white deal. For common floor-boarding, Swedish or Norwegian yellow or white deal.

Floor Joists.—Russian deals make the best joists, as they are straight-grained and free from knots, sound and tough. Baltic fir is cheaper and next best. Swedish and Norwegian not reliable.

Half-timber Framing.—Oak is best, as it resists decay the longest, and can be obtained naturally shaped in curves or straight, as may be required. The colour and texture are also suitable for architectural effect.

Teak is good, but does not weather quite so good a colour; it is apt to split with nailing. Larch is next best.

Pile Foundations.—Greenheart, oak, elm, creosoted Memel, alder. Greenheart is undoubtedly best, but the cost is prohibitive except for marine work, where it is sometimes essential, as sea-worms will not attack it. Oak is next best when it can be afforded. Memel fir (*Pinus sylvestris*) in 13 in. to 14 in. whole timbers, creosoted or in its natural state, is the most suitable under ordinary circumstances, owing to its convenient size, length, and general character. Riga fir is generally too small, and Dantzic fir too large and coarse. Pitchpine is considered suitable by some; its chief advantage is the large size and great length in which it may be obtained. American elm and English elm, beech, and alder are suitable if wholly immersed, but not otherwise.

Planking to Earth Waggons.—Elm, with ash for shafts, if any.

Roof Trusses.—Oak, chestnut, pitchpine, Baltic fir (Dantzic, Memel, or Riga). For tie-beams to open timber roof 40-ft. span pitchpine is best, as it can be obtained free from knots, in long straight lengths, and the grain is suitable for exposure either plain or varnished. Oregon pine is suitable for similar reasons, but not so well marked in the grain. Riga fir is good material for roof timbers, but difficult to obtain in long lengths. For tie-beam of king-post roof truss, the same as above, or pitchpine, if it is to be wrought and varnished.

Shop Fronts.—Mahogany is the favourite material, and weathers well if kept French polished; black-walnut and teak are perhaps next in order.

Treads of Stairs.—Oak, pitchpine, Memel fir, ordinary yellow deal.

Weather-boarding.—Oak is best under all circumstances, but is expensive. Larch (*Larix Europœa*) perhaps stands next, as it resists the weather well and bears nails without splitting. Ordinary weather-boarding consists of yellow deal from various ports—say, four out of a 2½-in. by 7-in. batten or 3-in. by 9-in. deal cut feather-edged. For work to be wrought and painted, American red fir is clean-grained and cheap. For very common rough work white spruce deal may be used as being the cheapest.

Window-sills.—Oak or teak for best work ; occasionally pitchpine is used, but it is not so durable as either of the former.

Brands and Shipping Marks on Timber.

Simple Explanation.—A few brands and marks have already been illustrated, but the subject needs special explanation, there being a very general ignorance as to the reasons for, and meanings of, the great number of marks found on imported timber. The difficulty of identifying parcels of timber consigned in the same freight, or stored in the same place, but belonging to different owners, was no doubt the original reason for the introduction of a marking system ; the extension of the system to marks that indicate quality was the natural sequel to the marks of ownership. There is nothing of a mysterious or cryptic nature in this system of timber marks, nor should the various marks be regarded in the light of a secret code ; the great increase in the number of manufacturers and the consequent multiplication of brands are the only causes that have brought about any obscurity that may be thought to exist. There is also generally an entire want of organisation, each new manufacturer being absolutely at liberty to adopt any brand or mark that he may think fit to adopt ; and though, in most cases, respect is paid to old-established marks, plenty of examples of repetition and overlapping exist. Reduced to simple terms, the system (if system it can be called) resolves itself into a parallel of the imaginary case described below. John Brown is a sawmill proprietor and forest owner in Sweden. He manufactures sawn wood goods for the English market, and in order to distinguish the goods produced at his mills from the goods of other sawmillers he stamps or stencils on the end of each piece a more or less abbreviated form of his own name ; and, at the same time, uses variation in the arrangement of the lettering in order to indicate differences in quality. Thus he may export six grades or qualities of material :—

The 1st quality may have   J B   on the end,
2nd   ,,   ,,   ,,   J B N   ,,   ,,
3rd   ,,   ,,   ,,   J + B + N   ,,   ,,
4th   ,,   ,,   ,,   J * B * N   ,,   ,,
5th   ,,   ,,   ,,   J — B — N   ,,   ,,
unsorted   ,,   ,,   J N B N   ,,   ,,

or if, instead of firsts and seconds, a *mixed* grade is substituted (consisting of mixed firsts and seconds), the mark will probably be J w B. John Brown makes no secret of these marks, and would gladly inform any inquirer of the significance (as to quality) *of any given brand.* In fact, he is at much pains to advertise the fact that these classes of material are manufactured by him, and that the above arrangement of initials is to be taken as an indication of the comparative qualities of the stuff. The two real examples given below will show how the *matter works out in practice.*

Holmsunds Marks.—The Holmsunds Aktiebolag (Holmsunds Share Company) manufacture and export sawn goods and planed goods from Holmsunds, Sweden, and the following is their advertised quality code :—

SAWN GOODS.

Mixed	...	...	...	H	D
Thirds	...	...	...	H D	D
Fourths	...	...	...	H N	D
Fifths	...	...	...	H F	D
Sixths	...	...	...	H M	S
Inferior Sixths	...	...	H S U	S

PLANED GOODS.

Firsts	...	...	...	H S w N	D
Seconds	...	...	...	H S * N	D
Thirds	...	...	...	H L N	D
Fourths	...	...	...	H L	D
Unsorted (Sawn or Planed)	...	H S U N	D

Here, obviously, the word Holmsunds has been made use of as the base for quality variations.

Wifsta Warfs Marks.—The Wifsta Warfs Bolag, a sawmilling firm in the Sundswall district of Sweden, exports under the following marks (also, very clearly, derived from the name) :—

Mixed	...	...	...	W W B
Thirds	...	...	...	W S W
Fourths	...	...	...	W T W
Fifths	...	...	...	W F W
Sixths	...	...	...	W W W
Unsorted	...	...	...	W W W W

A fact that should be noted with respect to Swedish goods is that where a mixed grade is shipped usually no separate firsts and seconds are exported, as these best qualities are not sorted from one another.

List of Marks.—In the same way nearly all other firms in the Baltic and Norway trade make use of some simple method of signifying qualities, in which the initials of the head of the firm or of the company (where a company is in proprietorship) form the chief distinguishing features. Obviously, therefore, no universal key can exist that will at once make clear all details as to qualities, port of shipment, etc., except it be in the nature of a long list of names and addresses of manufacturers, and of the initials and symbols that are peculiar to the productions of each. Such a list has been compiled, and is in general use by timber merchants and all connected in any way with the timber trade ; it contains upwards of two thousand marks and brands. One of the essentials of such a work is that it should be kept up to date, as new firms and symbols are constantly appearing on the market, while others fall off from time to time. Lastly, the marking, when applied to logs, assumes several new characters ; it may be said that frequently group numbers, cutting numbers, private sub-owner numbers, and marks, contents marks, and even dates, are sometimes placed on the ends and sides of logs.

JOINTS.

Introduction.—Full instructions on setting out, cutting, and fitting most of the joints used in carpentry and joinery are given in the companion volume, "Woodworking," to provide a collection of illustrations handy for reference, so that the present treatment of technical woodworking may not be incomplete.

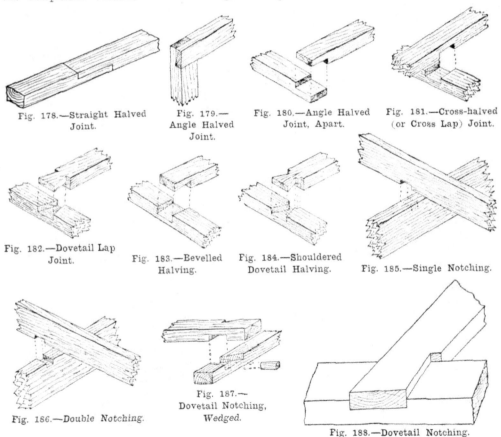

Fig. 178.—Straight Halved Joint.

Fig. 179.—Angle Halved Joint.

Fig. 180.—Angle Halved Joint, Apart.

Fig. 181.—Cross-halved (or Cross Lap) Joint.

Fig. 182.—Dovetail Lap Joint.

Fig. 183.—Bevelled Halving.

Fig. 184.—Shouldered Dovetail Halving.

Fig. 185.—Single Notching.

Fig. 186.—Double Notching.

Fig. 187.—Dovetail Notching, Wedged.

Fig. 188.—Dovetail Notching.

and the reader is assumed to be familiar with all these processes. The object of the present chapter is merely to present brief particulars of the joints in general use and

Joints in Carpentry.

Halved Joints.—The simplest joints used in carpentry are the various forms of halving: simple halved joints (Figs. 178

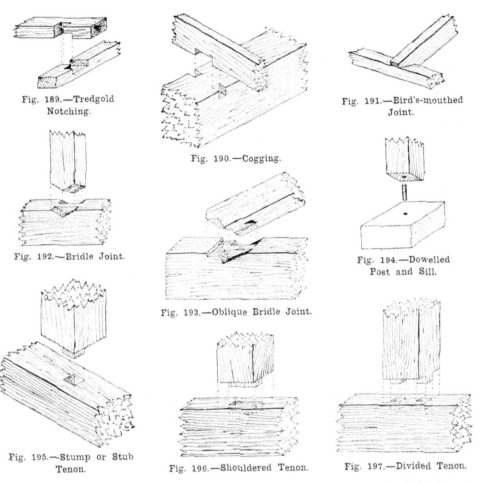

Fig. 189.—Tredgold Notching.

Fig. 191.—Bird's-mouthed Joint.

Fig. 190.—Cogging.

Fig. 192.—Bridle Joint.

Fig. 194.—Dowelled Post and Sill.

Fig. 193.—Oblique Bridle Joint.

Fig. 195.—Stump or Stub Tenon.

Fig. 196.—Shouldered Tenon.

Fig. 197.—Divided Tenon.

Fig. 198.—Inserting Tenon in Chase Mortise.

to 181), dovetail halving (Fig. 182), bevelled halving (Fig. 183), and shouldered dovetail halving (Fig. 184).

Notched and Other Joints.—Of the many forms of notching there are : single notching (Fig. 185), double notching (Fig. 186), dovetail notching (Figs. 187 and 188), and Tredgold notching (Fig. 189). Cogging is shown by Fig. 190, the bird's-mouthed joint by Fig. 191, the bridle joint by Figs. 192 and 193, and dowelling of wood to stone by Fig. 194.

Tenon Joints (Carpenters').—Of tenon joints there is the stump, or stub tenon (Fig. 195) ; the shouldered tenon (Fig. 196) ; the divided tenon (Fig. 197) ; the chase mortise (Fig. 198), in the side of a timber, with one cheek cut away and the depth gradually tapering out to the face of the

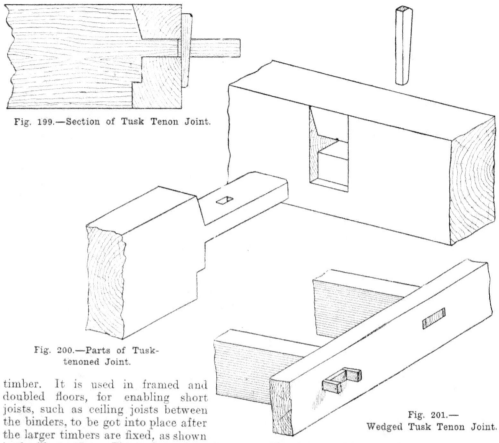

Fig. 199.—Section of Tusk Tenon Joint.

Fig. 200.—Parts of Tusk-
tenoned Joint.

Fig. 201.—
Wedged Tusk Tenon Joint.

timber. It is used in framed and
doubled floors, for enabling short
joists, such as ceiling joists between
the binders, to be got into place after
the larger timbers are fixed, as shown
in the illustration. The tusk tenon is shown
by Figs. 199 to 201 ; struts tenoned into
the heads of king- or queen-posts are
shown by Figs. 202 to 204.

Toe Joints.—Simple toe joints are shown
by Figs. 205 and 206, and a toe joint with
tenon by Fig. 207.

Gantry Strut Joints.—Bird's-mouth and

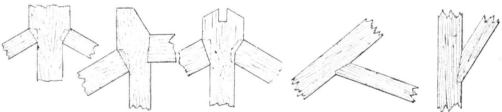

Fig. 202.   Fig. 203.   Fig. 204.

Fig. 202.—Strut Tenoned into King- or Queen-Post.
Fig. 203.—Principal Rafter Tenoned into Queen-
Post, Straining Beam Joggled into same.
Fig. 204.—Principal Rafters Tenoned into King-
Post.

Fig. 205.—Toe Joint
between Principal
Rafter and Strut.

Fig. 206.—Toe Joint
between Vertical
Post and Strut.

mitred butt joints for a gantry strut are
shown by Figs. 208 and 209 respectively.

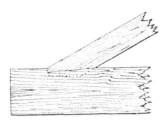

Fig. 207.—Toe Joint with Tenon.

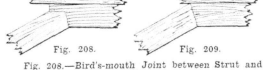

Fig. 208.   Fig. 209.

Fig. 208.—Bird's-mouth Joint between Strut and Straining Piece, or Head.

Fig. 209.—Mitre Butt Joint between Straining Piece and Strut.

Fig. 210.—Dovetailed Halving Bolted.

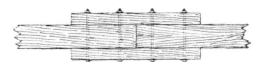

Fig. 211.—Common Fished Joint.

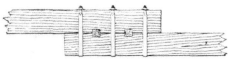

Fig. 212.—Lapped Joint with Keys and Straps.

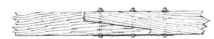

Fig. 213.—Raking Scarf with Butt End.

Fig. 214.—Tabled Joint.

Fig. 215.—Tabled Scarf with Folding Wedges.

Fig. 216.—Tabled and Splayed Scarf.

Fig. 217.—Indented Beams for Lengthening and Strengthening.

Fig. 218.—Splayed Scarf with Folding Wedges.

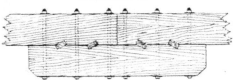

Fig. 219.—Fished Joint with Oblique Keys.

Fig. 220.—Fished and Tabled Joint.

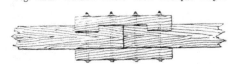

Fig. 221.—Fished and Tabled Joint.

Joints for Lengthening Beams and Posts.
—A joint suitable for *tension only* is the dovetailed halving (Fig. 210). A joint suitable for *compression only* is the common fished joint (Fig. 211). Joints suitable for *cross strain only* are as follows: Lapped, with keys and straps (Fig. 212), and the raking scarf with butt end (Fig. 213). Joints suitable for *tension and compression* are as follows: Tabled (Fig. 214), and the tabled scarf with folding wedges (Fig. 215).

Joints suitable for *tension and cross strain* are as follows: Tabled and splayed scarf (Fig. 216), indented beams for lengthening and strengthening (Fig. 217), and the splayed scarf with folding wedges and iron plate covering joint on tension side (Fig. 218). A joint suitable for *compression and cross*

3*

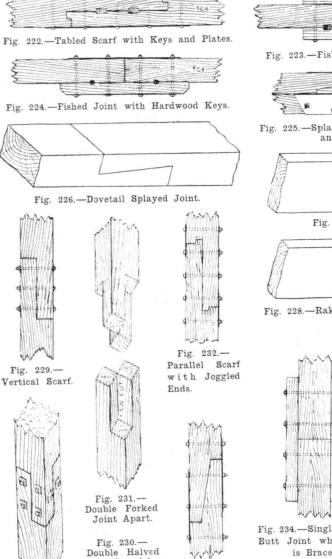

Fig. 222.—Tabled Scarf with Keys and Plates.

Fig. 224.—Fished Joint with Hardwood Keys.

Fig. 226.—Dovetail Splayed Joint.

Fig. 229.—
Vertical Scarf.

Fig. 232.—
Parallel Scarf
with Joggled
Ends.

Fig. 231.—
Double Forked
Joint Apart.

Fig. 230.—
Double Halved
or Double
Forked Joint
Together.

Fig. 230.

Fig. 233.—
Splayed Scarf.

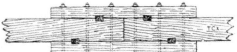

Fig. 223.—Fished Joint, Keyed and Bolted.

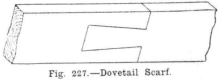

Fig. 225.—Splayed Scarf with Folding Wedges
and Iron Fish Plates.

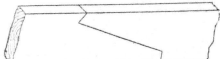

Fig. 227.—Dovetail Scarf.

Fig. 228.—Raking Scarf used for Ridges, etc.

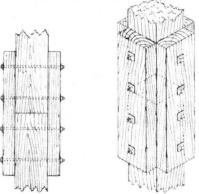

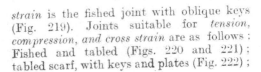

Fig. 234.—Single Fished   Fig. 235.—Double Fished
Butt Joint when Post   Butt Joint for Detached
is Braced.   Post.

strain is the fished joint with oblique keys (Fig. 219). Joints suitable for *tension, compression, and cross strain* are as follows: Fished and tabled (Figs. 220 and 221); tabled scarf, with keys and plates (Fig. 222); fished, keyed, and bolted (Fig. 223); fished, with hardwood keys (Fig. 224); and splayed scarf with iron fish plates and bolts (Fig. 225), which is used in the warehouses at the South-West India Dock, London. Other joints used for lengthening plates and ridges are shown at Figs. 226, 227, and

228. Joints for beams and posts are : the vertical scarf—a halved joint (Fig. 229), double halved joint (Figs. 230 and 231), parallel scarf with joggled ends (Fig. 232), splayed scarf (Fig. 233), single fished butt joint when the post is braced (Fig. 234), and the double fished butt joint (Fig. 235) when the post is detached.

say 10 in. by 3 in. The joint may also shear across B C or G F, therefore section at B C or G F must equal $\frac{360}{1\cdot3} = 277$, say 28 in. by 10 in. The joint may also be crushed at B D or G H, therefore section at B D or G H must equal $\frac{360}{10} = 36$, say 10 in. by $3\frac{1}{2}$ in.

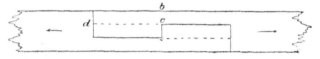

Fig. 236.—Tabled Scarf Joint.

Rule for Proportioning Parts of Scarf.— Tredgold gives the following practical rules for proportioning the different parts of a scarf according to the strength possessed by the kind of timber in which it is formed, to resist tensional, compressional, or shearing forces respectively. In Fig. 236 cd must be to cb in the ratio that the force to resist detrusion bears in the direct cohesion of the material—that is, in oak, ash, elm, cd must be equal to from eight to ten times cb ; in fir and other straight-grained woods cd must be equal to from sixteen to twenty times cb. The sum of the depth of the indents should be equal to one and one-third depth of beam. The length of scarf should bear the following proportion to the depth of the beam :—

Thus the beam should be about 10 in. by 10 in., with wedges as shown ; but in ordinary practice the folding wedges do not exceed one-fourth the depth of the beam, and are usually placed square to the rake of the scarf, the scarf being further strengthened by bolts and plates.

Strength of Joints in Struts and Beams.— If two deals are bolted together, with distance pieces between, they will be stronger than a solid timber strut of the same sectional area, because the dimension of "least width" in the formula for calculation of strength will be increased. There

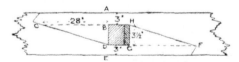

Fig. 237.—Splayed Scarf with Folding Wedges.

Wood Used	Without bolts	With bolts	With bolts and indents
Hardwood (oak, ash, elm)	6 times	3 times	2 times
Fir and other straight - grained woods	12 ,,	6 ,,	4 ,,

Calculation of a scarfed joint with folding wedges as Fig. 237 :—

			Per sq. in.
Working resistance to tearing ..		$=$	12 cwt.
,, ,,	compression	$=$	10 ,,
,, ,,	shearing..	$=$	1·3 ,,

Load equals, say, 360 cwt. direct tension beyond that taken by bolts or plates. The joint may tear across A B or D E (Fig. 237), therefore section at A B must equal $\frac{360}{12} = 30$,

would be no appreciable advantage in making the distance pieces of different thicknesses, to swell or reduce the middle diameter ; they should be all alike, and enough to make the combined thickness not less than three-fourths of the width of the deals, and the distance apart in feet should be equal to the length of the deal in feet multiplied by its thickness in inches and divided by the width in inches. Single $\frac{1}{2}$-in. bolts are of no use in rough carpentry, except for very small work ; instead, two $\frac{5}{8}$-in. bolts should be placed diagonally through each block. Horizontal connecting rods in machinery are sometimes swelled in the middle to allow for the cross strain upon

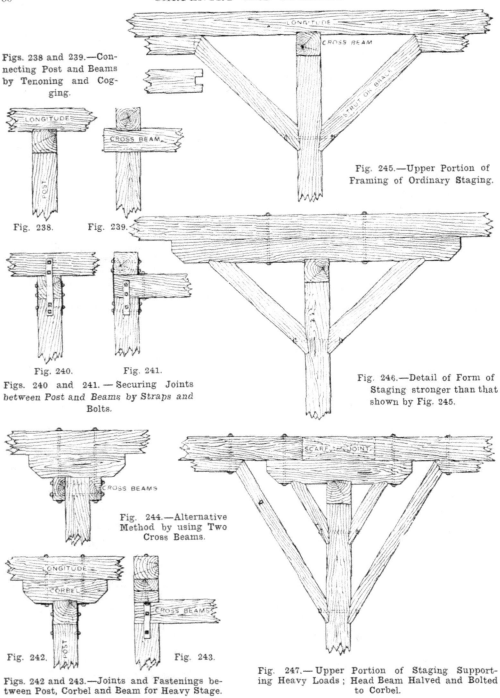

Figs. 238 and 239.—Connecting Post and Beams by Tenoning and Cogging.

Fig. 238.

Fig. 239.

Fig. 245.—Upper Portion of Framing of Ordinary Staging.

Fig. 240.

Fig. 241.

Figs. 240 and 241. — Securing Joints between Post and Beams by Straps and Bolts.

Fig. 246.—Detail of Form of Staging stronger than that shown by Fig. 245.

Fig. 244.—Alternative Method by using Two Cross Beams.

Fig. 242.

Fig. 243.

Figs. 242 and 243.—Joints and Fastenings between Post, Corbel and Beam for Heavy Stage.

Fig. 247.— Upper Portion of Staging Supporting Heavy Loads ; Head Beam Halved and Bolted to Corbel.

them in addition to the end-long strain, while vertical struts have no cross strain to meet.

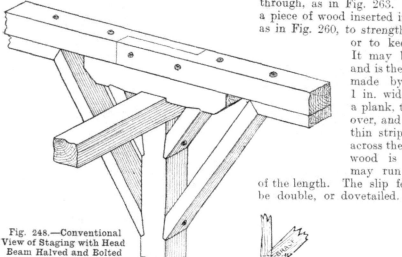

Fig. 248.—Conventional View of Staging with Head Beam Halved and Bolted to Corbel.

Jointing Beams to Posts and Struts.— The usual methods of forming joints between beams, posts, struts, and braces as used in the construction of gantries, stagings, jetties, bridges, etc., are illustrated by Figs. 238 to 254. The inscriptions to the illustrations make the methods quite clear to understand.

Joints in Joinery.

Edge Joints.—Eleven joints used in connecting boards edge to edge are shown by

Fig. 251. Fig. 252.

Fig. 251.—Mitred Butt and Tenoned Joint between Brace and Straining Piece.

Fig. 252.—Double Abutment Joint between Strut, Head, and Straining Piece.

Figs. 255 to 265. Matchboarding is thin stuff with a tongue and bead worked on one edge and a groove on the other, so that when

the pieces are put together the joint is masked by the bead, and the tongue prevents dust and draught from passing through, as in Fig. 263. A slip feather is a piece of wood inserted in plough grooves, as in Fig. 260, to strengthen a glued joint, or to keep out the dust. It may be of soft wood, and is then in short lengths, made by cutting pieces 1 in. wide off the end of a plank, turning the pieces over, and cutting them into thin strips, with the grain across their length. If hard wood is used, the grain may run in the direction of the length. The slip feathers may also be double, or dovetailed.

Fig. 249. Fig. 250.

Fig. 249.—Strut and Post Joint Supported by Cleat Spiked to Post.

Fig. 250.—Brace and Post Joint. Brace Tenoned into Post : Cleat Joggled in and Spiked to Post.

Right Angle Joints.—Fourteen styles of angle joints are shown by Figs. 266 to 279.

Obtuse Angle Joints.—Four kinds of these joints are illustrated by Figs. 280 to 283.

Fig. 253. Fig. 254.

Fig. 253.—Treble Abutment Joint between Strut and Straining Piece.

Fig. 254.—Tenoned and Bird's-mouth Shouldered Joint between Strut and Straining Piece.

Dovetail Joints.—These are known in great variety, but it will be sufficient to show a few kinds only : the ordinary dove-

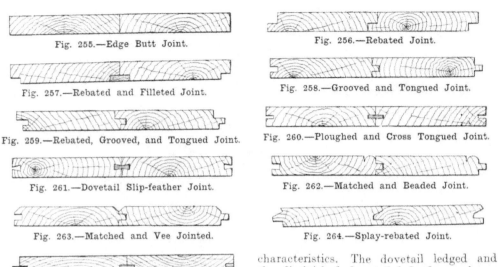

Fig. 255.—Edge Butt Joint.

Fig. 256.—Rebated Joint.

Fig. 257.—Rebated and Filleted Joint.

Fig. 258.—Grooved and Tongued Joint.

Fig. 259.—Rebated, Grooved, and Tongued Joint.

Fig. 260.—Ploughed and Cross Tongued Joint.

Fig. 261.—Dovetail Slip-feather Joint.

Fig. 262.—Matched and Beaded Joint.

Fig. 263.—Matched and Vee Jointed.

Fig. 264.—Splay-rebated Joint.

Fig. 265.—Dowelled Joint.

tailing (Figs. 284 and 285), lapped dovetail (Fig. 286), two secret or double-lap or rebated dovetails (Figs. 287 and 288), and the secret mitred dovetail (Fig. 289). The box pin joint (Fig. 290) is not a dovetail joint, but has some of the latter's characteristics. The dovetail ledged and the diminished dovetail ledged are shown respectively by Figs. 291 and 292.

Dowelled Joint.—The ordinary dowelled joint is represented by Fig. 293; sections showing a dowel fitted incorrectly and correctly are represented by Figs. 294 and 295 respectively. A right angle dowelled joint is shown by Fig. 296. Allied to the dowel joint is the screwed straight joint (Figs. 297

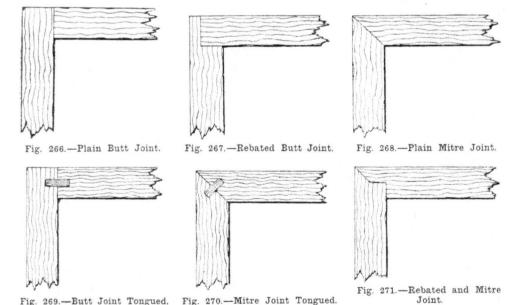

Fig. 266.—Plain Butt Joint.

Fig. 267.—Rebated Butt Joint.

Fig. 268.—Plain Mitre Joint.

Fig. 269.—Butt Joint Tongued.

Fig. 270.—Mitre Joint Tongued.

Fig. 271.—Rebated and Mitre Joint.

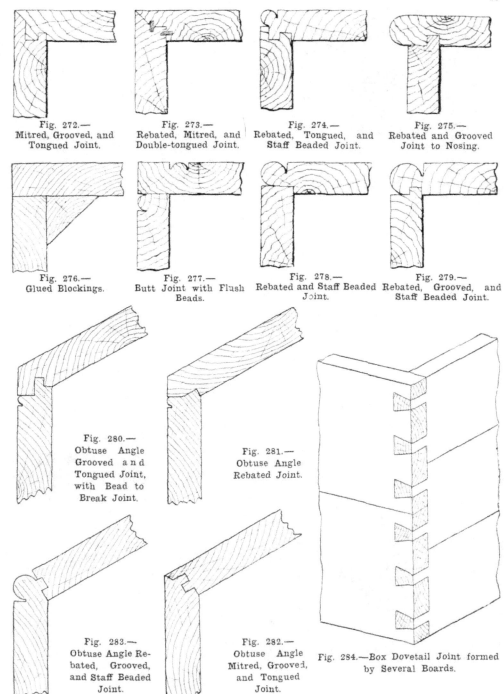

Fig. 272.—
Mitred, Grooved, and
Tongued Joint.

Fig. 273.—
Rebated, Mitred, and
Double-tongued Joint.

Fig. 274.—
Rebated, Tongued, and
Staff Beaded Joint.

Fig. 275.—
Rebated and Grooved
Joint to Nosing.

Fig. 276.—
Glued Blockings.

Fig. 277.—
Butt Joint with Flush
Beads.

Fig. 278.—
Rebated and Staff Beaded
Joint.

Fig. 279.—
Rebated, Grooved, and
Staff Beaded Joint.

Fig. 280.—
Obtuse Angle
Grooved a n d
Tongued Joint,
with Bead to
Break Joint.

Fig. 281.—
Obtuse Angle
Rebated Joint.

Fig. 283.—
Obtuse Angle Re-
bated, Grooved,
and Staff Beaded
Joint.

Fig. 282.—
Obtuse Angle
Mitred, Grooved,
and Tongued
Joint.

Fig. 284.—Box Dovetail Joint formed
by Several Boards.

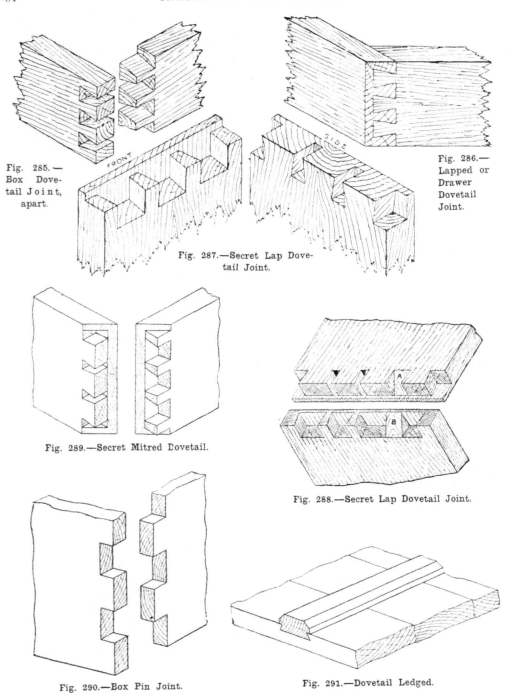

Fig. 285.—
Box Dove-
tail Joint,
apart.

Fig. 286.—
Lapped or
Drawer
Dovetail
Joint.

Fig. 287.—Secret Lap Dovetail Joint.

Fig. 289.—Secret Mitred Dovetail.

Fig. 288.—Secret Lap Dovetail Joint.

Fig. 290.—Box Pin Joint.

Fig. 291.—Dovetail Ledged.

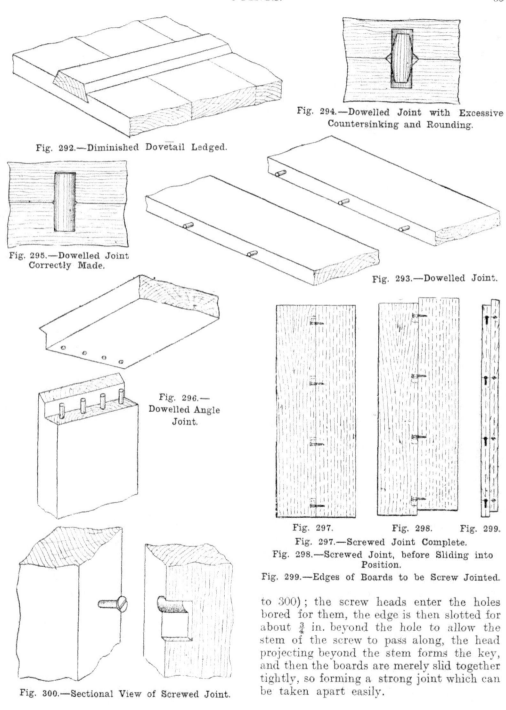

Fig. 292.—Diminished Dovetail Ledged.

Fig. 294.—Dowelled Joint with Excessive Countersinking and Rounding.

Fig. 295.—Dowelled Joint Correctly Made.

Fig. 293.—Dowelled Joint.

Fig. 296.—Dowelled Angle Joint.

Fig. 297.   Fig. 298.   Fig. 299.

Fig. 297.—Screwed Joint Complete.

Fig. 298.—Screwed Joint, before Sliding into Position.

Fig. 299.—Edges of Boards to be Screw Jointed.

Fig. 300.—Sectional View of Screwed Joint.

to 300); the screw heads enter the holes bored for them, the edge is then slotted for about $\frac{3}{4}$ in. beyond the hole to allow the stem of the screw to pass along, the head projecting beyond the stem forms the key, and then the boards are merely slid together tightly, so forming a strong joint which can be taken apart easily.

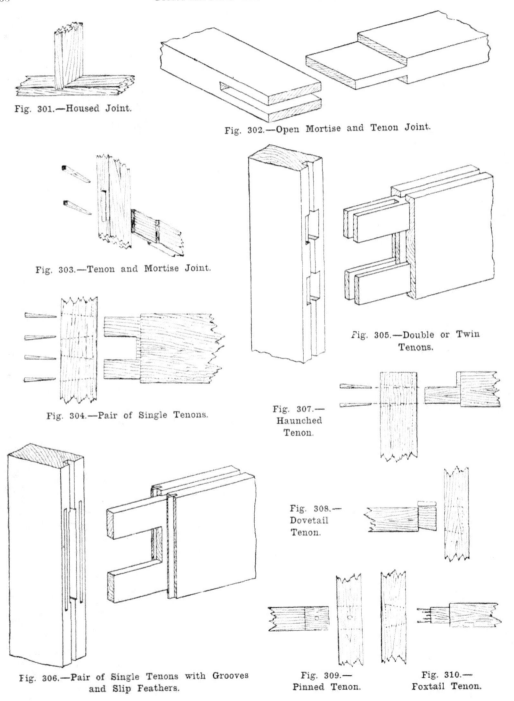

Fig. 301.—Housed Joint.

Fig. 302.—Open Mortise and Tenon Joint.

Fig. 303.—Tenon and Mortise Joint.

Fig. 305.—Double or Twin
Tenons.

Fig. 304.—Pair of Single Tenons.

Fig. 307.—
Haunched
Tenon.

Fig. 308.—
Dovetail
Tenon.

Fig. 306.—Pair of Single Tenons with Grooves
and Slip Feathers.

Fig. 309.—
Pinned Tenon.

Fig. 310.—
Foxtail Tenon.

Housing.—The simple housing joint is shown by Fig. 301.

Tenon Joints (Joiners').—Some tenon joints have already been shown under the heading, "Joints in Carpentry" (p. 55). Further tenon joints, more especially used in joinery, are : the simple open tenon and mortise (Fig. 302); closed mortise and tenon (Fig. 303); pair of single tenons, commonly called "double" tenons (Fig. 304); double or twin tenons (Fig. 305); pair of single tenons, with grooves and slip feathers (Fig. 306); haunched tenon (Fig. 307); dovetail tenon (Fig. 308); pinned tenon (Fig. 309). Stump or stub tenons and tusk tenons are also used in joinery, and have already been illustrated (Figs. 195

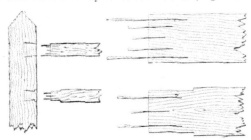

Fig. 311.—Foxtail Tenons with and without Housing.

and 200, pp. 55 and 56). The foxtail tenon (Fig. 30) is a good joint; alternative methods (with and without housing) of applying this in fitting rails into an oak gate-post are shown by Fig. 311.

Proportioning Tenons.—There is no universal rule for proportioning tenons, but the practice is to give from half to the whole of the width of the rail, when this does not exceed 5 in., for the width of the tenons. If more space than half were given to a haunched tenon, the end of the stile would be liable to be driven out in wedging up, and there is no reason why more space should be given. Wide tenons are objectionable, owing to their liability to shrink from the wedges or the sides of the mortises.

Applications of Tenon Joints.—With regard to the application of the various tenon joints, a few of these are noted below : A simple tenon, one-third thickness of the

stuff, is used in framing together pieces of the same size, the mortise being just long enough to allow of a wedge being driven in on each side of the tenon to secure it. A pair of single tenons, usually called a double tenon, is used for connecting the middle rail of a door to the stiles. A haunched tenon for connecting the top rail of a door to the stiles; the tenon being half the width of the top rail leaves a haunch or haunching to prevent the rail from twisting. A stump or stub tenon is used at the foot of a post to prevent movement. A tusk tenon is used in framing trimmers to trimming joists, to

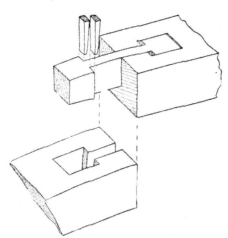

Fig. 312.—Hammer-headed Key Joint.

obtain the maximum support with the minimum reduction of strength. A tenon with only one shoulder is used in framed and braced batten doors, and in skylights, when the rail requires to be kept thin for other parts to pass over; this is known as a barefaced tenon. A pair of double tenons is used for the lock rail of a thick door, to receive a mortise lock.

Hammer-headed Key Joint. — A conventional view of a hammer-headed key joint apart is presented by Fig. 312.

Special Joints.—Many other joints adapted to particular purposes are described in subsequent sections. Reference to these may easily be found by consulting the index.

FLOORS.

General Considerations.—The remarks in this paragraph will be found applicable to all sorts of floors. The joists should be laid across the narrowest part of the room, and girders and binders should be so arranged as to take a bearing on a solid pier or wall, wet, for as long a period as possible before they are required for use. Where such an arrangement is possible it is well to have the boards laid face downwards for some months in the position they are to occupy before they are finally nailed.

Fig. 313.—Method of Supporting Joists round Brickwork Fender in Basement.

and not over door or window openings. In cases where a long distance has to be traversed by a joist, which is supported by one or more girders in the length, it should be made as long as possible. By this means the strength of the joist is greatly increased, as also is its usefulness as a tie to the walls. Flooring-boards should be cut and prepared, and stacked in the open air, with free ventilation all round, with proper protection from

Basement or Ground Floors.

The floor in a basement storey, or on the ground level where there is no basement, is formed of joists laid on wooden sleepers. themselves bedded on dwarf walls (Figs. 313 and 314). The walls and sleepers are usually 4 ft. or 5 ft. apart, and the joists 4 in. to 6 in. deep. Occasionally the walls and sleepers are further apart, and then joists 6 in. or even 8 in. deep are used. Fig.

313 is a conventional view, and Fig. 314 a section through a floor of this description, clearly showing how the joists are supported by the brick fender round the fireplace. Oak is considered best for sleepers, and to ensure of its being thoroughly seasoned,

Single Floors.

The simplest floor consists of a row of beams or joists, varying in thickness and depth with the width or bearing between the walls on which they are supported. To the upper sides of these joists is nailed

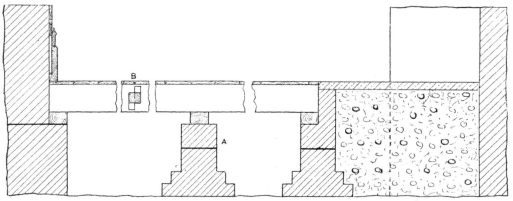

Fig. 314.—Section through Basement or Ground Floor.

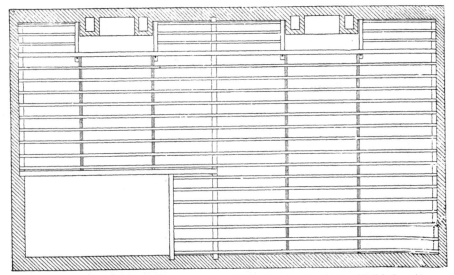

Fig. 315.—Plan of Single Floor showing Trimming to Fireplaces, Well-hole, etc.

ship oak is sometimes specified. Formerly it was the practice, more so than at present, to notch or cog the joists to the sleepers. When the joists are deep enough, rows of herringbone strutting are introduced, as indicated at B (Fig. 314), cut and fixed as shown later by Fig. 317.

the floor-boarding, and to the under side the laths which carry the ceiling. These joists should not be placed at a greater distance than 15 in. from centre to centre. An ordinary example of a single floor is shown at Fig. 315, this figure being the plan of the timber of a floor of two rooms, and well-hole for staircase for a dwelling-house, 34 ft. from front to back, and 20 ft. wide in the clear ; it shows also the trimming for two 6-ft. chimney-breasts in flank-wall, and for well-hole at opposite flank next the back wall. The well-hole is 7 ft. wide and 12 ft. long. The floor is constructed to carry two framed partitions, one 18 ft. from, and parallel to, the front wall, and the other extending from this partition to the back wall along the well-hole. The middle bearing is required to be under the

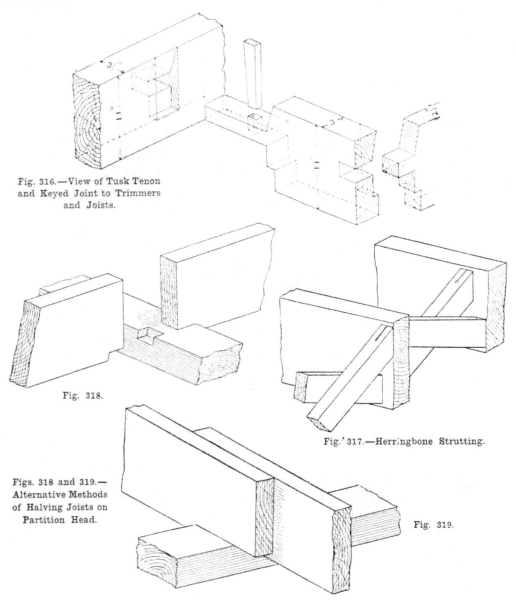

Fig. 316.—View of Tusk Tenon and Keyed Joint to Trimmers and Joists.

Fig. 318.

Fig.' 317.—Herringbone Strutting.

Figs. 318 and 319.— Alternative Methods of Halving Joists on Partition Head.

Fig. 319.

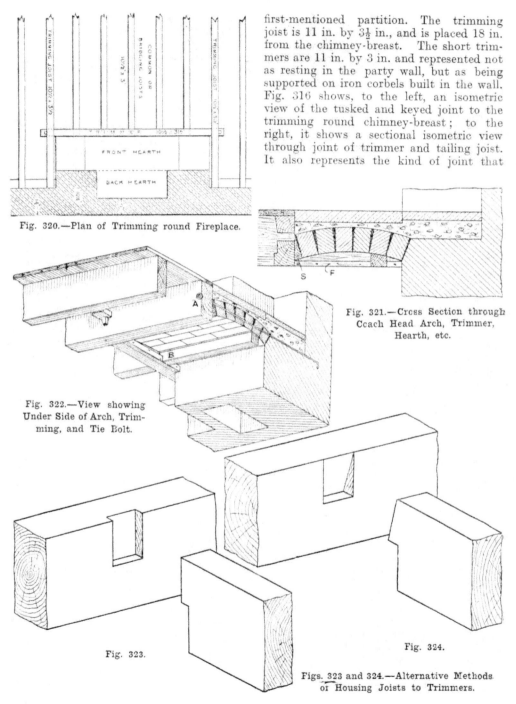

first-mentioned partition. The trimming joist is 11 in. by $3\frac{1}{2}$ in., and is placed 18 in. from the chimney-breast. The short trimmers are 11 in. by 3 in. and represented not as resting in the party wall, but as being supported on iron corbels built in the wall. Fig. 316 shows, to the left, an isometric view of the tusked and keyed joint to the trimming round chimney-breast; to the right, it shows a sectional isometric view through joint of trimmer and tailing joist. It also represents the kind of joint that

Fig. 320.—Plan of Trimming round Fireplace.

Fig. 321.—Cross Section through Coach Head Arch, Trimmer, Hearth, etc.

Fig. 322.—View showing Under Side of Arch, Trimming, and Tie Bolt.

Fig. 323.

Fig. 324.

Figs. 323 and 324.—Alternative Methods of Housing Joists to Trimmers.

would be used to connect the staircase trimmer and joists shown in well. Fig. 317 (p. 70) gives a view of the herringbone strutting (2 in. by 1½ in.), four rows of which are indicated on the plan. The joists going from back to front are required to be

parallel with the chimney-breast, and the trimmers which carry the joists are against the sides of the breasts. Fig. 320 is a reverse case, there being only one trimmer, which is parallel to the breast, but two trimming joists, these being at right angles to

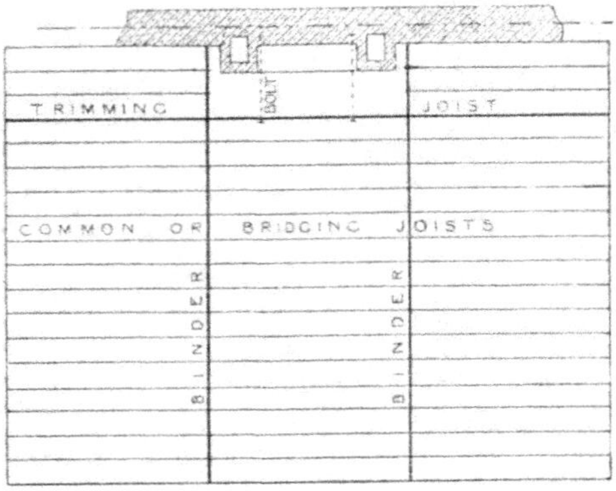

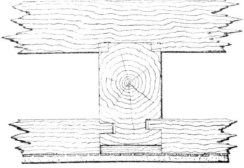

Fig. 327.—Section through Binder showing Bridging Joists Cogged, and Alternative Methods of Connection with Ceiling Joists.

Fig. 325.—Plan of Binder or Double Floor.

34 ft. 9 in. long ; therefore all, or the greater part, would have to be formed of two lengths and halved on the middle bearing ; alternative methods of doing this are shown by Figs. 318 and 319.

it. Fig. 321 is a section through the trimmer, hearth, coach head brick arch, etc., shown in plan at Fig. 320. s (Fig. 321) is a feathered-edge piece of board (a springing piece) nailed to the trimmer for the arch to

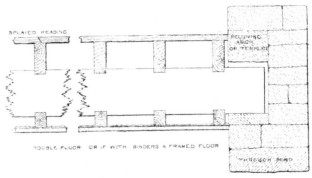

Fig. 326.—Section through Joists, showing Side of Binder supported by Wall.

Trimming Round Openings.

In projections where fireplaces and flues (usually known as chimney breasts) occur in walls it is necessary to trim round them, so that the nearest timber in front shall be at least 18 in. distant, whilst that at the sides may be only an inch or so. In the plan (Fig. 315) the trimming joist runs

butt against ; F is a fillet nailed to the trimming joists so as to support the piece of scantling to which the laths are nailed. This construction is clearly shown at B (Fig. 322). When a trimmer has to support an arch, to prevent any likelihood of the arch forcing it back, one or two iron bolts are inserted, one end being bedded and hooked into the brickwork, the other having a screw or nut, as indicated at A (Fig. 322). Figs. 323 and 324 show alternative simple methods of housing short joists into trim-

mers. These are generally adopted in positions where there is not sufficient space to allow of their being inserted with the usual tusk tenon. Trimmers and joists to

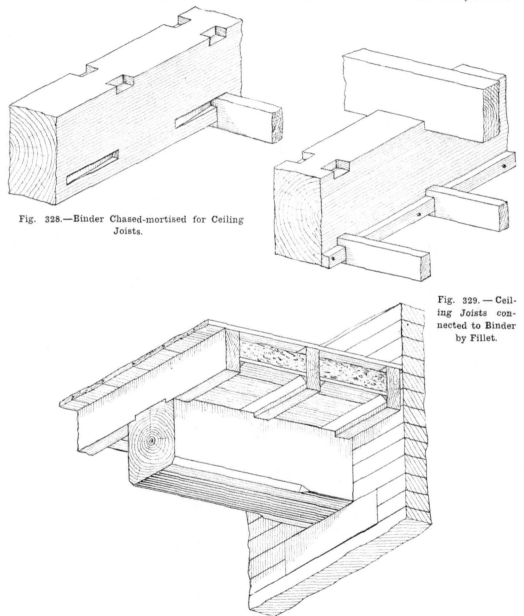

Fig. 328.—Binder Chased-mortised for Ceiling Joists.

Fig. 329. — Ceiling Joists connected to Binder by Fillet.

Fig. 330.—Under Side of Floor with Wrought Binder.

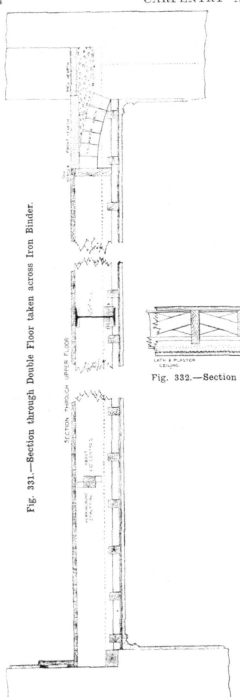

Fig. 331.—Section through Double Floor taken across Iron Binder.

which they are connected should always be thicker than the ordinary joists. A common rule is to make the trimmers and trimming joists $\frac{1}{8}$ in. thicker for each joist carried. Single floors may span as great a distance as 18 ft. by using 11-in. by 3-in. joists stiffened with two or three rows of herring-bone or solid strutting.

Double Floors.

When the distance between the supporting walls exceeds 14 ft. or 15 ft., it is usual to place binders or girders of wood or iron at intervals of from 6 ft. to 10 ft., and on these to support the bridging joists. Floors so constructed are known as double floors, having two sets of joists, the lower set (ceiling joists) being smaller, and used solely to support the ceiling. Thus the ceiling, being supported independently of the floor joists above, is not liable to be jarred by

Fig. 332.—Section across Bridging Joists showing Method of Fixing Ceiling Joists.

the traffic overhead, and the connection between the ceiling and floor being broken by the space between the two sets of joists, sound from above is not so audible below as when the floor is single.

Wooden Binders.—The outline plan of a double floor is given at Fig. 325, and Fig. 326 is a section through the joists, flooring, and ceiling, showing the side of the binder and also the method of supporting it. Fig. 327 is a section through the binder showing alternative ways of connecting the ceiling joists with the binder by mortise and tenon joints. Ceiling joists which have to be got into position after the binders are built in have their tenons inserted at one end into an ordinary mortise, whereas the tenon at the other end has to slide into a chase mortise as indicated at Fig. 328. To avoid weakening the binder, sometimes a fillet is nailed on so as to support the ceiling joists, which are notched to it as shown at Fig. 329. Fig. 330 illustrates the case where ceiling joists are not used. The binder is wrought and stopped chamfered ; the laths for the ceiling would be nailed to the under edges of the bridging joists. The transmis-

sion of sound would be lessened by sound boarding and pugging as shown.

Iron Binders.—Two sections through a double floor are presented by Figs. 331 and

Fig. 333.—Conventional View of Double Floor.

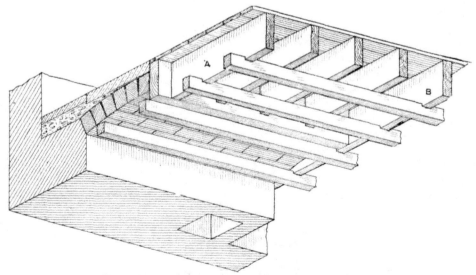

Fig. 334.—View of Part of Under Side of Floor adjacent to Chimney Breast.

Fig. 335.—Method of Fitting Oak Border to Floor Boards.

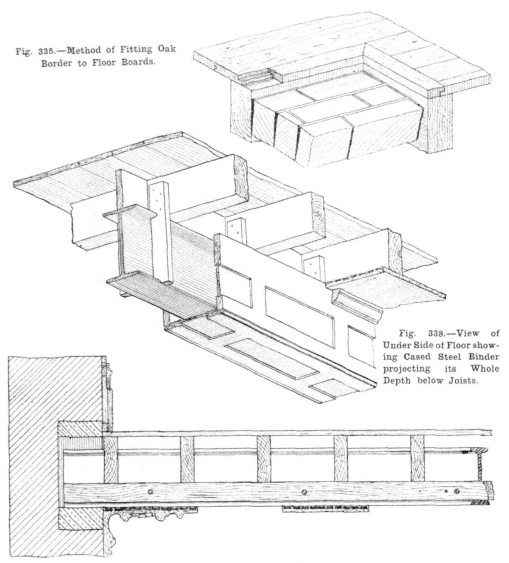

Fig. 338.—View of Under Side of Floor showing Cased Steel Binder projecting its Whole Depth below Joists.

Fig. 336.—Section taken Parallel to Steel Binder in Double Floor.

Fig. 337.—Section taken at Right Angles to Section—Fig. 336.

332. Just above the lath-and-plaster ceiling are the ceiling joists, and running parallel with these is a 10-in. by 5-in. rolled-iron

Fig. 335 shows a method of mitreing and fitting an oak border to the floor-boards ready to receive the hearth. Figs. 336 and

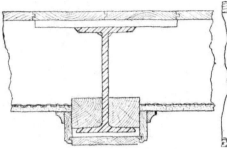

Fig. 339.—Steel Binder Projecting Part of its Depth below Joists.

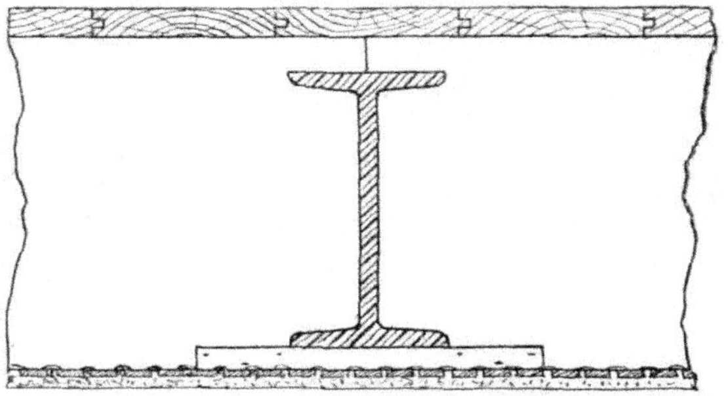

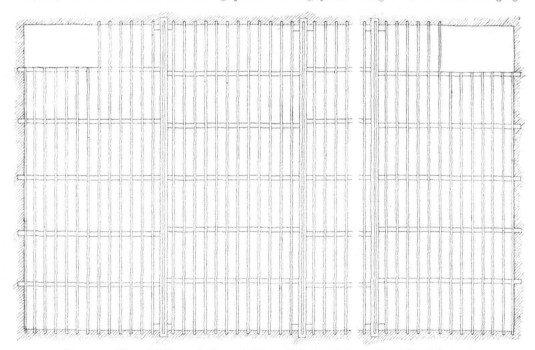

Fig. 340.—Section showing Arrangement to avoid Binder Showing.

joist (the binder). Fig. 333 shows the general construction of this floor, the special feature of which is that the ceiling joists

337 are sections through a somewhat similar floor, but of a more ordinary character, the ceiling joists being fixed to each bridging

Fig. 341.—Plan of a Framed Floor, showing Girders, Binders, Joists, Trimming, etc.

are notched to and supported by every fourth bridging joist, which are stouter and deeper, as shown at A and B (Fig. 334).

joist. The binders are of rolled-iron or steel 11 in. deep and 4½ in. wide in the flanges and 10 ft. apart. Fig. 338 illus-

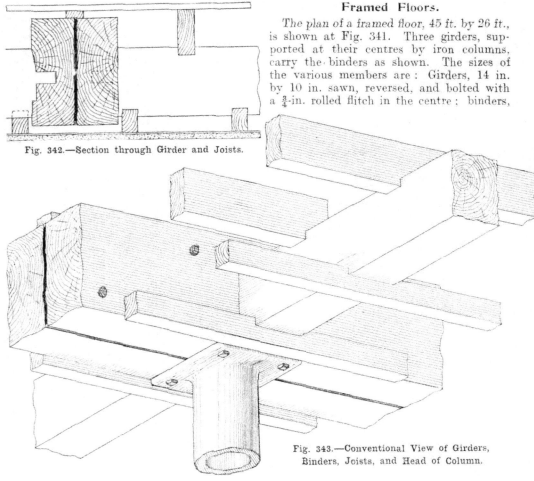

Fig. 342.—Section through Girder and Joists.

Framed Floors.

The plan of a framed floor, 45 ft. by 26 ft., is shown at Fig. 341. Three girders, supported at their centres by iron columns, carry the binders as shown. The sizes of the various members are : Girders, 14 in. by 10 in. sawn, reversed, and bolted with a $\frac{3}{4}$-in. rolled flitch in the centre : binders,

Fig. 343.—Conventional View of Girders, Binders, Joists, and Head of Column.

trates a case where the bridging joist rests direct on the iron binder, solid strutting being inserted between the joists to keep them vertical. The ceiling is formed of either lath and plaster or match-boarding fixed direct to the joists. the binder being cased round as shown. Fig. 339 illustrates an arrangement of casing the under side of a girder or binder when it is deeper than the joists. If constructed as shown at Fig. 340 a flat ceiling can be obtained under the binder : but this construction cannot be adopted when the iron member has to serve as a girder for floors having heavy loads to carry, as a single binder would not be deep enough.

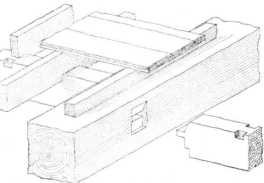

Fig. 344.—General View of Part of Framed Floor.

9 in. by 6 in. ; bridging joists, 6 in. by 2½ in. ; ceiling joists, 3 in. by 2 in. Figs. 342 and 343 will make the construction clear. Figs. 344 to 346 are details of a double floor for a smaller span. Figs. 345 and 346 are views taken at right angles to each

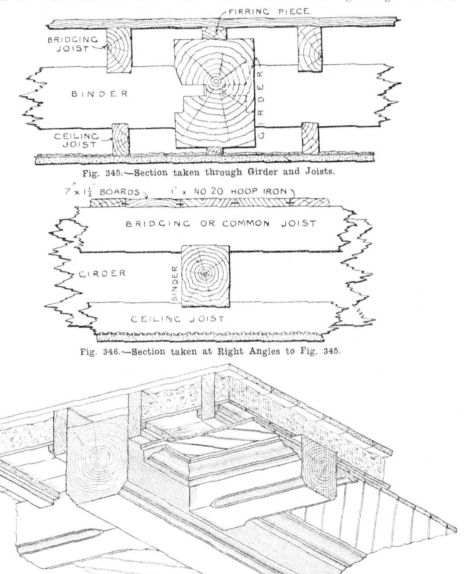

Fig. 345.—Section taken through Girder and Joists.

Fig. 346.—Section taken at Right Angles to Fig. 345.

Fig. 347.—View of Under Side of Framed Floor, with Wood Ceiling and Beams Wrought and Moulded.

other. Fig. 347 is a conventional view showing girder, 12 in. by 10 in. ; binders, 8 in. by 6 in. ; bridging joists, 8 in. by 2¼ in ; and matchboard ceiling. There ends of the binders, and thus they are well supported without the girder being weakened. Two different forms of malleable iron stirrups are illustrated by Figs. 348

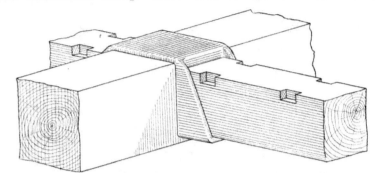

Fig. 348.—Binders supported on Girders by Malleable Iron Stirrup.

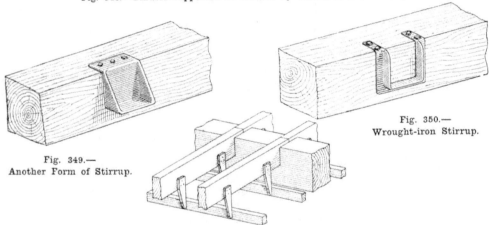

Fig. 349.—
Another Form of Stirrup.

Fig. 350.—
Wrought-iron Stirrup.

Fig. 351.—Method of Hanging Ceiling Joists from Bridging Joists.

being no ceiling joists, the girders and binders have their under-edges moulded. To intercept sound, the floor may be pugged as shown. The strength of wooden girders often being weakened to the extent of one-

and 349, and one of wrought iron by Fig. 350. A system of supporting ceiling joists by connecting them to the bridging joists by nailing them to strips of wood is shown at Fig. 351, but it has become obsolete.

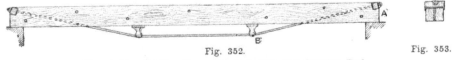

Fig. 352.   Fig. 353.

Figs. 352 and 353.—Beam Trussed with One Tension Rod.

eighth by being mortised and housed to receive the binders, various forms of stirrup irons have been introduced to carry the

Floors with Trussed Beams.

In warehouses and factories where there are heavy loads and vibration the girders are sometimes strengthened by trussing. Various methods are adopted. Two ways of trussing by wrought-iron rods are shown by Figs. 352 to 358. In the case of Fig. 352 the beam is sawn down the middle, ends reversed, and bolted together with blocks

Fig. 354.—Enlarged View of End A (Fig. 352).

between, so as to allow of the iron rod passing through the iron heel plate at each end (Fig. 354), so that it can be tightened. Figs. 356, 357, and 358 illustrate a very strong form of trussing by using a solid beam and a tension rod on each side.

(see Fig. 317) for the insertion of the nails. A great advantage in this form of strutting is that, although *the joists may shrink in* thickness and depth, the strutting remains firm owing to the greatest shrinkage taking place in depth. This will be made clear by Fig. 359. Let *a, b, c, d* represent the original position of the strutting ; then upon shrinkage taking place, the struts move about

Fig. 355.—View of Cast-iron Strut B (Fig. 352).

their centre O, and tend to the positions indicated by the dotted lines *a' h'* and *c' d'*, the greatest movement being produced by the depth shrinkage ; thus the greater this is the more the compression on the struts, which would produce greater distances

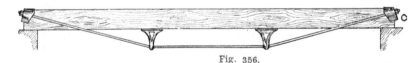

Fig. 356.   Fig. 357.

Figs. 356 and 357.—Beam trussed with Two Tension Rods.

Strutting.

Herringbone Strutting.—Cross-pieces of wood, about 2 in. by 1½ in., or 2 in. by 2 in., are frequently fixed between joists, as already shown by Figs. 315, 317, and 332, with

Fig. 358.—Enlarged View of Cast-iron Shoe C (Fig. 356).

the view of strengthening and increasing the rigidity of the whole floor. To prevent splitting at the ends by boring, it is usual to make a saw kerf at each end of the struts

between the joists, were it not for the floor boards being nailed to the joists.

Solid Strutting.—When pieces of board are cut and simply driven in tightly between the joists and nailed, they often become loose some months after the floor is completed, owing to the shrinkage of the joists

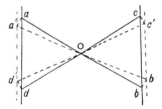

Fig. 359.—Movement of Herringbone Strutting produced by Shrinkage of Joists.

in thickness, and thus they are of very little use for the purpose for which they were intended. Solid strutting is a most valuable form for stiffening and strengthening floors

4*

of warehouses, etc., if a wrought-iron bar or *tube is passed through each joist* a little above its centre. The bar must have a thread and nut at each end working against an iron plate, so that the struts and joists may be tightened perfectly close to each other. A view of this arrangement is given at Fig. 360.

Supporting Joists by Walls.

Joists are now often supported direct by the brickwork or masonry, or they may take their bearing on a tar and sanded or galvanised iron bar. Figs. 361 to 364 show four general methods of bedding plates for *joists in or upon the walls. Fig. 364 shows* the plate supported by iron corbels built in the walls. So that the plate may not project below the ceiling, sometimes the joists are notched down to bring their lower edges level with the under side of the plate; but, of course, this weakens the joists.

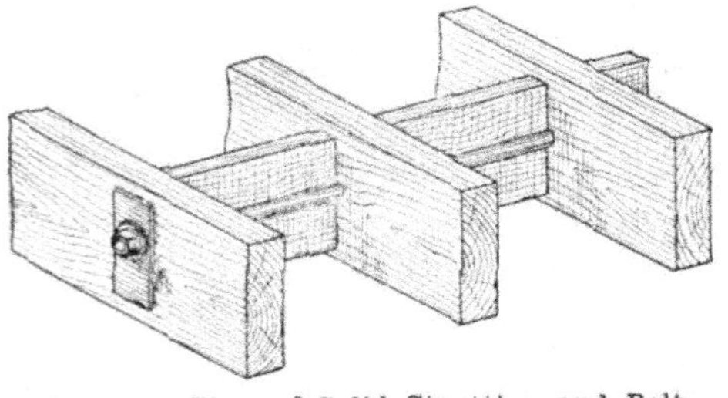

Fig. 360.—View of Solid Strutting and Bolt.

Determining Sizes of Joists.

Common joists are spaced 12 in. apart, with herringbone strutting every 4 ft. Dimensions for common joists are as follow:

Span or Length of Bearing in Feet.	Depth in Inches.			
	$1\frac{3}{4}$ in. thick.	2 in. thick.	$2\frac{1}{2}$ in. thick.	3 in. thick.
6	6	$5\frac{3}{4}$	$5\frac{3}{8}$	5
8	$7\frac{1}{2}$	7	$6\frac{1}{2}$	$6\frac{1}{4}$
10	$8\frac{1}{2}$	8	$7\frac{1}{2}$	7
12	$9\frac{3}{4}$	$9\frac{1}{2}$	$8\frac{1}{4}$	8
14	$10\frac{1}{2}$	10	$9\frac{1}{2}$	9
16	$11\frac{1}{2}$	11	$10\frac{1}{2}$	10

The nearest available size should be used, and 2-in. ceiling joists should be $\frac{1}{2}$ in. deep per foot span. The trimming joist is made $\frac{1}{8}$ in. thicker for every common joist carried by the trimmer. A rough rule used some years ago was to fix the depth of the joists at one-sixteenth of the clear span, or $\frac{3}{4}$ in. to each foot between the bearings. The Ecclesiastical Commissioners prescribe the size of joists to be 9 in. by $2\frac{1}{4}$ in. for 12-ft. spans and 12 in. by 3 in. for 18-ft. spans. A metropolitan authority has fixed upon $8\frac{1}{2}$ in. by $2\frac{1}{2}$ in., and $11\frac{1}{2}$ in. by $2\frac{1}{2}$ in. for the

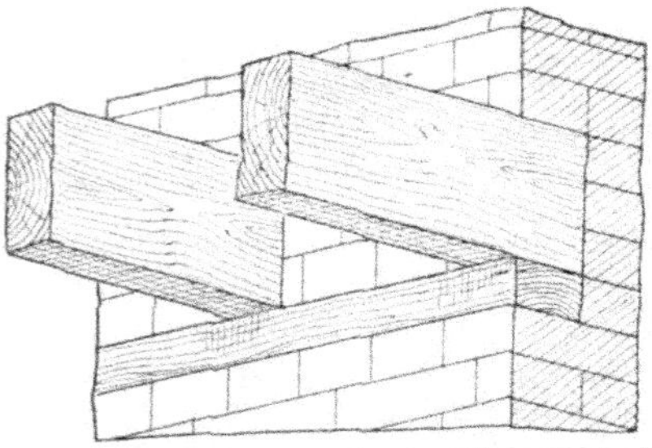

Fig. 361.—Joists supported by Wall Plate built in Wall.

same respective bearings. By the rough rule of one-sixteenth the distance between the bearings, the depth for an 18-ft. span should be:—

$$\frac{18 \times 12}{16} = 13\tfrac{1}{2} \text{ in.}$$

If, however, the thickness of the joist is taken to be 3 in., *the strength of the joist* will allow for

$$\frac{13{\cdot}5^2 \times 3 \times 2{\cdot}5}{18} = 76 \text{ cwt.}$$

central breaking load, or

$$\frac{76}{6} = 12\tfrac{1}{2} \text{ cwt.}$$

central safe load, which is considerably more than is required (see the calculation given below).

Weight on Joists.

The weight on ordinary joists of, say, 18-ft. span, 12 in. deep, and 3 in. thick, and 1 ft. 3 in. centres, may be taken to be as follows:—The superficial space carried on *the joist is 18 ft.* by 1 ft. 3 in. = 22·5 sq. ft., and this covered with people at, say, 84 lb. per square foot amounts to 22·5 ft.

by 84 lb.   = 1,890 lb.

The sound-boarding and pugging may be taken at 100 lb. per yd. super., and the lath, plaster, etc., at 80 lb., giving a total weight of 180 lb. per yd., or per ft. super. $\frac{180}{9} = 20$ lb. This multiplied into the area gives

22·5 × 20 lb.   450

The floorboards will be 18 ft. by 15 in. by 1½ in. = 2·81 ft. cube the joists 18 ft. by 1 ft. by 3 in. = 4·50 ft. cube
        7·31

and the total weight of timber will be 7·30 ft. by 35 lb.   =   257

Thus the total distributed weight is   2,597

This is equal to $\frac{2597}{2} = 1,299$ lb. central load, or 11·6 cwt. The strength of the joists under this load will be, by the formula already given, $\frac{12^2 \times 3 \times 2\cdot5}{18} = 60$ cwt. breaking load, or $\frac{60}{6} = 10$ cwt. safe load.

Estimating Load on Floors.

Floors should be estimated for according to the nature of the building and the probable load. A crowd of persons is variously estimated to weigh from 41 lb. to 147·4 lb. per square foot of the surface covered. Probably a safe average would be 1 cwt. per ft. super. considered as a live load. Dwelling houses are usually designed for a dead load of 1¼ cwt. per ft. super., churches and public buildings 1½ cwt., and warehouses 2¼ cwt. The weight of the structure must be allowed for in addition to the above loads, and this is most important to bear in mind in connection with fireproof floors. For dwelling houses the 1¼ cwt. is usually made to include the weight of the floor itself.

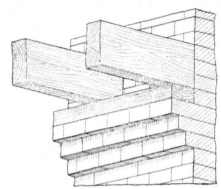

Fig. 363.—Wall Corbelled Out to carry Wall Plate.

Bridging Joist for 18-ft. Span, Load 1 cwt. per ft. super.

Let it be required to determine the size of a bridging joist suitable for a span of 18 ft. and capable of carrying a load of 1 cwt. per ft. super., the joists fixed 12 in. centre to

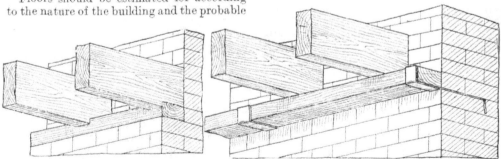

Fig. 362.—Joists supported by Wall Plate bedded on Set-off.

Fig. 364.—Plate carried by Wrought-iron Corbels built in Wall.

centre. The preliminary calculation will be as follows : (1) The total weight on one joist is equal to the load on the half space on either side of the joist—that is, 6 in. on each side. Then the total load = 18 × 1 ft. × 1 cwt. = 18 cwt. (2) The load that may be safely carried on the joist is a certain fraction of the breaking weight —that is, of the load that would break the joist. This fraction varies, for the different purposes for which the scantling is to be used, from one-fifth to one-tenth. In the case of floor timbers, where the joist has to sustain a live load, it should not exceed one-seventh or one-eighth the breaking weight. In the example given above, the joist has to carry a load of 18 cwt. Hence the breaking weight is equal to $18 \times 8 = 114$ cwt. (3) The breadth or thickness of the joist must bear a certain proportion to the depth so as to be satisfactory as regards strength and economy. Let this proportion for a bridging joist be decided by the formula $B = \cdot 3 \, D$, where $B =$ the breadth and D the depth— all in inches. It is evident that the joist in such a case must be considered as strutted. The preliminary calculations as regards the joist having been made, a formula applicable to every case for calculating the strength of timber, no matter where or for what purpose the scantling may be required, must be decided on. A piece of wood of the same kind as that used for the joist, and 1 ft. long by 1 in. square, loaded at the centre till it breaks, will be the constant for all purposes of calculation when dealing with the same material. It will be found that the strength varies directly as the breadth, directly as the square of the depth, and inversely as the length; this may be proved by increasing the breadth, length, and depth, and carefully noting the difference in the loads required to break the beam in each case. Briefly, the formula may be stated thus: $B.W. = \dfrac{c\,b\,d}{L}$ —that is, for a central load. But a floor-joist carries a distributed load, and this load will be found to be equal to twice the load it will carry when centrally loaded. Then the formula will be :—

$$B.W. = \frac{2\,c\,b\,d^2}{L}$$

$$114 = \frac{2 \times 4 \times \cdot 3d \times d^2}{18}$$

and $d^3 = \dfrac{144 \times 18}{2 \times 4 \times \cdot 3} = 1080.$

$\therefore d = \sqrt[3]{1080} = 10$ in. nearly,

and $b = \cdot 3 \times d = \cdot 3 \times 10 = 3$ in.

Let c be the constant $= 4$ cwt.; b the breadth in inches; d the depth in inches; L the length in feet; B.W. the breaking weight $= 114$ cwt. Therefore a joist 10 in. by 3 in. would be suitable for a span of 18 ft., and would carry a load of 1 cwt. per ft. super. The following rule is given by Tredgold for fir joist :—

$$D = \sqrt[3]{\frac{L^2}{B}} \times 2 \cdot 2$$

In this case a breadth must be assumed, which is, in most cases, a difficult and very uncertain proceeding; however, assuming for the present example the breadth to be 3 in.,

Then

$$D = \sqrt[3]{\frac{L^2}{B}} \times 2 \cdot 2,$$

$$D = \sqrt[3]{\frac{18 \times 18}{3}} \times 2 \cdot 2,$$

$$D = \sqrt[3]{108} \times 2 \cdot 2 = 4 \cdot 5 \times 2 \cdot 2 = 9 \cdot 9 \text{ in.}$$

The result is very much the same as in the previous example, but the advantage of the first method will be obvious when dealing with further calculations, as it is applicable to other beams than floor timber.

Determining Size of Binder.

Say it is required to determine the size of a binder 10 ft. long and fixed 6 ft. apart, capable of carrying a floor weighing 1 cwt. per ft. super. Make, as before, the necessary preliminary calculation. (1) Total load carried by the binder $= 10 \times 6 = 60$ cwt. (2) Breaking weight (say) seven times safe load $= 60 \times 7 = 420$ cwt. (3) Let the ratio of the breadth and the depth be as 6 is to 10, that is $\cdot 6 \, D$, which is a very suitable ratio for all purposes where stiffness is required. (4) Let c the constant $= 4$ cwt. Then, using the same formula as before,

breaking weight $= \dfrac{2\,c\,b\,d^2}{L}$

$$420 = \frac{2 \times 4 \times \cdot 6d \times d^2}{10}$$

$$d^3 = \frac{420 \times 10}{2 \times 4 \times \cdot 6} = 875$$

$$\therefore d = \sqrt[3]{875} = 9.5 \text{ nearly,}$$
$$\text{and } b = .6d = .6 \times 9.5 = 5.7 \text{ in.}$$

Therefore, a binder 9·5 in. × 5·7 in. would carry a floor weighing 1 cwt. per ft. super. over a span of 10 ft. The following rule is given by Tredgold :—

$$D = \sqrt[3]{\frac{L^2}{B}} \times 3.42.$$

In this case, again, the breadth must be assumed. Let this be taken as 5·5 in.,

$$\text{then } D = \sqrt[3]{\frac{10 \times 10}{5.5}} \times 3.42$$
$$\therefore d = \sqrt[3]{18} \times 3.42$$
$$\therefore d = 2.7 \times 3.42$$
$$= 9.2 \text{ nearly,}$$

which corresponds very nearly with the first case.

Determining Size of Girder for Supporting Floor.

Girders 10 ft. apart from centre to centre carry a floor weighing $1\frac{1}{4}$ cwt. per ft. super. Required, the breadth and depth for strength ; span 20 ft. (1) The total load carried by the girder is $20 \times 10 \times 1.25 = 250$ cwt.—that is, the length multiplied by half the bay on either side multiplied by the load per ft. super. (2) Let B.W. = 7 times the safe load $= 250 \times 7 = 1,750$ cwt. (3) Let breadth be ·6 d. (4) Let c the constant be 4 cwt.

Then
$$B.W. = \frac{2\,c\,b\,d^2}{L}$$
$$1750 = \frac{2 \times 4 \times .6d \times d^2}{20}$$
$$\therefore d^3 = \frac{1750 \times 20}{2 \times 4 \times .6} = 7262.5$$
$$\therefore d = \sqrt[3]{7262.5} = 19.25 \text{ in. nearly,}$$
$$\therefore b = .6d = .6 \times 19.25 = 11.550 \text{ in.}$$

Therefore, the breadth and depth of a suitable girder for the required purpose must be 11·5 in. wide and 19·25 in. deep. It is needless to remark that a wooden girder 20 in. deep is impracticable, and a wrought-iron girder would be substituted for it ; but as the above is merely an illustrative example, the construction of the girder need

not be discussed. Tredgold's rule for fir girder is :

$$D = \sqrt[3]{\frac{L^2}{B}} \times 4.2.$$

Let the breadth (which must be assumed) be 12 in. Then :—

$$D = \sqrt[3]{\frac{20 \times 20}{12}} \times 4.2$$
$$= \sqrt[3]{\frac{400}{12}} \times 4.2$$
$$= \sqrt[3]{33.3} \times 4.2$$
$$= 3.2 \times 4.2$$
$$= 14 \text{ in. nearly.}$$

It is evident from this that a girder 20 in. deep is by far too large, or that a girder of 14 in. is much too small. If the formulæ in each case are examined it will be found that the first is based on the strength of a small beam determined by trial, while the second is doubtful. It is certain, however, that a girder, 12 in. by 14 in., and 20 ft. long, is not capable of carrying a load of 250 cwt., as determined by the recognised formulæ. It may be mentioned further that the loads are considered as distributed loads, while in reality they are loads placed at certain fixed points, namely, the points where the binders are connected to the girder ; consequently the dimensions obtained by the formulæ are slightly less than they ought to be.

Thus
$$B.W. = \frac{2\,c\,b\,d^2}{L}$$
$$= \frac{2 \times 4 \times 11.5 \times 19.25 \times 19.25}{20}$$
$$= 1704.58 \text{ cwt.}$$

which is less than the actual breaking weight calculated for, namely, 1,750 cwt. The strongest floor, for the quantity of timber used, is given in the first case, while the apparent strength shown in the second and third cases results in actual weakness. But single floors should not be used for spans exceeding 16 ft. ; and though they are sometimes used for spans up to 24 ft., in such cases deflection is considerable, resulting in cracked ceilings, etc. It may, nevertheless, be stated that each floor has its advantages and its disadvantages. The above calculations are not worked out exactly, the nearest fraction being taken in each case.

Floor-Boards.

Timber Used for Flooring.—Many varieties of wood are manufactured into flooring. For elaborate purposes wainscot oak, teak, etc., are employed; but for less expensive work coniferous timber is used, of which there are several kinds. Pitchpine is the most elegant and durable; other kinds are: Riga (red and white), Bjorneborg (red and white), Swedish and Norwegian white wood, and Quebec red. White deal is a timber that is seldom if ever kept under cover; it is generally sawn and wrought direct from exposed stacks. Red deal—especially flooring battens and deals—is invariably kept under cover. Pitchpine is never imported in battens, but in logs, deals, and irregular sized scantlings, which can be with safety stacked, when pinned, in an exposed place. For flooring, etc., white deal is used to a far greater extent than red deal or pitchpine, on account of its cheapness and adaptability.

List of Shippers' Marks on Floorings.— The following is a list of floorings with the shippers' marks, showing the quality and port :—

A + A	3rd.	Iggesund
B S S C	1st.	Sandarne
B C	2nd.	,,
B B B	3rd.	,,
B – Co.	u/s.	,,
C + K	2nd.	Domigo
D M	3rd.	Skutska
D O M	2nd.	,,
E E	1st.	Sundswall
E E E	2nd.	,,
F A S	3rd.	Gothenburg
GLINGE	3rd.	Fredrikstad
N A S	2nd.	Gefle
K H B	1st.	,,
H A B	2nd.	,,
H + B	1st.	Fredrikstad
H + B + T	3rd.	,,

Shippers' Marks on Floorings (*continued*) :—

H H	1st.	Hudikswall
H T A B	3rd.	,,
J C K	1st.	Domsjo
J D & Co.	2nd.	Söderhamn
J F J	u s.	Fredrikstad
K H B	Extra 1st.	Gefle
K + K	3rd.	Domsjo
M H	1st.	Söderhamn
N + W	1st.	Sundswall
N = W	2nd.	,,
N W	3rd.	,,
P A T	1st.	Hudikswall
P P	2nd.	,,
S A F	2nd.	Gothenburg
S A L	2nd.	,,
S B	Extra 1st.	Skonvik
S B S	1st.	,,
S A B	2nd.	,,
S F	1st.	Gothenburg
S F A	1st.	Fredrikshald
S F B	2nd.	,,
S + W	3rd.	,,
S K B	Extra 1st.	Skutska
S K B	1st.	,,
S U N D	1st.	Sundswall
S V VIK	1st.	Swartwik
S V ★ VIK	2nd.	,,
S D D	3rd.	,,
W & Co.	2nd.	Fredrikshald
W D & Co.	1st.	,,
WIESE & Co.	1st.	Fredrikstad
	3rd.	Gefle
	u/s	,,

Sizes of Floor-Boards.—The sizes of flooring generally taken from white deal are 3 in. by $\frac{7}{8}$ in., 3 in. by $1\frac{1}{4}$ in., 3 in. by $1\frac{3}{8}$ in., $5\frac{1}{2}$ in. by $\frac{7}{8}$ in., 6 in. by $\frac{7}{8}$ in., $6\frac{1}{2}$ in. by $\frac{7}{8}$ in., $5\frac{1}{2}$ in. by $1\frac{1}{4}$ in., 6 in. by $1\frac{1}{8}$ in., $6\frac{1}{2}$ in. by $1\frac{1}{8}$ in., and $6\frac{1}{2}$ in. by $1\frac{3}{8}$ in. These sizes do not include the tongue, or feather, consequently the stuff is $\frac{1}{2}$ in. broader in the rough than when wrought. The most suitable battens

for flooring are 6 in. by 2 in., 6 in. by 2½ in.,
6½ in. by 2½ in., 7 in. by 2½ in., and 7 in. by
3 in. When 3-in. by ⅞-in. flooring is being
cut and wrought, the most suitable sized
batten is 7 in. by 3 in., which gives six
pieces, three saw cuts being sufficient—

Fig. 365.—Ordinary Direction of Grain in Floor-
Boards.

namely, two deep and one flat. This is
when wrought single with the flooring
machine. When run double with the
machine two saw cuts through the depth
are sufficient. The flat cutting in this
instance is done with the flooring machine.
The double working of flooring and lining
with machinery, though much the quicker
way, is not so satisfactory as the single
method, for each alternate board has to be
reversed, besides the further disadvantage.
if the battens are waney, of the groove being
always on the waney edge. Similar sized
flooring (⅞ in.) can also be cut from 7-in.
by 2½-in. battens. Two boards may be
cut ⅞ in. in thickness and one ½ in., thereby
utilising the whole batten; 5½ in. by ⅞-in.
boards are taken from 6-in. by 2-in. material ;
3-in. by 1¼-in. from 7-in. by 2½-in., 3-in.
by 1⅜-in. from 7-in. by 3-in.

Operation of Floor-Board Planing Machine.
—The fixed cutters or face irons of a flool-
ing machine produce the best and smooth-
est work. These tools operate on the under
side of the boards ; therefore the freshly
sawn side should be placed downwards to
receive the finishing, which the face irons
accomplish. The revolving top scutching-
block is not so much used for dressing as
for bringing the boards to an exact thickness. So long as one side of the board is
well dressed and of accurate thickness, it is
not important to have the other side so well
done. Some machines have fixed planers
on the upper side, but such cannot bring
the stuff to an accurate thickness like the
revolving scutch-block. It is heavy work
for fixed cutters to reduce boards ¹⁄₁₆ in. ;
the scutching-block, however, can easily

take ¼ in. off. With evenly sawn wood
heavy cutting has seldom to be resorted to.
The leading advantage of the scutching-
block compared with fixed cutters is that
the block makes an irregular surface parallel,
whereas fixed cutters follow the uneven
nature of the board, and do not alter any
irregularity which it may have. There are
many cutter heads for the formation of the
tongue and groove. The face-iron side of
the groove and tongue should project a little
more than the scutched side ; by this means
the faced side of the flooring, when driven
home and placed in position, has a better
joint than it otherwise would have. The
" Shimer " patent heads make the finest
work ; with a feed speed of 60 ft. or 80 ft.
per minute undue chipping is very rare with
the " Shimer " patent. A good machine
can run from 9,000 to 11,000 sup. ft. of 6-in.
or 6½-in. by 1¼-in. flooring per day, or 4,000
sup. ft. of narrow flooring. All 1¼-in.
material is taken from 2½-in. battens,
whether broad or narrow. Flooring above
1½ in. thick is sometimes run with two
grooves instead of one, and slip feathers
are employed in place of the solid formed
tongue. This plan saves ½ in. on the
breadth of each board.

Stacking Floor-Boards.—Finished flooring.
no matter how well it may be stacked and
pinned, is always liable to become twisted
whilst being seasoned. To obviate this,
the material should be sawn, pinned, and
stacked in the rough. Let it season for six
or eight weeks ; then finish it with
machinery. Work done in this way can
be stored in bulk under cover without
being pinned or ventilated. Flooring

Fig. 366.—Direction of Grain for Least Shrinkage
of Floor-Boards.

wrought on this principle does not twist,
cast, or shrink like material finished and
stacked at one operation ; it is, moreover,
much more easily laid. This rule applies
also to lining. Red deal flooring is not so
generally wrought for stock as white, for

the reason that red deal battens are, as a rule, kept under cover; orders can be executed and despatched without the necessary seasoning that white deal requires.

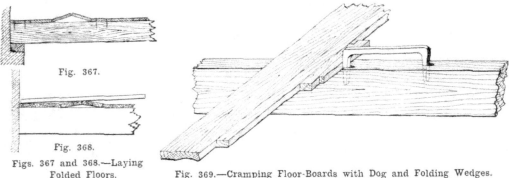

Fig. 367.

Fig. 368.

Figs. 367 and 368.—Laying Folded Floors.

Fig. 369.—Cramping Floor-Boards with Dog and Folding Wedges.

Red deal is more easily manufactured than white. It is to a certain degree softer and not so tough in the reed as spruce.

Direction of Grain in Floor-Boards.— If a specification does not insist on any particular position of the grain of the wood, it will be complied with by either of the examples shown in Figs. 365 and 366. If the grain is intended to show "annual rings parallel with the edges," words to that effect should be inserted in the specification, or it should be stated that "all boards are to be cut radially from the tree." No doubt the plank shown in Fig. 366 would be less liable to warp than that shown in Fig. 365;

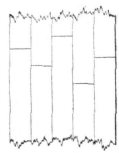

Fig. 370.—Floor with *Joints broken* at 3-ft. Intervals.

but to obtain all like this would mean picking over *a very large parcel of boards in* order to get the quantity required, and it may be looked upon as impracticable.

Laying Floor=Boards.

Folded Floor.—"Floors to be laid folding with the joints broken" means that the heading joints of the boards are not to be in line when laid, but are to be crossed in as long lengths as possible from joist to joist. The system of laying the boards *with a succession of joints in line causes* unevenness when the boards shrink, and weakens the floor. The term "laid folding" is an old one, and was applied when mechanical means were not available for bringing the joints tightly together. In the absence of a floor cramp the boards *may be laid with fairly tight joints by* jumping them in, as shown in Fig. 367. The first board next the wall is laid and nailed in its place; then other boards (say five), to make a width of about 3 ft., are

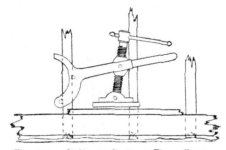

Fig. 371.—Ordinary Pattern Floor Cramp.

laid down. The final position of the fifth board having been ascertained, the fifth board is nailed down ¼ in. inside the line it takes when only hand tight. The four other boards are then jumped in and nailed.

A board placed over the loose boards, as seen in Fig. 368, will be found of assistance in getting the floor-boards down to the joists, but there will still be some difficulty

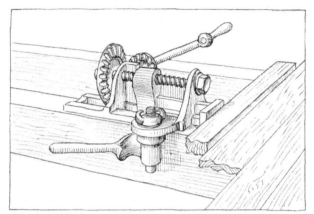

Fig. 372.—Improved Form of Floor Cramp.

unless the four boards are kept loose—that is, none of the intermediate boards between the first and sixth must be nailed until all of them are tight home. Another simple method of cramping is shown in

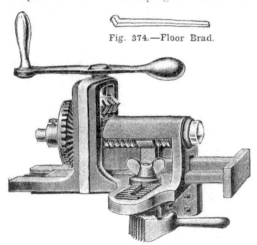

Fig. 374.—Floor Brad.

Fig. 373.—Another Improved Floor Cramp.

Fig. 369. An iron timber dog (Fig. 47, p. 11) is driven into the top edge of the joist, allowing about 3 in. from the edge of the floor-board. A piece of rough timber

2 in. thick is then laid next the board, and a pair of hardwood folding wedges is driven between the timber and the dog until the joints of the board are close;

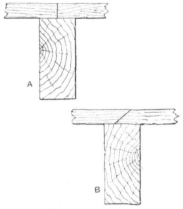

Fig. 375.—Butt and Splayed Heading Joints.

then the boards are nailed, the dog is removed, and more boards laid in the same manner. Both the methods above mentioned are usually adopted for the commoner kinds of work only.

Laying Floor - Boards with Aid of Cramp.—Floors laid with the heading joints crossed, as in Fig. 370, need a special cramp to bring up the joints ; three kinds of cramps are shown by Figs. 371 to 373, but a variety is available. For instance, batten-width tongued and grooved common Baltic flooring would be laid in the following manner. The joists would be tried over and brought to a level. A batten, or line of battens, would be laid down next the wall to line true at the outer edge, and then be nailed to the joists. The remaining rows are laid two or three at the time with the tongues inserted, then cramped into place, nailed, and the next lot of battens applied. If the battens are already tongued, they can be laid either way, as the block, or saving piece, between the cramp and batten can be grooved to clear the tongue. Figs. 371 and 372 show the modes of using floor cramps. When the floor has been finished so far that there is not sufficient room for the cramp, the remaining battens can be wedged in from the wall, or forced together by using a piece of quartering as a lever.

Floor Brads.

Nails used in flooring are called floor brads (Fig. 374), and they are driven through

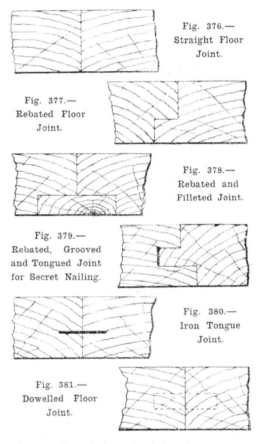

Fig. 376.—
Straight Floor
Joint.

Fig. 377.—
Rebated Floor
Joint.

Fig. 378.—
Rebated and
Filleted Joint.

Fig. 379.—
Rebated, Grooved
and Tongued Joint
for Secret Nailing.

Fig. 380.—
Iron Tongue
Joint.

Fig. 381.—
Dowelled Floor
Joint.

the floor-boards into the joists, two at each passing, about 1 in. from the edge.

Joints for Floor-Boards.

Heading Joints.—The points of contact between the ends of two floor-boards are called heading joints (Fig. 375). A (Fig. 375) shows the section of a butt heading joint, but slightly less simple than the splayed heading joint shown in section by B (Fig. 375). These joints should always be arranged to occur over a joist, and in floors laid with the aid of a cramp, contiguous boards should have their heading joints on different joists— that is, should break joint. The actual joint is made in different ways. In common floors the boards simply butt up against each other A (Fig. 375) ; in better work the heading joints are splayed B (Fig. 375). Even with plain headings it is usual slightly to undercut the ends so as to present as close a surface joint as possible. Sometimes the heading joints are grooved and tongued in a similar fashion to the longitudinal joints described below. In very expensive work the ends of the boards are cut into a series of sharp, salient and re-entering notches, whose ridges are parallel to the surface of the floor. These notches fit one another, and form a tight joint. Such joints are sometimes used in oak floors ; they are extremely troublesome and expensive to make, and the point nearest the surface of the floor is very liable to break away even in hard wood.

Edge Joints.—The ordinary straight joint for the longitudinal edges of floor-boards is shown in section by Fig. 376 ; the rebated joint (Fig. 377) is another common method, a joint requiring more work being the rebated and filleted (Fig. 378). The rebated, grooved, and tongued joint (Fig. 379) is useful for secret nailing. The joint shown in Fig. 380 has an iron tongue, and Fig. 381 shows the dowelled joint. The ploughed and cross-tongued joint with slip feather (Fig. 260, p. 62) is also used. In all floors which are ceiled underneath, means should be taken to prevent dust or particles of any kind from falling between the boards. Any accumulation of organic matter on the upper surfaces of the plaster is certain to decompose. The ceiling being, moreover, always more or less porous, these particles gradually work their way to the under surface, and produce a stained appearance, which no amount of whitewashing or scraping will remove. The usual method of preventing this is to form a ploughed and tongued floor. Each board is grooved on each edge, and thin slips, or tongues, either of wood or of galvanised iron, are then inserted (see Figs. 260 and 380). If of iron, the tongue should be galvanised. The tongue should be fixed nearer to the lower edge of the board than to the upper, so that as much wear as possible can be had out of the floor before the tongue is exposed. Another method of attaining the same object is known as rebating and filleting (see Fig. 378); a rebate is cut on the lower edge of each board, and a fillet of oak or some other hard wood fixed in the space thus formed. For superior work, a dowelled floor (Fig. 381) has the advantage of showing no nails on the surface; the boards are pinned together between the joists with oak dowels, and nailed obliquely on one edge only. Dowelled boards should not be more than 3 in. wide, and not less than $1\frac{1}{4}$ in. thick when finished. The "Pavodilos" joint is as shown by Fig. 382, a slightly modified form being that shown by Fig. 383, which, although the second key is lost, may possibly be preferred on account of the danger, when nailing down the flooring jointed as in Fig. 382, of damaging the feather-edge of the board that is being fixed.

Double-boarded Floor.

An upper layer of thin oak boards is sometimes fixed over a rough deal floor for the sake of appearance, and also in some cases to obtain an almost impervious surface. A floor of this kind, wax-polished and well laid, is much to be commended for the ease with which it can be cleaned, and for its non-absorbent nature.

Sound-proof Floors.

One method of preventing the sound from one room being audible in another room immediately below is to nail fillets to the joists, and on these nail a layer of rough boards, and to fill in on the top of these boards a stratum of lime-and-hair mortar. Slag felt, a preparation of slag wool, which is a material produced by blowing off waste steam into the slag of iron furnaces, is also used for this purpose. In the case of the slag felt the process is as follows: On the under side of the joists, fillets are nailed to wooden blocks 1 in. thick, and to these fillets the lathing for the plaster ceiling is affixed. The slag wool (known as "pugging") is then laid on the upper surface of the laths, and is felted by a patent process, this process of

felting removing entirely the property which the slag wool possesses of emitting sulphuretted hydrogen, and also reducing the weight of the material. Slag material, being fireproof, is to be preferred to sawdust and other combustible materials sometimes

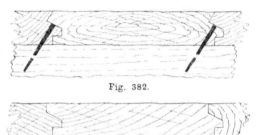

Fig. 382.

Fig. 383.

Figs. 382 and 383.—"Pavodilos" Joint in Flooring.

used. Fig. 384 shows the section of part of a common floor, showing 9-in. by 3-in. joists, and $1\frac{1}{2}$-in. boarding with a rebated heading joint. In addition, "pugging" and a lath-and-plaster ceiling are shown. The object of the pugging is to reduce the transmission of sound. The fillets for supporting the pugging need not be of the shape indicated in Fig. 384. Another means of attaining the desired end is to nail strips of felt on the upper edges of the joists, under the floorboards. By this means the connection between the joists and boarding is broken. This arrangement creates some difficulty in fixing the boards, which can be overcome by nailing a lath along the top of the felt.

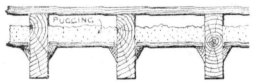

Fig. 384.—Section of Sound-proof Floor with Pugging.

Fireproof Wooden Floors.

Protected Wooden Floors.—One of the simplest and most economical methods of constructing a fire-resisting floor is to protect an ordinary wooden floor with slabs of asbestic plaster or of slag wool (silicate cotton), both of which can be obtained commercially in slabs, as cloth, or in the form of loose fibre or wool. The loose wool is useful for filling up the spaces between the joists as a pugging to deaden sound (as already described), as well as affording protection against fire. A convenient method

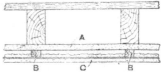

Fig. 385.—Asbestos Slabs under Wooden Floor.

of attaching the slabs is shown in Fig. 385. The slabs are formed by enclosing silicate cotton between sheets of galvanised wire netting, and are made of thicknesses varying from 1 in. to 3 in. They are secured to the under side of the joists, as shown at A, by wooden fillets B B nailed underneath, the nails passing through the slabs. To these fillets are secured the laths, when a lath-and-plaster ceiling c is desired. Additional security can be obtained by placing other slabs between the joists, resting on triangular fillets as shown in Fig. 386. Owing to the comparative cheapness of these methods of construction, and the measure of security they afford, they are worthy of more general adoption in dwelling-houses and office buildings.

Solid Wooded Floors.—Woodwork, when used in solid masses, is an excellent material for fireproof construction. It is extremely difficult to destroy timber in bulk by fire, and in America, partly on this account, and also on account of the cheapness of timber,

Fig. 386.—Asbestos Slabs between Joists.

floors and walls are constructed of planks nailed together side by side. The walls of many of the large grain elevators and station buildings are constructed in this way. The system of forming floors by close timbering instead of the ordinary use of joists and flooring boards, was introduced into England

by Messrs. Evans and Swain between 1870 and 1880. The joists, instead of being placed at some distance from each other, were laid close together, so that air could not penetrate between them, the planks being then spiked as shown in Fig. 387. As an alternative method, the spikes could be driven in diagonally, and, if thought necessary, the under side of the planks could be protected with a plaster ceiling keyed into grooves formed in the planks. As a test of the capability of this system, a building was erected 14 ft. square inside of 14-in. brick walls, and measuring 7 ft. from the ground to the ceiling. The flooring was laid as described above, of deal battens 7 in. deep by 2½ in. thick, spiked together side by side. One-third of the under side was plastered, the joists being grooved for this purpose; one-third was plastered on nails partly driven into the planks, and the remaining third was left unprotected. The

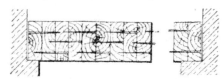

Fig. 387.—Floors of Solid Wood.

chamber underneath was packed almost full of timber, which was then lighted, and it was not until after five hours' continuous exposure to the flames that the unprotected portion of the floor gave way. The system was afterwards adopted in large warehouses for the East and West India Docks, London, and in other buildings.

Other Systems.—A modification of the system just described has been patented by Messrs. Hinton and Day, and is illustrated in Fig. 388. The joists are spaced apart in the ordinary way, but the spaces are filled in with solid blocks, having the grain placed vertically, tongued and grooved together in such a manner that the passage of air between them is prevented. The blocks are carried by fillets nailed to the sides of the joist. A test of this system of flooring was made at Westminster. Four walls of 9-in. brickwork were erected, and the under side of the floor to be tested was 9 ft. 6 in. from the ground. The lower part of the building was filled three parts full with inflammable material (no petroleum or grease, however), and a fierce fire maintained for more than two hours, after which it was extinguished, and the under side of the floor was found to be charred to a depth of $\frac{3}{4}$ in. In American factory and workshop buildings a layer of mortar D is often introduced between two thicknesses of flooring, as shown in Fig. 389. Here 8-in. by 4-in. wooden joists E support the flooring planks, which are 3 in. thick, on which a layer of mortar, $\frac{3}{4}$ in. thick, is spread. Floor-boards $1\frac{1}{2}$ in. thick, laid on the top of this, form the working surface of the floor. Sometimes the floor-boards are laid in two thicknesses, crossing each other diagonally, as shown in Fig. 390, in which F indicates the layer of mortar. The beams carrying the floors have air spaces round each end, and to avoid the danger of the wall being pulled down by a falling

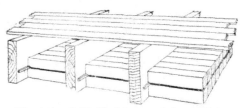

Fig. 388.—Solid Blocking carried on Fillets.

beam in case the latter should be burnt through, the upper end of the beam is cut away at both ends so that it can fall freely.

Wood=Block Floors.

Solid wood-block floors are now much used in the basements of dwelling-houses, on the ground floors of public buildings, and for covering certain forms of fireproof constructions in the upper floors of warehouses, etc. The advantages they possess over the ordinary boarded floor are : damp-proofness, freedom from dry rot, greater lasting properties, and freedom from vibration, and they do not transmit sound nor harbour vermin ; they are more sanitary, through the absence of shrinkage, and consequent open joints of the older system ; and the absence of nails is also a great advantage, as the holes made by these are always unsightly, and when

the boards wear down the heads project, to the discomfort of the users.

The Wood Blocks.—Wood blocks are generally made from 9 in. to 18 in. long by 3 in. wide, and from $1\frac{1}{2}$ in. to 3 in. thick, of yellow deal, pitchpine, oak, birch, maple, or beech. They should be prepared from thoroughly seasoned and sound stuff. The firms who make a speciality of this work usually dry the blocks in hot-air chambers after working, and afterwards store them in a dry building. Precautions should therefore be taken, when receiving a consignment from the factory, to store them under cover until they are required ; and it is wise not to order them until the place is ready, because their storage for any length of time

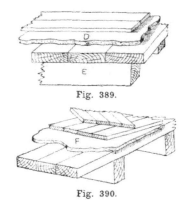

Fig. 389.

Fig. 390.

Figs. 389 and 390.—American Systems of Wooden Floors.

in a damp building will defeat the object of the previous drying, and for this the purchaser has to pay. The smaller sized blocks are sometimes made with square joints, and are held in place by the cement or mastic with which the foundation is covered, but in superior work the blocks are also connected by grooves and tongues or dowels. Several patented systems are on the market, some of the best of which are here illustrated ; these combine an interlocking of the blocks with the substance of the bed, by means of dovetailed grooves or inserted keys, and a connection with each other by means of pins or tongues.

Preparing Basement for Wood-Block Floor. —In preparing a basement to receive a wood-block floor, the ground should be taken out from 8 in. to 11 in. (according to the thickness of the blocks) below the intended floor-line ; the surface should be roughly levelled and rammed solid ; 1-ft. 6-in. stakes are then driven into the bottom about 6 ft. apart, and levelled off to 6 in. above the ground ; the site is then filled in with concrete to the depth of the stakes, and the surface beaten smooth. A blue lias lime, or Portland cement, should be used for the concrete, in the proportion of 1 cement to 6 aggregate. The concrete bed should be allowed to settle and dry before proceeding with the next step, which is the floating of the top with a $\frac{3}{4}$-in. layer of Portland cement and sand, 5 to 1 ; preparatory to

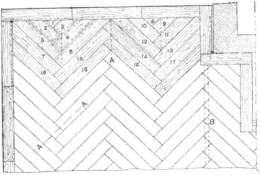

Fig. 391.—Herringbone Pattern of Wood-Block Floor with 18-in. Blocks.

this screeds of cement about 3 in. wide should be run around the margins, and across the room every 6 ft. or 8 ft. ; these should be accurately levelled and struck straight with a long float, and when set will become levelling points from which to strike off the surface of the cement ; before the latter has become hard it should be brushed over with a birch broom to score the surface ; it must then be allowed time to become perfectly dry, as any trace of moisture will be fatal to the adhesion of the bitumen coat next to be laid. From seven to fourteen days, according to the state of the atmosphere, will be required for this purpose ; and as an additional precaution just before laying the bitumen, or matrix, as it is termed, dust a little fresh lime or some fine dry ashes over the surface ; these must,

however, be swept thoroughly off before running on the mastic. The bitumen is sometimes laid in two coats, the first being allowed to set before proceeding with the second ; the purpose of this is to ensure a substantial layer of bitumen between the blocks and the cement, but this is only necessary on very damp sites. The objects of the three different layers under the floor are : The concrete is to form a substantial and unyielding foundation, and also to prevent the ground-air arising ; the cement layer is to form a hard and regular surface to which the matrix can adhere ; and the matrix is a damp-proof layer that will effectually prevent any moisture that may pass through the cement from reaching the blocks, and also, being strongly adhesive, it keeps the blocks attached to the cement. Various mixtures are used for matrices, the best having mineral bitumen as a base ; but frequently a simple mixture of Stockholm tar and pitch, in the proportions of 2 of tar to 1 of pitch, is used (note, gas tar is unsuitable). When this is laid in a single coat, screeds of wood about $\frac{1}{4}$ in. or $\frac{3}{8}$ in. thick are nailed lightly to the cement to form bays about 4 ft. or 5 ft. square ; two of these should be filled in with the melted matrix, which should be boiling hot, and the first filled in will be ready for laying the blocks by the time the second is filled. The best consistency of the matrix for laying is when it is thick enough to receive the weight of the block without allowing it to sink in, and yet warm enough to amalgamate properly with the mixture adhering to the blocks. Scaffold boards should be laid across the bays, resting on the screeds, for the men to kneel on whilst laying.

Laying the Wood Blocks.—The mastic, as the fumes are suffocating, should be heated in a large iron cauldron in the open air, and brought into the building in iron pails. The blocks should be stacked in the room near the doorway, each cut to its proper size and each series stacked by itself. To do the work properly two men at least will be required to lay, working into each other's hands, and one to deliver the blocks as required. The order of delivery and of laying will depend on the design, as will be mentioned presently. The blocks are dipped to half their depth into the pail of mixture, care being taken not to allow any to get on the surface, and lightly tapped into place; when a bay is completed a piece of quartering about 5 ft. long, with one side planed straight, should be struck on the face of the blocks to

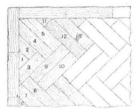

Fig. 392.—Double Herringbone Design with 12-in. Blocks.

bring them to a uniform level. In laying the herringbone design (Fig. 391), begin with the margin, laying this as far as the mastic runs; then taking two blocks, place them in the left-hand angle, and make a mark on the margin where the edge of the second block reaches. This will be the point for fixing the small triangular piece, marked No. 1; next fix the block marked 2, and then Nos. 3, 4, 5, 6, 7, in due order. This arrangement makes the insertion of the tongues or pins easy. Having reached No. 7, either move to the right, or let the second man take up the running with block No. 8, whose position is found by measuring from No. 6 with two blocks as before; then let

Fig. 393.—Tile Design with 12-in. Blocks.

him follow on with Nos. 9 to 14 consecutively, when the first man will lay Nos. 15 and 16, and the second Nos. 17 and 18, and so on. The shaded portion in Fig. 391 represents the recess between a chimney-breast and the wall. If a beginning were made against a straight wall all the three

pairs of contiguous blocks should be laid first right along that side—that is, all of those having mitred ends, as these provide the starting points of the pattern, then follow on alternately left and right as described. A beginning should always be made at the

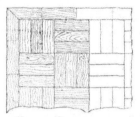

Fig. 394.—Chequer Design with 9-in. Blocks.

wall opposite the door, working towards the latter so that no traffic may pass over fresh-laid work; and after all the blocks are down, sawdust should be freely strewn over their surface to absorb any mastic that may have dropped thereon, and scaffold boards laid on spare blocks from the doorway, should it be necessary to pass that way. At least twenty-four hours should elapse before beginning the cleaning off, to allow the mastic time to set hard, and in cleaning off plenty of tallow should be used. It will be found an advantage to the workmen to have a pail of whiting handy, whilst they are laying the blocks, into which they can occasionally dip their hands, as the tar

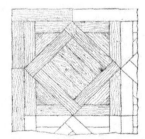

Fig. 395.—Panel and Frame Design with Mixed Blocks.

burns severely the unprotected skin. As before mentioned, the blocks should be all cut to size before beginning, and this necessitates the setting out of one "repeat" of the design full size upon a large board or a clean floor. The actual blocks should be used for this purpose, fixing down the margins, and cutting and fitting in a bay as shown by the dotted line A (Fig. 391). Once the spread of a bay is known, it is easy to space out the quantity for a room and

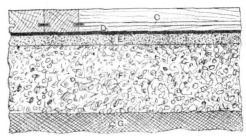

Fig. 396.—Section of Herringbone Patterns shown in Fig. 391.

ascertain how many of each length and shape are required. It is best to lay down all recesses like the one shown, and cut in all the blocks, specially marking them. To obtain the size of the recess, lay down the margin blocks tight between the walls, or frame a rough template to the opening. The herringbone pattern must always be laid square—that is, cut ends must be a mitre of forty-five degrees.

Designs of Wood-Block Floors.—Design Fig. 392 is laid similarly, beginning with the blocks No. 1 and following on with 2 and 3, etc. Fig. 393 is an easy design to lay when once the corner is passed ; the numbers indicate the order of laying the blocks. Care must be taken to keep the sides of each tile in a straight line, and they should be tested occasionally with a straightedge. Fig. 394 is an easy design to lay, and looks very well in pitchpine. Fig. 395 is more elaborate, but very effective in two coloured woods, the darker one for the frames and the lighter for

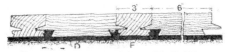

Fig. 397.—Turpin's Patent Block Floor.

the panels. All of these designs are based on the right-angled triangle, and, given the size of the block, they can be readily set out to fit any room ; each pattern being a repeat, one bay multiplied by the length and

width of the room will show the quantity required. It may be mentioned that these blocks are usually sold by the hundred.

Jointing and Fixing Wood Blocks.—Fig. 396 shows the section of a wood-block base-

Fig. 398.—Duffy's Patent Block Floor.

ment floor with grooved and tongued joints. Fig. 397 represents a section of Turpin's patent interlocking system ; here a tapering tongue with an undercut shoulder on the lower side is stuck on the solid all round one block, and a corresponding groove in the other, and when the two come together they form a dovetail groove into which the mastic is pressed when laying, thus forming a solid key with the bed. Duffy's patent is shown in Fig. 398, and consists in the connection of the blocks by means of dowels ; these are supplied with the blocks and driven in as the blocks are laid. The holes are bored by machinery and are at exactly the same

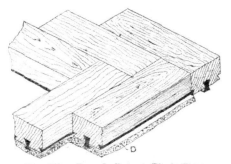

Fig. 399.—Geary's Patent Block Floor.

distance apart, whether on the end or side, and therefore the blocks can be laid in several combinations. In Geary's patent (Fig. 399) each block is fixed to the mastic by means of two metal keys driven into the

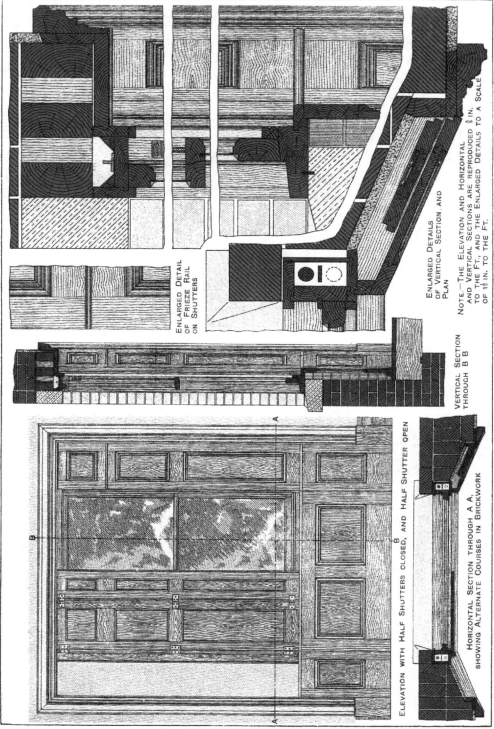

CARPENTRY AND JOINERY.

ENLARGED DETAIL OF FRIEZE RAIL ON SHUTTERS

ENLARGED DETAILS OF VERTICAL SECTION AND PLAN

NOTE.—THE ELEVATION AND HORIZONTAL AND VERTICAL SECTIONS ARE REPRODUCED ⅛ IN. TO THE FT., AND THE ENLARGED DETAILS TO A SCALE OF 1½ IN. TO THE FT.

VERTICAL SECTION THROUGH B B

ELEVATION WITH HALF SHUTTERS CLOSED, AND HALF SHUTTER OPEN

HORIZONTAL SECTION THROUGH A A, SHOWING ALTERNATE COURSES IN BRICKWORK

BOXING SHUTTERS TO SASH WINDOW

3

ends of the block; these project from the bottom, and are buried in the bed material. The key is drawn to enlarged scale in Fig. 400; it is easily knocked out when a block has to be cut, and is re-inserted in a small mortise. A half dovetail groove is also worked on the side of each piece, which forms an additional key to the block. In Fawcett's system, shown in plan at B (Fig. 391, p. 94), and in isometric projection by Fig. 401, the ends of the blocks have a $\frac{1}{8}$-in. groove cut across them at an angle of forty-five degrees, and these, when the blocks are laid in herringbone pattern, lie in a continuous straight line. Into these grooves a $\frac{3}{4}$-in. by $\frac{1}{16}$-in. steel tongue is inserted as shown in Fig. 401, the succeeding row of blocks fitting over and completing the

larger portions of the patterns, the natural colours of the wood afford sufficient contrast, but for bands in the borders, and for edgings for the geometric figures, more vivid colours are sometimes desirable, and these are obtained by dyeing some light-coloured wood, such as ash or sycamore, to the required tint. The three forms of parquetry in ordinary use are known respectively as thin, medium, and solid. The two former, which are respectively out of $\frac{1}{4}$-in. and $\frac{1}{2}$-in. stuff, are glued to $\frac{1}{2}$-in. or $\frac{3}{4}$-in. deal backings in squares or panels from 10 in. to 18 in. square, and these panels are grooved and tongued all round, or sometimes dowelled, and are attached to the counter-floor either with screws, which are afterwards pelleted, or by gluing down. The former method is

Fig. 400.—View of Metal Key.   Fig. 401.—Fawcett's Patent Block Floor.

groove. This system is very effectual in preventing the rising of individual blocks, and is much used on fire-resisting concrete floors. The letter references in Figs. 391 to 399 not mentioned in the text are: C groove, D mastic, E cement, F concrete, G ground.

Parquet Floors.

Parquetry is a method of covering a floor with hard and richly coloured woods, arranged in various fanciful and geometric patterns, the effect of the design being brought out by the various colours, and by the direction of the grain in the component pieces, which are selected chiefly for their differences in this respect. Usually, for the

employed when it is intended to remove the parquet at some future time; and the latter, when the parquet is to be permanent. The solid parquet is about 1 in. thick, and the various pieces are usually glued direct to the counter-floor and to each other in one operation, the design being formed as the work proceeds. In this method, all pieces more than $1\frac{1}{2}$ in. wide are dowelled, or, in a cheaper class of work, are nailed to each other with wire nails. Borders are fixed first, and, as far as possible, these are made wide enough to bring all small recesses and projections into line, so as to cause no interruption in the pattern; but large openings must have the borders broken and returned around them.

5

TIMBER PARTITIONS.

Common Stud Partitions.

This chapter will consist chiefly of illustrations showing the construction of timber partitions. Such partitions are built in a variety of styles, the simplest being the common stud partition, which is supported by a wall, as shown by Fig. 402. This is built with quartering, and is not braced or trussed in any way, but is stiffened by nogging pieces being notched into the edges of the studs and nailed as indicated. Fig. 403 shows a somewhat similar form, but the nogging pieces and the brick nogging shown add to the stability. The partition can be finished with lath and plaster.

Braced and Trussed Partitions.

Fig. 404 illustrates a form of partition the sill of which rests on floor joists whilst the head serves as a middle bearing for the floor above. The joists under the sill are notched on a plate and supported by a $4\frac{1}{2}$-in. brick partition in which it is assumed there is at least one opening in the middle, on account of which the braces and king-post are introduced into the partition shown. When the sill overhangs the joist, as shown at A, it is housed into the post, and the latter is supported on a bearer fixed between the joists, *as shown in Fig.* 405. Another method is shown at B (Fig. 404), the post and sill being mortised and tenoned or dovetailed together and supported by a bearer, which rests on fillets, nailed to the joists, as shown by Fig. 406. Sometimes the partition is framed and fixed with the sill *running through the openings, as indicated* by the dotted lines at B (Fig. 404) ; just before the floor is laid the sill is cut out between the posts. A trussed partition is usually so built as to carry its own weight, and often that of one or more floors as well, and to distribute the weight to particular *points of support, as will be made clear* by the following examples. Fig. 407 shows a partition which has to carry its own weight over the greater part of the span and also that of the floor above ; the head of the partition serves as a girder, the joists being

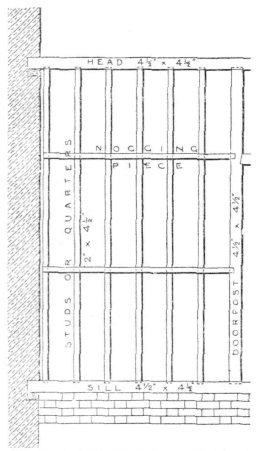

Fig. 402.—Part Elevation of Common Stud Partition supported from below.

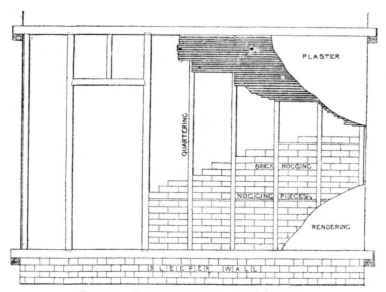

Fig. 403.—Elevation of Brick-nogged Partition.

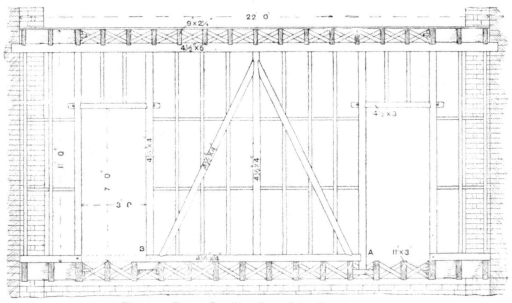

Fig. 404.—Braced Partition Framed for Two Doorways

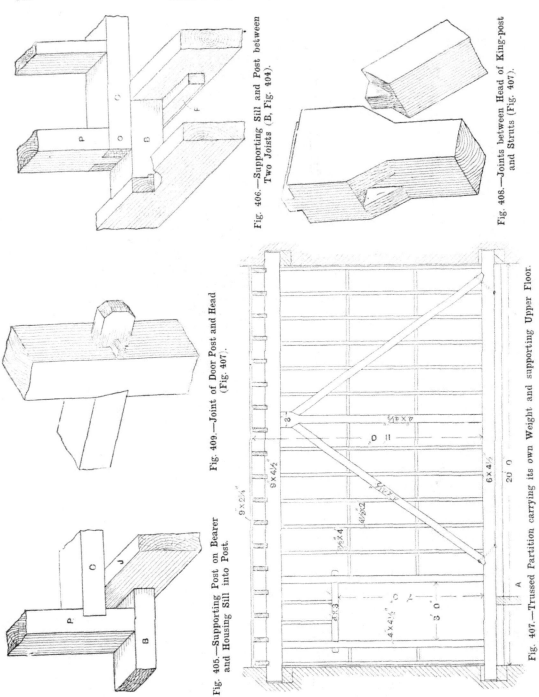

Fig. 405.—Supporting Post on Bearer and Housing Sill into Post.

Fig. 406.—Supporting Sill and Post between Two Joists (B, Fig. 404).

Fig. 408.—Joints between Head of King-post and Struts (Fig. 407).

Fig. 409.—Joint of Door Post and Head (Fig. 407).

Fig. 407.—Trussed Partition carrying its own Weight and supporting Upper Floor.

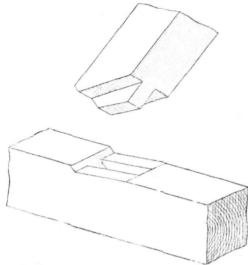

Fig. 410.—Joint between Strut and Sill in Trussed Partition (see Fig. 407).

notched or cogged to it. The ends of both head and sill are supported by stone corbels built in the walls. The sill has an intermediate support on a passage wall A. The foot of the king-post should be connected to the sill by a bolt or strap. Figs. 408 to 410 show the form of the three principal joints. The trussed partition illustrated by Fig. 411 is designed to answer the following requirements: A trussed "framed" partition between the front and the back room and the landing of the same house, providing a door opening on to the landing 7 ft. high by 3 ft. 4 in. wide, and opening for folding doors to back room, 9 ft. high by 9 ft. wide; the storey is assumed to be 11 ft. high clear of the joists. Particulars of the various joints are given in Figs. 412 to 416. This being an example of carpentry that requires a fair amount of judgment to design properly, it will probably serve as an example for reference if it is fully worked out, because

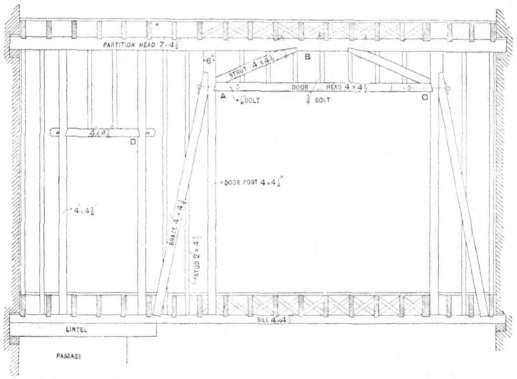

Fig. 411.—Trussed Cross Partition Frame with Two Openings and to support Upper Floor.

this kind of partition frequently forms the support of floors, as shown in Fig. 411. It has been assumed that the sill is supported

Fig. 413, joints between partition head and top of door post and strut (see B, Fig. 411). Fig. 414, joints between door post, door

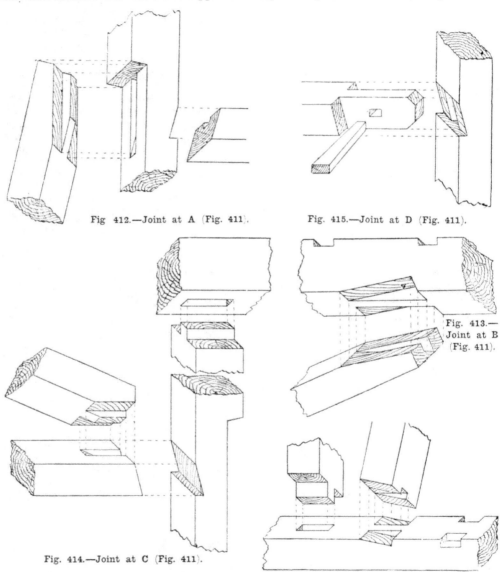

Fig 412.—Joint at A (Fig. 411).

Fig. 415.—Joint at D (Fig. 411).

Fig. 413.—Joint at B (Fig. 411).

Fig. 414.—Joint at C (Fig. 411).

Fig. 416.—Joints at Foot of Strut and Door Post (Fig. 411).

on a brick partition wall, except across the passage, where the sill is shown resting on a lintel. The enlarged details are explained as follows : Fig. 412, joints between door post, door head, and brace (see A, Fig. 411).

head, and strut (see C, Fig. 411). Fig. 415, joint at D, Fig. 411. Fig. 416, joint between

Fig. 417.—Conventional View of Main Members of Cross and Staircase Partitions through Two Storeys.

sill, brace, and door post (see Fig. 411); also showing sill notched to receive the joist. Figs. 417 to 421 illustrate a practical example of partitioning to the upper storeys over a ground floor which is used for business

419). Fig. 421 shows the plan of the cross and staircase partition. The staircase partitions are 6 ft. from the flank wall, so that the upper staircases may be formed of two flights. The cross partition (Fig. 418) has

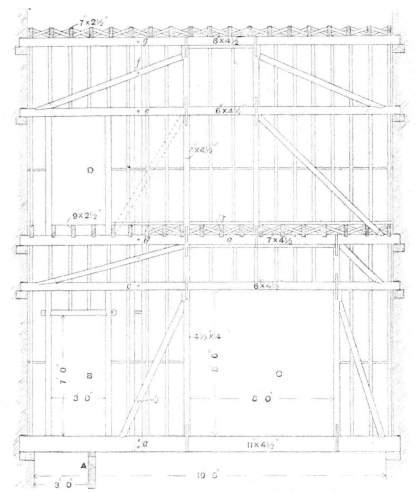

Fig. 418.—Elevation of Cross Partitions.

purposes. Sketch plans of the ground and second floors are given by Figs. 420 and 421. In the ground plan (Fig. 420), a private entrance, 3 ft. wide, and the staircase are enclosed by a 4½-in. brick and studded partition, which is indicated in section and elevation at A (Figs. 418 and

a doorway B leading from the staircase landing, and an opening C is provided for folding doors. The head of this partition is prepared to act as the middle bearing for the second floor joists, and it serves also as a sill for the main members of the cross partition to the second floor, which in turn supports the third or garret floor joists as shown. Fig. 419 shows the staircase partitions to the first and second floors, having door openings E and F. It must be noted that these partitions are not directly over supported by them. One end of these partitions is carried by the back wall, and the other is connected to the cross partitions by means of ⅝-in. bolts, which are indicated at a, b, c, d, e, f, and g (Fig. 418). Owing

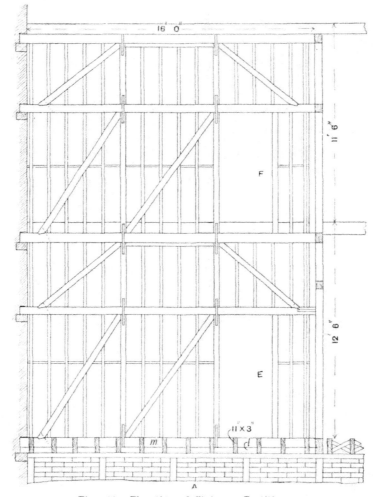

Fig. 419.—Elevation of Staircase Partitions.

the ground floor partition A (Fig. 418), and therefore do not receive any direct support from it. These partitions are designed to carry their own weight. The lower one supports one end of the first floor joists of the back room and landing, whereas only the landing of the second floor has to be to the sill m (Fig. 419) of the lower cross partition having to carry the ends of the joists, the strongest method doubtless would be to fix them to fillets as shown at l (Fig. 419). The fillets would be spiked or bolted to the sill; mortising and housing for tusk tenoning, etc., of the joists would greatly

5*

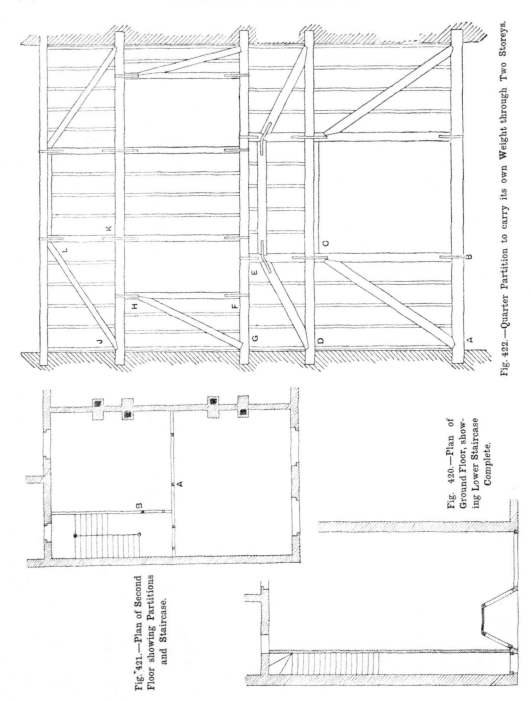

Fig. 422.—Quarter Partition to carry its own Weight through Two Storeys.

Fig. 421.—Plan of Second Floor showing Partitions and Staircase.

Fig. 420.—Plan of Ground Floor, showing Lower Staircase Complete.

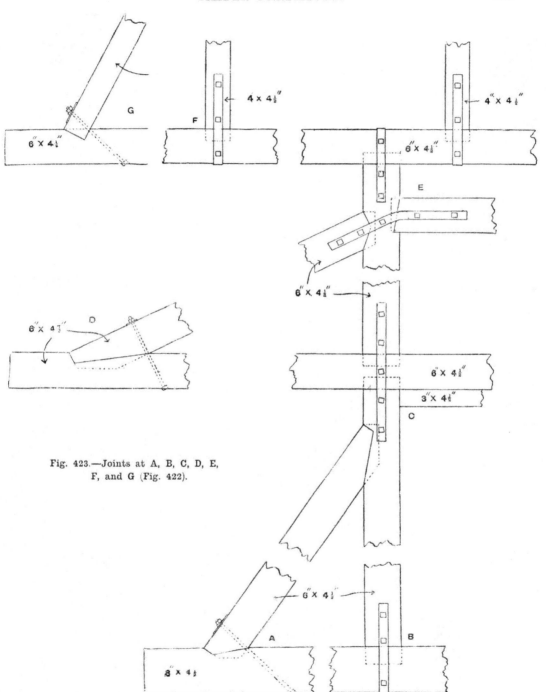

Fig. 423.—Joints at A, B, C, D, E,
F, and G (Fig. 422).

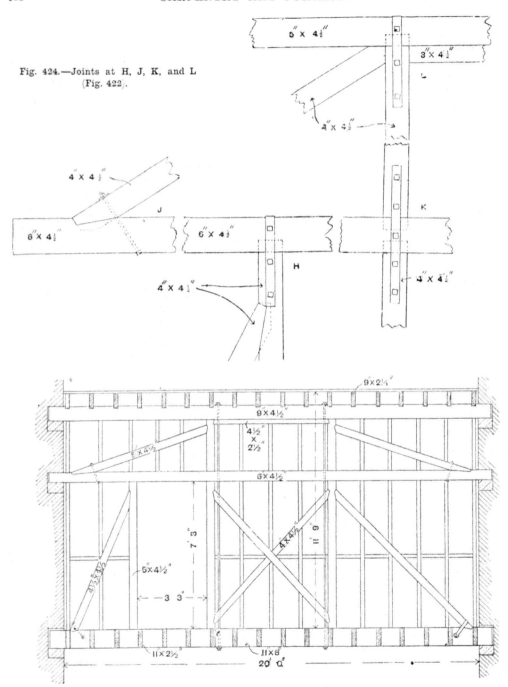

Fig. 424.—Joints at H, J, K, and L
(Fig. 422).

Fig. 425.—Trussed Partition framed for One Doorway and to support Two Floors.

weaken the sill. The feet of the studs of the upper partition (Fig. 418) may run down and be fixed to the head *o* or to a thin sill *p* secured to the tops of the joists as shown. In the conventional view (Fig. 417) the studs have been omitted, so that the main timbers of the framing may be clearly seen. The front second floor joists have also been omitted for a similar reason.

Quarter Partition Through Two Storeys.—Fig. 422 is the elevation of a quarter partition 18 ft. wide and 24 ft. high, running through

are tapped at each end for nuts. Fig. 426 shows a partition which supports similar loads, but having two openings. The sill of the partition has to answer as a girder also, and may have the joists connected to it by means of tusk tenoning, housing, etc.; formerly that was the general method, but, of course, the beam is thus greatly reduced in sectional area and strength. A much better way is to fix a fillet, either by nails or coach screws, to support the ends of the joists, as clearly shown in Fig. 427.

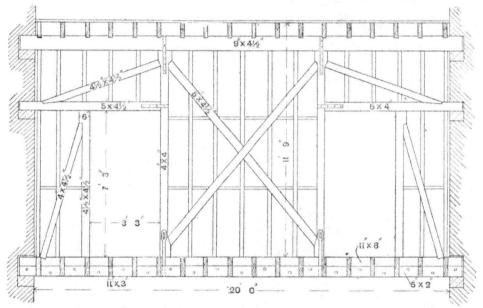

Fig. 426.—Trussed Partition framed for Two Doorways and to act as Middle Bearing for Two Floors.

two storeys and self-supporting over the ground floor. On the first floor is a central doorway 6 ft. 6 in. wide by 7 ft. 6 in. high; on the second floor is a doorway 3 ft. wide and 6 ft. 6 in. high, 3 ft. 6 in. from one side wall; and another 4 ft. wide and 6 ft. 6 in. high, 2 ft. from the other wall. Figs. 423 and 424 give details of joints, and show the necessary ironwork, joints A to G in the lower storey being shown by Fig. 423, and joints H to L in the upper storey by Fig. 424. A partition with one opening and supporting two floors is shown by Fig. 425. It is strengthened by two ¾-in. iron rods, which

Further Designs of Trussed Partitions.—Assume that a room 15 ft. wide and 11 ft. high is to be divided by a quarter partition having a central opening for a folding door 7 ft. wide and 8 ft. high. A suitable trussed partition would be the one shown by Fig. 428, in which all necessary scantlings are given, and the members named. When a timber partition in a storey 12 ft. high has a bearing of 21 ft., and has to carry itself and the floor above, the design may be as in Fig. 429, which shows provision for a door in the centre, and takes consideration of the fact that the binders of the floor above will rest on the top of the partition. All scantlings and names of members are indicated.

combustible material, the object being to prevent fire passing through the partition from one room to another. In a case where

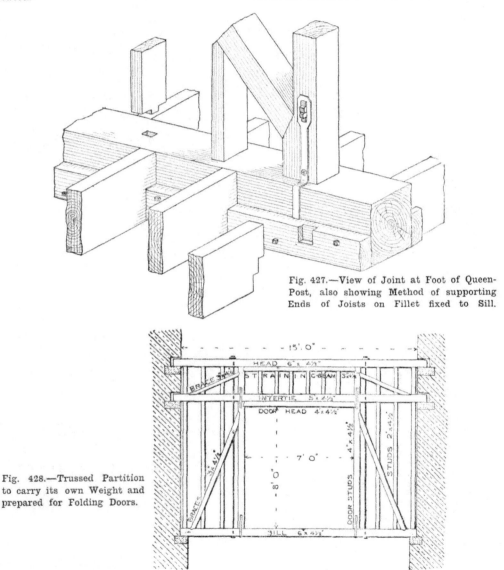

Fig. 427.—View of Joint at Foot of Queen-Post, also showing Method of supporting Ends of Joists on Fillet fixed to Sill.

Fig. 428.—Trussed Partition to carry its own Weight and prepared for Folding Doors.

Fireproof Partitions.

Some bye-laws render it compulsory to fill in the spaces in a timber partition with brickwork, concrete, pugging, or other in-

the studdings are of 3-in. stuff, bricks would be laid on edge. The partition can be covered in any suitable way—lath and plaster, wainscoting, etc.

Sound-proof Partitions.

Timber partitions are rendered more or less sound-proof by filling in with sawdust, substance because of the greater mass that has to be set in motion. For this reason clean dry earth, or, preferably, sand, is better than sawdust. But the weight of the

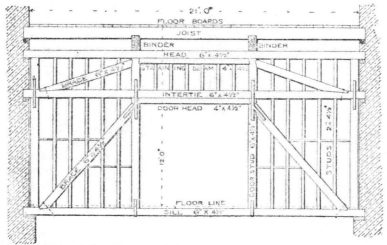

Fig. 429.—Trussed Partition framed for Central Opening and to support Binders.

which, however, does not answer the purpose so well as a heavier material. Sound or vibration—the same thing—is more readily damped or absorbed by a heavier sand needs to be taken into consideration in designing the floor supports. There is a number of patented systems available for building sound-proof partitions.

TIMBER ROOFS.

Roof Pitch.

For the roofs of ordinary buildings, either 30° or 26½° is adopted for the pitch, the former having a rise of half the length of rafter, and the latter having a rise of one-fourth the span, known also as square

lap should also be increased. Where a 2½-in. lap would do for a pitch of 60°, a 4-in. lap would be desirable for a pitch of 22½°. When the span and rise are given, the pitch (a) will be $\dfrac{\text{rise}}{\text{span}}$ (for example,

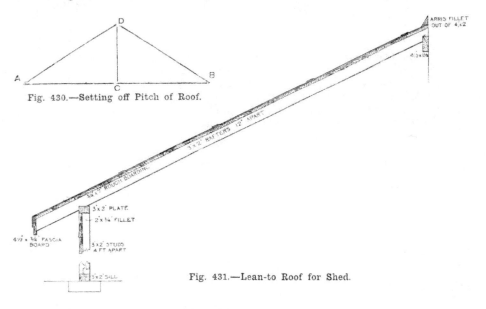

Fig. 430.—Setting off Pitch of Roof.

Fig. 431.—Lean-to Roof for Shed.

pitch. Sometimes, for large sheds with iron roof trusses, the pitch is reduced still more—to, say, a minimum of 22½°. For Gothic work and for exposed positions, high-pitched roofs are used, say 45° or 60°, and occasionally more, covered with shingles, slates, or plain tiles. The flatter the roof, the heavier the slates should be, and the

24-ft. span 6-ft. rise $= \dfrac{6}{24} = \frac{1}{4}$ pitch);

or (b) will be a slope of $\dfrac{\frac{1}{2}\ \text{span}}{\text{rise}}$ to 1 (for example, in the given case $\dfrac{\frac{1}{2} \times 24}{6} = 2$ to 1); or (c) the pitch in degrees will be the

112

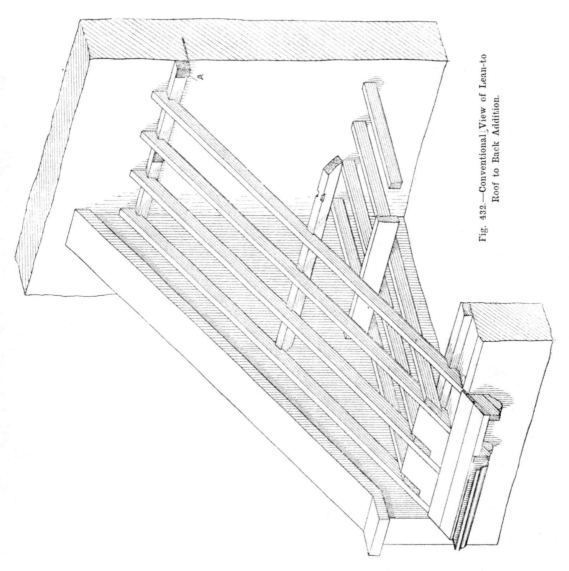

Fig. 432.—Conventional View of Lean-to Roof to Back Addition.

angle whose tangent is $\dfrac{\text{rise}}{\frac{1}{2}\text{span}}$ (for example,

in the given case $\dfrac{6}{\frac{1}{2} \times 24} = \cdot 5$), which is

the tangent of an angle of 26° 33′. To set off the slope of a roof, say, at one-third pitch draw the span A B (Fig. 430), divide it into three equal parts, and at the centre

of the span C set up the perpendicular C D equal to one part; join A D. Then A D will be the required pitch. This is the flattest pitch at which tiles should be laid.

Lean-to Roofs.

Lean-to roofs in their simplest forms are used for covering sheds and for temporary purposes shown in section by Fig. 431.

They are also largely used for covering the back and side additions to all kinds of buildings, the spans varying from a few feet up to 20 ft. or more. Fig. 432 is a conventional view of an ordinary lean-to roof over a back addition. The head of the rafters may fit on to a plate fixed into the main wall, or the plate may be supported on iron corbels, as shown at A. When the span is more than 8 ft., a purlin should be

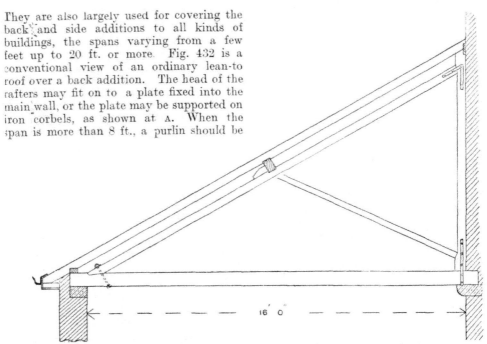

Fig. 433.—Section through Trussed Lean-to Roof for Span of 16 Feet.

ntroduced. The span for these is unlimited within reason, but, as in other forms of roofs, the rafters should be supported by purlins at about every 7 ft. In the larger spans for important work, framed trusses from 8 ft. to 10 ft. apart would be introduced. A section through a roof of this description is given at Fig. 433, which, it will be

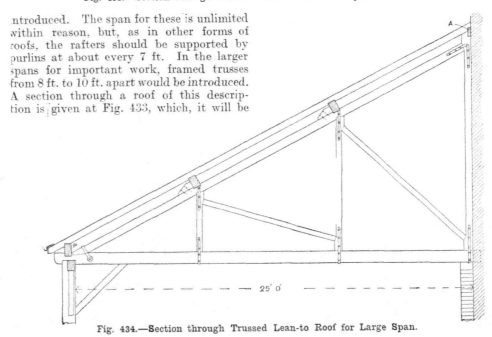

Fig. 434.—Section through Trussed Lean-to Roof for Large Span.

seen, is a half king-post truss. The example shown at Fig. 434 is a form often adopted for sheds attached to main buildings, where it is desirable to have a covered-in space with as little obstruction in the lower part as possible. One end of each truss is sup-

case is a proper application of the pole plate, which is so named because it has intermediate supports between the trusses. The conventional view (Fig. 436) will make clear the construction at the foot of the rafters.

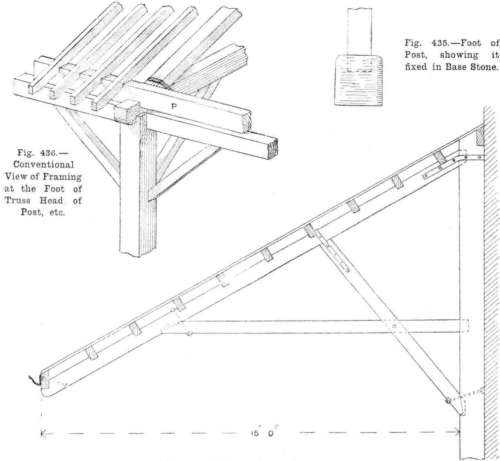

Fig. 435.—Foot of Post, showing it fixed in Base Stone.

Fig. 436.— Conventional View of Framing at the Foot of Truss Head of Post, etc.

Fig. 437.—Self-supporting Shed Roof.

ported by a pier bonded to the main building, and the other by a head and posts which are braced. The foot of each post is sometimes fixed into a stone or iron base (Fig. 435). This is to prevent damage by vehicles, etc., and to prevent decay. The common rafters are supported by a wall plate A, by purlins, and at their feet by a pole plate P. This

Fig. 437 illustrates a case where it is desirable to roof over a space adjacent to a building, and to leave the front of the covered space clear, and at the same time not to fix the members of the roof to the wall of the buildings. The boarding is supported by small purlins, or, as they are sometimes called, horizontal rafters. For this roof to be entirely self-supported the post would have to be well bedded in the ground.

Span, Couple-close, and Collar-beam Roofs.

Next to the lean-to, the simplest form of roof construction is that known as the span

But the more common method is to introduce purlins, one on each side of the roof, so as to support the centre of the rafters. If the purlins are sufficiently strong and bedded in the gable walls at each end, and the rafters notched on to them, there is very little outward thrust on the walls at the feet of the rafters. A section

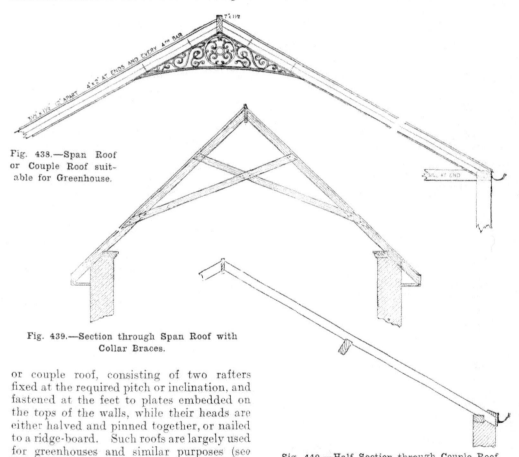

Fig. 438.—Span Roof or Couple Roof suitable for Greenhouse.

Fig. 439.—Section through Span Roof with Collar Braces.

Sig. 440.—Half Section through Couple Roof with Purlins.

or couple roof, consisting of two rafters fixed at the required pitch or inclination, and fastened at the feet to plates embedded on the tops of the walls, while their heads are either halved and pinned together, or nailed to a ridge-board. Such roofs are largely used for greenhouses and similar purposes (see Fig. 438). To remedy the obvious tendency of such a roof to spread at the foot and thrust out the walls, which tendency increases with the increase of span, various means are adopted.

Where it is desirable to have as much space in the roof as possible, this spreading may be obviated to a great extent by fixing collar braces as shown at Fig. 439.

through a little more than one half of a roof of this description is given at Fig. 440, a conventional view being shown by Fig. 441. The number of purlins should be increased as the span is increased, so that the common rafters do not have a greater bearing than 6 ft. to 8 ft.

adopted for roofing small houses, and is an enlargement of the previous case. Alternative methods of fixing the purlins are shown. At A the sides of the purlin are vertical

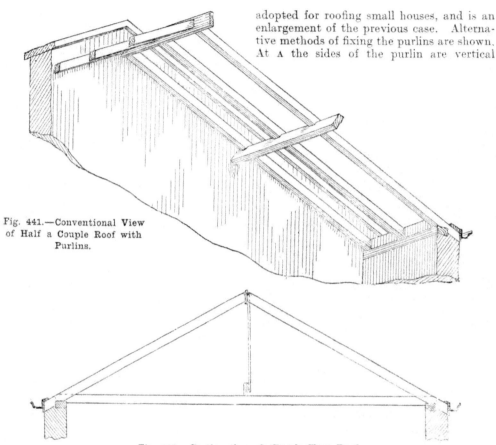

Fig. 441.—Conventional View of Half a Couple Roof with Purlins.

Fig. 442.—Section through **Couple-Close** Roof.

The couple-close roof (Fig. 442) over a small building consists of rafters, which are bird's-mouthed and fixed to the wall plates ; ceiling joists are fixed to the wall plates and act as ties and counteract the outward thrust on the walls. The ceiling joists are usually supported at the centre by being nailed up to a beam which is tied to the ridge by pieces of board 5 ft. or 6 ft. apart (Fig. 443).

A collar-beam is a horizontal beam or brace, generally of the same scantling as the rafters, placed from one-third to half-way up a span-roof and connected to the rafters at each side, the roof now becoming a collar-beam roof.

Fig. 444 is a form of roof frequently

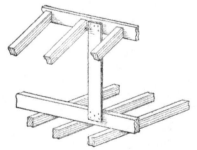

Fig. 443.—View of Method of Tying Ceiling Joist to Ridge.

and the rafters are notched on ; whereas at B the purlin is fixed with its sides at right angles with the pitch of the rafters, this being probably the stronger method. This roof is additionally strengthened by inserting a collar to every third or fourth rafter as shown by dotted lines. Fig. 447 (p. 119) shows a common application

Fig. 450 shows the joint between the principal rafter and tie-beam, which are additionally secured by an iron strap, the ends of which are prepared for bolts and nuts securing a heel-plate A. It also shows

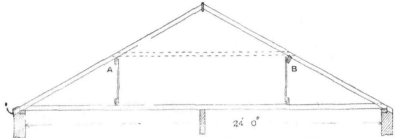

Fig. 444.—Section through Span Roof for a Small House.

of this kind of roof, largely adopted for dwellings, from the cottage to the villa class; the collars not only serving as ties to strengthen the roof, but also as ceiling joists. In this figure the collars are shown dovetail notched. Fig. 445 shows another form of dovetail notching. Fig. 446 shows a form of notching. When there is no

a 4½-in. wall-plate B, on which the common rafters are bird's-mouthed. The rafters project beyond the wall, and bearers c are fixed to their ends, and also into the wall, so that the soffit boarding and fascia board may be fixed. A cast-iron gutter is shown fixed to the fascia board, and a tilting fillet D is also shown.

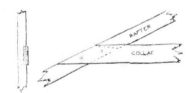

Fig. 445.—Form of Dovetail Notching.

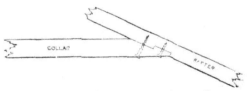

Fig. 446.—Notching Collar into Under Edge of Rafters.

ceiling a collar is usually fixed to every third pair of rafters.

King-post Trusses.

A king-post truss is suitable for any span up to 30 ft., and the sizes or scantlings of its members are shown in the table given on p. 121. A cross section of a little more than one-half of a 28-ft. span roof resting on stone template on 14-in. brick walls with 9-in. piers is presented by Fig. 448, and part longitudinal section at Fig. 449. The common rafters measure 4 in. by 2 in. The eaves overhang and are finished with fascia and eaves boarding. Certain details of construction require to be shown separately on a larger scale.

Fig. 451 shows the joint between principal rafter and strut; purlin, cleat, and common rafter are also illustrated. The cleats are usually fixed with spikes or coach bolts.

Fig. 452 shows the joints and three-way iron strap at the head of the king-post; also the ridge and its junction with the common rafters. Fig. 453 is the joint at the foot of the king-post, with stirrup-iron, gib, and cotters. Fig. 454 is a vertical section through this joint, showing straps, gibs, and cotters, clearance in mortise at C, and clearance in strap at D.

Fig. 455 shows the foot of a principal rafter and tie-beam connected with a wall having a cornice and parapet, with a gutter

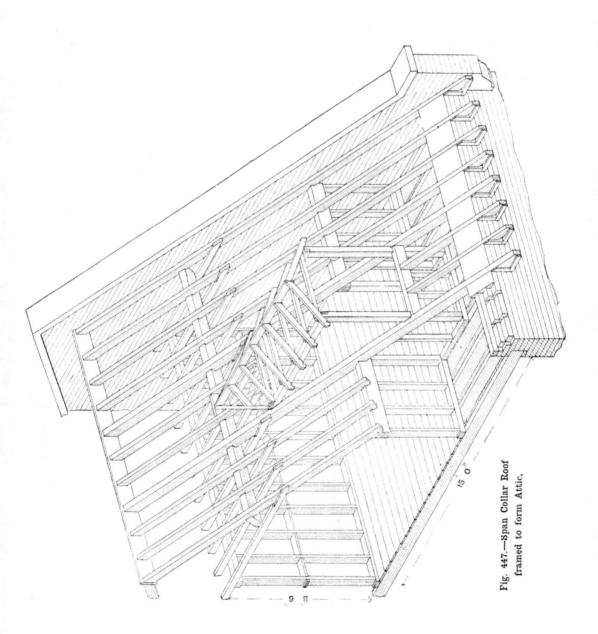

15' 0"

9 8

Fig. 447.—Span Collar Roof
framed to form Attic.

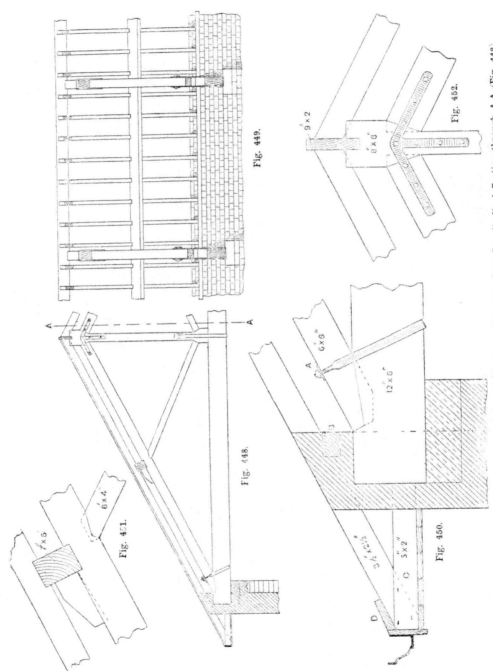

Fig. 449.

Fig. 452.

9 × 2

8 × 6"

Fig. 448.

A   A

6 × 4

7 × 5

Fig. 451.

6 × 6"
A

12 × 6"

3½ × 2½"
D
5 × 2"
C

Fig. 450.

Fig. 448.—Half Transverse Section through King-post Roof.   Fig. 449.—Part Longitudinal Section through A A (Fig. 448).
Fig. 450.—Detail of Foot of Principal Rafter.   Fig. 451.—Detail of Joint between Strut and Principal Rafter, Support
of Purlin, etc.   Fig. 452.—Detail of Head of King-post.

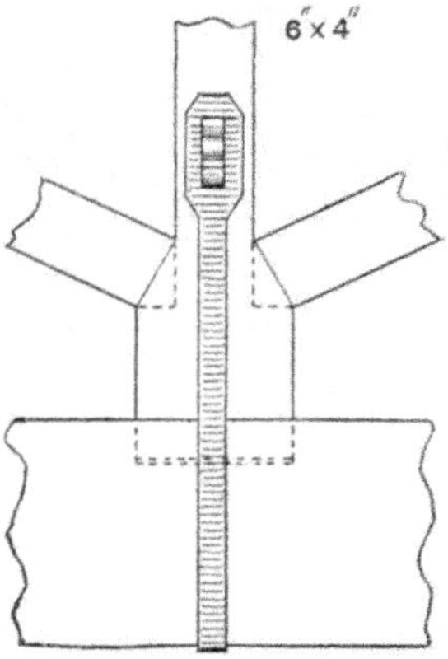

Fig. 453.

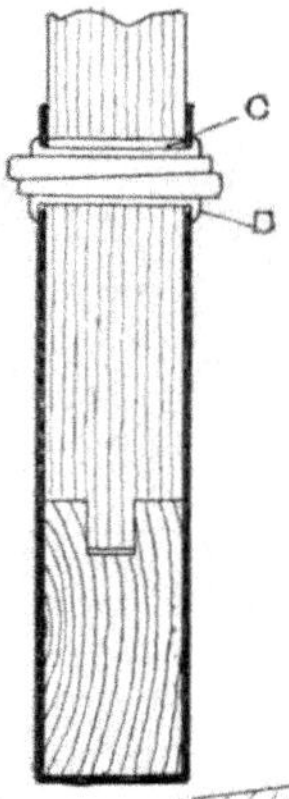

Fig. 454.

Fig. 453.—Detail at Foot of King-post.

Fig. 454.—Section taken through Centre of Strap and King-post.

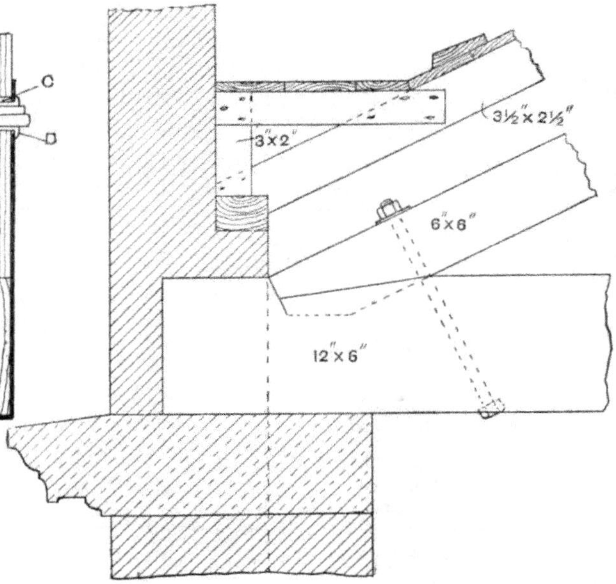

Fig. 455.—Detail at Foot of Truss in Connection with Parapet.

SCANTLINGS FOR TIMBER ROOFS.

The table below shows at a glance the respective scantlings for collar, king-post, and queen-post roofs.

Description of Roof.	Span.	Tie-beam.	Principal Rafter.	King-post.	Queen-post.	Strut.	Straining Beam.	Purlin.	Straining Sill.	Common Rafter.	Collar.	Ridge.
	Ft.	Ins.	Ins.	Ins.	Ins.	Ins.	Ins.	Ins.	Ins.	Ins.	Ins.	Ins.
Collar-beam Roof.	8									3 x 2	2 x 2	7 x 1¼
	10									3¼ x 2	2¼ x 2	7 x 1½
	12									3¼ x 2¼	3½ x 2	7 x 1½
	14									4 x 2	4½ x 2	9 x 1½
	16									4 x 2	5 x 2	9 x 1½
	18									4 x 2¼	5½ x 2	9 x 1½
King-post Roof.	18	7 x 3	4½ x 3	4½ x 3		3½ x 2		7 x 3		3½ x 2		
	20	9 x 4	4 x 4	5 x 4		4 x 2¼		7 x 4		4 x 2		
	22	9 x 4	6 x 3	6 x 3½		4 x 2¾		8 x 4		4½ x 2		
	24	9½ x 4	6 x 3½	6 x 4		4½ x 2		8 x 5		4½ x 2		
	26	9 x 5	6 x 4	6 x 4		4 x 3		8½ x 5		4½ x 2¼		
	28	10 x 5	6 x 4	6 x 6		4½ x 3		8¾ x 5¼		4½ x 2½		
	30	11 x 6	6 x 5	7 x 6		6 x 3		8 x 6		4½ x 2½		
Queen-post Roof.	30	9 x 4	5½ x 4		4½ x 4	4 x 3	7 x 4	8 x 4	4 x 4	4 x 2		
	32	10 x 4	6 x 4		5 x 4	4 x 3¼	7½ x 4	8 x 4	4½ x 4	4 x 2		
	34	10 x 5	6½ x 4		6 x 4	4½ x 3½	8 x 4	8½ x 5	5 x 4	4¼ x 2		
	36	10 x 6	6½ x 5		7 x 4	5 x 3½	8 x 4½	8½ x 5	5 x 4	4¼ x 2		
	38	10 x 6	6 x 6		7 x 5	5 x 4	8 x 5	8¾ x 5¼	5 x 4½	4½ x 2		
	40	11 x 6	7 x 6		7 x 6	6 x 4	8½ x 5	8 x 6	5 x 4½	4½ x 2		
	42	11½ x 6	7 x 6		8 x 4	6 x 5	8 x 6	9 x 5	6 x 4	4¾ x 2		
	45	12½ x 6	7¼ x 6		8 x 6	6 x 6	8 x 6	9 x 6	6 x 4	5 x 2		

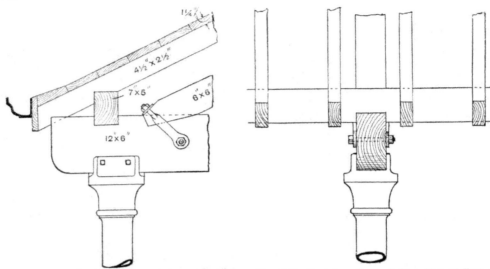

Fig. 456.—Elevation of Foot of Truss supported
by Iron Column.

Fig. 457.—Part Longitudinal Elevation of Foot of
Truss, Pole Plate, etc., Boarding removed.

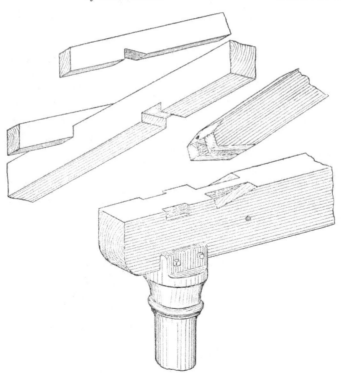

Fig. 459.—Heel Strap
and Bolt.

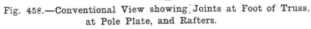

Fig. 458.—Conventional View showing Joints at Foot of Truss,
at Pole Plate, and Rafters.

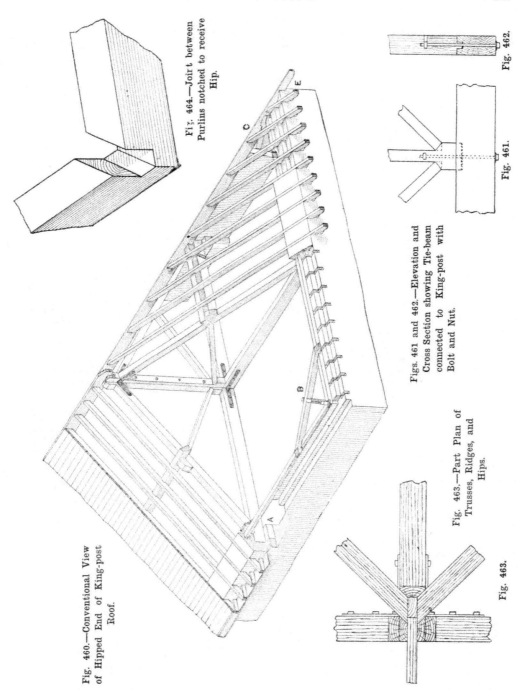

Fig. 460.—Conventional View of Hipped End of King-post Roof.

Fig. 464.—Joint between Purlins notched to receive Hip.

Figs. 461 and 462.—Elevation and Cross Section showing Tie-beam connected to King-post with Bolt and Nut.

Fig. 463.—Part Plan of Trusses, Ridges, and Hips.

Fig. 461.

Fig. 462.

Fig. 463.

formed behind the latter. In this case the joint between the tie-beam and principal rafter is fastened by means of a bolt.

of fastening for trusses where half-trusses have not to be attached to them ; but when this latter is the case, the stirrup - iron

Fig. 465.—Section through Main Truss showing Method of connecting Tie-beam, King-post, and Hips to Ridge.

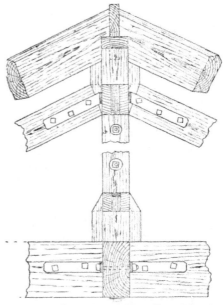

Fig. 466.—Part Elevation of Main Truss showing King-post and Section of Tie-beam of Half Truss.

Figs. 456 to 459 show the end of a truss supported by an iron column. This case illustrates the use of the pole-plate and also the oblique bridle joint. The pole-plate serves two purposes, viz. to connect the ends of the trusses longitudinally and to support the rafters. The tie-beam and principal rafter are secured together by the strap shown at Fig. 459. Obviously this form of strap, having an adjustable plate at the top which can be forced close to the heel of the principal rafter, is rather a better kind than when simply in the form of a stirrup-iron.

Hipped End of King - post Roof.—A conventional view of a portion of a king-post roof with hipped end is shown at Fig. 460. The method of constructing the truss and half-truss as shown at Fig. 460 is illustrated by Figs. 461 to 467. Undoubtedly, for most cases, the stirrup-iron with gibs and cotters is the best form

leads to rather a clumsy connection ; therefore the bolt and nut method shown at Figs. 461 and 462 is adopted for securing the tie-beam and king-post together.

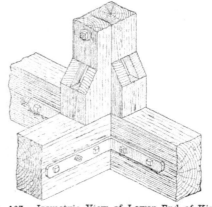

Fig. 467.—Isometric View of Lower End of King-posts and Portion of Tie-beam.

Fig. 463 shows part plan of trusses, ridge, and hips. Fig. 464 shows the meeting of the purlins, which are notched out for the hip. When this is deep there is no notching, the full ends of the purlins butting against it. Fig. 465 shows the tusk tenon joint between the tie-beams, with necessary straps and bolts; also the connection of king-posts and straps and bolts at head and at c. Fig. 466 is part elevation of main truss. Fig. 467 shows the lower ends of king-posts and portions of tie-beam.

Dragon Tie at Foot of Hip Rafter.

A dragging tie, or dragon tie or beam (Figs. 468 and 469), is a framework at the lower end of a hip rafter, in the angle of the building, connecting it with the wall-plates in such a way as to resist the thrust of the hip rafter. The foot of the hip rafter is halved, notched, stepped, or tenoned into

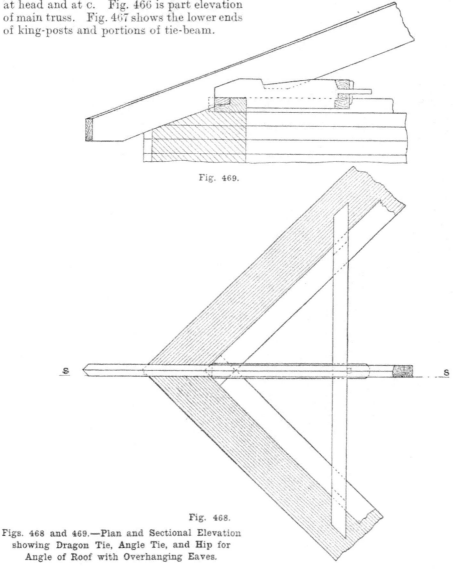

Fig. 469.

Fig. 468.

Figs. 468 and 469.—Plan and Sectional Elevation showing Dragon Tie, Angle Tie, and Hip for Angle of Roof with Overhanging Eaves.

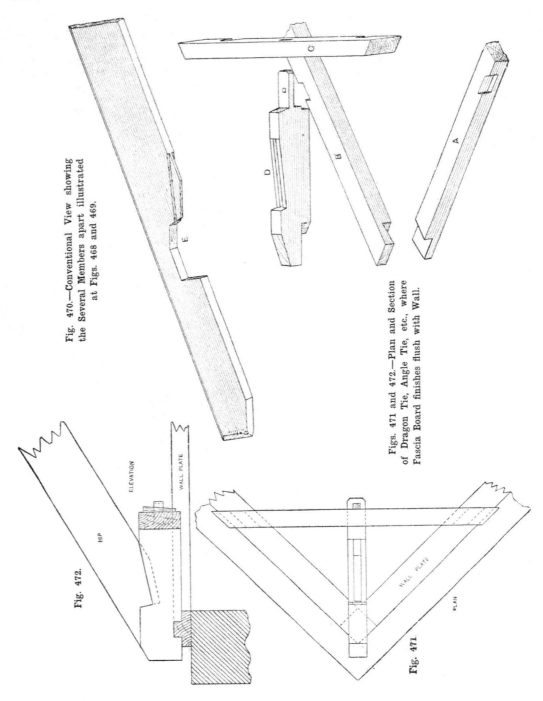

Fig. 470.—Conventional View showing the Several Members apart illustrated at Figs. 468 and 469.

Figs. 471 and 472.—Plan and Section of Dragon Tie, Angle Tie, etc., where Fascia Board finishes flush with Wall.

Fig. 472.

Fig. 471.

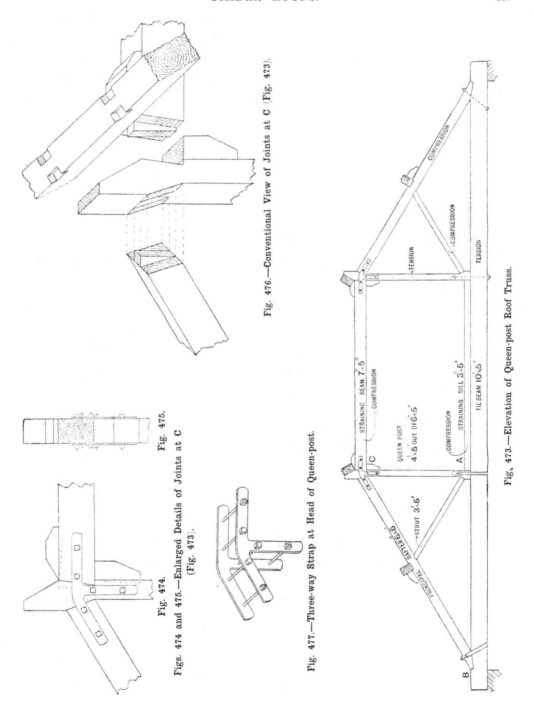

Fig. 476.—Conventional View of Joints at C (Fig. 473).

Fig. 475.

Fig. 474.

Figs. 474 and 475.—Enlarged Details of Joints at C (Fig. 473).

Fig. 477.—Three-way Strap at Head of Queen-post.

Fig. 473.—Elevation of Queen-post Roof Truss.

the dragging tie, which is notched at one end on to the wall-plates, at the angle where they are halved together, and at the other end is attached to the angle tie or

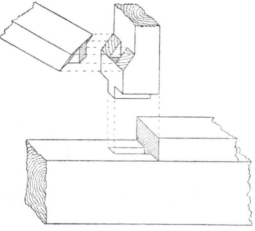

Fig. 478.—Conventional View of Joints at A (Fig. 473).

brace by means of a tusk tenon secured by a pin or wedge, the angle tie being notched over the wall-plates to keep it in place. There is more than one method of constructing this joint. The dragon piece and angle tie should be used in all hipped roofs, although in small roofs it may be of a simpler construction, such as a batten nailed diagonally across the plates, the hip rafter notching

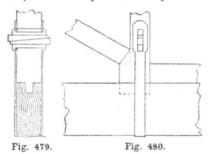

Fig. 479.   Fig. 480.

Figs. 479 and 480.—Enlarged Details of Joints at A (Fig. 473).

on to it. Fig. 470 is a conventional view showing the parts separated.

Figs. 471 and 472 illustrate a case where the hip does not overhang the walls. It

should be noted that in the plan the hip is not shown.

Queen-post Trusses.

The queen-post trusses are suitable for spans of 30 ft. and more. Suitable scantlings or sizes for the different members are

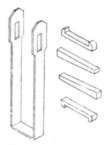

Fig. 481.—Conventional View of Stirrup-Iron, Gibs and Cotters at A (Fig. 473).

shown in the table on p. 121. Fig. 473 is the elevation of a queen-post truss for a clear span of 34 ft. Many of the details of this truss are the same as those of the king-post truss already fully illustrated. The principal differences are that in this case a horizontal straining beam has to be jointed to the queen-posts; the joint is clearly shown by Figs. 474 to 476, the ironwork being illustrated separately by Fig. 477. The joint and stirrup at the foot of the queen-post are illustrated by Figs. 478 to 480.

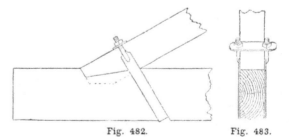

Fig. 482.   Fig. 483.

Figs. 482 and 483.—Enlarged Detail of Foot of Truss, B (Fig. 473).

Fig. 481 is a conventional view of the stirrup-iron, gibs, and cotters.

The joint of the principal rafter at its foot with the tie-beam is shown in elevation and section by Figs. 482 and 483, and conventionally by Fig. 484, the heel strap being shown by Fig. 485.

and 491), where the purlins are shown mitred together and also notched out to receive the hip. This latter would also be

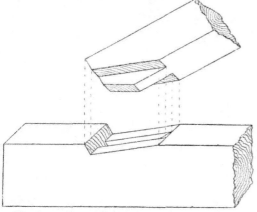

Fig. 484.—Conventional View of Joint between Principal Rafter and Tie-beam.

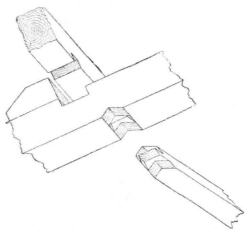

Fig. 486.—Conventional View of Joints at Head of Strut, Cogging of Purlin, etc.

The joints between principal rafter, strut, and purlin are illustrated by Fig. 486.

At Fig. 487 is shown the hipped end of a queen-post roof, and Figs. 488 and 489 show part sectional elevation and plan of same. The upper purlin at the end partly rests on

Fig. 485.—Conventional View of Heel Strap and Plate for fastening Principal Rafter to Tie-beam.

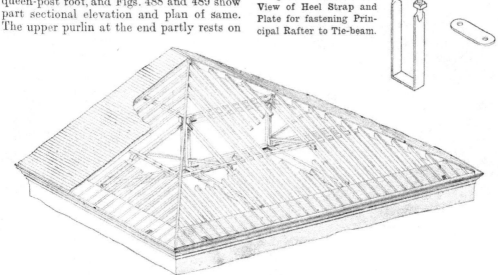

Fig. 487.—Conventional View showing general Construction of Queen-post Roof Truss with Hipped End.

the straining beam. The upper ends of the queen-posts are cut to receive the purlins, as shown by the conventional view (Figs. 490

notched out part of its depth so as to fit in with the purlins in this case.

The method of connecting the half-truss

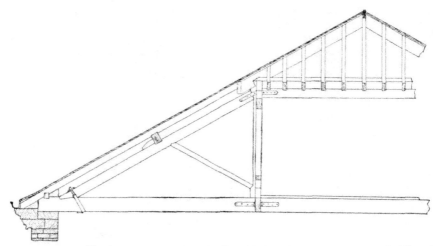

Fig. 488.—Part Sectional Elevation on Line D D (Fig. 489) of Queen-post Truss with Hipped End.

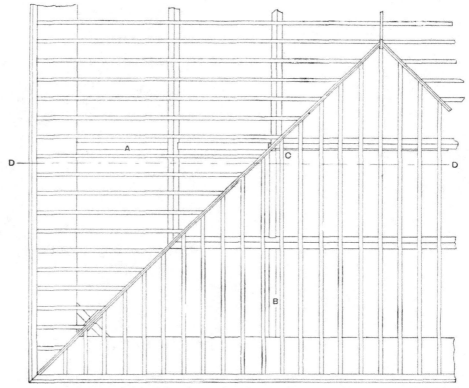

Fig. 489.—Part Plan of Queen-post Truss with Hipped End, showing Timbers—A, Main Truss,
B, Half Truss, both being connected at C.

to the main truss is shown by the part elevation and plan (Figs. 492 and 493); the construction will be more clearly understood

one of connecting the tie-beams by tusk mortise and tenon joints, and tightening up with keys. Other straps and connections

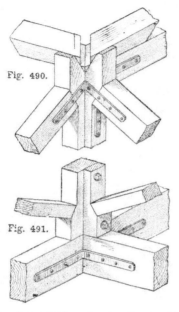

Fig. 490.

Fig. 491.

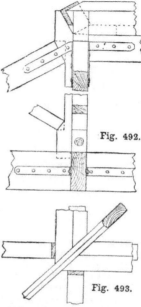

Fig. 492.

Fig. 493.

Fig. 490.—General View showing Construction at Heads of Queen-posts.

Fig. 491.—General View showing Construction at Feet of Queen-posts and Connection of Tie-beams.

Fig. 492.—Enlarged Sectional Elevation at Head and Foot of Queen-post.

Fig. 493.—Part Plan of Hip and Purlins.

from Fig. 491, the tie-beam of the former being connected to that of the latter by a

are clearly shown. At Fig. 494 suitable forms of joints are shown for the heads of the queen-posts, and at Fig. 495 a method

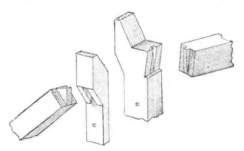

Fig. 494.—Forms of Joints at Head of Queen-posts.

Fig. 495.—Joints between Hips and Ridge.

short stub-tenon, and both being further secured by iron straps and bolts. No doubt this method is preferable to the old-fashioned

of fixing the hips to the ridge is illustrated. What is known as a king- and queen-post truss is shown by Fig. 496; this is suitable for a span of 50 ft.

Other forms of joints at head of queen-posts, etc., are shown by Figs. 497 and 498, the latter representing the better design, because the main stresses are

have only a two-way strap, as illustrated at Fig. 501.

Securing Principal Rafter to Tie-beam.
Two ways of securing the principal rafter

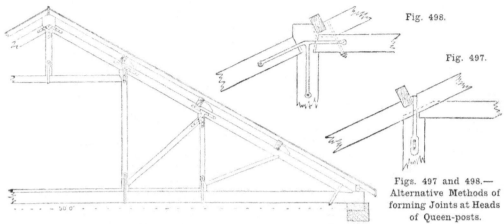

Fig. 498.

Fig. 497.

Figs. 497 and 498.—Alternative Methods of forming Joints at Heads of Queen-posts.

Fig. 496.—Half Elevation of King- and Queen-post Roof Truss for 50-Feet Span.

bounded by the tie-beam, by the principal rafters up to the straining beam, and by the straining beam itself. The portion of the truss above the straining beam has only a very small stress upon it, and it is therefore unnecessary to make the upper ends of the principal rafters of such large scantling as the lower ends, which form an essential part of the truss.

Figs. 499 and 500 illustrate a form of heel strap; this having been fixed with a bolt and nut, a hardwood or metal wedge is driven in between it and the principal rafter as

to the tie-beam are in occasional use. There may be a bolt as in Fig. 455, or a heel strap as in Fig. 499. The bolt and nut are better

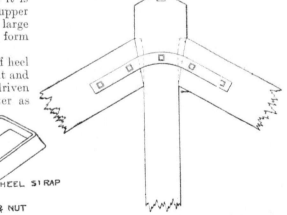

Fig. 499.

Fig. 500.

WEDGE

HEEL STRAP

HEEL STRAP

BOLT & NUT

Fig. 499.—Heel Strap connected to Tie-beam by Bolt, and tightened to Principal Rafter by Wedge.
Fig. 500.—General View of Strap and Bolt.

Fig. 501.—Two-way Strap for Head of King-post.

shown; but this is not quite such a good form as that illustrated at Fig. 459, page 122.

Occasionally the head of a king-post may

than this form of heel strap, which cannot be properly tightened; the bolt may be tightened, but it weakens the timbers a trifle by loss of sectional area, which, however, is not serious; the stirrup weakens the timber least, and can be tightened up readily.

Joint between Principal Rafter and Tie-beam.

In designing the joint between the principal rafter and the tie-beam, the object should be to obtain the best form of resistance, it being noted that the principal rafter and the portion of the tie-beam beyond that rafter are in compression. If through faulty roof design it were possible for the principal rafters to sag, in the case of Fig. 502 the

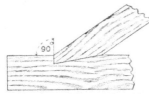

Fig. 502.—Improper Way of forming Joint at Toe of Principal Rafter.

rafter would ride on the heel A (Fig. 503), and the toe B would rise and split off the abutting piece C; not admitting the possibility of sagging, even then the greatest thrust would be to the point, and there would still be the danger of shearing or splitting. In the case of Fig. 504 (where the abutting surface is at right angles to the back of the

Fig. 503.—Result of forming Joint as at Fig. 502.

principal rafter), sagging, were it possible, would cause the rafter to ride on the heel D (Fig. 505) and the toe to slide along F G, and there would be no tendency to split off the abutting piece H. Regarding the thrust of a perfectly rigid rafter, the abutment shown in Fig. 504 is better than that shown in Fig. 502. The compromise (Fig. 506), in which the angle is bisected, is the best form for a properly designed roof, there being an equal abutment of fibres. Bolts and plates or straps affect the shape of the joint.

Cambering of Tie-beam.—The term "cambering," as applied to carpentry, means the

binding of a beam so that its centre is raised above the ends, causing it to assume an arc or arch-like form, the object being to prevent sagging of the middle below the

Fig. 504.—Usual Form of Joint at Toe of Principal Rafter.

straight line joining the ends when the beam is fully loaded. The following is a good example of the object to be obtained by cambering: When a king-post truss is being prepared, the king-post is made a little short, to the extent of about $\frac{1}{2}$ in. for every 10 ft. of span. Then, when the truss

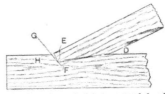

Fig. 505.—Opening of Joint caused by Sagging of Principal Rafter when made as at Fig. 504.

is put together, the tie-beam is forced up to the shoulders of the king-post and held fast by means of a bolt or stirrup strap with gibs and cotters, already shown. The object is twofold. This bending of the tie-beam shortens it to a slight extent, thus bringing the feet of the principal rafters a little

Fig. 506.—Best Angle for Toe of Principal Rafter.

nearer together, and tightening up the joints of the truss so that each shall take its proper bearing, and also making each respective member take its share of the load without sagging or distortion ; in short, making the truss rigid and firm.

Mansard Roof Trusses.

The Mansard form of roof takes its name from François Mansard, a French architect who was born in 1598 and died in 1666. It is essentially a roof with two pitches, and is usually employed as a means of economising space. There are one or two regular

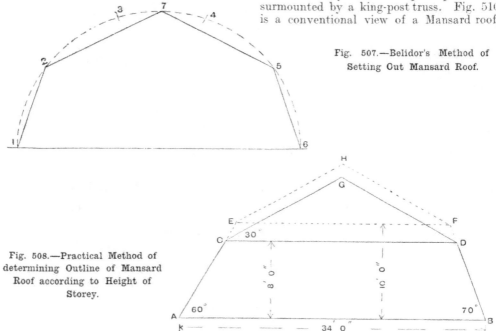

Fig. 507.—Belidor's Method of Setting Out Mansard Roof.

Fig. 508.—Practical Method of determining Outline of Mansard Roof according to Height of Storey.

desired height for the ceiling line, which is shown as 8 ft. From C D set off angles at 30 degrees meeting in G, which is the outline of upper portion of the roof. If it is desired to raise the height of the storey, as shown as 10 ft., the upper part of the roof becomes smaller, the main span remaining the same as does also the lower pitches. One of the usual ways of constructing a Mansard truss is illustrated by Fig. 509, which actually shows a queen-post truss surmounted by a king-post truss. Fig. 510 is a conventional view of a Mansard roof,

methods of getting the two slopes. Fig. 507 shows Belidor's system. On the line equal to the span describe a semicircle. Divide the circumference into five parts, numbering the points 1, 2, 3, 4, 5, 6, as shown. Join points 1 and 2 and 5 and 6. Divide the space between 3 and 4 equally, numbering the point 7 ; then join 2 to 7 and 7 to 5. The height of the storey is often the practical consideration, and therefore the above method is not always so applicable as that shown at Fig. 508. Set out the span and the outline for the lower part of the roof at an angle of 60 or 70 degrees (A and B) ; then draw the horizontal line C D at the

showing the complete timbering. It illustrates a case where the roof is designed to provide a room with as large a floor area as possible, this being often desirable for trade purposes. An enlarged detail of the foot of the main tie-beam, principal rafter and queen-post, with section through parapet, gutter and wall, is given at Figs. 511 and 512. There should be a stanchion or similar support A (Fig. 510) in the event of the floor being laden above the ordinary. A dormer is provided, the timbers of which are connected with the rafters, etc., as shown. The ceiling joists are fixed to the top edge of the upper tie-beam ; the main

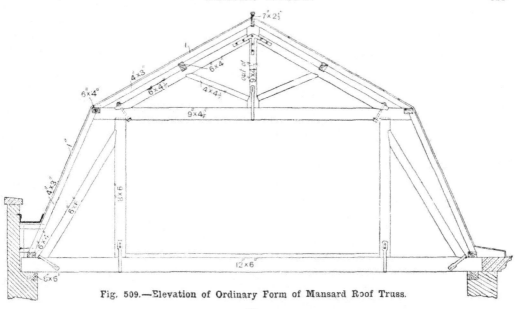

Fig. 509.—Elevation of Ordinary Form of Mansard Roof Truss.

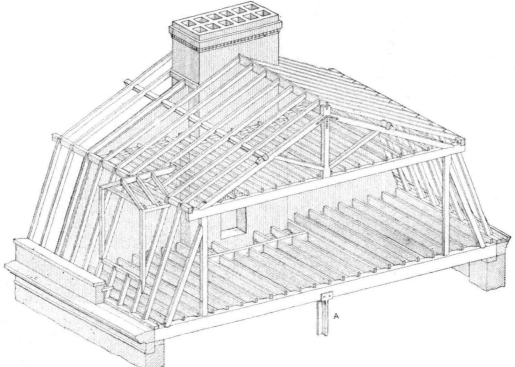

Fig. 510.—General View showing Timbering of Mansard Roof, with Main Tie-beam acting as Floor Binder.

tie-beam acts as a binder for the floor, the joists being supported by it. Two good methods of doing this are shown by Figs. 512 and 513; the joists at A are notched out to fit on the tie-beam, and their lower edges are further supported by a fillet fixed to the beam; fillet and

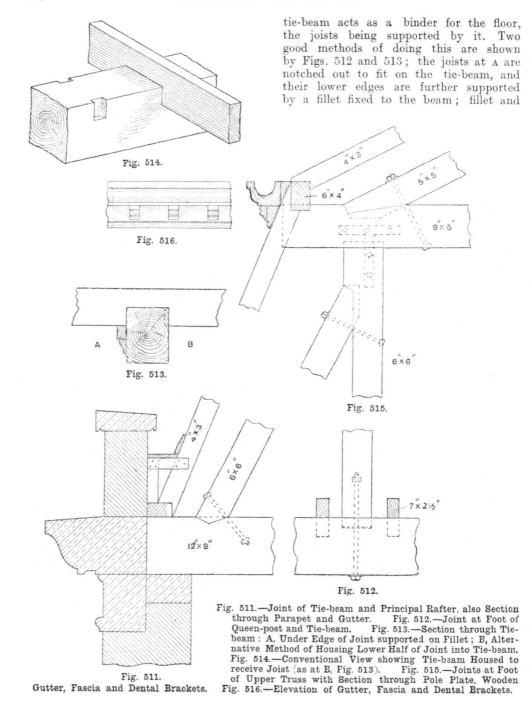

Fig. 514.

Fig. 516.

Fig. 513.

Fig. 515.

Fig. 512.

Fig. 511.

Gutter, Fascia and Dental Brackets.

Fig. 511.—Joint of Tie-beam and Principal Rafter, also Section through Parapet and Gutter.   Fig. 512.—Joint at Foot of Queen-post and Tie-beam.   Fig. 513.—Section through Tie-beam: A, Under Edge of Joint supported on Fillet; B, Alternative Method of Housing Lower Half of Joint into Tie-beam. Fig. 514.—Conventional View showing Tie-beam Housed to receive Joist (as at B, Fig. 513).   Fig. 515.—Joints at Foot of Upper Truss with Section through Pole Plate. Wooden Fig. 516.—Elevation of Gutter, Fascia and Dental Brackets.

beam may be finished off with a moulding as shown. At B the joists are notched on to the tie-beam and their lower halves tie-beam and principal rafter, also section through pole plate, wooden gutter, fascia and dental brackets; an elevation of a

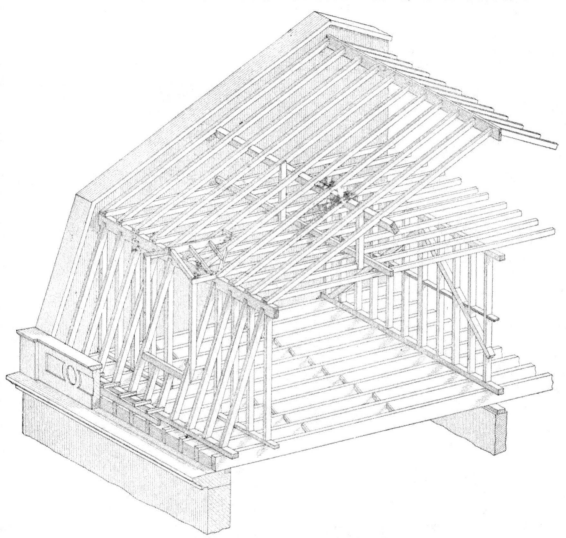

Fig. 517.—Mansard Roof constructed without Trusses.

housed in, as will be seen by reference to Fig. 514. The detail figures show general sizes, which, of course, would be increased or diminished according to the loading. Fig. 515 shows the connection at the head of the principal rafter, queen-post, upper portion of the latter is given by Fig. 516. The head of the queen-post and upper tie-beam may be strapped together as indicated by the dotted lines in Fig. 515, or the arrangement as shown by Fig. 520 (p. 138) may be adopted.

Mansard Roof without Trusses.—Of late years the custom has increased to dispense with trusses in roofs of moderate spans. The ends of the purlins 15 ft. to 35 ft. long can be carried on party or division walls, and the top storey is divided into rooms by partitions. Fig. 517 represents such an example ; the floor joists are fixed to plates at each end, and rest at

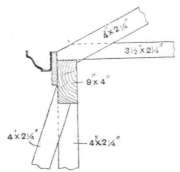

Fig. 518.—Detail of Junction at Lower and Upper Rafters.

their centres on the partition wall below, thus acting as ties to the back and front walls. A plate which receives the lower ends of the rafters is fixed to the top edges of the joists. The top ends of these rafters are

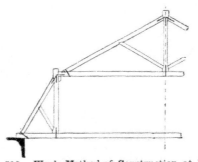

Fig. 519.—Weak Method of Construction at Head of Queen-post in Mansard Roof.

fixed to a plate, upon which the ends of the upper rafters and the ends of the ceiling joists are fixed, the latter tying in the plates and thus preventing any forcing-out action of the rafters. The heads of the studs are also fixed into this plate, the lower ends of the studs being connected to a sill

fixed to the joists. The upper rafters have a central bearing on purlins, and the ceiling joists are connected to a binding piece which is tied to the purlins as shown. the joists also being fixed to the partition head. Although there is no truss, clearly the whole is triangulated and supported to form a substantial roof.

Principles in Designing a Mansard Truss. —In considering the stresses borne by the members of a Mansard truss, it might be thought that the queen - posts are nothing more than mere vertical posts

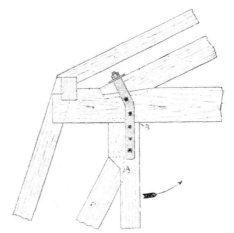

Fig. 520.—Construction at the Head of Lower and Upper Truss.

carrying a king-post truss, whereas the object of a king- or queen-post is to support the tie-beam and prevent it from sagging, thereby sustaining tensional stress. Actually, however, no compression whatever is thrown on the queen-posts, however they may be placed, as the lower principal rafter takes all the thrust from the load above, whether that load be a lead flat, an ordinary purlin roof, or the king-post truss of a Mansard roof ; and as a matter of fact, in many Mansard roofs of small span the queen-posts are omitted altogether, as they are only necessary when the lower tie-beam has to be supported near its middle for the purpose of carrying a floor, etc. Queen-posts are omitted, for instance, when the purlin or plate which takes the top ends of the lower common rafters and the bottom ends

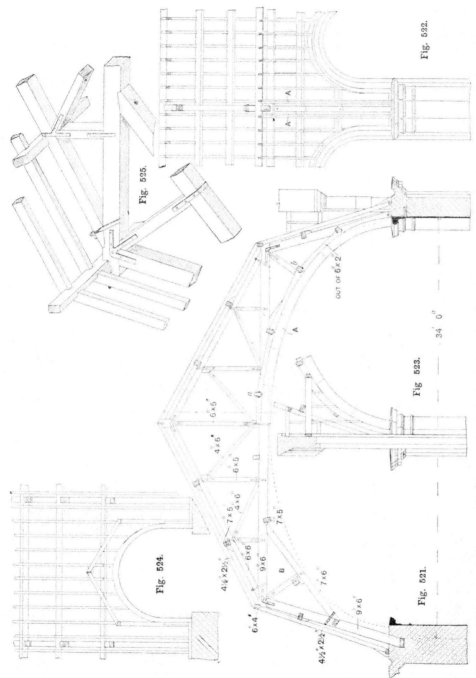

Fig. 521.—Transverse Section through Mansard Roof with Arched or Coved Ceiling, intersected by Circular-headed Opening for Windows. Fig. 522.—Part Longitudinal Section of Mansard Roof showing Curved Ribs and Framing to Circular-headed Window Openings. Fig. 523.—Transverse Section through Circular-headed Opening showing Angle Ribs, Head and Ceiling Ribs. Fig. 524.—Part Elevation of Outside Framework to Head of Window Openings behind Masonry. Fig. 525.—Conventional Detail View showing Construction of B, Fig. 521.

of the upper rafters is supported by main walls, intermediate walls, or partitions not more than about 15 ft. apart, this arrange-

the whole of the load carried by the upper truss would have to be supported by the cleat spiked to the queen-post and the stub tenon on the end of the upper tie-beam (Fig. 519), a most inefficient support; the effect or stress of the load would, of course, be withstood by the strut or principal rafter so long as the connections held good, as the head of the queen-post merely forms a convenient abutment for connecting the two trusses. The designer of the truss shown

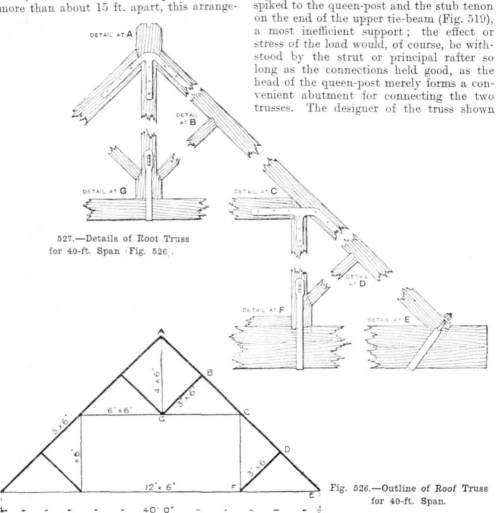

527.—Details of Roof Truss for 40-ft. Span (Fig. 526).

Fig. 526.—Outline of Roof Truss for 40-ft. Span.

ment rendering the construction of the lower truss unnecessary. They, however, are an advantage constructively in another way, as they afford a means of triangulating the enclosed figure, and thus rendering the truss immovable by wind pressure. Lest any student should be tempted to copy a form of truss (Fig. 519) which has been proposed as an improvement on the ordinary Mansard truss, it may be pointed out that

by Fig. 519 proceeded on the false assumption that the queen-posts of the ordinary Mansard were in compression, and his system was an attempt to avoid this. However, in an ordinary Mansard truss, the head of the queen-post is, of course, a direct support to the king-post truss, and therefore carries the weight of that truss and roofing, and thus, from the head of the queen-post to the bottom of the joint (see A, Fig. 520),

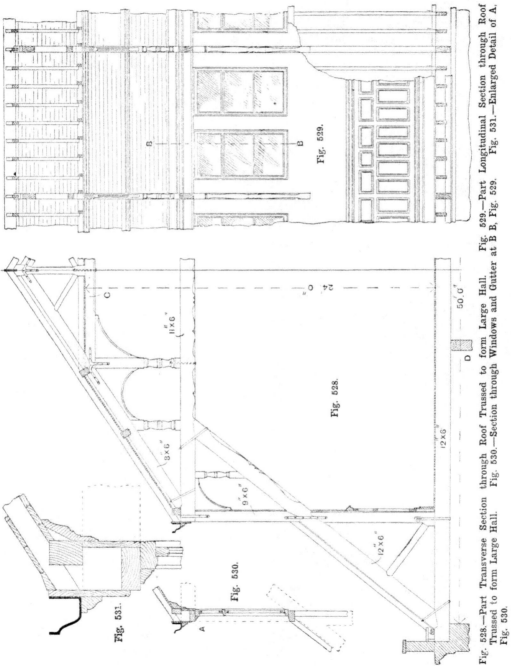

Fig. 528.—Part Transverse Section through Roof Trussed to form Large Hall. Fig. 529.—Part Longitudinal Section through Roof Trussed to form Large Hall. Fig. 530.—Section through Windows and Gutter at B B, Fig. 529. Fig. 531.—Enlarged Detail of A, Fig. 530.

Fig. 528.

Fig. 529.

Fig. 530.

Fig. 531.

where it is connected with the lower principal rafter, is in compression—but only this portion. The compressional stress is then transmitted to the lower principal rafter, the remainder of the queen-post

short; then the tie-beam is forced to the shoulders of the queen-posts, and secured by straps or bolts, and this clearly produces tension in the queen-posts from their connection with the tie-beam to the shoulders of the principal rafter (A, Fig. 520). It is also to be noticed that the connection between the queen-post and tie-beam prevents an inward turning action which would

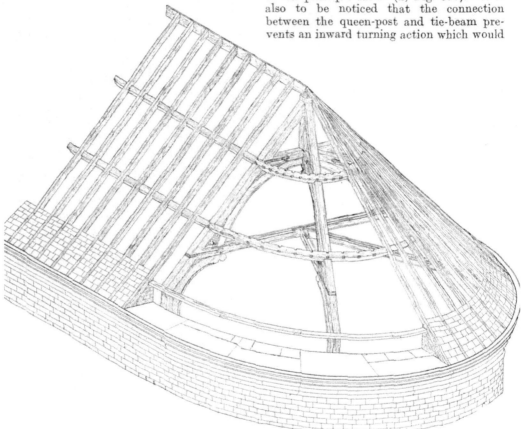

Fig. 532.—General Outside View of Apse End of Collar Beam Truss Roof.

being in tension; this may be curious, but is nevertheless a fact. The object of the tie-beam is to extend from wall to wall, taking the ends of the principal rafters, and thus preventing the outward thrust of the principal rafters on the walls, which would occur if a tie-beam were not used. Then, to prevent the tie-beam sagging, the queen-posts run down to it and support it. When these trusses are properly made, it is usual to have the queen-posts a little

otherwise take place (as indicated by the arrow in Fig. 520) about the top of the queen-post (B, Fig. 520), by the inward thrust on it by the lower principal rafter. This quite refutes any statement that the queen-posts, instead of supporting the tie-beam, add their load to it.

Mansard Roof over a Room with an Arched Ceiling.—Figs. 521 to 525 show a Mansard roof designed for a span of 34 ft., the room having an arched or coved ceiling which is intersected by circular-headed openings for windows. The figures show the construction very fully, and the chief dimensions are figured. The ribs for the curved ceiling are cut to shape out of 6 in. by 2 in. stuff, and are notched to fillets fixed on the sides of the purlins as shown at b (Fig. 521). A rib built up of two thicknesses out of 11 in. by 1 in. is fixed on each side of the truss as shown at A (Figs. 521 and 522) to take the laths

50-ft. span, allowing of a large well-lighted room being formed within it, 34 ft. wide and 24 ft. high from the floor to the ceiling. The ceiling is level with the top edge of the collar C. The main tie-beam is supported by two corridor walls, one of which is indicated at D, some of the leading dimensions being given.

Collar Beam Roof.—A collar beam roof over a small church or similar building with a circular apse end is illustrated by

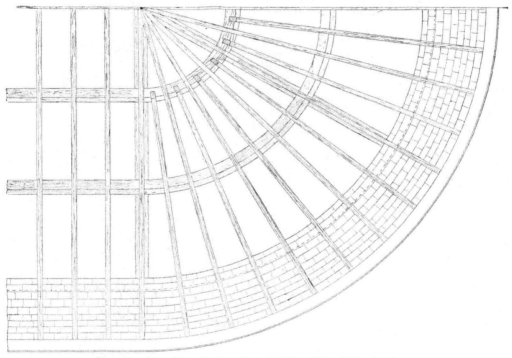

Fig. 533.—Part Plan of Apse End of Collar Beam Truss Roof.

and plaster, so that in the finished ceiling a panelled and moulded rib would be formed. The construction of the circular-headed window opening in connection with the main ceiling is clearly shown in the illustrations.

Open Timber Roofs.

The open timber roof shown by Figs. 526 and 527 is suitable for covering a room 40 ft. wide. Figs. 528 to 531 illustrate a good example of construction for a roof

Figs. 532 to 538. The circular end has two half trusses framed and fastened to the main truss; the king-post is common to the three, being cut to the form shown at F (Fig. 537) to receive the heads of the principal rafters. Only each alternate common rafter is carried up to the vertex. The main collar beam and the two half beams are mortised and tenoned together as shown at G (Fig. 537), and further secured by a strong three-way strap bolted to the top surfaces (Fig. 538). The curved purlins would be made out of two thicknesses with joints breaking. The leading dimensions will be found clearly stated in the figures.

Hammer Beam Roofs.—Figs. 539 to 554 fully illustrate the roofing to a church. The roof to the nave is supported by beam trusses of simple construction. The *dimensions of the main members are as* follow :—Principal rafters, hammer beam, purlins and collar, 10 in. by 7 in. ; ribs out of stuff 4 in. thick. A hammer beam truss of a good ornamental design is shown

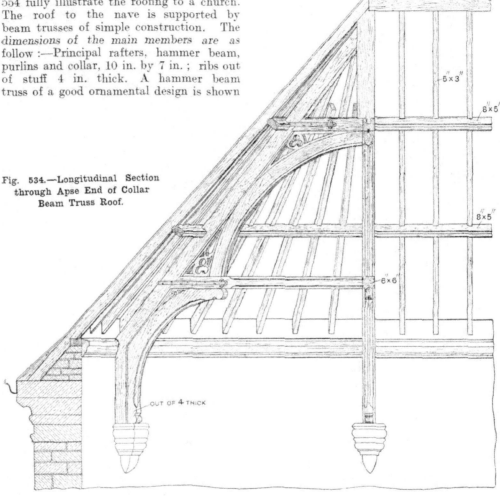

Fig. 534.—Longitudinal Section through Apse End of Collar Beam Truss Roof.

OUT OF 4 THICK

by Fig. 555. Details of construction are illustrated by Figs. 556 to 560.

Raising Roof Principal.

For raising into position a hammer beam principal weighing about 1 ton, two upright poles may be used ; and a horizontal pole, as indicated in Fig. 561, should be

securely lashed to the principal to prevent its being strained during the operation of hoisting. Treble-sheave blocks will be ample for the purpose, seeing that each rope will have to lift only about half a ton ; if double-sheave blocks are used, the time occupied in hoisting will be lessened, but more than one man must pull at each rope. A 3½-in. good quality hemp rope will lift 11 cwt. easily with an ample margin of safety. A method recommended by an experienced carpenter for raising into position a hammer-

CARPENTRY AND JOINERY.

HALF SIDE ELEVATION

CONVENTIONAL SECTIONAL VIEW

HALF LONGITUDINAL SECTION THROUGH A A

NOTE.—THE ELEVATIONS AND SECTIONS ARE REPRODUCED ½ IN. TO THE FT., AND THE ENLARGED DETAILS TO A SCALE OF 1½ IN. TO THE FT.

ENLARGED DETAIL AT E

ENLARGED SECTIONS THROUGH F F AND G G

HALF END ELEVATION

HALF TRANSVERSE SECTION THROUGH B B

ENLARGED SECTION THROUGH D D

ENLARGED DETAIL AT C

LANTERN LIGHT WITH GABLE ENDS.

4

beam truss of 50-ft. span is as follows :—
A derrick is erected and held in nearly an
upright position by guy ropes. A block and
tackle are secured to the top of the derrick,
at the lower end of which a single block is

required by men working the crab. To
prevent the truss swaying and doing damage
during ascent, it is guided by workmen
holding ropes tied to it. The truss is next
placed as nearly as possible in its position,
then plumbed, adjusted, and stayed temporarily with pieces of timber attached to
the plate, or other convenient fixing, until

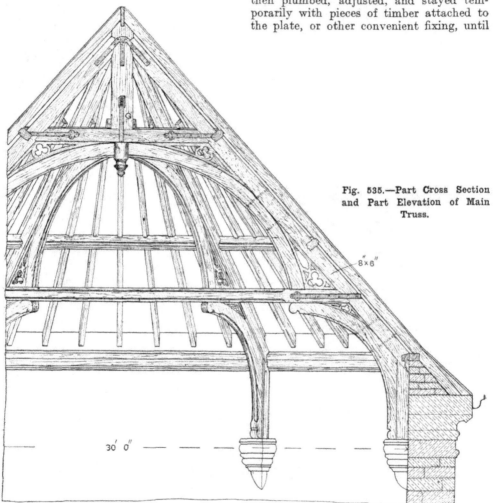

Fig. 535.—Part Cross Section
and Part Elevation of Main
Truss.

8″×6″

30′ 0″

fixed. The other end of the tackle is connected to the truss, and the free end of the
cord passes through the top block and down
through the single pulley fixed at the lower
end of the derrick, from which the cord is
continued and connected to a crab. The
truss is now gradually raised to the height

it can be connected to others by purlins,
ridge, etc. At the present time derrick
cranes are frequently used for hoisting, in
which cases the trusses would be raised,
guided, and placed in position more speedily
than by the method above described of
using a block and tackle.

7

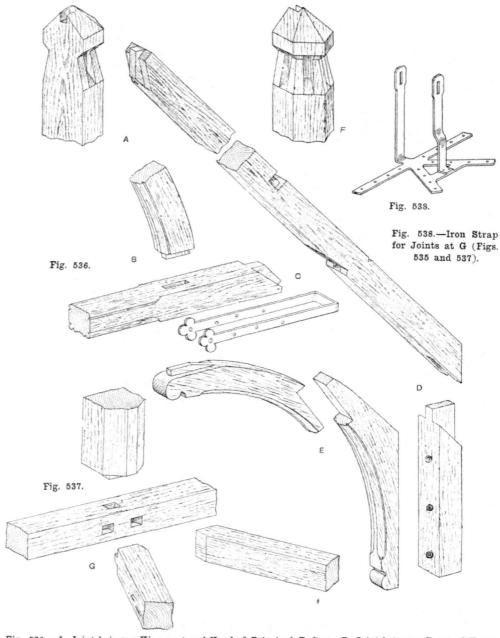

Fig. 536.

Fig. 537.

Fig. 538.

Fig. 538.—Iron Strap
for Joints at G (Figs.
535 and 537).

Fig. 536.—A, Joint between King-post and Head of Principal Rafter. B, Joint between Front of Upper Rib and Collar Beam. C, Joint between Collar Beam and Principal Rafter and Iron Strap. D, Joint at Front of Principal Rafter. E, Jointing of Lower Ribs. F, View of Mortising and Joggling to receive Principal Rafters to Upper End of King-post to Main and Half Trusses. Fig. 537.—Joints at Bottom End of King-post and Half Collars with Collar to Main Truss (G).

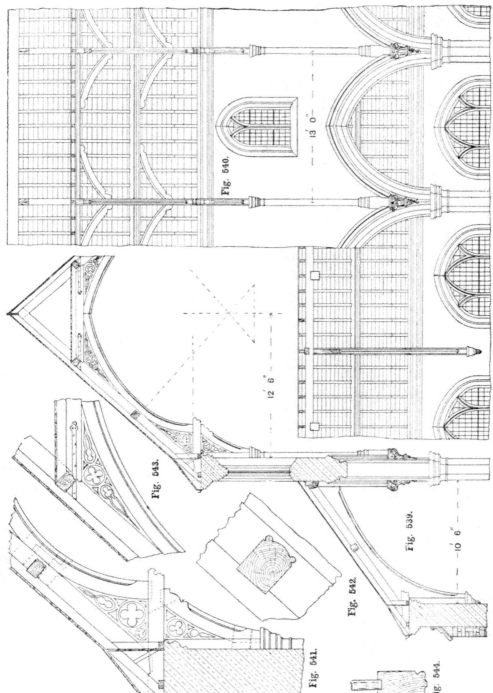

Fig. 539.—Transverse Section through Roof of Nave and Aisle. Fig. 540.—Part Longitudinal Section showing Under Side of Roof over Nave and Aisle. Fig. 541.—Detail of Connections with Hammer Beam. Fig. 542.—Section of Purlin. Fig. 543.—Detail of Connection of Collar Beam and Principal Rafter. Fig. 544.—Section of Moulded Rib Grooved for Spandril.

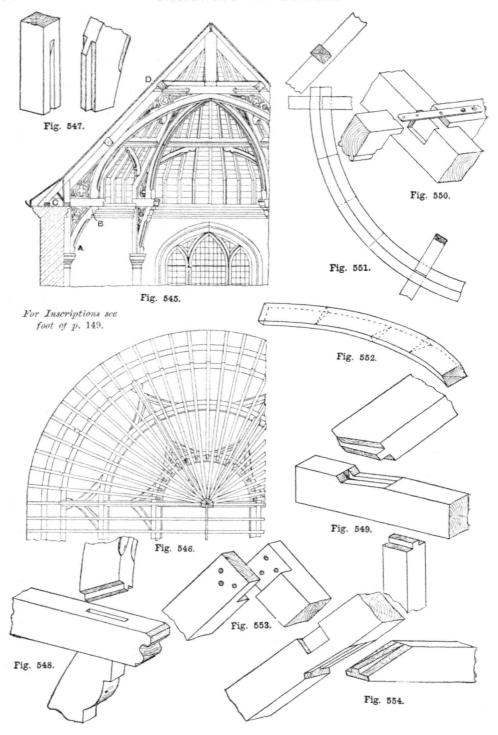

Fig. 547.

Fig. 545.

For Inscriptions see foot of p. 149.

Fig. 546.

Fig. 548.

Fig. 549.

Fig. 550.

Fig. 551.

Fig. 552.

Fig. 553.

Fig. 554.

Composite Roof Trusses.

A composite roof truss usually has wooden rafters and tie-beam and iron bolts with wooden struts. Sometimes it has an iron bent tie-rod instead of a tie-beam. The rod replaces the tie-beam, the construction is as shown in Figs. 564 to 568. A king-bolt truss with struts (Fig. 569) is suitable for a span of from 20 ft. to 30 ft. A queen-bolt truss (Figs. 570 to 572) is adapted for any

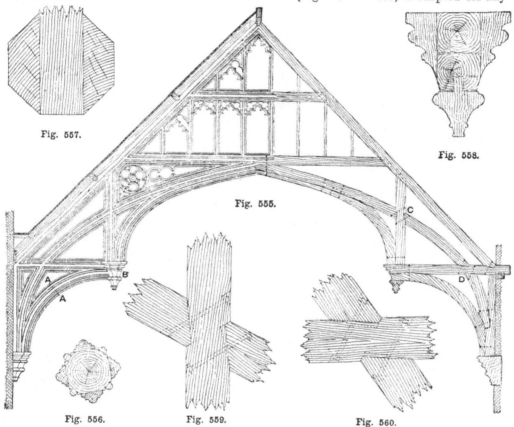

Fig. 557.

Fig. 555.

Fig. 558.

Fig. 556.   Fig. 559.   Fig. 560.

Fig. 555.—Ornamental Design for Hammer-Beam Truss.   Fig. 556.—Section of Hammer-Beam Truss through A A.   Fig. 557.—Plan of Pendant.   Fig. 558.—Section through Pendant.   Fig. 559.—Detail of Joint at C (Fig. 555).   Fig. 560.—Detail of Joint at D (Fig. 555).

simplest type is the king-bolt truss (Figs. 562 and 563), a simple span roof with tie-beam and king-bolt; this is suitable for a span between 15 ft. and 20 ft. When a tie-span between 30 ft. and 40 ft., as is also the king- and queen-bolt truss shown by Figs. 573 and 574. When a composite roof having an iron tie-rod is strutted, the

---

(For illustrations see previous page.)

Fig. 545.—Transverse Section showing Inside of Circular End Roof over Apse.   Fig. 546.—Plan of Timbers to Circular End over Apse.   Fig. 547.—Joint at Foot of Bracket at A (Fig. 545). Fig. 548.—Joints at End of Hammer Beam, B (Fig. 545).   Fig. 549.—Joints between Principal Rafter, Hammer Beam and Wall Piece.   Fig. 550.—Joints between Purlins and Principal Rafters. Fig. 551.—Plan of Part of Curved Purlin indicating Method of Building Up.   Fig. 552.—Conventional View of Curved Purlin showing Method of Construction.   Fig. 553.—Halving at Heads of Principal Rafters.   Fig. 554.—Joint between Collars and Principal Rafters.

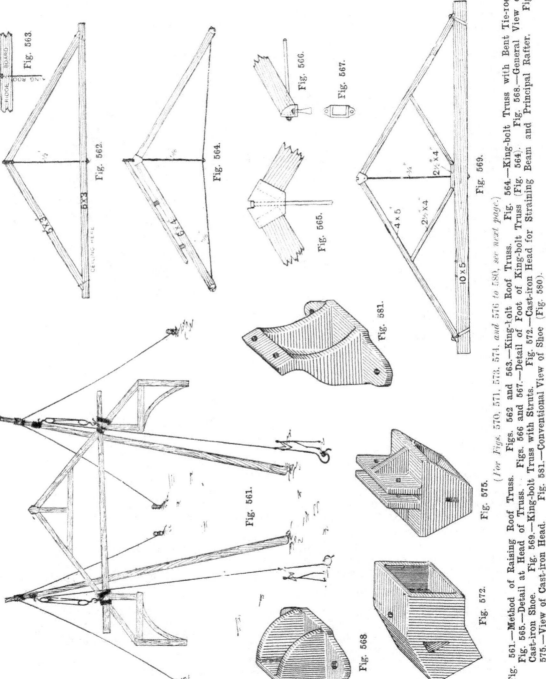

Fig. 563.

RIDGE BOARD
KING ROD

Fig. 562.

5 × 3

CEILING PIECE

Fig. 564.

6 × 4

Fig. 566.

Fig. 567.

Fig. 565.

Fig. 569.

4 × 5

2½ × 4

2½ × 4

10 × 5

Fig. 581.

Fig. 561.

Fig. 575.

Fig. 572.

Fig. 568.

(For Figs. 570, 571, 573, 574, and 576 to 580, see next page.)

Fig. 561.—Method of Raising Roof Truss.   Figs. 562 and 563.—King-bolt Truss.   Fig. 564.—King-bolt Roof Truss.   Fig.
Fig. 565.—Detail at Head of Truss.   Figs. 566 and 567.—Detail of Foot of King-bolt Truss (Fig. 564).   Fig. 568.—General View of
Cast-iron Shoe.   Fig. 569.—King-bolt Truss with Struts.   Fig. 572.—Cast-iron Head for Straining Beam and Principal Rafter.   Fig.
575.—View of Cast-iron Head.   Fig. 581.—Conventional View of Shoe (Fig. 580).

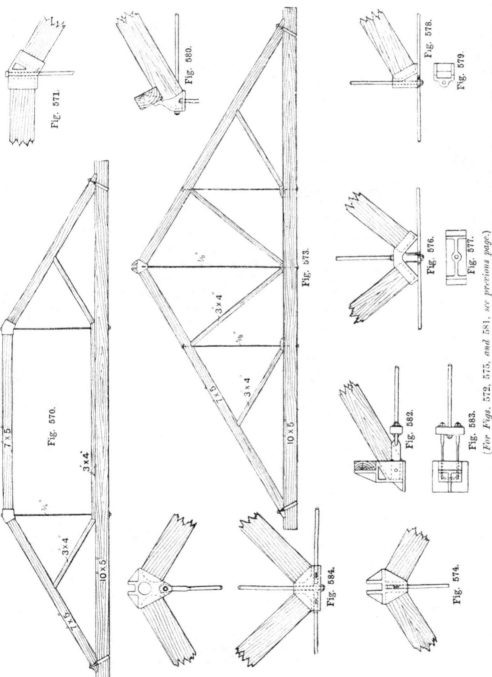

(For Figs. 572, 575, and 581, see previous page.)

Fig. 570.—Queen-bolt Truss. Fig. 571.—Detail at Head of Queen-bolt. Fig. 573.—King- and Queen-bolt Truss. Fig. 574.—Detail at Head of Truss. Figs. 576 and 577.—Cast-iron Shoe for Feet of Struts. Figs. 578 and 579.—Form of Shoe for Strut and Queen-rod. Fig. 580.—Shoe to receive Principal Rafter, Tie-rod and Plate. Figs. 582 and 583.—Detail at Foot of Principal Rafter, showing Method of Tightening Tie-rod. Fig. 584.—Details at Head and Foot of King-rod.

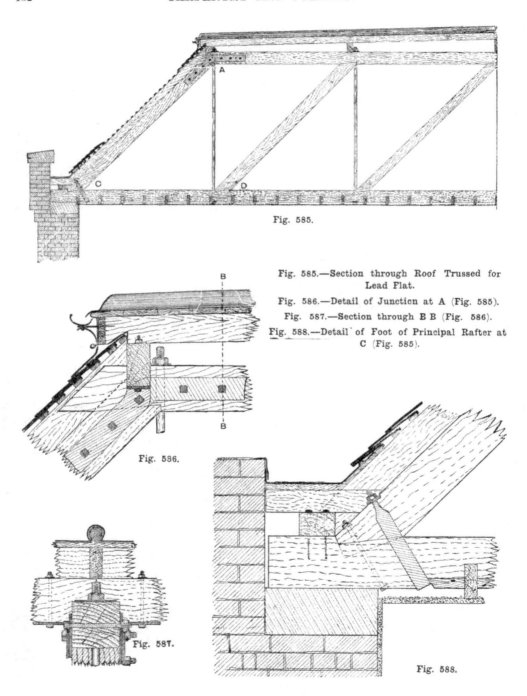

Fig. 585.

Fig. 585.—Section through Roof Trussed for Lead Flat.

Fig. 586.—Detail of Junction at A (Fig. 585).

Fig. 587.—Section through B B (Fig. 586).

Fig. 588.—Detail of Foot of Principal Rafter at C (Fig. 585).

Fig. 586.

Fig. 587.

Fig. 588.

connections are designed as in Figs. 576 to 584.

Composite Truss for Flat Roof.

For the span of 47 ft. between pier walls shown in Fig. 585, the " Howe " form of truss illustrated is quite as suitable as the queen-post truss sometimes adopted. It is proportioned to carry a 6-lb. lead flat, with side slopes battened and tiled, as well as a plaster ceiling supported on 2-in. by 6-in. ceiling joists, suspended from the lower chord or tie-beam, which, without over-stressing the truss proper, can be of an enriched class of decoration, suitable for a

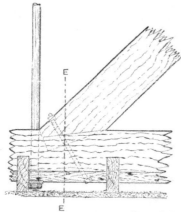

Fig. 589.—Detail at D (Fig. 585).

trusses are spaced 10 ft. apart, and the camber is ¾ in. for every 10 ft. between the wall and the centre rod. This will represent, roughly, a 1½-in. rise at the centre, which can be obtained by springing the lower chord and marking shoulder lines on the braces when framing. The material may be Memel fir or pitch-pine. The stuff is framed up from the saw to 8 in. by 8 in. for principal rafters, top chord, and second panel braces, with 8-in. by 6-in. middle braces, and 8-in. by 8½-in. or 9-in. lower chord or tie-beam. The round iron tie-rods have diameters as follow : Outside rod, 1⅜ in. ; second rod, 1¼ in. ; and centre rod, 1 in. The rods are threaded at both ends for hexagon or square nuts, with 2½-in. by ½-in. plates, 7 in. long, to each nut, the lower plates being let into the under side of the tie-beam, as shown at Fig. 589. The braces are bolted to the chords with ⅝-in. bolts. The feet of the principal rafters are further secured

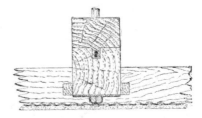

Fig. 590.—Section through E E (Fig. 589).

public hall or concert room. Allowance has been made for a snow-load of 5 lb. per square foot, and for a wind pressure of 26 lb. per square foot on the side slopes. Fig. 585 is a half-elevation ; Fig. 586, a detail at A (Fig. 585) at the top of the principal rafter ; Fig. 587, a section on the line B B (Fig. 586) ; Fig. 588, a detail at C (Fig. 585) at the junction of principal rafter and tie-beam ; Fig. 589, a detail at D (Fig. 585) at foot of brace ; and Fig. 590, a section on the line E E (Fig. 589). The details are as follow : Width between pier walls, 47 ft. ; extreme length of lower chord (tie-beam), 50 ft. 1 in. ; between external points of principal rafters, 48 ft. 4 in., which is divided into six panels 8 ft. wide and 8 ft. high on the centre lines ; hence all braces are inclined at 45°, approximately, and the tie-rods are vertical. The

7*

with double bolt-ended straps, forged out of 2½-in. by ½-in. flat plate-iron, with top plates (of the same dimensions) drilled to receive the ⅝-in. bolt ends. At the head of the principal rafters, an angle plate of ¾-in. iron, 4 in. wide, is drilled for ⅝-in. bolts and nuts, an auxiliary angle plate being placed at right angles, to line with the purlins, and bolted up with the corner plate mentioned above. This plate is bolted to the under side of the purlins with two ½-in. bolts, each 10 in. long. The purlins are notched ½ in. on to the top chord. In the half-elevation (Fig. 585), the outer purlin is shown partly removed, in order that the angle plate on the truss may be seen. Details of the lead flat, the tiled slope, and the plaster ceiling are clearly shown in the illustrations. A gutter is provided to drain the lead flat, with down pipes leading to cesspools at the lowest levels of the bottom gutters. The lead apron is copper-nailed under the lap of the roof-sheeting. This board may be rounded on the edge and rebated out to the thickness of the lead apron; or, alternatively, a thinner outside board may be used. A fascia board (1¼ in.) is nailed to the end of the rafters to serve as backing when dressing the lead apron, which

Belfast, Irish, or Bowstring Truss.

The Belfast, Irish, or bowstring roof (Fig. 591) is a cheap form of construction extensively used for large spans. The truss consists of the sole-piece A, made double, as shown, of two pieces of 9-in. by 1½-in. pitch-pine; the bows B, also double, are of 3-in. by 1½-in. pitch-pine, bent to the curve, and struts or lattices C, of 3-in. by ⅞-in. spruce, passing between the sole-pieces and

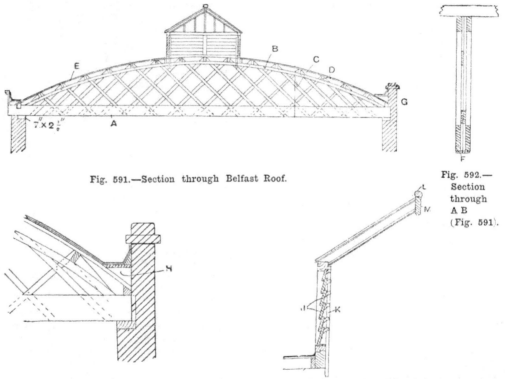

Fig. 591.—Section through Belfast Roof.

Fig. 592.—Section through A B (Fig. 591).

Fig. 593.—Enlarged Detail of Foot at G (Fig. 591).

Fig. 594.—Enlarged Detail of Ventilator of Belfast Roof.

comes down on the tiles a sufficient distance to cover the joints at the lap. The side rafters, 2½ in. by 5 in., are bird's-mouthed on to the pole plate, which is secured to the tie-beam with four coach screws (each 9 in. long), and has a bearing on the brickwork of the main wall. The scale of Fig. 585 is 3/16 in. to 1 ft., and the other illustrations (Figs. 586 to 590) are reproduced to a scale of ¾ in. to 1 ft.

bows, and clipping the purlins. They are usually put together with wire nails, one-half being laid out and nailed together, and the other half of bows and tie-beam or sole-piece put on top and nailed together. Rough timber is generally used. The purlins D, at 2-ft. centres, are usually double, about 3½ in. by 1½ in., and are covered with ¾-in. rebated or tongued and grooved boarding E, and roofing felt well lapped at the joints,

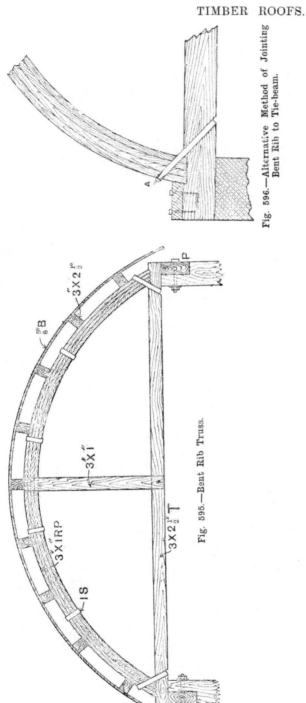

Fig. 596.—Alternative Method of Jointing Bent Rib to Tie-beam.

Fig. 595.—Bent Rib Truss.

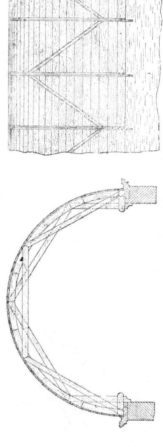

Fig. 597.—Transverse Section and Part Longitudinal Section through Circular Roof constructed of Boards.

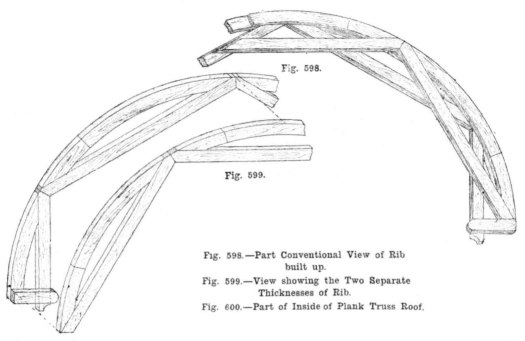

Fig. 598.

Fig. 599.

Fig. 598.—Part Conventional View of Rib
built up.
Fig. 599.—View showing the Two Separate
Thicknesses of Rib.
Fig. 600.—Part of Inside of Plank Truss Roof.

Fig. 600.

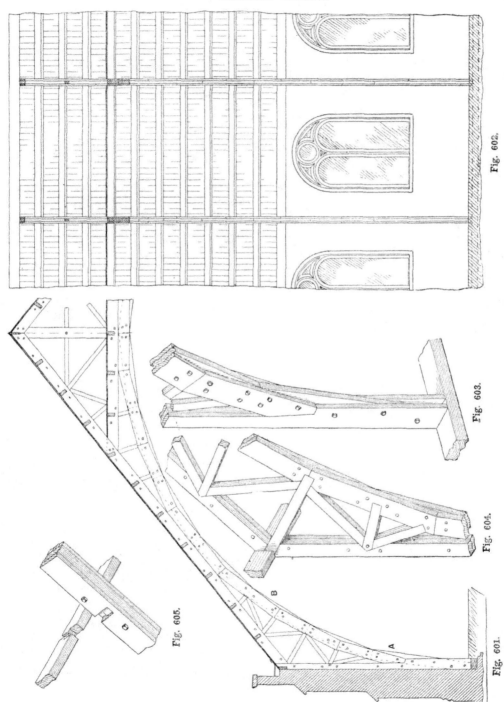

Fig. 602.

Fig. 603.

Fig. 604.

Fig. 605.

Fig. 601.

Fig. 601.—Cross Section through Roof and Part Elevation of Truss. Fig. 602.—Part Longitudinal Section of Roof. Fig. 603.—Foot of Truss connected to Sill. Fig. 604.—Jointing of Struts and Braces with Centre Planks. Fig. 605.—Joint between Purlins and Principal Rafters.

well coated with varnish, and sanded. The trusses are most usually made by the local felt manufacturers, of which there are several large firms in Belfast. The trusses

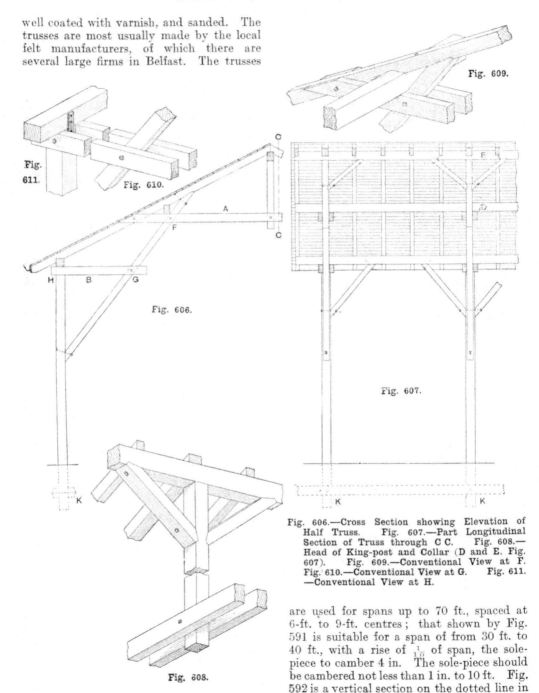

Fig. 606.—Cross Section showing Elevation of Half Truss. Fig. 607.—Part Longitudinal Section of Truss through C C. Fig. 608.— Head of King-post and Collar (D and E, Fig. 607). Fig. 609.—Conventional View at F. Fig. 610.—Conventional View at G. Fig. 611.—Conventional View at H.

are used for spans up to 70 ft., spaced at 6-ft. to 9-ft. centres; that shown by Fig. 591 is suitable for a span of from 30 ft. to 40 ft., with a rise of $\frac{1}{6}$ of span, the sole-piece to camber 4 in. The sole-piece should be cambered not less than 1 in. to 10 ft. Fig. 592 is a vertical section on the dotted line in

Fig. 591, showing the beaded cover board F underneath the sole-piece. Fig. 593 is an enlarged detail at G (Fig. 591), H being a wood block supporting the gutter. Fig. 594 is an enlarged detail of the ventilator, the louvres J being of 6-in. by 1-in. stuff; K is an oak rod for opening the louvres; L, the ridge roll of 2¾-in. by 1¼-in. stuff; and M the ridge of 6-in. by 1½-in. stuff.

Light Truss with Bent Rib.

The principle of the truss shown in Fig. 595 was applied in the Bristol Exhibition Buildings in 1893. The bent rib is built up with stuff of 1 in. or 1¼ in. thickness, each bent separately into position; the whole is held together with iron bands. This method

8 in., centre to centre. Fig. 597 illustrates the method of bracing the trusses together. If a neat appearance inside is desired, grooved and tongued boarding 1 in. thick would be most suitable as an inner covering. The outer covering should be of felt, corrugated iron, or similar material. Fig. 598 is a conventional view of a little more than half of a completed truss, Fig. 599 being a conventional view of the two separate thicknesses, and of the necessary breaking of the joints, etc. The several parts of the trusses must be nailed together.

Plank Truss Roof to Cover Large Area.

A plank truss roof to cover a large area such as a drill hall or similar building is

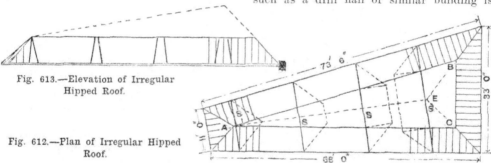

Fig. 613.—Elevation of Irregular Hipped Roof.

Fig. 612.—Plan of Irregular Hipped Roof.

is superior in many respects to bending a solid rib, distributing any weakness, in the shape of cross-grain, knots, etc., that otherwise would endanger the solid rib, and counteracting any excessive strain on the convex and concave surfaces of any timber so bent, the surfaces being in length as required by the sweep. For the same reasons the rib would have little or no tendency to revert to a straight position.

Circular Roof constructed of Boards.

At Fig. 597 is shown a transverse section and also a part of a longitudinal section through a circular roof made cheaply of boards. The ribs of the trusses are made out of two thicknesses of 9-in. by 1¼-in. boards, and are finished to a parallel width of 6 in. The truss braces also are made of 6-in. by 1¼-in. stuff. For a building about 47 ft. long, eleven would be a suitable number of trusses, spacing them out at about 4 ft.

illustrated by Figs. 600 to 605. This roof has been designed to span 60 ft., the trusses being 12 ft. apart. The principal rafters, collars and ribs are built up in three thicknesses, the centre planks being 11 in. by 3 in. and the two outer ones 11 in. by 2 in.; the braces and struts are 4½ in. by 3 in. and are notched into the central planks as shown by the conventional view (Fig. 604). The whole is bolted together by 7-in. by ⅝-in. bolts and nuts, as illustrated. The feet of the trusses are fixed into an 11-in. by 4-in. oak sill which runs the whole length of each side of the building. The purlins are 8 in. by 3 in., placed about 3 ft. apart, and are connected to the principal rafters by being housed and notched as shown in Fig. 605; 1¼-in. grooved and tongued and beaded boarding is fixed to the purlins to form the ceiling, and so is carried across level with the top of the collar.

Roof for Large Open Shed.

The construction of a roof for a large span shed is illustrated by Figs. 606 to 611. In constructing such a roof, the aim should

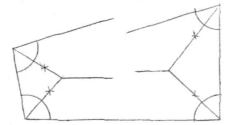

Fig. 614.—Method of Obtaining Positions of Hips.

be to avoid complicated joints. Therefore, the collar A and tie-beam (Fig. 606) are made of two thicknesses, the former being bolted to the king-post, brace and principal rafter, and the latter being bolted to the brace and head of the post. The general further construction is clearly shown by the detailed views (Figs. 608 to 611). To prevent the shed being lifted by wind, the post should be let some distance into the ground and there connected by ties as indicated at K (Figs. 606 and 607).

Irregular Hipped Roof.

There are several ways of covering a building erected upon an irregular site, such as that shown in the plan (Fig. 612). First the ridge may be kept level, and the rafters thrown into winding; secondly, the planes of the sides of the roof may be kept true, and the ridge thrown out of level, as indicated by the dotted outline in Fig. 613;

Fig. 615.—Detail of Irregular Hipped Roof Truss.

thirdly, the ridge may be kept level, and the inclinations of the various sides of the roof made to differ; and, lastly, the method shown in the accompanying illustrations,

which is perhaps the one that is most generally useful, may be adopted. In this case the roof is truncated—that is, it is treated as if the upper portion contained within the

Fig. 616.—Elevation of Corbelled Wall.

dotted outline in Fig. 613 were cut off; then a flat is formed on the top, as shown in the plan by the triangle A B C; the inclination of the lower portion being everywhere alike, and the lengths of the rafters being the same on each of the four sides.

Setting Out the Roof.—In setting out a roof of this description, the first process, after the plan of the roof is drawn in outline, is to ascertain the position of the hips.

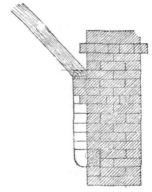

Fig. 617.—Method of Corbelling Wall to carry Plate.

The method is shown in a separate diagram to larger scale (Fig. 614). Each of the angles formed by the wall plates is bisected as shown, and the bisectors (which are the seats of the hips) are produced until they intersect, the point of intersection being the centre line of the ridge, or ridges. Having drawn the plans of the hips, from the point of intersection at A draw lines A B and A C parallel to the respective walls, and from the points of intersection of these lines, with the hip lines at the wide end, draw the line B C, which, if the construction is correct, will be parallel to the wall at that end. The triangle so formed is the outline of the flat to be covered with lead or zinc. Next, to obtain the shape of the trusses, determine their position and number, which would depend on the size of the roof and the nature of the covering. In the illustration, four are shown. Draw the centre line A E, which would represent the ridge if the roof was carried up to a single ridge, and draw the seats of the trusses S, S, S, S, at right angles to this central line. From the points in the plan where the seats

of the trusses cross the outlines of the flat, draw perpendiculars to the seat lines, as shown; make these equal in length to the height of the roof, as given, and join these points by straight lines to each other, and the intersection of the seat line with the wall. The outlines so obtained will be the shape of the respective trusses; or rather,

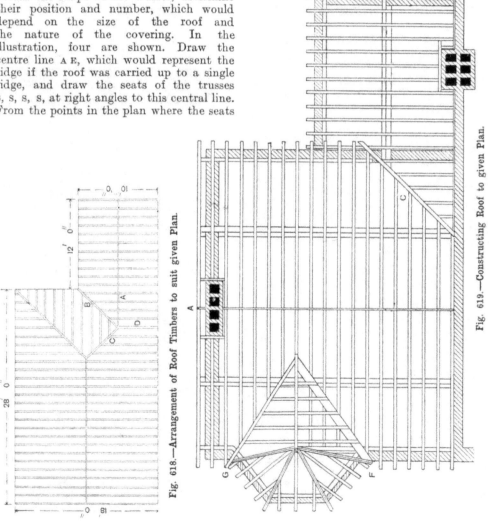

Fig. 618.—Arrangement of Roof Timbers to suit given Plan.

Fig. 619.—Constructing Roof to given Plan.

the outline of common rafters and bearers, as shown in Fig. 615 ; the truss is drawn to the same shape, but within the outlines, as shown. It will be noticed that the truss at the narrow end is a king-post truss, the remainder being queen-post trusses. The common rafters, a few only of which are shown, should be drawn at right angles to the walls ; if laid otherwise, their edges will not lie in one plane, and the boards or battens will not sit solidly on them. The jack rafters are cut against the hips in the usual manner. The thick lines in Fig. 613 represent the trusses, the thin lines the

bearers framed into it, and a central bearer at B, to carry the other ends of the cross

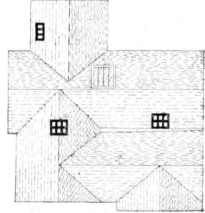

Fig. 620.—Plan of Completed Villa Roof.

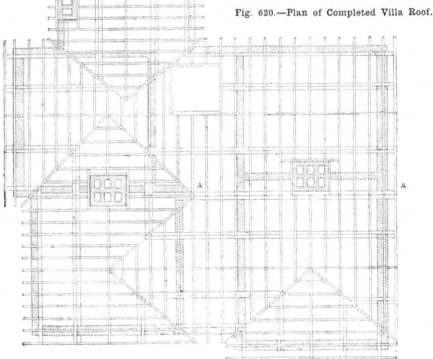

Fig. 621.—Timbering for Villa Roof.

rafters. Fig. 615 represents an enlarged detail at the head of one of the trusses, showing one of the ridge boards R, with cross

bearers ; this is sometimes raised to give a fall to either side of the lead flat. Figs. 616 and 617 show a method of carrying the wall plate at the wide end of the building when the roof is pitched from an existing wall. Three courses of bricks are corbelled out to take the wall plate, and about every

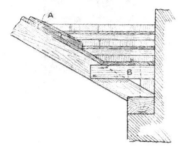

Fig. 622.—Section through Gutter behind Parapet.

8 ft. a short pier is built out resting upon a stone corbel. This is done in order to obtain the necessary weight to prevent the upper courses overturning.

Arrangement of Roof Timbers to Suit Given Plan.

First Example.—The arrangement of the timbers for a roof of the plan shown in

continued. A stouter rafter as shown at D, to meet on the opposite side of the ridge where the valley rafter and the hip rafter B and C meet, would be advantageous.

Second Example.—In Fig. 619 A is gable end. B lean-to, C valley rafter. The arrangement of the various members of the roof is shown. The sizes are as follows : Common rafters, 4 in. or 4½ in. by 2 in. ; ridge, 11 in. by 1½ in. ; purlins, 9 in. by 4 in. ; valley rafter, 11 in. by 1½ in. ; wall plates, 4½ in. by 3 in. At G is shown a valley rafter ; at F is an alternate method, in which the common rafters continue to the wall plate, and a valley board is nailed to them, to which the feet of the short rafters of the covering to the bay are nailed as indicated.

Third Example.—For the villa roof shown by Fig. 620 the arrangement of timbers will be as in Fig. 621. The outside dimensions of the whole roof are, roughly, 42 ft. by 45 ft. Suitable sizes for the roof timbers are as follows : Purlins, 7 in. by 4 in. ; valley rafters, 11 in. by 2 in. ; common and jack rafters, 3½ in. by 2¼ in. ; ridges, 7 in. by 1¼ in. The thickness of the external walls should be brick and a half (13½ in.). The internal walls, which are connected with the

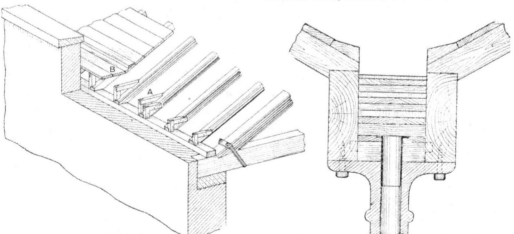

Fig. 623.—Arrangement of Bearers behind Parapet Wall.

Fig. 624.—Cross Section through Trough Gutter supported by Iron Column.

Fig. 618 will be as shown in that illustration. The valley rafter B and the hip rafter C would fit together against the ridge A, and the hip rafter C need not be

chimney breasts, must be at least 9 in. ; the other internal walls should be 9 in. for good substantial work, but for ordinary purposes the walls are more frequently only half a brick thick (4½ in.). Of course, the best method constructionally is to build up the internal walls to carry the purlins, but this is seldom done, as the purlins can be supported by struts resting on plates bedded on the walls. The feet of the rafters along A A can be carried by bird's-mouthing them on to two 9-in. by 3-in. deals bolted together, then the **V** gutter is formed between the rafters in the usual manner with two drips and falls as indicated on the plan (Fig. 620).

Gutters behind Parapet, behind Chimney, and to "M" Roof.

Fig. 622 shows a section through a gutter behind a parapet: A is the tilting fillet, B the bearer; drips and fall are shown. The conventional view (Fig. 623) clearly shows a method of constructing and fixing the bearers; those at A are arranged to form a drip. Gutter boarding is shown at B. Fig. 624 is a cross section through a trough gutter in an "M" roof showing falls, drips and cesspool, also outlet in a cast-iron column; the column assists in supporting the roof.

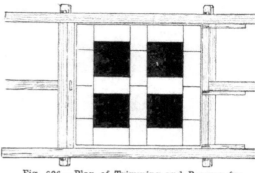

Fig. 626.—Plan of Trimming and Bearers for Gutter to Chimney.

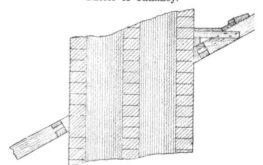

Fig. 627.—Vertical Section of Trimming and Gutter behind Chimney.

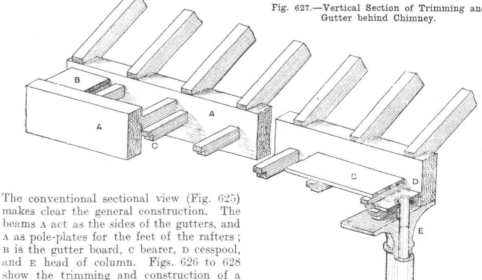

The conventional sectional view (Fig. 625) makes clear the general construction. The beams A act as the sides of the gutters, and A as pole-plates for the feet of the rafters; B is the gutter board, C bearer, D cesspool, and E head of column. Figs. 626 to 628 show the trimming and construction of a gutter behind a chimney. The gutter in Fig. 628 is formed to discharge the rainwater on both sides.

Fig. 625.—Conventional Sectional View of Trough Gutter supported by Iron Column.

Ridges and Purlins Trimmed to Chimneys.

Usually the ridge simply butts up against the chimney, being held in position only by

rafter next to the wall is generally fixed to a trimming piece A (Fig. 629), which rests a little way in the wall, and is nailed to the rafter at the other end ; sometimes these two latter timbers are stub mortised

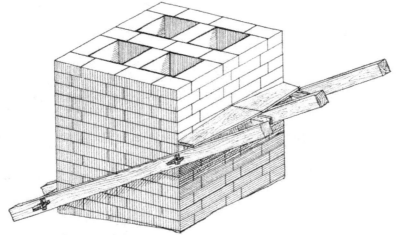

Fig. 628.—Conventional View of Trimming and Gutter.

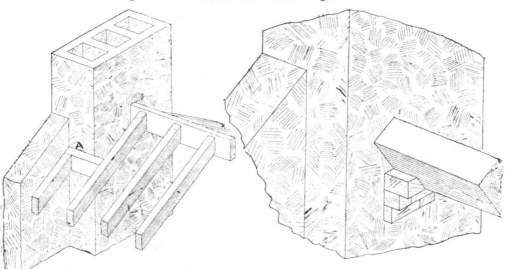

Fig. 629.—Trimming for Rafter against Chimney.

Fig. 630.—Supporting Purlin against Chimney by Corbelling.

the rafters If desired, the ridge can be supported on an iron corbel built into the chimney, but this is seldom considered necessary. When trusses are provided the ridge is further supported by them. The

and tenoned together. Of course, a purlin must not be allowed to enter a chimney breast, but only butt against it, therefore a purlin must be supported by brick or stone corbelling built in the chimney as illustrated at Fig. 630. In the case when the position for the purlin has not been anticipated, and thus no corbelling has been

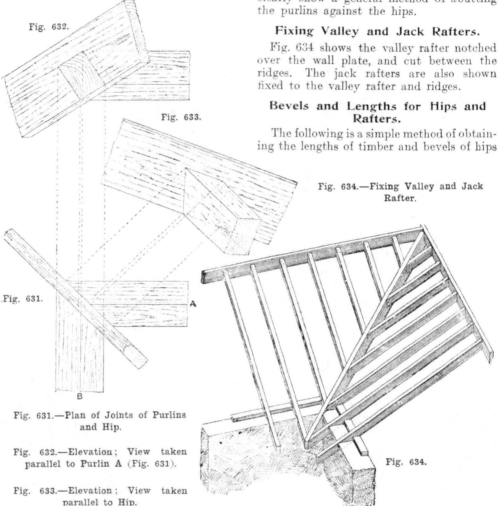

Fig. 632.

Fig. 633.

Fig. 631.

B

Fig. 631.—Plan of Joints of Purlins and Hip.

Fig. 632.—Elevation; View taken parallel to Purlin A (Fig. 631).

Fig. 633.—Elevation; View taken parallel to Hip.

A

with two purlins, and the latter being a view of the hip and purlins taken parallel to the purlin A (Fig. 631). Fig. 633 is a view of the hip and purlin B. These views clearly show a general method of abutting the purlins against the hips.

Fixing Valley and Jack Rafters.

Fig. 634 shows the valley rafter notched over the wall plate, and cut between the ridges. The jack rafters are also shown fixed to the valley rafter and ridges.

Bevels and Lengths for Hips and Rafters.

The following is a simple method of obtaining the lengths of timber and bevels of hips

Fig. 634.—Fixing Valley and Jack Rafter.

Fig. 634.

provided, a good plan is to rake out a joint and fix an iron corbel with cement.

Joints between Purlins and Hips.

The proper way to support the purlins at the hipped end of an ordinary roof is shown by Figs. 631 and 632, the former being a plan of the hip and the meeting of the hip

and rafters. First set up the elevation of pitch as shown in Fig. 635, and in the plan in Fig. 636. To obtain the length of the hip set up C E at right angles to B C, and, making C E equal to the height C' D, join B E, which gives the length required. The bevels for application to the side of the hip are shown at 1 and 2. The bevel for the edge is shown

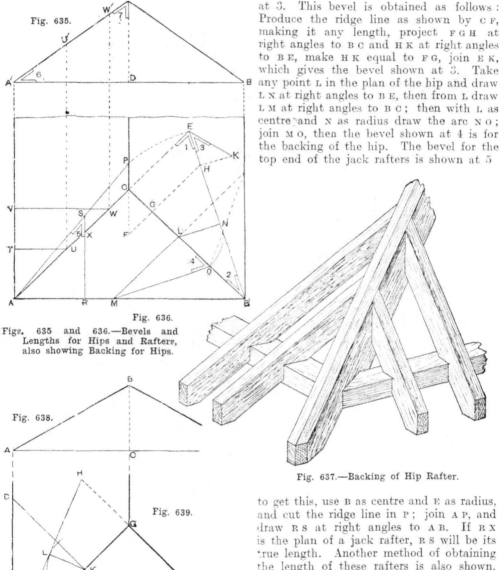

Fig. 635.

Fig. 636.

Figs. 635 and 636.—Bevels and Lengths for Hips and Rafters, also showing Backing for Hips.

Fig. 638.

Fig. 639.

Fig. 638.—Line of Pitch of Roof at A B C.
Fig. 639.—Plan of Roof at D E F G.

Fig. 637.—Backing of Hip Rafter.

at 3. This bevel is obtained as follows: Produce the ridge line as shown by C F, making it any length, project F G H at right angles to B C and H K at right angles to B E, make H K equal to F G, join E K, which gives the bevel shown at 3. Take any point L in the plan of the hip and draw L N at right angles to B E, then from L draw L M at right angles to B C; then with L as centre and N as radius draw the arc N O; join M O, then the bevel shown at 4 is for the backing of the hip. The bevel for the top end of the jack rafters is shown at 5

to get this, use B as centre and E as radius, and cut the ridge line in P; join A P, and draw R S at right angles to A B. If R X is the plan of a jack rafter, R S will be its true length. Another method of obtaining the length of these rafters is also shown. Let T U and V W be the plans of two rafters, project U U′ and W W′ as shown, then A′ U′ gives the length of the rafter shown in plan T U, and A′ W′ that of V W. The bevels to apply to the sides of rafters are shown at 6 and 7.

Backing of Hip Rafter.—The bevel shown at 4 (Fig. 636) is for the backing of the hip—

that is, planing the upper edge into two surfaces, so that each is in the same plane as the adjacent top edges of the rafters ; this will be clearly understood by referring to Fig. 637. The object of the backing is to

is the formation of the top edge into two planes, as shown and described above. The following method can be adopted for finding backing to hips. Set out to scale the line of the pitch of the roof as shown at

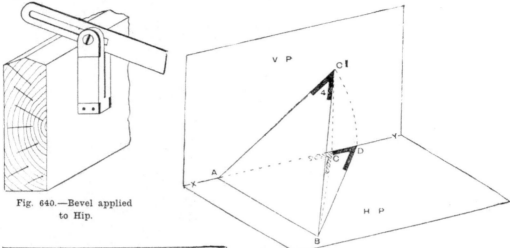

Fig. 640.—Bevel applied to Hip.

Fig. 642.—Diagram showing pictorially how to obtain Bevels for Jack Rafters.

Fig. 641.—Geometrical Method of obtaining Bevels for Jack Rafters.

A B C (Fig. 638), and a portion of the plan D E F G (Fig. 639) ; E G will be the plan of the hip. At right angles to E G set up G H, making it the same length as the height B C, then E H is the pitch of the hip. In E G take any point, as K, and at right angles to this line draw D F through K. With K as centre draw the arc L M tangent to E H as shown, join M F, which is the angle of the backing. Set the bevel to the drawing as shown. Fig. 640 is a sketch showing the bevel being applied to the hip. A drawing as shown at Figs. 638 and 639 can be sketched on a board to about 1-in. scale on a building, and it will be found to take up much less time than the rule-of-thumb method of guess and trial. If work is to be done properly, and without mistakes, time must be allowed to set it out. There is no other proper way.

Bevels for Jack Rafters.

The making of a cardboard model, as described in this paragraph, will greatly help to make clear the method of determining the bevels for jack rafters. On a piece

prepare a direct bearing for the boarding or slating battens. The term backing of hip is sometimes used to denote the distance from the back of the hip to the edge of the plate, but the proper meaning of the term

of card, say about 6 in. square (Fig. 641), draw a line x y; then set up A c equal to the pitch of the roof at right angles to x y. Draw A B, which corresponds to the wall plate. Now draw B c as the line plate of

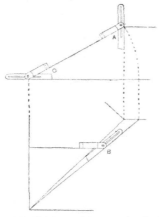

Fig. 643.—Taking off Bevels for Rafters.

the hip; then from c raise a projection to c¹. There is now a line plan of a portion of the roof, and also a line elevation. In the elevation at 1 and 2 are the bevels for the feet and head of the rafters, and at 3 is shown the plan of the bevel required for the top edges of the rafters. Now, to get the true shape of this bevel, rotate the plane A B C into the horizontal plane. To do this, with centre A describe the arc c¹ D, and join B D; then A B D is the true shape of A B C, and the bevel shown at 4 is the one re-

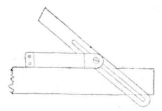

Fig. 644.—Applying Bevel to Edge of Rafter.

quired. To make the construction quite clear, cut out a piece of card the shape of A B D; fold the horizontal plane and the vertical plane at right angles to each other, and place the pieces of card in position, as

8

shown at Fig. 642. It will then be seen that the bevel at 4 is the one required, and that it stands over its plan as shown at 3 in Fig. 642.

Taking off Bevels for Rafters.

Bevels for rafters are taken off the drawing and put on the stuff to be cut in the way

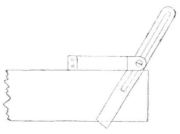

Fig. 645.—Applying Bevel to Side of Rafter.

described below. Set out for the bevels as shown at Fig. 643, the bevel at A being for the vertical cut, and that at B for the bevel to be applied at the edge of the rafter. The bevels can be set from the drawing as shown at Fig. 643. Fig. 644 shows the bevel B (Fig. 643) applied to the top edge of the rafter, and Fig. 645 shows bevel A (Fig. 643) applied to the side of it. This

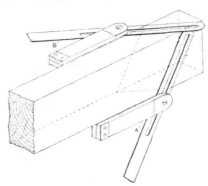

Fig. 646.—Application of Bevels to Side and Edge of Rafter.

will perhaps be more clearly understood from the isometric view given at Fig. 646, which shows the application of the bevels. The form of the cut c (Fig. 643) is the bevel for feet of rafters.

Bevels for Hips, etc., to Roofs over Obtuse or Acute Angles.

The foregoing cases of obtaining bevels for hips and rafters have been for roofs with the wall plates at right angles and the plan of the hip bisecting the angle between them. Many students are able to deal readily with such examples; but when a case is presented as shown by Fig. 647, where the plan of the plates at A makes an obtuse

bevel shown at M is for application to the edges of the jack rafters. To obtain the development of the hip end A G B, produce A B and draw $x\,y$ at right angles; project from G, and make x G' equal to E F. Join G' to H, so obtaining the inclination of the hipped end of the roof. With H as centre and G' as radius, obtain the point K'. From G project down at right angles to the line A B; this gives the point K. Then the bevels shown at K will be for the jack

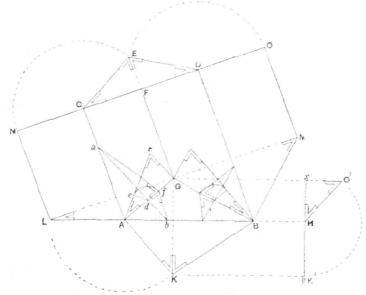

Fig. 647.—Method of obtaining Bevels for Feet and Side Cuts, and also the Backing of Hips.

angle, and at B an acute angle, and the plan of the hips does not bisect the angles, they often cannot get a correct result, although the geometrical principles are similar, as will be shown. At C D E set up the pitch of the roof taken at right angles to the plates; produce F C and F D indefinitely; from C with C E as radius obtain point N; from D, with D E as radius, obtain point O; draw N L parallel to A C, obtaining the point L. Join A L; then clearly A L N C is the development of the side of the roof A C F G, and the bevel at L is for the top cuts of the jack rafters. The development of the side of the roof B G F D has been obtained in exactly similar manner at B D O M, and the

rafters on each side. The bevels for the side top cuts are shown at E and G', and for the feet at C, D, and H. To obtain the bevels for the backing of the hip at A G, at any point in its plan draw $a\,b$ at right angles to A G. Set up G c at right angles to A G, making it equal to E F. Join A c. On the line A c set up $e\,d$ at right angles to A c, cutting A G in d. With d as centre, draw an arc tangent to A c, with radius $d\,e$, and so obtain point f. Join $a\,f$ $b\,f$; then the bevels shown at f will be for the backing of the hip, that at c will be for the vertical cut of the hip, whereas that at A will be for the foot. If this working has been carefully followed no difficulty will be found in setting out the bevels for the acute angle at b. The plan A C D B has been reproduced in Fig. 648 with the complete setting out of the timbers, and also the development of the sides of the roof which give the true length of each timber and the bevel for the top cut against the hip. A cardboard model set out on this principle, and folded up, would prove the working. The bevels shown at 1, 2, 3, and 4 are similar to the corresponding ones at Fig. 647.

then from d, $d\,c$ at right angles to $a\,c$, from b and c set up the pitches of the roof surfaces, as shown by lines $b\,e$ and $c\,f$. Set out the sections of the purlins from b and c and at right angles to the lines $b\,e$ and $c\,f$. Project down from the section at g' and h' to g, and so obtain the plan of the top edge and inner surface of the purlin; from $p'\,n'$ to 1 and 2 the plan of the edge and side of the adjacent purlin is obtained. To make the working clearer, assume that the end of the

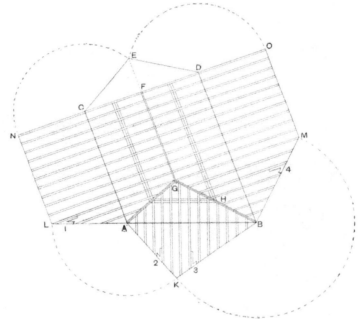

Fig. 648.—Plan of Timbering of Hipped End Roof, with Wall Plates making Obtuse and Acute Angles; also Development of each Side of Roof.

Bevels for Purlins.

The following will be found a good general method for obtaining the bevels for purlins. Assume that the plans of the wall plates are at right angles to each other, and that the plan of the hip bisects the angle between the plates. To an enlarged scale set out a portion of the plan as at A B C (Fig. 649). Set out the plan of the hip, its centre line being B D as shown. Fix on any point a in the centre line of the hip, and draw $a\,b$, $a\,c$ parallel to the plan of the line of wall plates A B, B C. At any convenient point in $a\,b$ draw $b\,d$ at right angles to $a\,b$;

purlin which fits against the hip has been removed to the position $n\,o\,p$ (which, of course, is parallel to 1 and 2). Continue $d\,b$, taking this line as an X Y; then, with b as centre, and p' and n' as radii, obtain points $q'\,r'$, thus constructing the edge and side of the purlin into the horizontal plane. Then project down from q' and r' parallel to B C, and from p and n parallel to X Y, thus obtaining points q and r. Then the true shape of the end of the side of the purlin is shown by the bevel E, and that for the edge at F. The application of these bevels to the purlin is shown by Fig. 650. The case just

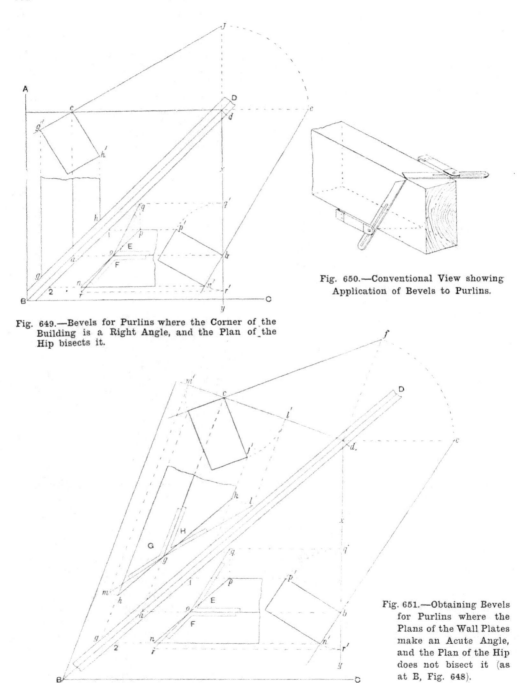

Fig. 649.—Bevels for Purlins where the Corner of the Building is a Right Angle, and the Plan of the Hip bisects it.

Fig. 650.—Conventional View showing Application of Bevels to Purlins.

Fig. 651.—Obtaining Bevels for Purlins where the Plans of the Wall Plates make an Acute Angle, and the Plan of the Hip does not bisect it (as at B, Fig. 648).

dealt with is soon mastered by the average student, but a difficulty is usually met with in obtaining bevels for purlins where the plates are at obtuse or acute angles, and the plans of the hips do not bisect the angle between them. Therefore take the case of the purlins fitting against the hip at E (Fig. 648). By carefully working out this case, it will be seen that the geometrical principles are those involved in the preceding example. Set out A B, B C, for the lines

pitch of that side of the roof; continue the lines $e\,b$ and $f\,c$, and set out the sections and plans of the purlins as shown. Then with b as centre and p' and n' as radii obtain points q' and r', projecting down from these and p and n respectively parallel to X Y, points q and r are obtained, and thus the bevels E and F. Although the bevels G and H are not the same angles as E and F, they are obtained in exactly similar manner, as will be seen. It should be noted that as far as practicable the same lettering has been adopted as in the case of Fig. 649.

Bevels for Hips.

At Fig. 652 is shown a method of obtaining the bevels for the head of a hip cutting full against the side of a ridge (see Fig. 654). Let A represent the plan of the end of the

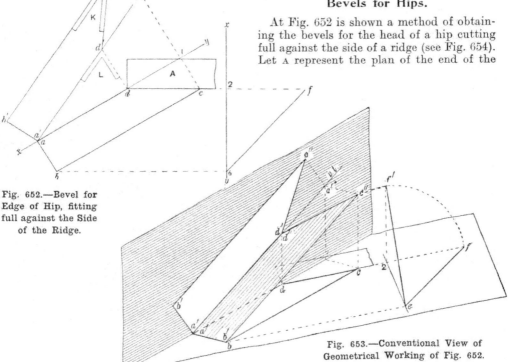

Fig. 652.—Bevel for Edge of Hip, fitting full against the Side of the Ridge.

Fig. 653.—Conventional View of Geometrical Working of Fig. 652.

of wall plates (Fig. 651), and B D the centre line of the plan of the hip; fixing upon any point a in it, draw $a\,b$, $a\,c$, parallel to B C, A B respectively. At any convenient point b draw $b\,d$ at right angles to B C, and from d draw $d\,c$ at right angles to $a\,c$. From b set up the pitch of the roof and draw $d\,e$ at right angles to $b\,d$, and $d\,f$ at right angles to $d\,c$. Make $d\,f$ equal to $d\,e$; joining $c\,f$ should give the

ridge and $a\,b\,c\,d$ the plan of the hip (not backed), cut to fit against the ridge. At right angles to the plan of the ridge draw $x\,y$ from b; project up to it, obtaining point e; then set up the angle of the roof; then project up from c, so obtaining the point f. By continuing $a\,d$ each way, use it as an $x\,y$; project up from c, and make $l\,c'$ equal to the height $2f$. Join $a'\,c'$; consider this the

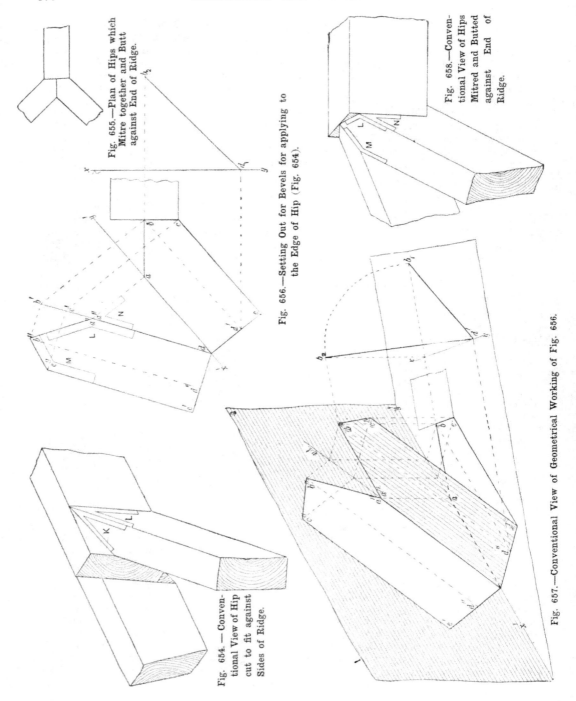

Fig. 655.—Plan of Hips which Mitre together and Butt against End of Ridge.

Fig. 656.—Setting Out for Bevels for applying to the Edge of Hip (Fig. 654).

Fig. 654.—Conventional View of Hip cut to fit against Sides of Ridge.

Fig. 657.—Conventional View of Geometrical Working of Fig. 656.

Fig. 658.—Conventional View of Hips Mitred and Butted against End of Ridge.

vertical trace ; project from d and obtain d' ; from c' at right angles to the vertical trace obtain c'', which is equal to c'. From a, set out at right angles to the vertical trace, making $a' b'$ equal to $a b$; join b' to c'' and c'' to d'. Then the bevel K is for the edge of the hip, that at L, of course, being for the side. If cuts be made along the lines $a' b'$, $b' c''$, $c'' d'$, and folds made along the line $a' d'$, and also along the line X Y, the

ing. Fig. 654 is a conventional view of the hips in their positions against the ridge, and indicates the application of the bevels to the top edge and side of hip, where the hips are mitred together, and also butt

Fig. 660.—Plan of Octagonal
Pyramidal Roof.

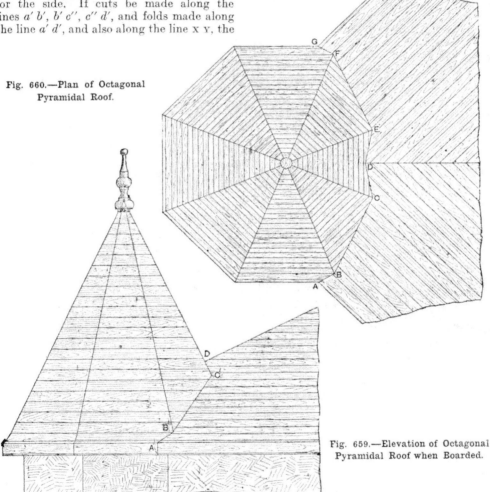

Fig. 659.—Elevation of Octagonal
Pyramidal Roof when Boarded.

true shape of the edge can be made to stand over its plan, as indicated by the conventional sketch (Fig. 653), which can be easily followed, as the same lettering is adopted as in Fig. 652. A model made as suggested in thick paper or card would prove the work-

against the end of the ridge, as shown by Fig. 655. The geometrical principles of this are exactly the same as explained in the previous case, and the working is shown at Fig. 656, where M and L are the required bevels for the edge, and N for the side. Fig.

657 is a conventional view of a model made to prove the working of Fig. 656. If the method shown by Fig. 652 has been mastered, no difficulty will be found in this example. Fig. 658 shows the hips mitred and butted against end of ridge.

Octagonal Pyramidal Roof at the Angle of a Building.

The method of obtaining the intersections, and the method of construction of an octagonal pyramidal roof intersecting at the angle of a hipped roof, will now be described. Figs. 659 and 660 show, respectively, the

is the line of boarding. The other half, A 3 4 5 6 B, it will be noticed, is a little less, this being the line of rafters. To avoid confusing the diagrams with a number of lines, the line of the wall has been omitted ; it would, of course, form a smaller parallel octagon to those shown. Next set out line A B, which is the line of feet of rafters, and

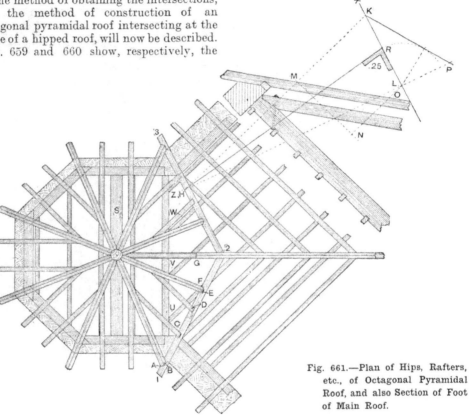

Fig. 661.—Plan of Hips, Rafters, etc., of Octagonal Pyramidal Roof, and also Section of Foot of Main Roof.

elevation and plan, and also the intersections where the pyramidal roof meets the hipped roof, as A, B, C, D, E, F, G. Before the intersections shown by Figs. 659 and 660, and the timbers shown by Figs. 661 to 663, can be properly set out, it will be necessary to obtain the intersections of the boarded surfaces geometrically. The method of doing this is shown by Fig. 664, and is as follows : Set out the half octagon A 2 1 8 7 B, which

E F, which is the line of face of the fascia board of the main roof, also the line of the main hip, as shown at G H. At right angles to 1 8, draw O P, and at right angles to this line set up O R, making it equal to the height. Join P R, which is the true inclination of the sides of the pyramidal roof. At any point along E F draw $x\,y$ at right angles to it, and set up the pitch of the main roof as shown by x s. Now take any point T on this pitch line, and project down at right angles to $x\,y$, meeting it in a, as shown. From o mark off o v equal to the height T U. From v project across to w, parallel to o P. From w project down at right angles to o P, meeting it in d. Then from d draw $d\,a$ parallel to 1 8, which will meet T U in a. Join E a, which gives the intersection of the surface o 1 8, and the main roof. For the next intersection, from where w a cuts o 8 in e, draw a line parallel to 7 8. Now produce T U, which meets the last line in f. Then from b draw through f to meet o 7 in g. This gives the intersection of the main roof with the triangular portion o 7 8. The side B 7 should be continued so as to meet E F in h. Join $h\,g$, and produce to G. Then q G is half of the intersection of the surface 7 o 6. Workers having a knowledge of geometry will see that the principle of working has been based on a problem in horizontal projection, the specific problem being : Given the horizontal traces and the inclinations of planes, find their intersections.

Developments of Surfaces.—If it is desired to obtain the developments of the several surfaces, they can be obtained in the following manner :— Draw P o at right angles to 1 8, and o R at right angles to o P. Measure on o R the height. Join P R, which gives the inclination and true length of the centre line of the full surfaces. Bisect line 2 3, and at right angles to it draw A K, making A K the same length as P R. Join 2 K and 3 K, which gives the true shape of each of the full surfaces. This development can be used to show the true shape of the surfaces which intersect with the roof. From

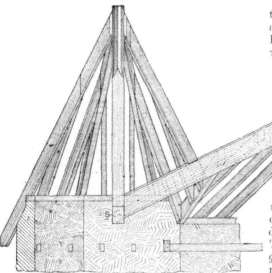

Fig. 662.—Vertical Section of Octagonal Pyramidal Roof showing Principal Timbers.

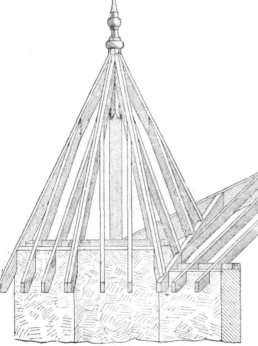

Fig. 663.—Elevation of Octagonal Pyramidal Roof ready for Boarding.

8*

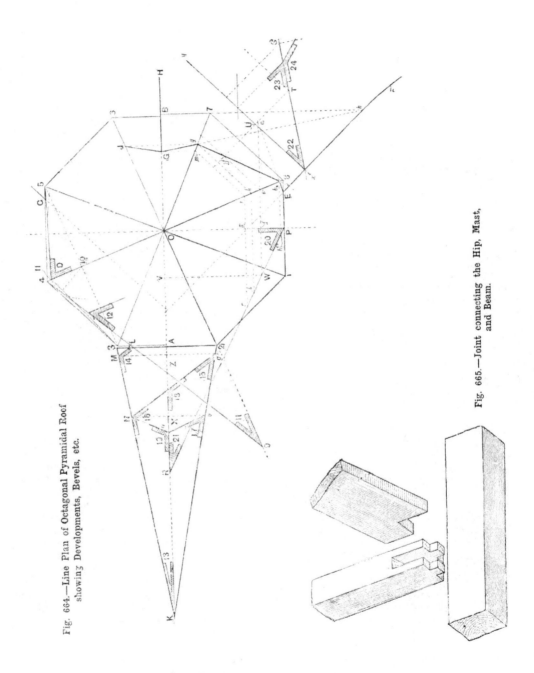

Fig. 664.—Line Plan of Octagonal Pyramidal Roof
showing Developments, Bevels, etc.

Fig. 665.—Joint connecting the Hip, Mast,
and Beam.

b project up to meet the line P R in b', make
A Z equal to P b', then through Z draw line
g' Z M parallel to A 3; next make 3 L equal to
E 8. Then 2 L M K is the true shape of the
side 1 E b O. From g draw a parallel to 7 8,
meeting 8 O in k, then from the point k draw
a line parallel to 1 8, and continue it to
meet P R in l. Now measure off on
the line A K a distance A X equal to P l.
Through X draw a parallel to 2 3, meeting 3 K
in point N. Join N q, then N q K is the true
shape of the surface g b O. From G draw
G m n s parallel to B 7 8 P. From s project
up to meet P R in r. Make A u equal to r P.
Join N u and u v; then N u v K is the true
development of the surface J G g O.

Bevels of Parts, and Backing of Hips.—
The method of obtaining the bevels of the
several parts may now be described, the
means of obtaining the backing of the hips
being first shown. At right angles to O 4
set up O 9, and make it equal to the height;
join 4 9, which gives the true rake of the
hips. At 11 is shown the bevel for the
vertical cut of the top, and at 12 that for
the foot. As will be seen, one edge of this
bevel is adjacent to the pitch line; the
other, being horizontal, is drawn parallel to
O 4. For the backing of the hips, join 3 5,
and from where this line meets O 4 in point
10, draw an arc tangent to the pitch line
4 9. From where the arc meets 4 10 in point
11, join to 5 and 3 as shown; then D is the
bevel required. The bevel for where the
hips meet each other is shown at 13. Reference to Fig. 662 will show where this bevel
will be required, and also that the upper
part of the mast or central post is octagonal.
This allows the upper cuts of the hips to be
made square through their thickness, and
therefore no bevel is required. The development of the intersection shown at
3 M N K v q 2 gives us the bevels for the feet
of the hips and rafters A, B, C, D, E, F, and
G (Fig. 661). The bevels 14, 15, 16, and 17
(Fig. 664) are for the feet of the hips A, B, E,
and F respectively (Fig. 661). These bevels
are for application after the hips have been
backed. The bevels to apply to the backs
of the feet of the jack rafters at C and D
(Fig. 661) are shown at 18 (Fig. 664), whilst
the bevel for the foot of G (Fig. 661) is shown
at 19 (Fig. 664). It will be noticed that

the valley rafters shown by 1, 2, 3 (Fig. 661)
have their upper edges in the same plane as
the main roof; therefore it will be necessary
to obtain a bevel for the preparation of these
edges. The geometrical construction for this
is as follows :—From any point in the plan
of the valley, as H (Fig. 661), draw the horizontal line H K and draw H L at right angles
to 2 3; then at any point in H L draw X Y at
right angles to it, and cutting line H K in K.
From where H K cuts the pitch of the roof
as shown at M, draw M N at right angles to
H K. Then from L drop a perpendicular to
M N as shown. Next project L P at right
angles to X Y, and make it equal in length to
N O. Now join K P; then with L as centre,
draw an arc tangent to K P, meeting X Y
in R. Join R H, which will give the bevel
required as shown at 25. The bevels for
application to the sides of the jack rafters
are shown at 20 and 21 (Fig. 664), whilst the
bevel at 13 is for application to the tops of the
jack rafters. The methods of obtaining the
bevels for the jack rafters for the main roof
are shown by 22, 23, and 24 (Fig. 664)
respectively.

Further Constructional Details.—The hip
of the main roof, as will be noticed, requires
supporting at the lower end. This is done
by placing a beam across the octagonal
space, as shown at S and S' (Figs. 661 and
662). Then the end of the hip is bird's-
mouthed on to this beam, an isometric
detail of which is shown at Fig. 665; this
figure also shows how the mast forks over
the end of the hip, and has two stub tenons
fitting into mortises made in the beam so as
to keep the whole secure. The ceiling
joists in the octagonal space are built into
the walls as shown in plan (Fig. 661) and
section (Fig. 662). Of course, as is usual,
the ceiling joists under the main roof run
parallel to one of the front walls. The
ends of four of these (U, V, W, and Z) cannot
be carried to any wall, therefore a trimmer
is provided of stouter scantling to carry
these ends, as shown in the plan and section
(Figs. 661 and 662). The boarding is clearly
shown in Figs. 659 and 660, and therefore
does not require further description. There
are other little points which are fully shown
in the illustrations, but it is not necessary
to enlarge upon them here.

FRAMEWORK OF DORMER WINDOWS.

Introduction.—A dormer window is a window formed in a sloping roof, the window piercing through the incline and having its framing erected vertically on the rafters. Dormer windows are much in use in modern buildings, to give adequate light to rooms

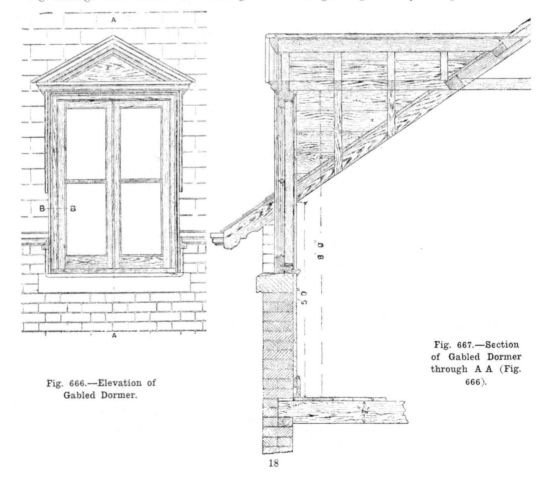

Fig. 666.—Elevation of Gabled Dormer.

Fig. 667.—Section of Gabled Dormer through A A (Fig. 666).

18

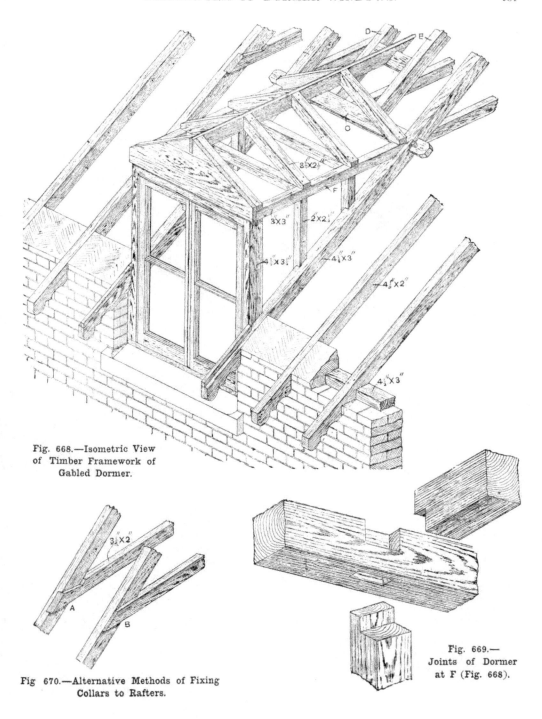

Fig. 668.—Isometric View of Timber Framework of Gabled Dormer.

Fig 670.—Alternative Methods of Fixing Collars to Rafters.

Fig. 669.—Joints of Dormer at F (Fig. 668).

formed in the roof, and to afford a good view of the surrounding district. They are also used to add to the architectural effect of

Gabled Dormer.

Figs. 666 to 672 show the leading details of construction of a dormer window, the

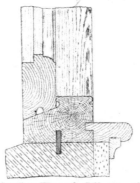

Fig. 671.—Section Through Sill, etc., of Dormer.

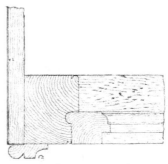

Fig. 672.—Enlarged Section of Dormer through B B (Fig. 666).

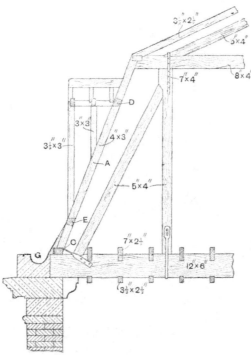

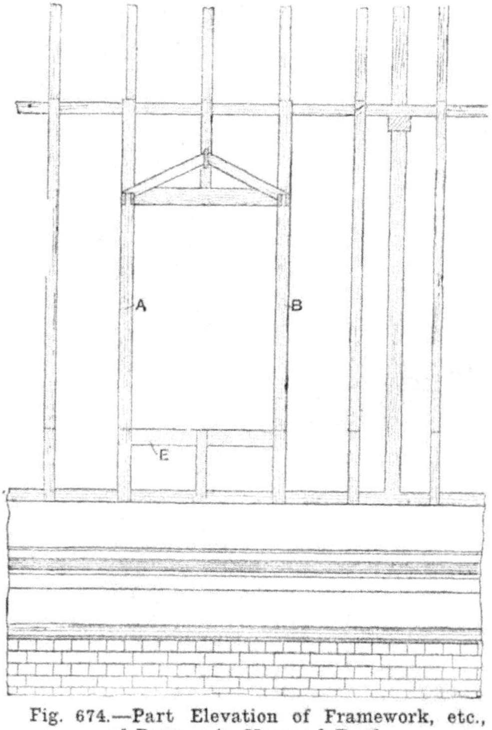

Fig. 673.—Section of Dormer through Portion of One End of Roof.

Fig. 674.—Part Elevation of Framework, etc., of Dormer in Mansard Roof.

the roof. The construction differs in various parts of the country, as will be gathered from the typical illustrations given in this section.

lower part of which is in the front wall of the house; the oak sill of the casement frame rests on the stone sill, connected to it by a water bar, as shown. The upper part of the window is constructed mainly of wood, the exposed woodwork being moulded as shown. The roof and the sides of the dormer are prepared for covering with lead or zinc. The advantage claimed for this kind of dormer, as compared with those situated wholly in the roof, is a larger window

view (Fig. 668) clearly shows the relation and arrangement of the several timbers. The trimming piece shown at c is tenoned through the two main rafters and keyed. The two rafters at D and E are stub-tenoned into the trimmer. Fig. 669 shows the joints at F (Fig. 668). At A in Fig. 670 the collar

Fig. 675.—Oblique Projection of Framework of Dormer in Mansard Roof.

opening and a higher ceiling to the roof next the front wall (see Fig. 667). The roof has overhanging eaves, with the feet of the rafters moulded as shown, an ogee gutter being fixed. The form of the exposed woodwork is shown in the elevation given (Fig. 666). Fig. 667 is a section through A A (Fig. 666), which exhibits the construction of most of the parts. The isometrical

is shown dovetail-lapped into the rafter. At B another method is shown, the collar being halved and lapped on to the rafter. Fig. 671 shows the oak sill, bottom rail of casement, nosing connection to stone sill, etc. At Fig. 672 is given an enlarged detail through B B (Fig. 666), showing the joint between the frame and the stile of the casements, also of the boarding nailed to the frame and covered with lead or zinc. After this boarding has been fixed, the moulding G is attached as shown. The dimensions shown of some of the principal parts should

to 675. In constructing the framework, the following are the principal points requiring attention. The wall is 18 in. thick, and is finished with a stone cornice. The top

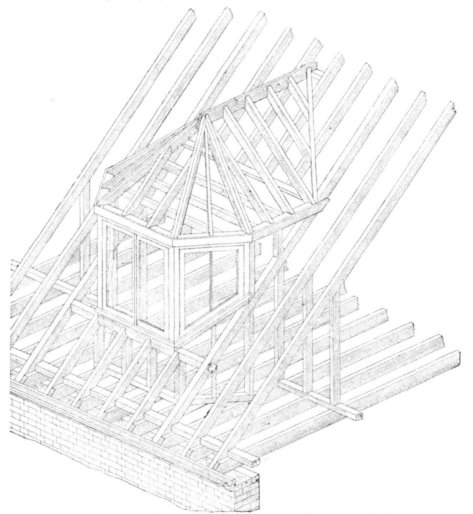

Fig. 676.—General View of Complete Framing of Bay Dormer Window.

be regarded as suggestive rather than absolute.

Small Dormer Window in Mansard Roof.

The bare framework of a small dormer in a Mansard roof is illustrated by Figs. 673

surface of this is hollowed out of the solid stone, and afterwards lead is dressed in so as to form the gutter (see G, Fig. 673, which shows the brickwork, stonework, and timber framing). The main tie-beam is supported by a stone corbel as shown. This beam also acts as a girder to support the floor and

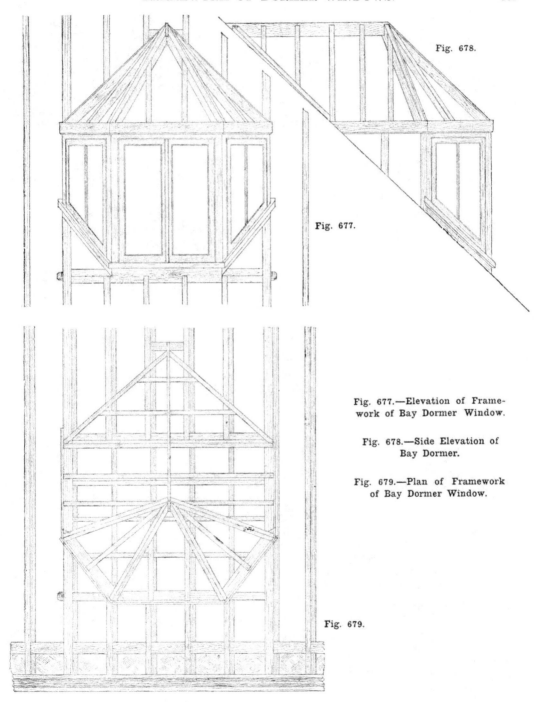

Fig. 678.

Fig. 677.

Fig. 677.—Elevation of Framework of Bay Dormer Window.

Fig. 678.—Side Elevation of
Bay Dormer.

Fig. 679.—Plan of Framework
of Bay Dormer Window.

Fig. 679.

ceiling joists. The former are cogged to the tie-beam, but the latter need only be notched on. The two rafters A and B (Fig. 674), to which the framing of the dormer is attached, are thicker than the other rafters. The opening between these two rafters is formed by the trimming pieces D and E. The lower wall plate is held fast to each tie-beam by iron straps as shown at C (Fig.

681 sufficiently show the construction, and therefore only the leading points need be mentioned. The joists rest on a wall plate in the usual manner, and a second plate, supporting the rafters, is fixed to the upper edges of the joists, the ends of the rafters also being attached to the joists. The gutter is of wood, supported on wooden bearers built into the wall as shown at Figs.

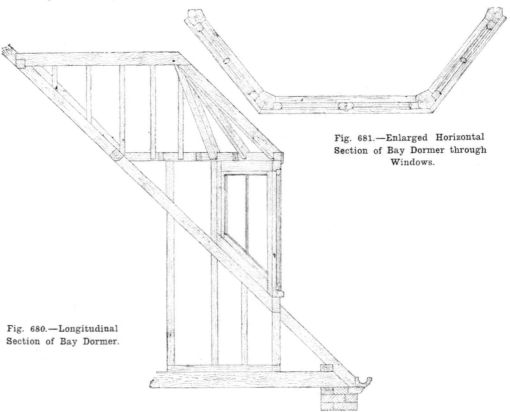

Fig. 681.—Enlarged Horizontal Section of Bay Dormer through Windows.

Fig. 680.—Longitudinal Section of Bay Dormer.

673). Suggestive sizes are figured on the different members, but of course these would vary according to circumstances.

Bay Dormer Window.

At Fig. 676 are shown conventionally the timbers connected with the framing of a bay dormer window, ready for boarding, battening, slating, leadwork, etc. The sidelights of the window are fixtures, the front casement opening outwards. Figs. 676 to

676 and 680. At Fig. 676, in order that the construction may be clearly shown, the brickwork is carried up level with the lower plate only ; but, of course, the wall when completed would be finished flush with the tops of the rafters. The first trimmer for the dormer would be formed by the two stout rafters, having a trimmer fixed at the lower end of the window and another at the ceiling level of it, the short rafters being fixed to these trimmers as shown.

North Country Style.

The arrangement next shown is that most generally favoured in the North of England, roof timbers are trimmed to give the necessary opening, and two stout raking pieces, 9 in. by 3 in., are provided to carry the sides of the dormer where the latter is of large size. These rakers are notched (prefer-

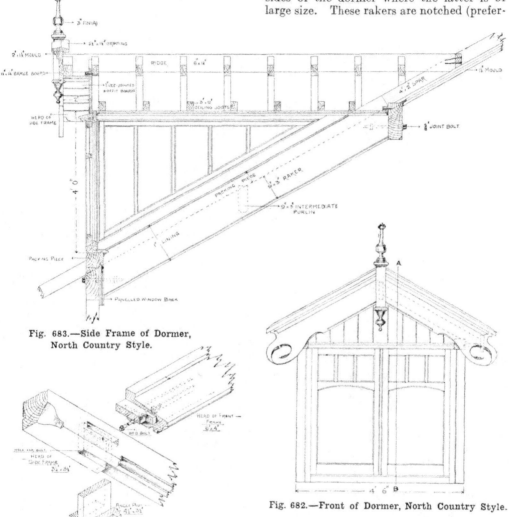

Fig. 683.—Side Frame of Dormer, North Country Style.

Fig. 684.—Joints in Angle Post.

Fig. 682.—Front of Dormer, North Country Style.

where side-lights and casement sashes are required. In constructing the window, the ably dovetailed) to purlins as will be illustrated, and are fixed with joint bolts at their head and foot. Any intermediate purlin can be trimmed into these rakers where necessary, and the design, as well as the dimensions, can be easily adapted to suit local conditions or the special requirements of any given case.

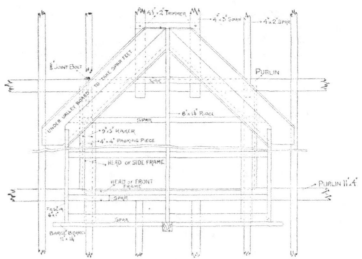

Fig. 685.—Construction of North Country Dormer Window.

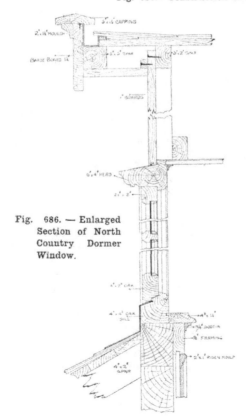

Fig. 686. — Enlarged Section of North Country Dormer Window.

Construction of North Country Dormer Window.

The dormer consists of a front (Fig. 682) and two side frames (Fig. 683), the latter being made to the slope of the roof, and being rebated for the insertion of a 2-in. sash from the outside. The stile of the side frame serves as the angle post, and in a small dormer the head and sill of the front frame are tenoned directly into it. In a larger dormer, however, the method here shown is much more convenient to adopt, especially where there are muntins in the front frame, as the latter is made as a separate frame, with light stiles tenoned to the head and sill, which are tenoned to the angle posts and secured and drawn tight with bed bolts as shown in detail (Fig. 684). When fixed, the stile of the front frame should also be screwed through the rebate to make a perfectly close joint between the angle post and the front frame. The roof in this case is made to overhang all round (see Fig. 685). The ridge and heads of the side frames are allowed to project beyond the gable, to carry the overhanging spars, bargeboards, and finial. The overhanging sides are formed by the spars projecting to the required amount; the spar feet are covered on the soffit by a soffit board tongued into the head of the frame and into a fascia board

nailed to the ends of the spars. The bargeboards are shaped and pierced, and are
provided with a double-splayed capping, into
which the bargeboards and bed mould are
housed, and which projects over the gutter

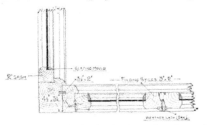

Fig. 687.—Detail Horizontal Section through
Angle and Sashes.

behind the gable (see detail, Fig. 686). The
gable and the overhanging portion of the
front of the dormer are boarded with 1-in.
tongued, grooved, and V-jointed boards.
The front frame is fitted with casements that
open outwards (see Fig. 687), and are hung
with 3½-in. brass butts. The interior is

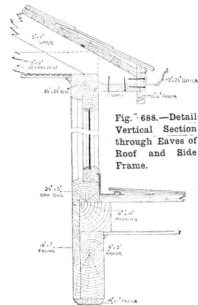

Fig. 688.—Detail
Vertical Section
through Eaves of
Roof and Side
Frame.

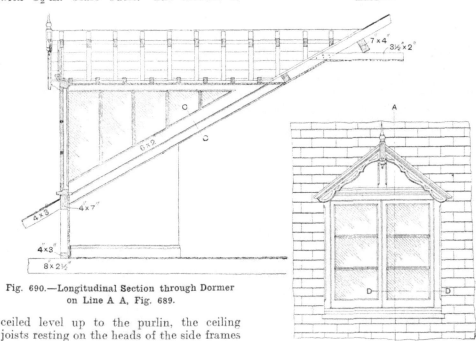

Fig. 690.—Longitudinal Section through Dormer
on Line A A, Fig. 689.

ceiled level up to the purlin, the ceiling
joists resting on the heads of the side frames
and nailed to them, the whole being lathed
and plastered and a small scotia mould fixed
in the angle (see detail, Fig. 688). All the

Fig. 689.—Front Elevation of Dormer.

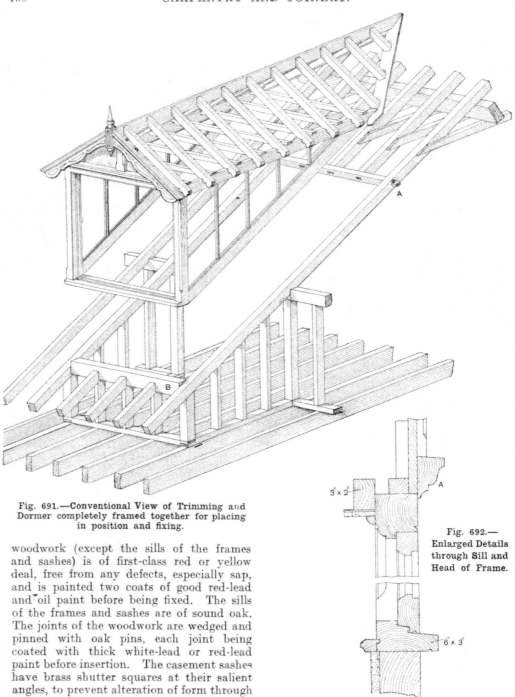

Fig. 691.—Conventional View of Trimming and
Dormer completely framed together for placing
in position and fixing.

Fig. 692.—
Enlarged Details
through Sill and
Head of Frame.

woodwork (except the sills of the frames
and sashes) is of first-class red or yellow
deal, free from any defects, especially sap,
and is painted two coats of good red-lead
and oil paint before being fixed. The sills
of the frames and sashes are of sound oak.
The joints of the woodwork are wedged and
pinned with oak pins, each joint being
coated with thick white-lead or red-lead
paint before insertion. The casement sashes
have brass shutter squares at their salient
angles, to prevent alteration of form through

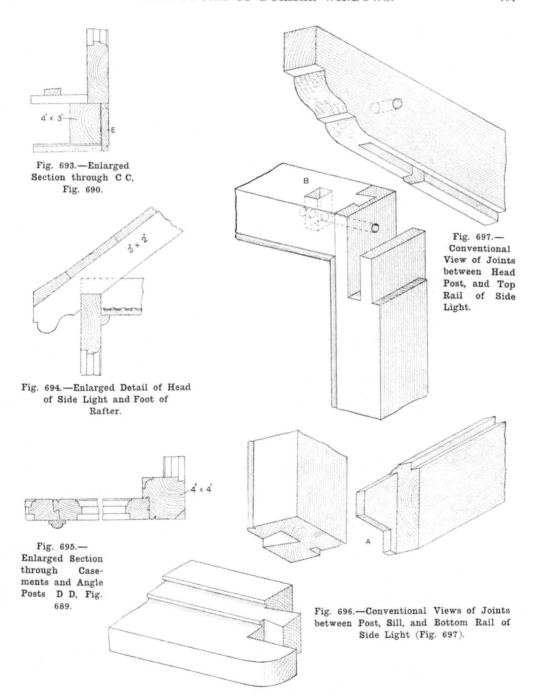

Fig. 693.—Enlarged
Section through C C,
Fig. 690.

Fig. 694.—Enlarged Detail of Head
of Side Light and Foot of
Rafter.

Fig. 695.—
Enlarged Section
through Case-
ments and Angle
Posts D D, Fig.
689.

Fig. 697.—
Conventional
View of Joints
between Head
Post, and Top
Rail of Side
Light.

Fig. 696.—Conventional Views of Joints
between Post, Sill, and Bottom Rail of
Side Light (Fig. 697).

the weight of the glazing. It is not in- tended to describe the construction of sashes in this chapter, that subject being reserved for exhaustive treatment later.

represents the elevation, Fig. 690 being a longitudinal section. It will be seen that the front consists of a frame with casement —sashes opening outwards—and the side is

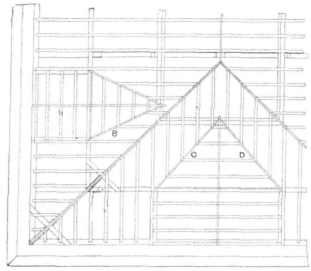

Fig. 699.—Plan of Naked Framework.

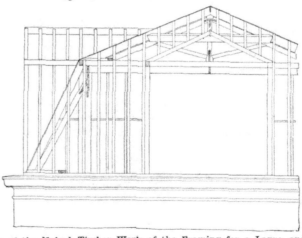

Fig. 698.—Elevation of the Naked Timber Work of the Framing for a Large and a Small Dormer, in the Side and End of Hipped End Mansard Roof.

Another North Country Dormer Window.

The kind of dormer that is used where it is desired to gain as much light as pos- sible is shown by Figs. 689 and 690. Fig. 689

framed with bars for glazing. At Fig. 691 the trimming is clearly shown; on each side of the opening a stout rafter, 4 in. by 3 in., is provided, and at the top of the opening a trimmer is tenoned through the rafter and keyed (Fig. 691). This trimmer is mortised

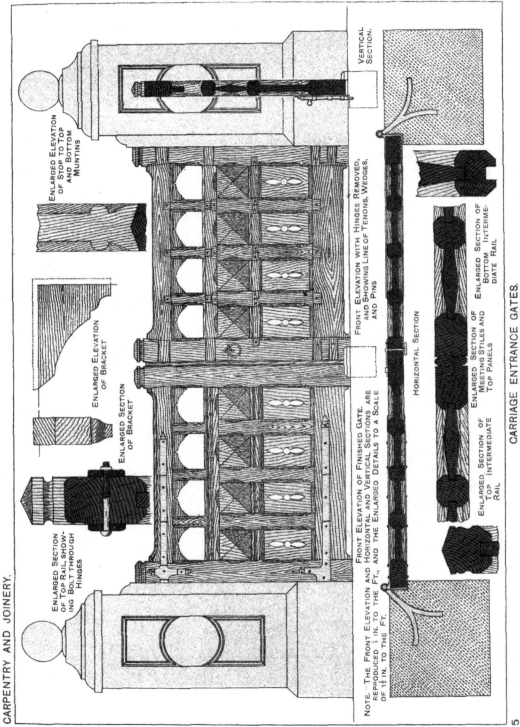

CARPENTRY AND JOINERY.

CARRIAGE ENTRANCE GATES.

ENLARGED SECTION OF TOP RAIL SHOW-ING BOLT THROUGH HINGES

ENLARGED SECTION OF BRACKET

ENLARGED ELEVATION OF BRACKET

FRONT ELEVATION OF FINISHED GATE.

ENLARGED ELEVATION OF STOP TO TOP AND BOTTOM MUNTINS

VERTICAL SECTION.

FRONT ELEVATION WITH HINGES REMOVED, AND SHOWING LINE OF TENONS, WEDGES, AND PINS

HORIZONTAL SECTION

ENLARGED SECTION OF TOP INTERMEDIATE RAIL

ENLARGED SECTION OF MEETING STILES AND TOP PANELS

ENLARGED SECTION OF BOTTOM INTERME-DIATE RAIL

NOTE.—THE FRONT ELEVATION AND HORIZONTAL AND VERTICAL SECTIONS ARE REPRODUCED ½ IN. TO THE FT., AND THE ENLARGED DETAILS TO A SCALE OF 1⅛ IN. TO THE FT.

5

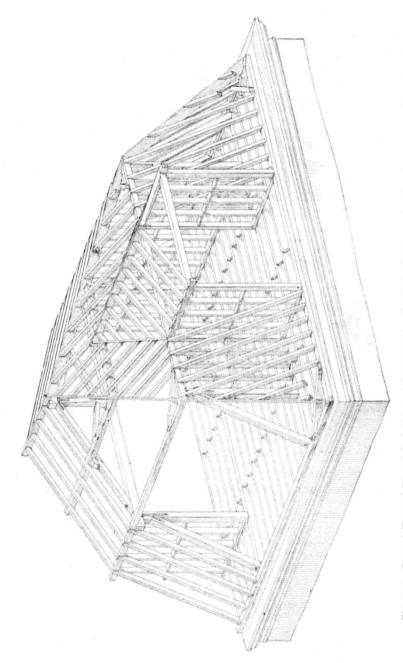

Fig. 700.—General View of Hipped End Mansard Roof, showing the Main Members, also Framing, etc., for Large Dormer in the End of Roof.

to receive the upper rafters as shown at A. A deeper piece of stuff is used for the trimmer, B, against which the lower rafters butt. This trimmer is fixed to the stout rafters, and is allowed to project as shown so as to support the sill of the dormer. Fig. 691, which is a conventional view of the dormer framed together ready to be fixed in its proper position, also shows how the stout rafters are supported by ashlering studs which are fixed to them.

or bed screws may be used if desired. Fig. 697 shows the connections for the post, the head, and the top rail of light, the joints being firmly held together by a bolt inserted from the outside of the top rail. This bolt is passed through into the head in which has been made a mortice to receive a nut, as indicated at B, Fig. 697. The bottom and top rails of the side lights are connected by mortice and tenon joints at their upper end The top rails are made to

Fig. 701.—View of Stone Dormer in End of completed Mansard Roof.

Constructional Details.

The preparing of a dormer of this description is the work of a joiner rather than of a carpenter. A few leading particulars of the construction will now be given. The angle posts are rebated and beaded to receive casement sashes ; they are also chamfered on the outside and moulded on the inside as shown by the enlarged section, Fig. 695. These angle posts are connected to the double sunk oak sill in the way represented at Fig. 696, and to the bottom rail of the side light by a barefaced haunched tenon as shown at A in the same illustration. These joints may be held together more firmly by the insertion of stout screws 5 in. long,

project beyond the posts so that the lower ends of the bargeboards may be fixed to them. The ridge also projects, and is tenoned into the finial to which the upper ends of the bargeboards and moulding are butted and fixed. The top end of the top rail and end of ridge piece are connected by means of two pieces of $\frac{7}{8}$-in. boards. The ceiling joists of the dormer are notched down on the top rails of the side lights and nailed. The rafters or spars are cut to fit the ridge, and are notched on to the top rails as shown. The two pieces of board before mentioned receive the ends of the small jack rafters. The gabled part of the dormer is formed by a chamfered muntin, and is plain boarded on each side. Just above the head of the frame a moulding of the section shown at Fig. 691 is planted on as a finish. The joints of the framing should be well coated with white-lead and red-lead before they are put together. In better-class work immediately after fixing the dormer its roof is boarded as represented in section at Fig. 690. The bottom rails of the side lights

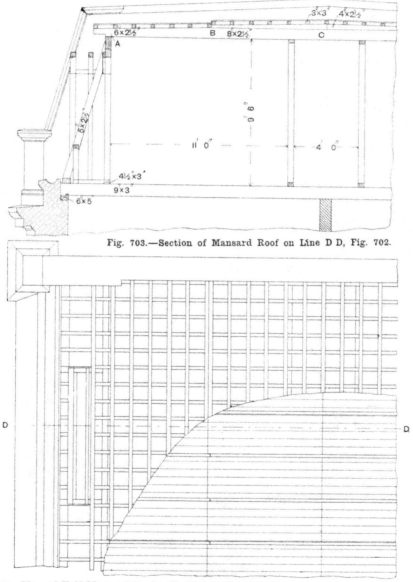

Fig. 703.—Section of Mansard Roof on Line D D, Fig. 702.

Fig. 702.—Plan of Half Mansard Roof with Flat (part boarded), showing Trimming to Dormer, Rolls, Joists for Boards, Bridging Joists, Rafters, etc.

rest upon and are fixed to the stout rafters, and the finish on the inside is formed by fixing a beaded lining as shown in section at E, Fig. 693. The leading dimensions of the various parts are figured on the illustrations.

Stone Gabled Dormer.

A more important case with regard to dormers is shown by Figs. 698 to 701. Reference to Figs. 698 and 699 will show that the side of the hipped end of a Mansard

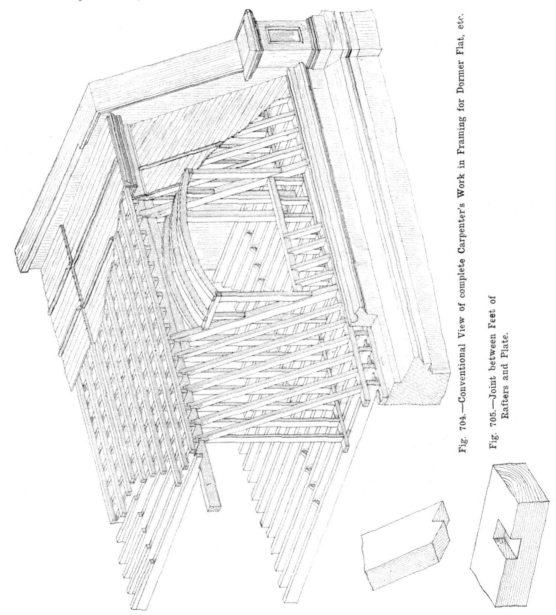

Fig. 704.—Conventional View of complete Carpenter's Work in Framing for Dormer Flat, etc.

Fig. 705.—Joint between Feet of Rafters and Plate.

roof has been designed to provide for a large stone-gabled dormer in the end, a general view of which is given at Fig. 701. A dormer of equal height but narrower is provided for on the side. Framing for these dormers is of such magnitude as to necessitate the provision of valley rafters, A, B, C, and D, Fig. 699. The framing for the smaller dormer is similar to that for the larger one; therefore it has not been

with one piece, or smaller sheets may be used with rolls, but these are not shown. This is an important example, both as regards roofing and dormers; but further description is thought unnecessary, as careful attention has been given in the preparation of the illustrations to show clearly all the essential points of construction, and therefore they should present no difficulty to the careful reader.

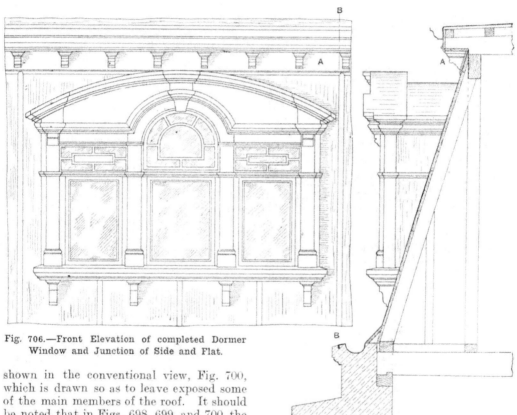

Fig. 706.—Front Elevation of completed Dormer Window and Junction of Side and Flat.

Fig. 707.—Sectional Elevation through B B, Fig. 706.

shown in the conventional view, Fig. 700, which is drawn so as to leave exposed some of the main members of the roof. It should be noted that in Figs. 698, 699, and 700 the masonry is only carried up level with the wall plate; the parapet and gutter bearers and boarding are omitted so as to make clear the more important construction. Fig. 701 is a conventional view of the stone gable, which also shows the two inclined surfaces of the main roof slated and with moulded fascia and guttering, with a lead apron under. The roof and side of the dormer are shown covered with lead. Each part may be covered

Dormer in Half Mansard with Flat Top Roof.

Unlike the Mansard roof proper, this kind has practically no upper pitch, nor trusses, rafters, etc., to the upper portion, and thus there is no loss of space. It is used largely where intermediate walls or partitions for support are available to assist in carrying the flat, and also where it is desired to get as large a room space as possible. Dormers are almost always framed, and form an important part of the construction of this kind of roof. Figs. 702 to 707 fully show the construction of this description of a roof with dormers, etc. The following are the general particulars. The bridging joists of the floor are cogged and nailed on to a rebated wall plate, or which has a fillet nailed on it. The plate to receive the lower ends of the rafters is notched and secured to the top edges of the joists as indicated, the rafters being notched into this plate as shown at Fig. 705. The curb plate is supported by the rafters, and also by the studding as shown, the latter of course tenoning into the under side of the plate. The bridging joists for the flat are out of 8-in. by 2½-in., and to produce the necessary fall the first 7 ft. is tapered from 6 in. to 8 in. (see A to B, Fig. 703), then the remaining 7 ft. as a firring piece 2 in. thick, increasing to 4 in. thick, nailed on as shown at B to C (Fig. 703), the drip being provided for at B. Scantlings are nailed on to the firring described, forming the joists on which to

nail the boarding for the flat. This boarding should be 1¼ in. thick, grooved and tongued, and cleaned off smooth to receive the lead. The boarding should always be fixed running parallel with the fall of the flat, so that in the event of any of the boards curling up and thus forming hollows in the lead, the rain will not be retained in puddles, as its flow is not interfered with. Grooved and tongued 1¼-in. boarding is nailed diagonally on the rafters to receive the lead, as shown at Fig. 704. The curb plate having to span 8 ft. over the dormer, it is strengthened by a 6-in. by 4½-in. lintel bolted to its under side and supported by the angle studs, as indicated at Fig. 704. The front elevation and side elevation of the completed Venetian dormer window are shown at Figs. 706 and 707. The ceiling of this is arched; the ribs to carry this and the boarding are shown at Fig. 704. The framework to receive the completed window is fully shown at Fig. 704. The junction between the side and flat is finished with an upper fascia and a lower fascia, and soffit boards, moulded modillions, guttering, etc., as shown in elevation and section (Figs. 706 and 707). The construction of the dormer frame, casements, etc., will be treated of in a subsequent chapter, where a number of detail illustrations will be given.

HALF-TIMBER CONSTRUCTION.

Introduction. — In substantial half-timber work, English oak is used, but sound, resinous pitchpine or Scotch pine is often

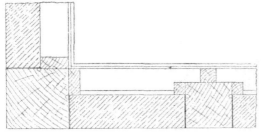

Fig. 708.—Section showing Front filled in with 4½-in. Brickwork Back-lathed and Plastered.

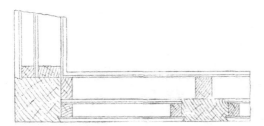

Fig. 709.—Section showing External Stucco, Middle Coat of Lath and Plaster, and Internal Lathing and Plastering.

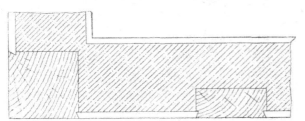

Fig. 710.—Section through Post filled in with 9-in. Brickwork —Front Stuccoed ; Inside Plastered.

substituted. Head and sill should run through and be framed with the angle posts. Mullions should be tenoned into the horizontal members, and secured by draw-boring, the pins being oak ¾ to 1 in. in diameter, split and then shaped, and allowed to project from the surface about ¾ in. All curved braces should be made from natural curved (compass) timber. Diagonal braces should be halved together at their centres

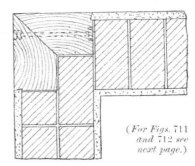

(*For Figs.* 711 *and* 712 *see next page.*)

Fig. 713.—Alternative Method of forming Angle Post of Two Pieces mitred and bolted together.

The angle posts are usually 8 in., 9 in., or 10 in. square, and the intermediate posts the same width, but 4 in. or 6 in. thick. The head pieces and sill pieces are of the same scantling as the angle posts, and are halved at the angles, and mortised for the tenons. The exposed faces of the timbers are wrought and oiled or

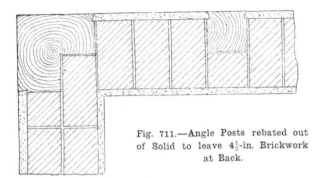

Fig. 711.—Angle Posts rebated out of Solid to leave 4½-in. Brickwork at Back.

Fig. 712.—Angle Post formed of Two Pieces butting and bolted together.

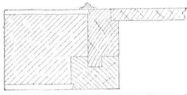

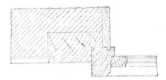

Fig. 714.—Method of fixing Door Frame in Half-timbered Work.

Fig. 715.—Method of connecting Window Frames in Half-timbered Work.

(For Fig. 713, see previous page.)

Fig. 716.—Front Portion of Half-timbered Cottage.

painted, red-lead mixed with boiled oil being used for all joints. Various forms of half-timber work are shown in section by Figs. 708 to 713 (scale = 1 in. to 1 ft.). In Fig. 708 the angle posts shown are 8 in. by 8 in., and the intermediate posts 8 in. by 4½ in., grooved at the sides, and

ensures greater warmth and dryness. The fillets and battens are fixed to the sides and backs of the timbers to receive the lath and plastering. The intermediate studs are shown to be rebated for the middle coat of plaster, but this rebating is not indispensable. In the example represented

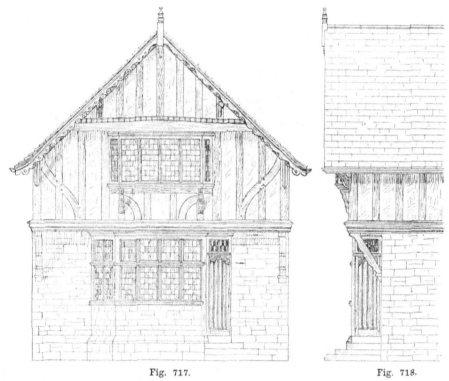

Fig. 717.

Fig. 718.

Fig. 717.—Front Elevation of Half-timbered Cottage.
Fig. 718.—Part Side Elevation of Half-timbered Cottage.

filled in with 4½-in. brickwork, which is rendered at the back with cement, while small fillets fixed to the sides of the uprights keep the brickwork in position. Battens are fixed to the backs of the timbers to take the lath-and-plaster work. Sometimes the horizontal timbers are covered on the upper side with sheet lead. Fig. 709 shows the angle posts 6 in. square, and the intermediate ones 6 in. by 3 in., grooved along the outer edge for external stucco or roughcast. A middle coat of plaster shown

by Fig. 710 the timbers are filled in with 9-in. brickwork, with roughcast face. The angle posts are 9 in. square, and the intermediate posts 9 in. by 4 in., splay-grooved along the outer edge. This form is very substantial, and some building bye-laws demand it. According to some bye-laws, there must be at least 4½ in. of brickwork behind all timber; then if the angle post is above 5 in. by 5 in., it must be rebated at the back (see Fig. 711), or it must be formed of two pieces bolted together as shown in

9*

section by Fig. 712, or mitred and bolted together as in Fig. 713 (p. 199).

Fixing Door= and Window=frames.

Figs. 714 and 715 (scale = 1 in. to 1 ft.) show common methods of fixing the door-

Half=timbered Cottage.

The application of half-timber work to a cottage is shown in Figs. 716 to 721. Figs. 717 and 718 are reproduced to a scale of ⅛ in. to 1 ft. The following would be the

Fig. 719.—Conventional View of naked Timber Work fitted together.

and window-frames in half-timber work. The posts are in each case rebated for the frame, and sometimes extend the full thickness of the wall.

leading points in the specification :—The timber (oak, fir, or pitchpine) to be of sound quality, without defects, thoroughly seasoned, and wrought on the exposed sides.

Angle posts tenoned to the head and sill, and the sills, 9 in. by 9 in., to be halved at the angles (see Figs. 719 and 720). All intermediate posts to be 6 in. by 5 in., or, as shown, every third post to be 6 in. by 4½ in. and the others 6 in. by 3 in. The front braces to be same thickness as the posts, as shown. The head to be from 9 in. by 4 in. to 9 in. by 9 in. The joists to project beyond the stonework below, and tusk tenon into the sill. The sills to be connected by wrought-iron angle plates and bolts. When oak pins are used, two are usually inserted at each joint. The gable overhanging the filling in must be of light character, as shown by Fig. 708 or Fig. 709, in the event of the window below being principally constructed of wood. In some districts it is compulsory for 9 in. of brickwork to be fitted in the gable as well as the sides, and

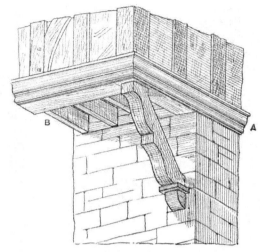

Fig. 721.—General View of Bracket at Angle and Under Side of Joists, etc.

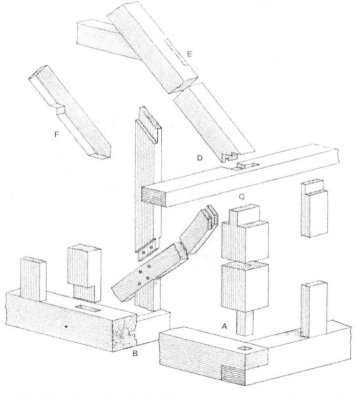

Fig. 720.—Conventional View showing Principal Joints in Fig. 719

therefore the window would have to be built of stone with mullions strong enough to give the necessary support. The sills (shown as moulded) are supported at the

the gable, a moulding being planted on the upper edge. The heads of the side framings project, and are supported by brackets. These heads and also the purlins and ridge

Fig. 722.—
South Elevation of
Half-timbered House.

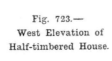

Fig. 723.—
West Elevation of
Half-timbered House.

angles by wooden bracket pieces whose bottom ends rest on stone corbels. In the best class work the sills are moulded on the solid, but more frequently the moulding is planted on (see A, Fig. 721); the meeting surfaces should be well painted. A projecting transom is shown in the upper portion of

support the first rafter, bargeboard, and finial. Fig. 719 shows a part of the naked framework fitted together, and Fig. 720 shows the various joints in the framework. Fig. 721 is a general view of the bracket supporting the angle, also the under side of the joists.

Design for Half-timbered House.

The south elevation (Fig. 722), west elevation (Fig. 723), sectional elevation (Fig.

wood framing, and the open panels plastered on to lathing, with grooves in the sides and top of the timber, so as to give a key for the plaster. The face of the plaster is kept

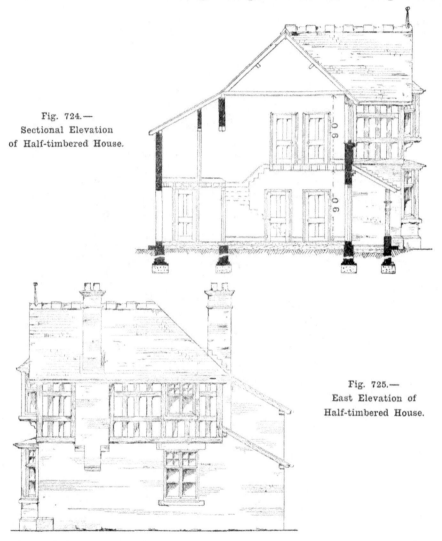

Fig. 724.—
Sectional Elevation
of Half-timbered House.

Fig. 725.—
East Elevation of
Half-timbered House.

724), and east elevation (Fig. 725) show a half-timbered house suitable for the country or suburbs. Red facing brick is used from the base to the first floor, the remaining portion being half-timbered work. The last-named can be carried out with strong

back about ¾ in. from the face of the wood-work ; the inside is lathed and plastered, or it may be of wood framing, with rough brick nogging between the panels, and then plastered on the front of the bricks, also on the outside of the framing as before.

Old-fashioned Half-timbered Gable.

Figs. 726 to 729 show part of the half-timber work for a house, the design being based on old-fashioned examples. The gable projects, and is supported by the joists, which overhang and are tenoned into the should be set back 1 in. to 1½ in. from the face of the posts. The members of the framing should be grooved to receive the plaster panel. The gable bargeboard is cut out of 2-in. stuff and chamfered. Fig. 726 is a general view, Fig. 727 a front elevation, and Fig. 728 a vertical section.

Fig. 726.—General View of Half-timber Gabled House based on Old Design.

sill, and some additional support is rendered by the three solid wooden brackets, two being bolted to the posts of the porch and the third built into the lower masonry. A small projecting oriel window is supported on brackets as shown. The woodwork should be cut from sound dry balk timbers, mitred and tenoned together, and secured with ¾-in. oak pegs, which should project about 1 in. from the face. The wood framing is shown backed with brickwork, which

Sham Half=timber Work.

Sham half-timber work (Figs. 730 and 731) is formed of pieces of scantling only 1½ in. or 2 in. thick, but as wide as the timbers used in real half-timber work. The pieces are mortised and tenoned together, and often pinned as shown. Often the whole framing is set up in position, the brickwork carried up against it, and strips of wood or wooden bricks, which are fixed to the backs of

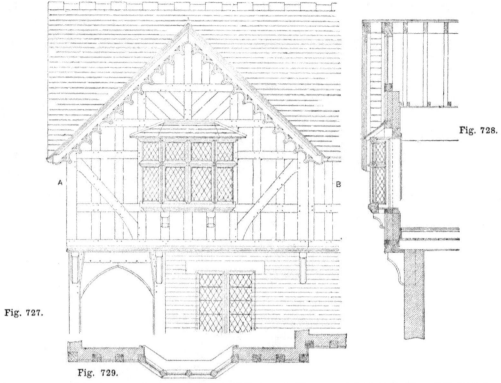

Fig. 728.

Fig. 727.

Fig. 729.

Fig. 727.—Front Elevation of Half-timber Gabled House.   Fig. 728.—Vertical Section of Half-timber Gabled House.   Fig. 729.—Section of Half-timber Gabled House through A B.

Fig. 732.—Method of supporting Angle by Ornamental Wooden Corbel.

the members, are bonded in. Probably this is the best method. An alternative is first

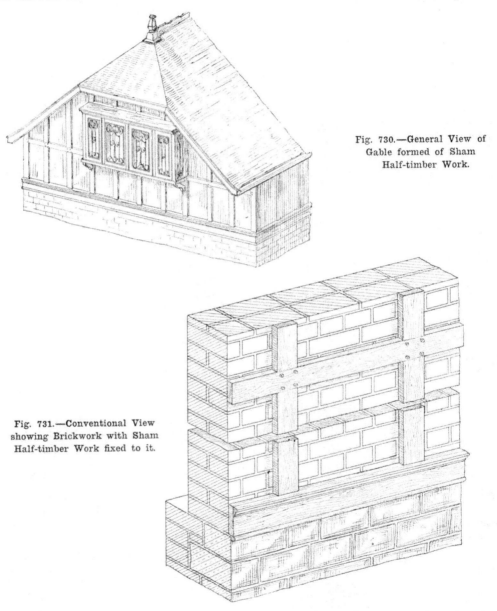

Fig. 730.—General View of Gable formed of Sham Half-timber Work.

Fig. 731.—Conventional View showing Brickwork with Sham Half-timber Work fixed to it.

the stucco or roughcast. The sham sill is often finished off with a moulding which

to build the brickwork up, and then to fix the wooden framing to wooden bricks or plugs. The edges are bevelled in to receive

is splayed on its upper edge for weathering. Fig. 732 illustrates a case where the side sill B runs forward and has its end carved. The

Fig. 733.—Conventional View of Projecting Window supported on Moulded Brackets.

front sill c tenons into this. The joists are notched out to project under the sill, and have their ends ornamented as shown. The angle is supported by the moulded wooden corbel A.

Supporting Upper Windows.

One of the general methods of supporting an upper projecting window by shaped brackets fixed to the posts is illustrated by Fig. 733. A carved bracket is shown in Fig. 734. The fixing for the corbels or brackets is obtained by housing the back edge into the posts about 2 in.; where there are no posts below the projecting window, the brackets are built into the wall.

Fig. 734.—Pierced and Carved Bracket for supporting Window.

Gable Treatment: Panelling, Bargeboards, etc.

Fig. 735 shows the upper portion of a gable and part of the side of a half-timbered in Fig. 736. In the case of the two upper storeys in the projecting gable of a house (Fig. 737), the first floor portion is formed of timber work, with brick or one of the other general fillings. The outside is covered with

Fig. 735.—Design for Gable End and Side of House based on Old Examples.

house, with posts, sill, transoms, and intertie, the panelling being partly filled in with ornamental woodwork. The design is based upon a good old example. The treatment of the upper portion of a small gable is shown tiles, which are fixed to oak laths (see Fig. 738). Half-timber work similar to what has already been explained forms the walling of the attic or second floor. Figs. 739 to 756 are designs for bargeboards.

Fig. 736.—Method of finishing Upper Portion of Small Gable.

Fig. 737.—Gabled Front of House with First Floor of Timber Work Tiled, and with Upper Storey of Half-timber Work.

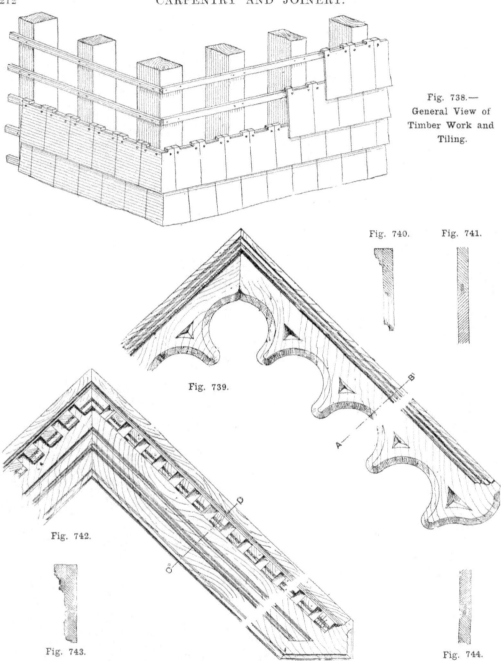

Fig. 738.—
General View of
Timber Work and
Tiling.

Fig. 740.   Fig. 741.

Fig. 739.

Fig. 742.

Fig. 743.   Fig. 744.

Fig. 739.—Bargeboard Ornamented and Chamfered.   Fig. 740.—Section of Bargeboard on A B
(Fig. 739).   Fig. 741.—Section showing Joint of Bargeboard.   Fig. 742.—Elevation of Bargeboard Moulded and Dentils.   Fig. 743.—Section of Bargeboard through D D (Fig. 742).   Fig.
744.—Section of Jointing of Bargeboard (Fig. 743).

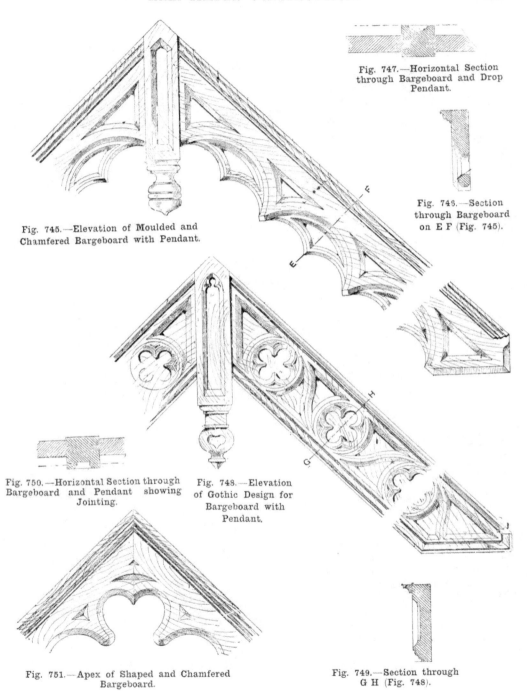

Fig. 747.—Horizontal Section through Bargeboard and Drop Pendant.

Fig. 746.—Section through Bargeboard on E F (Fig. 745).

Fig. 745.—Elevation of Moulded and Chamfered Bargeboard with Pendant.

Fig. 750.—Horizontal Section through Bargeboard and Pendant showing Jointing.

Fig. 748.—Elevation of Gothic Design for Bargeboard with Pendant.

Fig. 751.—Apex of Shaped and Chamfered Bargeboard.

Fig. 749.—Section through G H (Fig. 748).

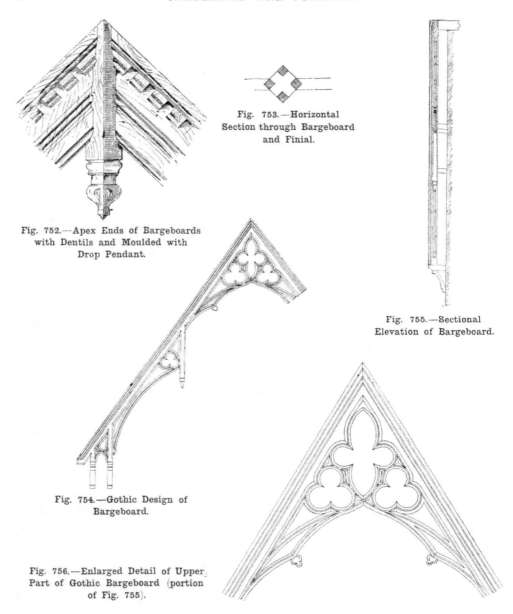

Fig. 752.—Apex Ends of Bargeboards with Dentils and Moulded with Drop Pendant.

Fig. 753.—Horizontal Section through Bargeboard and Finial.

Fig. 755.—Sectional Elevation of Bargeboard.

Fig. 754.—Gothic Design of Bargeboard.

Fig. 756.—Enlarged Detail of Upper Part of Gothic Bargeboard (portion of Fig. 755).

GANTRIES, STAGING AND SHORING.

Builder's Gantry.

A GANTRY, forming a temporary wooden staging, erected over a public footway, is an elevated basis from which building opera-

may be spaced out into spans ranging from 6 ft. to 10 ft. in the length of the gantry, and into one or two bays in the width from building line to kerb. The timber gener-

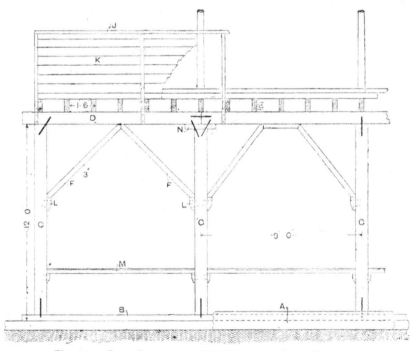

Fig. 757.—Front Elevation of Two Bays of Builder's Gantry.

tions are conducted. A gantry also has in many cases to carry all the front scaffolding of the building; such an instance is illustrated by Figs. 757 and 758. The bays

ally used for standards and heads is square, but commonly any section, from 9 in. by 3 in. up to 12 in. by 12 in., is used. In the accompanying illustrations the sections used are :—Fender A, 12 in. by 12 in. ; sole pieces B, 8 in. by 4 in. ; uprights C and heads D, 8 in. by 8 in. ; joists E, 9 in. by 3 in. ; struts F, 4 in. by 3 in. ; sheeting G, 9 in. by 3 in., or 9 in. by 1½ in. (see H) ; guard frame J, 4 in. by 2 in. ; guard boarding K, 6 in. by ¾ in. ; cleats L, 9 in. by 4 in. by 3 in. ; handrail M, 4 in. by 3 in. ; and impost N, 8 in. by 4 in. The dogs are out of ¾-in. square iron (see Figs. 760 and 761).

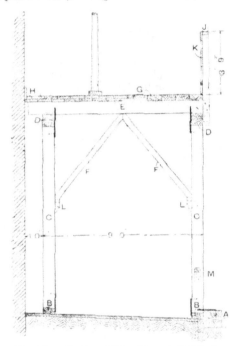

Fig. 758.—Sectional Elevation of Builder's Gantry.

Erecting the Gantry.—In erecting the gantry, the practice is to first lay down the sole pieces, then set out the position of the standards on them. These standards are then cut off to the required length, allowing for the difference in level owing to the fall of the footway. The uprights are now placed in position, dogged to the sole pieces, and temporarily braced with scaffold boards or any other handy material. The heads are next laid on the uprights and dogged to them ; the bridging joists are thrown across the heads and spiked at from 15-in. to 2-ft.

centres. Those coming immediately over the uprights are dogged to the heads with

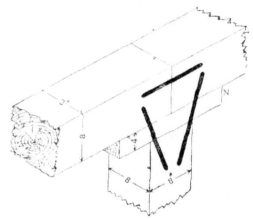

Fig. 759.—Enlarged View of Impost Piece of Builder's Gantry.

those shown at Fig. 761 (sometimes known as " bitches"). The gantry may now be braced as shown in Figs. 757 and 758, the

Fig. 760.— Dog used for   Fig. 761.—Bitch used for
Builder's Gantry.   Builder's Gantry.

latter showing three different methods of cutting the braces in general use. Fig. 759 illustrates an impost piece, used for the pur-

Fig. 762.

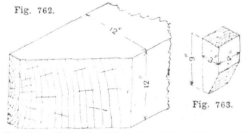

Fig. 763.

Fig. 762.—Method of Finishing End of Fender of Builder's Gantry.

Fig. 763.—Enlarged View of Cleat for Builder's Gantry.

pose of giving a greater bearing surface where a joint occurs in the head. Fig. 757

illustrates two methods of sheeting. If the 9-in. by 3-in. sheeting be used, then the whole area of the platform should, previous to laying the deals, be covered with tarred felt, to prevent water percolating through to the annoyance of the public. Or, if double sheeting scaffold boards be used, the contact with it will glide off; this is a very desirable precaution.

Dogs.—Figs. 760 and 761 are the types of dogs used in the above class of work. They run from 12 in. to 18 in. in length, and with points from 2 in. to 3 in. long. That shown at Fig. 760 is used for heading,

Fig. 764.—Conventional View of Gantry.

joints, both lateral and heading, should be lapped. The platform is then sanded, and the sand worked into the joints with a broom. The guard frame is then fixed and boarded to the height shown at Figs. 757 and 758. The fender may now be laid in the gutter and dogged to the uprights, and the handrail fixed to cleats between the upright, at from 3 ft. to 3 ft. 6 in. from the ground. Fig. 762 shows how the end of the fender should be cut so that any vehicle coming in lateral, and shoulder joints, and that at Fig. 761 (which is made with its points at right angles to each other, and, as already remarked, is sometimes known as a "bitch") is used in positions where it holds more effectively than the other, such as the fender to the uprights and the joists to the heads, etc. They are made rights and lefts, or, as it is often termed, in pairs. Fig. 763 is a view of a cleat as spiked to the upright to receive the thrust of the strut.

10

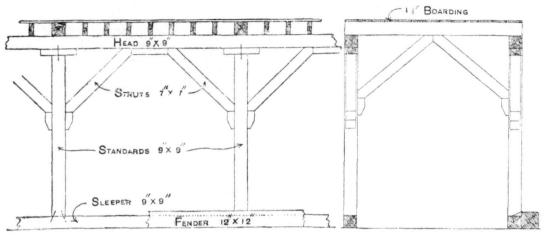

Fig. 765.—Elevation of Gantry.   Fig. 766.—Section through Gantry.

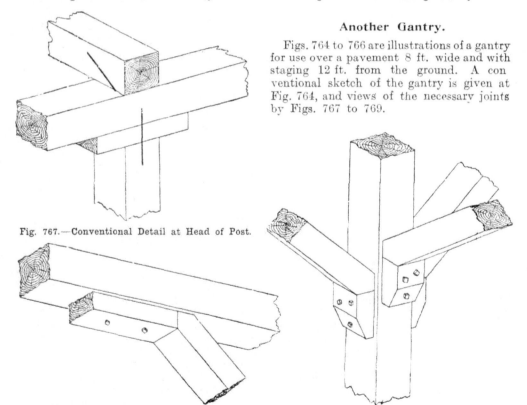

Fig. 767.—Conventional Detail at Head of Post.

Another Gantry.

Figs. 764 to 766 are illustrations of a gantry for use over a pavement 8 ft. wide and with staging 12 ft. from the ground. A conventional sketch of the gantry is given at Fig. 764, and views of the necessary joints by Figs. 767 to 769.

Fig. 769.—View of Straining Piece and Strut butting against it.   Fig. 768.—View showing Cleats supporting Posts.

Builder's Staging.

A portion of an important specimen of builder's staging is shown in front and side elevation respectively by Figs. 770 and 771. Staging of similar design has been used 9 in. by 9 in., according to the weight, number of stages, and strain brought upon it. The whole is braced by 7-in. by 2-in. to 9-in. by 3-in. scantlings. The upper part of one bay is framed out as shown at A in

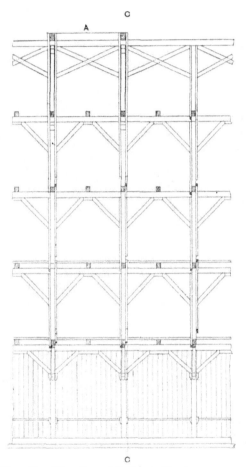

Fig. 770.—Elevation of Builder's Staging.

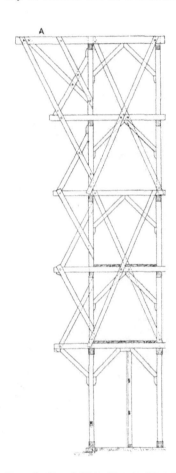

Fig. 771.—Sectional Elevation on Line C C (Fig. 770).

for many important buildings in different parts of the country. It is particularly serviceable for supporting heavy blocks of stone, girders, and other materials. Of course, the design in each case must be modified to meet requirements, but the illustrations will give a general idea of this kind of structure. The principal members would be of whole timbers from 6 in. by 6 in. to

the illustrations, for the purpose of supporting a travelling hoisting apparatus, so that materials, etc., can be taken direct from the vans in the street and delivered on to either of the projecting platforms below. The conventional view (Fig. 772) will make clear the general arrangement of the various members.

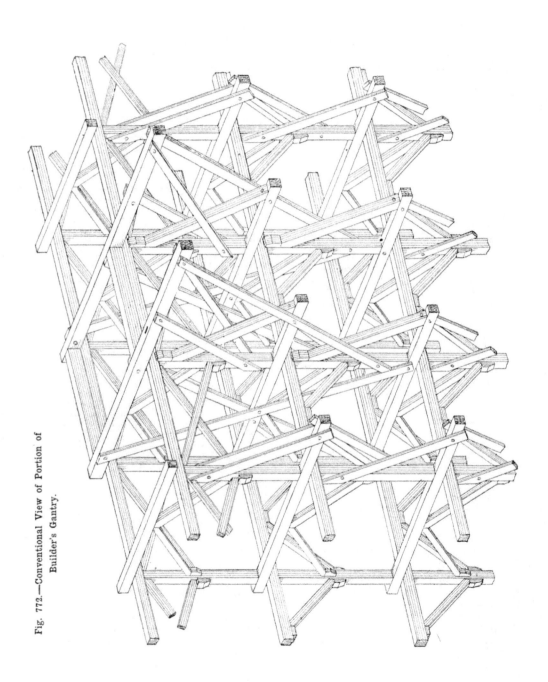

Fig. 772.—Conventional View of Portion of
Builder's Gantry.

Derrick Tower Gantry.

A general view of one of these is given at Fig. 773. There are now very few large buildings erected without the aid of this form of gantry, and it is gradually superseding other forms, on account of the following considerations: Occupying a small space, by the double movement of the jib, material can be raised from the ground on

usually built of baulk timber, whereas the towers, etc., are principally built of 7-in.

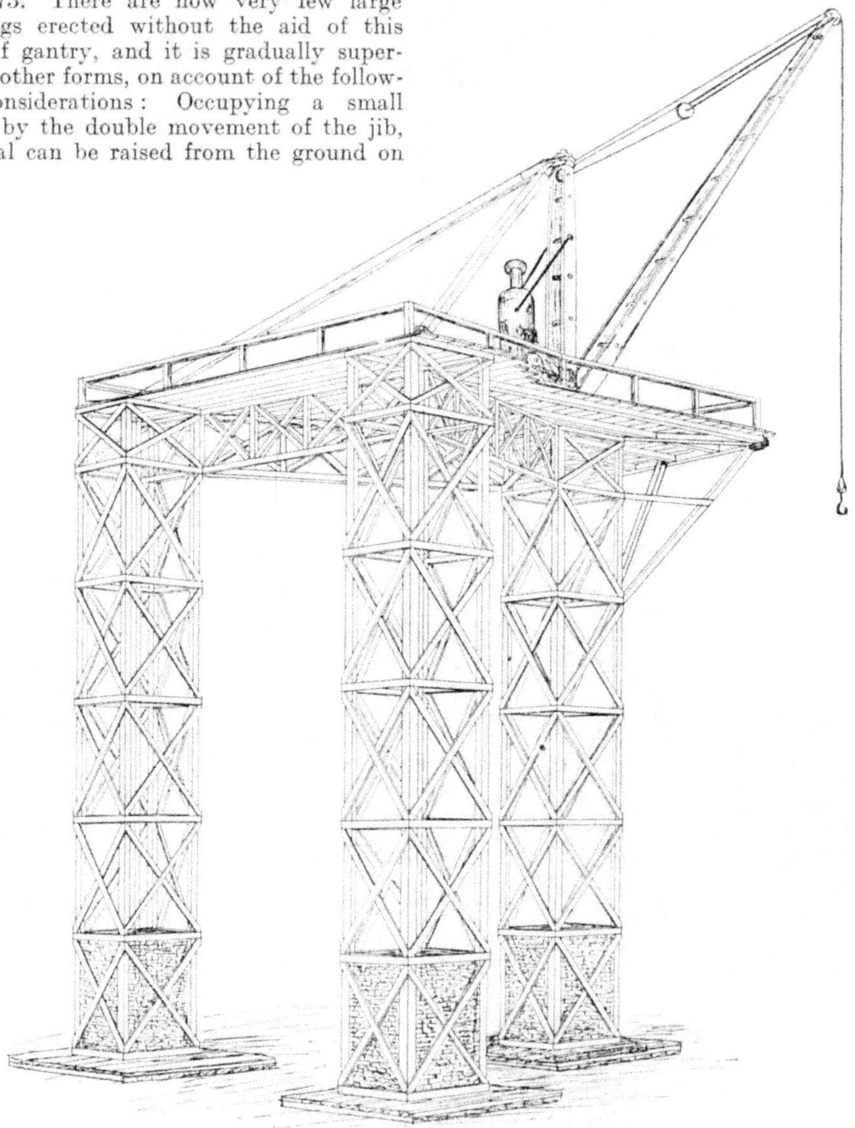

Fig. 773.—General View of Derrick Tower Gantry.

one side of the building, and deposited direct in its proper position on an opposite side. The cost of erecting is low compared with other kinds of stagings, which are

battens or 9-in. deals. There are three or four towers, but usually only three. They are about 6 ft. square, and are so arranged that lines joining at the centre of the plan of each tower form an isosceles triangle. Each tower has four posts, formed either of three 7-in. by 2½-in. battens, or three 9-in. by 3-in. deals (Fig. 771) bolted together, the layers, of course, breaking joint. Transoms, 8 ft. to 10 ft. apart, of similar scantlings, connect the posts

material; thus the back is anchored down. The front or king tower has a standard through the centre of its whole length, which is held to the posts by bracing. This standard is to give additional strength, for the support of the machinery of the crane, etc.

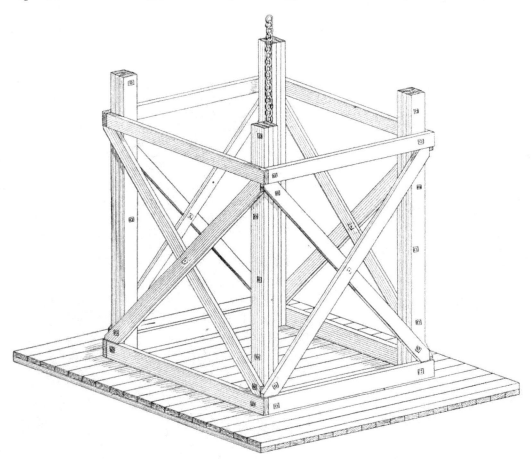

Fig. 774.—View of Timber Foundation and a Bottom Bay of an Anchor Tower.

by being bolted to them. The spaces between the transoms are braced as illustrated. Each tower rests on a double plank foundation (Fig. 774). In the two back or anchor towers the platforms and foundation planks (see Fig. 775) are connected together by means of a strong chain, the lower bay being loaded with bricks or other heavy

The upper ends of the towers are connected by trussed girders, as illustrated, the heads and sills being about 9 in. by 4 in. and the braces and struts 4 in. by 4 in., the whole being held together by ⅝-in. or ¾-in. bolts passing through the heads and sills, and thus connecting them. The towers are often tied together by bracing.

in connection with the construction of dock and other similar work. The illustration represents a gantry about 25 ft. high, 18 ft. clear between the sides of the framing.

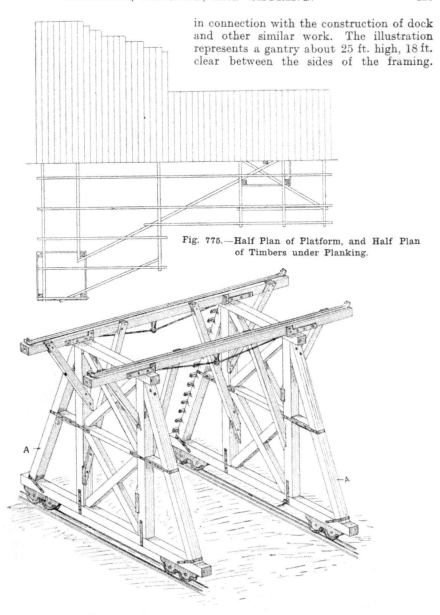

Fig. 775.—Half Plan of Platform, and Half Plan of Timbers under Planking.

Fig. 776.—Conventional View of a Movable Gantry to support Traveller.

Movable Gantry to Support Traveller.

A gantry of this description is illustrated at Fig. 776. These are used principally

The beams for supporting the rails are 34 ft., allowing the traveller to work on the outside of the side frames when required.

These are trussed with 1½-in. tension rods as shown. The main members are 12 in. by 12 in., braces A 10 in. by 12 in., internal braces 10 in. by 8 in., braces from the rail beams to the posts 8 in. by 6 in., sills 14 in. by 12 in. The principal joints are secured together by straps and bolts as illustrated. The traveller, being engineer's work, is omitted.

Gantry for Traveller.

Fig. 777 illustrates a form of gantry useful for lifting heavy blocks of masonry re-

Window Stand.

Figs. 778 and 779 show the construction of a stand for the window of a private house. This will seat thirty persons, in a space of 8 ft. by 7 ft. 3 in. Every precaution must be taken, in erecting these stands, to ensure absolute safety and stability. County Council and other officials are, quite justifiably, most stringent in their demands for proper and safe structures; and no fear need be entertained upon this head if the structures here described are carefully erected as shown.

Fig. 777.—General View of a Gantry for Traveller.

quired in building thick walls. To allow of free movement of the blocks, intermediate bracing is not obtainable, therefore this has to be arranged for at the outside as shown. This kind of gantry is usually built of baulk timber from 8 in. by 8 in. upwards, according to the height and strength required. The feet of the outer braces are frequently bolted to stakes driven firmly in the ground as indicated in Fig. 777, which gives a general view, in which the principles of construction are shown with sufficient clearness to render further description superfluous to the practical builder.

Stands for Spectators.

In Figs. 780 and 781, which illustrate a stand to accommodate about 1,000 persons, the lettering is explained as follows: A, Brace shouldered over half thickness; B, braces halved together; C, brace cut in between principals; D, short tenon; E, tenon mortised through and wedged; F, brace bolted on face of principal; G, bearers mortised and tenoned together and pinned; H, bearer mortised into raking beam and pinned; I, rail dovetailed to post; J, post mortised to receive tenon on raking beam

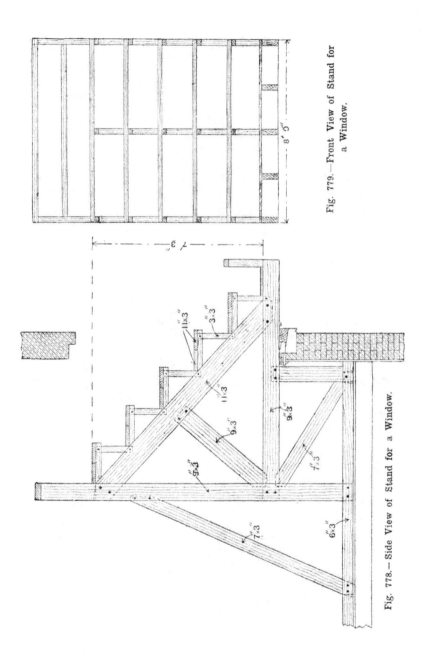

Fig. 779.—Front View of Stand for a Window.

Fig. 778.—Side View of Stand for a Window.

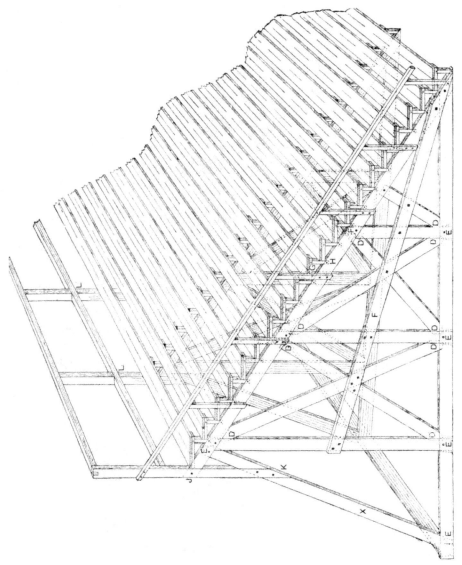

Fig. 780.—Elevation of Principal and Perspective View of Stand to seat 1,000 Persons.

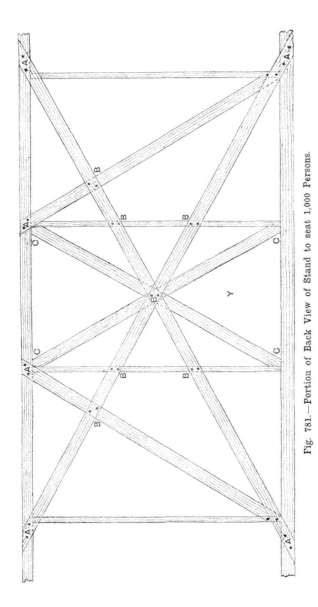

Fig. 781.—Portion of Back View of Stand to seat 1,000 Persons.

and wedged; κ, post halved on to raking strut; L, rail halved to centre posts.

The stand is 70 ft. long, and will contain twenty rows of seats, which are 18 in. high and 22 in. wide. The principals, which will be nine in number, equally spaced, are framed together, as shown, with cross and raking braces. The cross braces are shown in Fig. 781, in which the opening marked Y will occur between each pair of principals, and also between the intermediate principals situated nearer the centre of stand, which are cut in between in the same manner and bolted to the uprights. The other braces shown in the back view will be fixed on the raking strut X in Fig. 780. Where a stand is fixed between two walls, no side struts are required; but when it is fixed in the open, good strong raking struts are wanted at each end, firmly secured to the raking beam with bolts. The timber throughout should be of the soundest. The raking beam and the uprights should be 11 in. by 4 in.; the braces may be 9 in. by 4 in.; sole pieces, 9 in. by 4 in.; framed bearers, 4 in. by 4 in.; seats, two 11-in. by 3-in. planks, well spiked to bearers. The timbering at the back is run up to prevent spectators from falling, and an awning can be fixed to it. The handrail newels are bolted to the side of the raking beam. A similar but smaller structure, might be made to provide more comfortable seating accommodation, the seats being 26 in. instead of 22 in., thus giving more room for the feet.

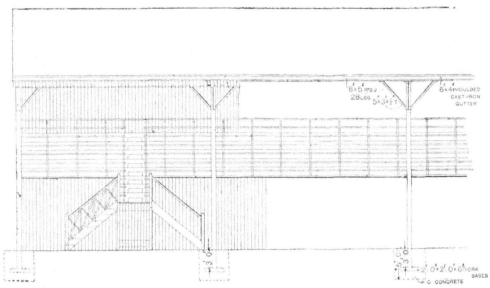

Fig. 782.—Part Elevation for Grand Stand for Sports Ground.

Fig. 783.—Plan of Grand Stand for Sports Ground.

Grand Stand for Sports Ground.

Fig. 782 shows part elevation, Fig. 783 a plan, and Fig. 784 a section of a stand, 192 ft. long by 19 ft. wide, with seating accommodation for one thousand spectators. The stand is constructed entirely of timber, trussed and braced as necessary to make a perfectly safe structure. Twenty trusses are framed as shown in Fig. 784, and tied together with raking braces, forming the entire length of the stand. Each post in the truss stands on a Portland cement concrete base, 2 ft. by 1 ft. 6 in. by 1 ft. 6 in. The seating and floor are carried upon 9-in. by 3-in. deals, spaced between the trusses, and with 4½-in. by 3-in. framed bracketing. The floor is composed of 1½-in. wrought-one-side boarding, the risers of 1-in. floorboards. The seats are of 11-in. by 1½-in. wrought pitchpine boards with rounded edges. The front of the stand is matchboarded on 3-in. by 2-in. wrought framing, with a moulded capping. The usual offices, with refresh-

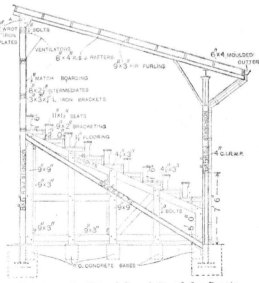

Fig. 784.—Section of Grand Stand for Sports Ground.

ment bar, etc., etc., are provided ; and a space called the press box is set apart for reporters. The seats are reached by flights of steps, as shown. The roof is

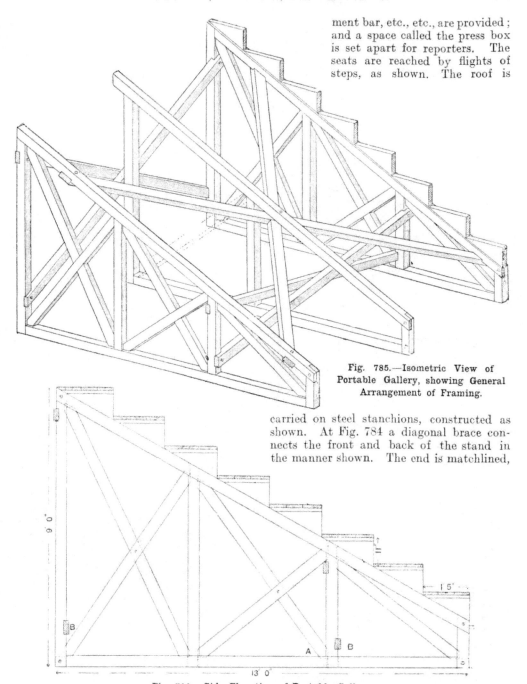

Fig. 785.—Isometric View of Portable Gallery, showing General Arrangement of Framing.

carried on steel stanchions, constructed as shown. At Fig. 784 a diagonal brace connects the front and back of the stand in the manner shown. The end is matchlined,

Fig. 786.—Side Elevation of Portable Gallery.

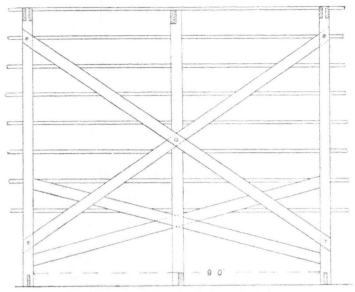

Fig. 787.—Back Elevation of Portable Gallery.

and finished with handrail to the rake of the seating. The stand is covered with corrugated-iron sheeting, No. 16 gauge, screwed into position. In Fig. 783, the letters R B O indicate reserved box over; and P B O, press box over.

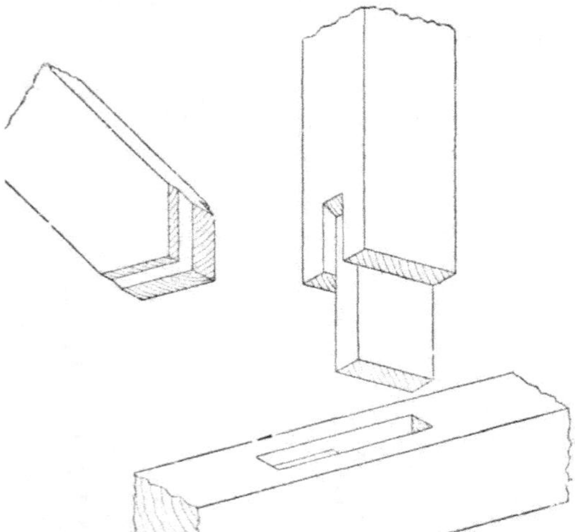

Portable Gallery.

A portable gallery, suitable for a public hall or other similar building, is illustrated by Figs. 785 to 791. The structure is one that can be erected and taken to pieces with little trouble, and the materials can be stored away until again required. The size and form of the gallery will, of course, vary with the requirements to be fulfilled; but the method of framing set out below should be strictly observed, for the safety of the structure depends on the care with which this part of the work is carried out. The framing should be properly mortised and tenoned, and in some parts lap-jointed, as shown in Figs. 785 to 789. The tenons that go through should be wedged into their respective mortices; in other cases, the joints should be secured by bolts and nuts. Butterfly nuts will be very useful, as they are readily adjusted. The joints at A and B (Fig. 786) are shown in detail by Figs. 788 and 789 respectively. Each piece of framing should be properly braced (see Figs. 785

Fig. 788.—Enlarged View of Mortice and Tenon Joints at A (Fig. 786).

and 786), and should be again well tied together by braces as shown in Figs. 785 and 787. It will be observed that these parts are notched and lapped, as little as possible of the wood being cut away, so that the framing may not be weakened more than is absolutely necessary. The main standards should be about 9 ft. apart, as shown; the standards between these need not, for ordinary purposes, have more than one brace, which is indicated by dotted lines (Fig. 785). This intermediate standard is intended to support the boarding forming

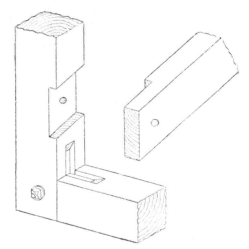

Fig. 789.—Enlarged View of Joints at B (Fig. 786).

the staging. If the gallery is for children, triangular pieces cut out of 11-in. by 2-in. stuff, and firmly secured by nails, will be suitable (see Figs. 785, 786, and 790); but if it is intended for adults, greater height and breadth will be required, and it will be necessary to frame the supports for boarding of 2-in. by 2½-in. stuff, as shown at Fig. 791. These supports should be halved together and stub-tenoned into shallow mortices (Fig. 791), and firmly secured by nails. A simple method of securing the boarding is shown at Fig. 790. On the under side, ledges are nailed so as to clip on each side of the support; then by boring holes, as shown at A and A', these ledges can be held together

with an iron pin or bolt. Suitable sizes of timber will be 4 in. by 2 in. for the smaller

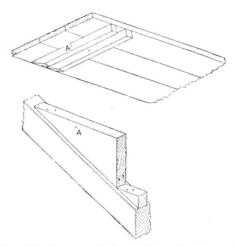

Fig. 790.—Enlarged View of Triangular Support, and Method of connecting it to Boarding.

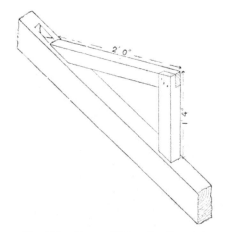

Fig. 791.—Frame Triangular Support.

braces, and 4 in. by 3 in. for the larger braces.

Shoring.

Shoring may be described briefly as temporary supports for walls that are considered

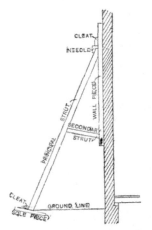

Fig. 792.—Single Strut Raking Shore.

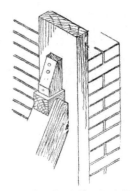

Fig. 793.—Detail of Head of Strut.

unsafe, or for girders, etc., in course of erection or repair. The three most typical kinds of shoring are raking, horizontal, and dead shores. Every other kind of shoring appears to be an adaptation of one or other of these three kinds. In shoring and underpinning, probably as much as in any other

Fig. 794.—Foot of Strut with Groove for Tightening Up.

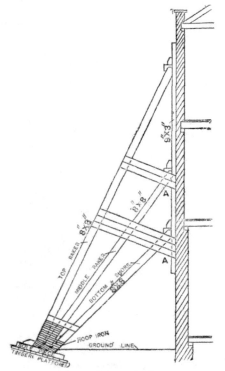

Fig. 795.—Triple System Raking Shore.

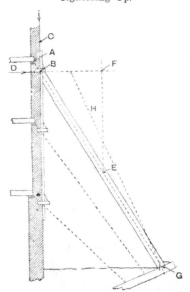

Fig. 796.—Raking Shore.

branch of the building trade, a ripe experience is essential to success. It is necessary to be thoroughly prepared for any emergency that may arise; for it is only when the cutting away is actually commenced that it becomes possible to find out exactly what circumstances have to be met. The first—and the most important—thing is to secure a solid base to shore from. If this is not obtained, the support given, or supposed to be given, is a deceit. There may be a cellar at the point where the shores have to be erected. If a strong wall of the cellar is not available at a suitable point, then the point of support must be found outside on the pavement or roadway. If it is earthy ground, try it with a crowbar. It may be solidified to a certain extent by ramming, or thick planks may be placed to form a solid platform.

Raking Shores.—The most simple type of raking shore is that consisting of only one principal strut, as shown in Fig. 792. It is erected thus :—At a little way down, usually at about 2 ft. from one end of the "wall-piece," a hole, rectangular in shape, is cut to take the "needle"; and when the wall-piece is in position, the needle fits into a hole in the wall made to receive it by removing a half-brick. The needle also projects on each side of the wall-piece to receive the head of the principal strut. To counteract the upward thrust of the shore, a cleat is nailed over the needle. These details are shown more clearly in Fig. 793. A secondary strut, as illustrated, is necessary. The sole piece or footing block is a timber balk let into the ground, and a cleat is nailed on that also to keep the foot of the shore from slipping. In soft soils a little timber platform is placed to receive the sole piece indicated in Fig. 795. Sometimes wedges are driven in at the foot of the principal strut, but the heavy hammering necessary to drive them home is likely to defeat the purpose for which the shore is being erected. The more approved method of tightening up is to cut a groove in the foot of the shore (Fig. 794), and gradually lever it into position. The most common type of raking shore is that shown in Fig. 795, which is really a triple system on the same principle as that shown by Fig. 792. The illustration therefore

explains itself in the light of the foregoing description. The top and middle shores are called top and middle rakers respectively; the underneath one of all is the bottom shore. This arrangement has to be strengthened by more than one secondary strut, on account of the length of the top raker, and for this purpose pieces of timber are brought right back to the wall and nailed to the shores and wall-piece as shown at A in Fig. 795. In the still more intricate system of four shores sometimes seen, the topmost strut is called the rider shore.

Scantlings of Shoring Timbers.—The following table of shores and scantlings has been found useful (taking the angle of the shore at about 65°) :—

Height of Wall up to	Number of Shores.	Scantling.
15 ft. to 30 ft.	2	6″ × 6″
40 ft.	3	8″ × 8″
50 ft. and beyond ...	4	9″ × 9″

Beyond 50 ft., if the distance apart between each system exceeds 12 ft., the scantling of each shore should be 12 in. by 9 in.

Erecting a Raking Shore.—Let it be assumed that a building requires support, and that raking shores are in this case most suitable. The work can be carried out according to the following directions given by Mr. H. A. Davey in a paper read in 1899 at the British Institute of Certified Carpenters. All the window openings must be strutted, and care must be taken that the brickwork is not jarred more than is absolutely unavoidable. Next find out the heights of the floors and the thickness of the wall, and make a rough sketch, to any scale, of a vertical section of the wall. The next step is to decide where to pitch the foot of the shores, and great care must be taken in making this selection; for the shores, should the footblock yield to their pressure, would become a source of danger instead of a support. Old drains and vaults will probably give most trouble in this respect, but everything must be made solid before the shore is put into position. In the case of a vault, Mr. Davey found the most satisfactory treatment was to run the shore through the crown to firm ground. Old drains can, as a rule, be either cleared away or filled in. The angle the shore should make with the horizon is generally decided by the width of the pavement ; but assuming that there is no distance given, then, to obtain the maximum thrust, the shore would have to be inclined at an angle of 45° with the ground ; but there are two reasons against so large an angle : (1) The shore would take up too much space ; (2) increased lengths of timber would be required. It has been decided, therefore, that in practice the

(this is quite near enough for all practical purposes), and from E draw a vertical line intersecting D produced at F, and the resultant H of the forces D and C will lie between B and F and in the direction of the foot of the shore ; so a line drawn from G to the centre of B F will be the mean of the directions the resultant H will take. At G draw a line at right angles to H, and this line will represent the face of the footblock. It will be seen by this that the footblock cannot be at right angles to the shore, owing to the resultant of the forces acting outside

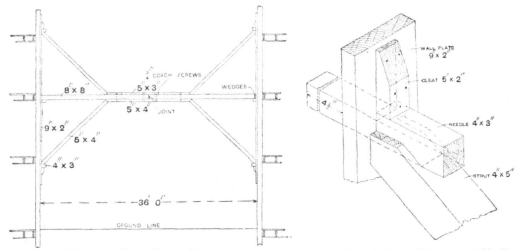

Fig. 797.—Single Flying Shore.

Fig. 798.—View of Head of Raker and Needle.

best angle for the top raker shall be between 60° and 70° (see Fig. 796). Draw the face of wall, position of joists and wall plate on face of wall, draw a line from A making an angle of 60° with the ground line ; on each side of this line set off half the thickness of the shore (assuming whole timber is being used, which should always be the case), and where the centre line intersects the wall plate at B draw lines to represent the needle, and cleat nailed above it ; this finishes the head of the shore for the present. Now discover the angle the footblock should make with the shore. Draw lines C and D to represent the horizontal and vertical forces acting at the back of the wall and opposite the head of the shore, assume the centre of gravity of the shore to be at E

the shore. The shore should be levered into its place with a crowbar, and fixed to the footblock with iron dogs. The practice of driving wedges in with a sledge-hammer is most dangerous, and no man understanding the nature of the work would run such a risk. Sometimes when the building is very high it is necessary to put up the top raker in two pieces. The top piece is then called a rider, but it is much better in one piece if it can be managed, on account of the objection to wedging. Three or four pieces of 1-in. boarding are nailed to the sides of the shores and wall plate, to hold them together and to act as struts and ties. For this reason all the shores in a system should be of the same size. The distance between the shores should not be more than about 12 ft. ; but this depends on the position of the piers, as there must always be a good abutment for the head of the shore. Mr. Davey does not recommend putting the shores close together at the bottom, for the

Horizontal or Flying Shores.—Horizontal or flying shores are used when a house is taken down in a terrace, and the adjoining

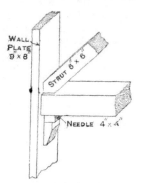

Fig. 801.—Detail of Joint of Flying Shore at B (Fig. 799).

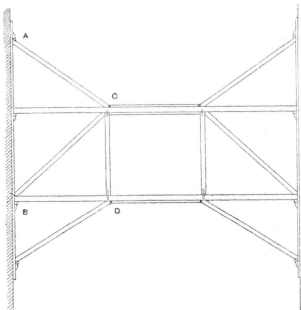

Fig. 799.—Elevation of Double Flying Shore.

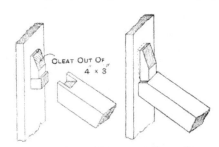

Fig. 800.—Details of Joints of Flying Shore at A (Fig. 799).

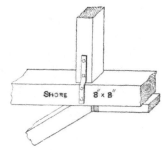

Fig. 802.—Detail of Joint of Flying Shore at D (Fig. 799).

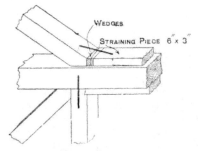

Fig. 803.—Detail of Flying Shore at C (Fig. 799).

reason that if the bottom shore has to be removed first, which is often the case, it is a difficult matter if they are close together, whereas by leaving a space of about 6 in. or 8 in. and cutting in a block, it is quite easy to remove any one shore without interfering in any way with the others in the system.

walls require supporting, a clear way being required underneath. Text-books generally limit the length to about 32 ft., because fir timber is not easily obtained longer ; but pitchpine may be obtained in 70-ft. lengths, so that if flying shores are the best for the purpose, there is no need to trouble about the length. Flying shores are superior to raking shores, because their thrust is immediately opposite the disturbing force. The most common type of flying or horizontal shore is that shown in Fig. 797. For detail of the joint between needle and strut see Fig. 798. In a case where two houses of 18-ft. frontage, each in a terrace, have been pulled down, and shoring is required for supporting the adjoining houses on each side, the strut just shown, or, preferably, the one shown by Fig. 799, would be suitable. The joint shown at Figs. 798 and 800 is to be preferred for this, and frequently the struts are simply butted against cleats, as shown at Fig. 800; these should be housed-in as well as being spiked to the wall-piece as indicated. The method of wedging and connecting the joints with dog-irons and straps is shown by Figs. 801 to 803. It is best, where practicable, to have the horizontal shores cut just tight between the wall plates, but should they be a little short, a pair of folding wedges may be driven between one end of each shore and the wall-piece.

Flying Shores for Buildings of Unequal Heights.

The examples of flying shores that have just been illustrated are in general use for buildings that are of equal, or nearly equal, heights. When the buildings are not about the same height the shores are usually of a special design more or less complicated to suit each particular requirement of the work to be executed. Figs. 804, 805, and 806 show three systems which are somewhat similar to those generally illustrated as suitable for cases where it is necessary to support a high house by means of a flying shore against a lower house. The spans would not be so great as is shown in Fig. 807, which is a typical case of shoring up the side of a five-storey house, standing in a narrow street, the traffic of which must not be obstructed by raking shores. On the opposite side of the street is a house that has to be used for support; this being two storeys lower, care

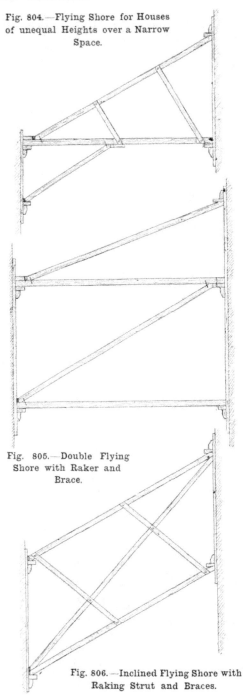

Fig. 804.—Flying Shore for Houses of unequal Heights over a Narrow Space.

Fig. 805.—Double Flying Shore with Raker and Brace.

Fig. 806.—Inclined Flying Shore with Raking Strut and Braces.

Fig. 807.—Flying Shores over Thoroughfare between Two Houses of unequal Heights.

must be taken to distribute as much as possible the thrusting force of the shores so as to prevent injury to the supporting house. It will be seen by examining the illustra-

ing points E, F, and G; in the same way, if there is a pressure against any or all of the points A, B, C, and D, the pressure is also transmitted to the same points. To carry out effectively this system of distributing the pressure, large cleats are bolted to the horizontal timbers to form a firm abutment for the feet of the struts. The timber for the

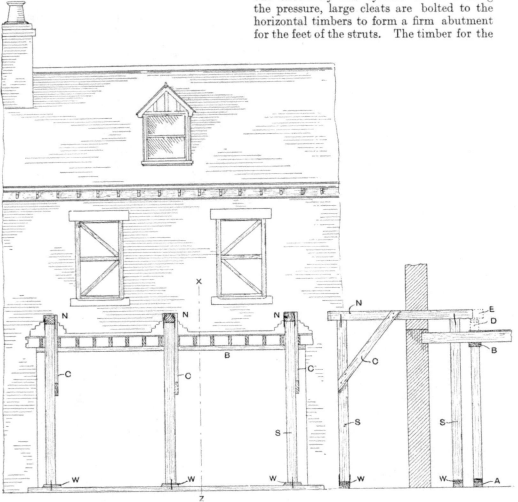

Fig. 808.—Elevation of Dead Shoring for converting Ground Storey of Small House into Shop.

Fig. 809.—Section on Line X Z (Fig. 808).

tion that the bracing and strutting has been arranged so that when the wall of the higher house exerts a pressure at either one of the points A, B, C, or D, this pressure against the lower house is transmitted to the three bear-

main members would vary from 5 in. to 9 in. by 8 in. according to the span, and the amount of the thrust would be a necessary factor for special consideration in each case. The braces and struts would be proportionate to the size of the other members.

Vertical or Dead Shores.

Vertical or dead shores are in general use for the following cases : (1) When the foundations of a building have given way and it is necessary to support the walls by shoring in sections so as to underpin them during the renewal of the foundations. (2) When the ground storey of a private house is converted into a shop, the upper part of the house is supported by shoring until the bressummer is fixed in its place and the new brickwork built upon it to support the old wall. (3) Where it is desired to raise the front of a shop, then it is necessary to support the upper part of the structure whilst a new bressummer or girder is being fixed in position so that it can carry the wall above. In larger and more important classes of buildings, which range from three storeys high upwards, it is usual, in addition to dead shores, to use raking shores, with the object of steadying the walls, as well as giving a certain amount of support, and thus to minimise the chance of accident to the building.

Dead Shoring for Converting Private Dwelling into Shop.

Figs. 808 and 809 represent a usual method of shoring in the common case of converting the ground storey of a small private dwelling house into a shop. In many cities, for such small jobs as this, raking shores are not used. The leading methods of procedure may be summarised as follows : The windows are strutted by pieces of timber about 3 in. by 3 in. or 3 in. by 4 in. The sill shown at A is supported on the ground floor, and a head is put plumb over this against the ceiling, as shown at B (Figs. 808 and 809) ; three or more vertical posts are cut to a length to fit tightly between the head and the sill. Sometimes the posts are cut a little short so as to allow of a pair of oak wedges to be driven between the post and sill ; in this way the dead shores or posts support the floor, and thus the front wall is relieved of its weight. Holes are then cut through the front wall about 6 in. or a foot above the floor, for the insertion of needles N (Figs. 808 and 809). It is usual to put a needle under each pier between

openings, but when the piers are very wide it is sometimes necessary to insert two needles, as shown in the illustrations. The needles are supported by dead shores at each end s (Figs. 808 and 809), both inside and outside the building. These shores rest upon continuous sills, and are fixed tight under the needles by the insertion of oak wedges W. It is usual to brace the outer dead shores to the needle by a piece of scantling as shown at C. The feet and head of the shores are also secured to the head and sill by iron dogs. Next, it is usual to remove sufficient of the brickwork to allow of the insertion of the bressummer which spans the opening and to which the floor joists are fixed in some one of the various ways. Next the walling is made good, with brick or stone laid in Portland cement, as far as this can be done without removing the needles ; after the new work has properly set the shoring is removed, and the making good of the wall is completed. When placing the shores on the ground floor it should be carefully noted whether this is sufficiently strong to support the shoring. If not, part should be taken up and the sills bedded firmly on the solid earth. In the case where there is a basement it would be necessary to support the ground floor by a sill-head and dead shores, as shown in Figs. 808 and 809. With small jobs, sometimes the inner standard supporting the needle is dispensed with ; pieces of square timber about 3 ft. long, D (Fig. 809), rest on the floor directly over the head B ; the inner end of the needle rests on these square timbers ; and if required, pairs of wedges are driven between the timber and the needle, as indicated at E.

Shoring Large Corner House for Converting the Ground Storey into Shop.

A familiar example, but not such a common one as that just described, and one that is of much greater magnitude, is illustrated by Figs. 810 to 815. A case of this description generally calls for the exercise of considerable skill and judgment, especially if the house is an old one. Figs. 810 and 811 show a five-storey corner house, with an area on the two fronts. It is shown shored up for the conversion of the ground floor into shop premises. The raking shores would first be erected, and it being a high house, there would be two rakers, and also a rider G to K (Fig. 811); these shores wouldvary in size from 6 in. by 6 in. to 7 in. by 7 in.,

should rest on a solid foundation formed by the solid earth. Standards marked A, B, would rest on this sill and carry the head B. In this way the ground floor would be supported, and in its turn this would support the

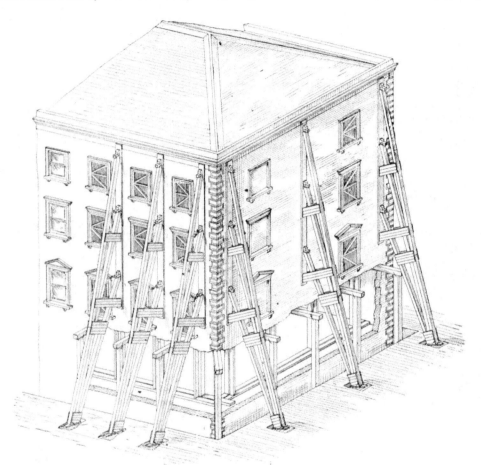

Fig. 810.—General View of Shoring to large Private House the Ground Storey of which is to be altered into Shop.

according to the special requirements of the particular case. To give the best support the needles at the head of the rakers should be inserted just below each floor. As there is an area, the feet of the shores would require to have a foundation to rest on, F (Fig. 811) placed at least 2 to 3 ft. from the area wall. The sill in the basement, A (Fig. 811),

sill C, the standards, the head, and, in a way, the whole of the first floor. Occasionally the shoring is continued through the floor above, and this would, to some extent, reduce the load on the needles. Next, holes would be cut in the walls for the insertion of the needles, then the first floor D (Fig. 811) would have a hole cut for the inner dead

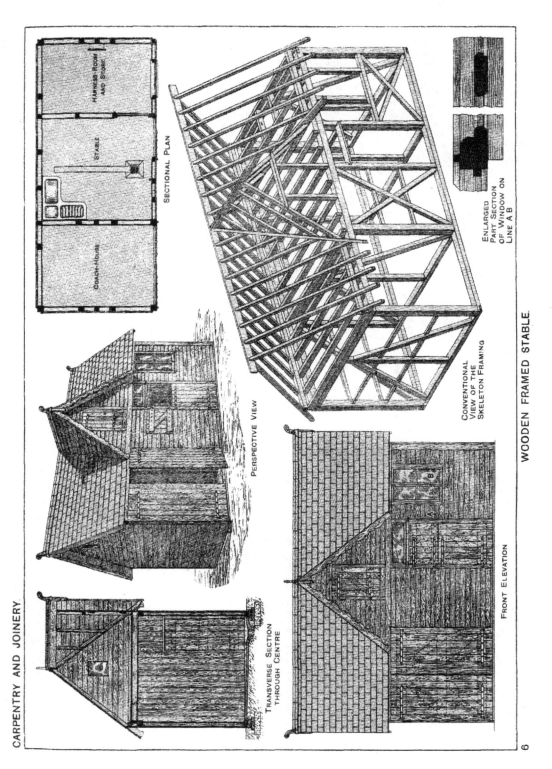

CARPENTRY AND JOINERY

SECTIONAL PLAN

HARNESS-ROOM AND STORE

STABLE

COACH-HOUSE

ENLARGED PART SECTION OF WINDOW ON LINE A B

CONVENTIONAL VIEW OF THE SKELETON FRAMING

PERSPECTIVE VIEW

TRANSVERSE SECTION THROUGH CENTRE

FRONT ELEVATION

WOODEN FRAMED STABLE.

6

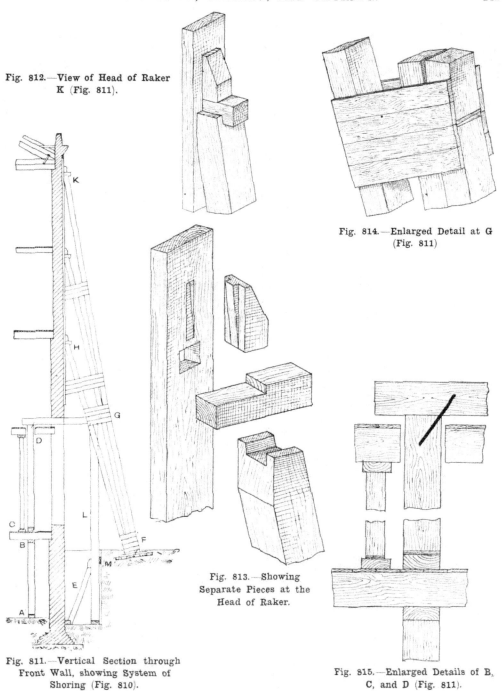

Fig. 812.—View of Head of Raker
K (Fig. 811).

Fig. 814.—Enlarged Detail at G
(Fig. 811)

Fig. 813.—Showing
Separate Pieces at the
Head of Raker.

Fig. 811.—Vertical Section through
Front Wall, showing System of
Shoring (Fig. 810).

11

Fig. 815.—Enlarged Details of B,
C, and D (Fig. 811).

shore to pass through, and this shore would carry the end of the needle. The outer dead shore L (Fig. 811) should be long enough to rest on a sill bedded firmly in the area as shown. As a stay to these standards,

angle of the building should be timbers of much greater sectional area, as they have to support a greater weight, and all chances of movement must be guarded against. Fig. 812 is the general view of the top end of a

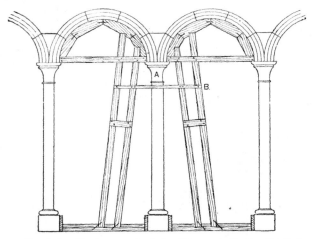

Fig. 816.—Shoring required for the removal of a Capital.

Fig. 817.—Shoring necessary when an Arched Stone has to be renewed.

a raking strut E could be birdsmouthed on to a plate and its head cut to fit against the shore, which would have a cleat fastened to it. When these do not fit tight against the walls of the area, blocks should be placed between as shown at M (Fig. 811). The needles and the standards carrying the

raker at its connection with the needle, the cleat, and the wall piece. Fig. 813 shows each one of these parts separated. Fig. 814 is an enlarged detail at G (Fig. 811), showing the wedges, etc., at the foot of the rider. Fig. 815 gives enlarged details at B, C, and D respectively in Fig. 811.

Shoring an Arcade.

Fig. 816 shows the method of shoring required when it is necessary to remove the capital of a column A. Sole plates are cut between the bases of the columns on each side of the one from which the cap is to be removed. Ribs cut from 6-in. planks are fitted into the arch, the edges fitting perfectly to the soffits; the butt joints, cut radiating with the centre, must also fit perfectly, and the planks are held together by iron dogs; or the joint may be crossed with pieces of $1\frac{1}{4}$-in. deal, as

vary according to the weight above the arches. If there are very high and heavy clerestory and roof, an additional shore might be necessary, but should be kept almost vertical. Fig. 817 shows the timbering necessary when one of the arch stones, as at C, is to be cut out. Fig. 818 shows an elevation, and Fig. 819 a section of the timbering necessary for the removal of the shaft D. The capital is kept in position by the collar E, which is fitted round it and bolted together, and in turn fixed to the shores. Struts F and shores G are cut tight under the collar. The centering in each

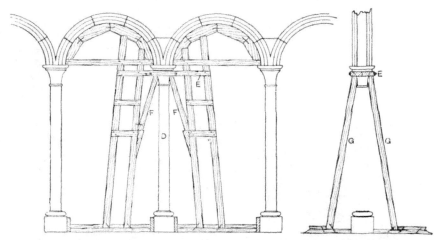

Fig. 818.—Elevation and Section of Shoring required for the removal of a Column.

in making an ordinary centre. Stretcher pieces are cut tight in between the arches at the springing. Shores of 12-in. by 6-in. timber are then cut up as shown. Folding wedges are not used, the shore being pinched up tight with an iron bar; and when it is home, pieces of timber are cut in between and spiked to the sole plate. The shores are pitched to about 85°, and sufficient room is allowed for the removal and the reinstatement of the capital. Horizontal pieces are fixed on each side of the shores at B; these are allowed wide enough to scribe around the shaft, and so hold it rigid during the process of removal and fixing. The free use of iron dogs is recommended, and the work must be well done to ensure success. The size of the timbers would

case may be made additionally secure and strong by the free use of braces and struts.

Shoring the Arcade of a Church.

This work calls for the highest skill and judgment. The one example here illustrated and described will serve to give some idea of the nature of this kind of work. In the portion of a church arcade shown by Fig. 819 there are cracks in the masonry, indicating a subsidence in the foundation of one or more of the pillars; it is assumed that it is found necessary to renew the foundation. The first thing will be to construct strong centres. In the case illustrated the arch mouldings can be supported on three centres; the middle one is constructed of stuff 7 in. thick, and those on each side of stuff 3 in.

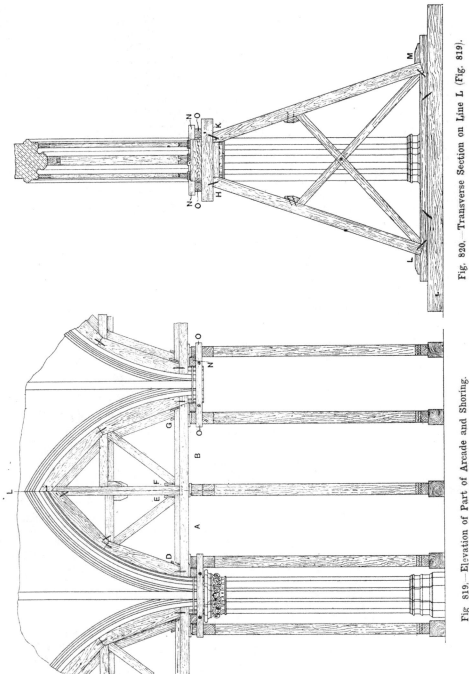

Fig. 820.—Transverse Section on Line L (Fig. 819).

Fig. 819.—Elevation of Part of Arcade and Shoring.

thick. So that the ribs may properly fit the intrados of the arches, templates made of $\frac{1}{2}$-in. boarding should be carefully scribed to fit ; then these should be used for making

Fig. 821). The raking shores L H and K M are made to spread at the bottom, the sill being notched out of the solid to receive the square ends of these ; they are further

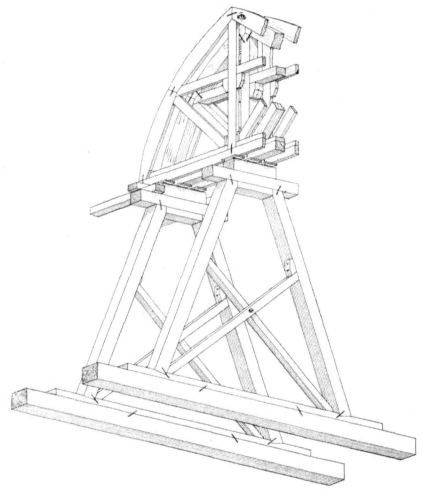

Fig. 821.—Conventional View of Central Trestle and one adjacent to Pillar, also of Part of Centres, Collars, etc.

the timber ribs, which are out of 7-in. by 12-in. and 3-in. by 11-in. stuff respectively. The joints between the ribs, tie, king-post, struts, etc., are clearly illustrated at A, B, and C (Fig. 823). Strong trestle shores are next made to the sizes and form clearly shown in the illustrations (especially see

secured by fixing on cleats. The whole of the trestle is supported by a large timber sleeper 12 in. by 12 in. or more, as the case may demand. The object of the above arrangement is to obtain the necessary support for the shoring, at sufficient distance from the pillar to allow ample room for its removal or the further shoring of it, and to allow sufficient space for the excavation necessary for the new foundation. The

the load on the shores adjacent to the pillars, and obviously lessen the weight on the ground near the foundation. In the case

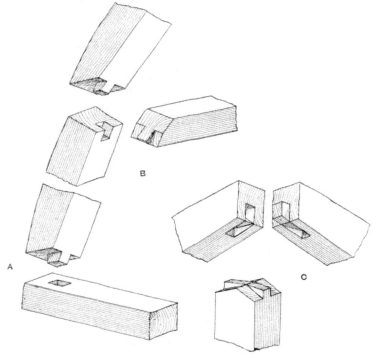

Fig. 823.—Conventional View of Joints of Centres.

centres are also supported in the middle by trestle shoring, which will take a full share of the load, and thus to a large extent reduce

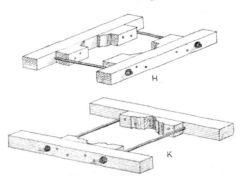

Fig. 822.—Conventional View of Collars to support the Springing of Arches.

of ordinary earths the excavation for the foundation would require to be timbered strongly, so as to prevent any movement. The shoring having been placed in position, the centres can next be erected and forced home so as to fit the arch, by wedging up from the head pieces of the shore as indicated in the illustrations, and also the several parts of the centres themselves, by the wedges shown at D, E, F, and G (Fig. 819). The centres should be connected together by fixing blocks between them, or by bracing them together with pieces of scantling. As the centres might not directly support some of the stones resting on the capital, two collars should be made ; these are shown in position in Figs. 819 to 821. A conventional view of them is given at Fig. 822, in which the collar at H shows pieces of timber which have been accurately scribed to fit the soffit and mouldings adjacent to it near the springing of the arches. The collar at K is at right angles to H, and has had blocks scribed to fit the moulding and the front and back immediately above the capital. These blocks are bolted to the main pieces of the collars as indicated. The collars are made to grip firmly to the

similar to those which have been treated, but of larger scantlings.

Shoring to Railway Arch.

The illustrations (Figs. 824 to 830) show the centering and strutting employed for shoring up two arches of a viaduct over a shallow river. A scaffold and staging to

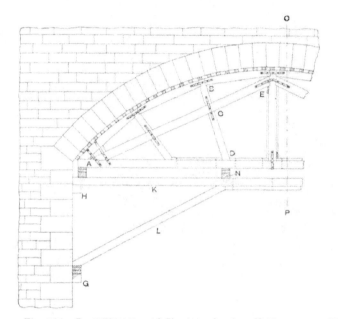

Fig. 824.—Part Elevation of Shoring, showing Half a Principal, One Main Strut, etc.

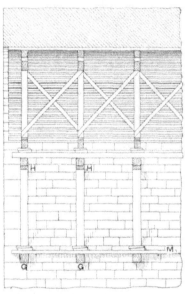

Fig. 825.—Part Transverse Section through Arch and Shoring on Line O P.

stonework by tightening up the iron bolts as shown. In heavy work, of course, it might be necessary to have two bolts at each end so as to prevent accident by the breaking of one. The lower collar is supported on the heads of the trestle shores, and it in its turn supports the upper collar. If it is desired not to remove the capital, the collars would be scribed to fit the pillars immediately below it, in which case the collar would be supported by transoms fixed to the trestles. In this kind of shoring dogs should be freely used. In some cases it might be necessary to support the wall of the arcade on each side; these shores would be somewhat

work upon are erected, holes are cut into the piers at about 6-ft. centres, and wood corbels 9 in. by 9 in. (G and H, Figs. 824 and 825) are set in firmly by wedging and filling in with Portland cement. On the upper corbels transoms K are placed, their centres being supported by the shores L. The corbels G support a plate M from which the shores are wedged. The radii and span of the arches being known, a full-size drawing is set out. The templates for the several pieces of ribs are next made; they are used for cutting out by the bandsaw pieces of timber stuff for the ribs. The struts, tie-beam, and other members are next set out from the drawings and the various joints made. The ribs are set out and jointed with the tie, heads of struts and king-post. Each principal is now fitted together, and when found satisfactory each of the parts

forced up as close as possible to the soffit of the arch by driving in the wedges on the plate N. If a lagging does not fit close, force it up by driving wedges under it on the top of the rib, the object being to make the

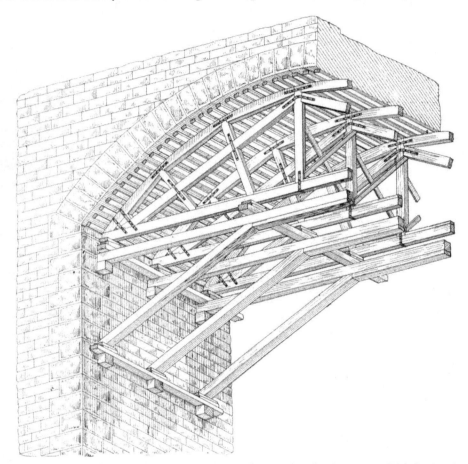

Fig. 826.—General View looking from the Under Side, showing the Arrangement of the Various Members of the Shoring.

joining together is worked, thus allowing each principal to be taken to pieces and sent to the job. Iron straps are provided to secure the principals when re-erected. Between the plates N and the tie-beams of the centres, wedges are inserted. Battens are next placed on the ribs to form the laggings. Each principal and the laggings are then

laggings take their proper share of bearing. To keep the centres upright, braces are fixed between each pair, as shown by Figs. 825 and 826. The former figure represents a part transverse section through arch and shoring on line O P (Fig. 824), the latter a general view (looking from the under side) showing the arrangement of the various members.

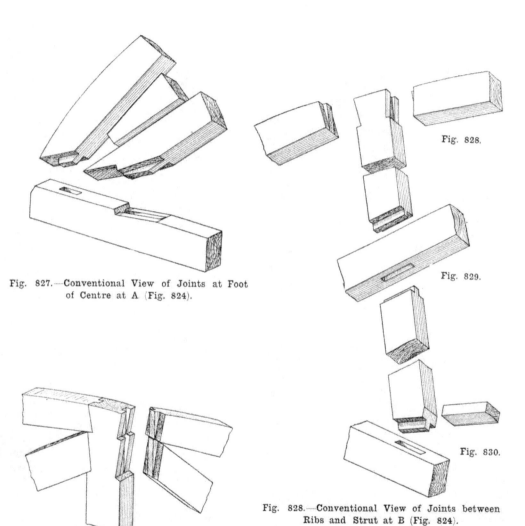

Fig. 828.

Fig. 827.—Conventional View of Joints at Foot of Centre at A (Fig. 824).

Fig. 829.

Fig. 830.

Fig. 831.—Conventional View of Joints at E (Fig. 824).

Fig. 828.—Conventional View of Joints between Ribs and Strut at B (Fig. 824).

Fig. 829.—Conventional View of Joints at C (Fig. 824).

Fig. 830.—Conventional View of Joints at D (Fig. 824).

ARCH CENTERINGS.

Setting Out for Segmental Arches.

THE wooden centerings or centres used for supporting brick or stone arches until the construction is complete, or until the mortar or cement is dry, are made by carpenters, who find it necessary to be able to set out the particular curve required before starting the actual construction of the woodwork. The curves for constructing the centres or turning-pieces, or for setting out

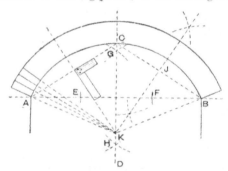

Fig. 832.—Drawing Curves of Arch from a Centre.

the voussoirs or members of segmental and cambered arches, can be obtained by several different methods. The practice, generally, among engineers, architects, and surveyors is merely to indicate the width of the opening to be spanned, and the height or rise above the level line of the springing at the abutments, and so long as the completed work is in conformity with such instructions the artisan is allowed to produce it by the method he is best acquainted with. Fig. 832 shows a segmental arch nearly approaching a semicircle, and on the left-hand side of the figure is indicated a very common method of producing it. The chord of half

the arc is drawn, and the centre of it measured off. The square is then applied to the chord, as shown, and by means of a straightedge a line is drawn at right angles to the chord, bisecting it at its centre, and intersecting the perpendicular line at K, which is the centre of the curve required. This method, while correct in theory, is unscientific in practice, being too dependent on crude mechanical aids, and is not creditable to an intelligent craftsman. On the right-hand side of Fig. 832 a better and more expeditious method is shown, as follows :—First draw the line A B, and mark off the span ; then with A and B as centres mark off equal distances at E and F. With E and F as centres and with any radius greater than half the distance between them, draw the intersecting arcs G and H ; through the points of intersection draw the lines C D. Mark off at C the rise above the line A B, and, with C and B as centres, draw intersecting arcs ; a line drawn through the points of intersection will bisect equally the line B C at J, and being continued will intersect the line C D at K, which is the centre required. A light deal rod and a fine bradawl for a pivot centre will make an excellent compass for this purpose, and is called a radius rod.

Formula for Arch Curves.—For various reasons, such as the absence of anything upon which it can be worked out, and the small round-up of the curve, it is not always convenient to adopt the method just described, and in cases of this kind the mathematical formula

$$\text{Diameter} = \frac{\frac{1}{2}\ \text{span}\ \times\ \frac{1}{2}\ \text{span}}{\text{rise}} + \text{rise}$$

is extremely useful.

Obtaining Radius of Segment of Circle.—

To get the radius of any segment of a circle, the following rule should be committed to memory :—Given the span or chord line, and versed sine (rise), square half the chord, divide by the rise, and to the quotient add the rise. This gives the diameter ; divide by 2 for radius. Referring to Fig. 832 as an example, half chord = 2 ft. 11 in. = 35 × 35 = 1225 ÷ 7 = 175 + 7 = 182 ÷ 2 = 91 in. = 7 ft. 7 in. radius.

in Fig. 833, and these divisions may be subdivided as indicated. The object of bisecting the arcs is to get the lines radial.

Setting Out Curves for Large Arches of Moderate Rise.

Arches of large span and moderate rise cannot conveniently be struck out from a centre, owing to the length of the radius, neither can small arches that have but very little rise. Fig. 834 shows a method

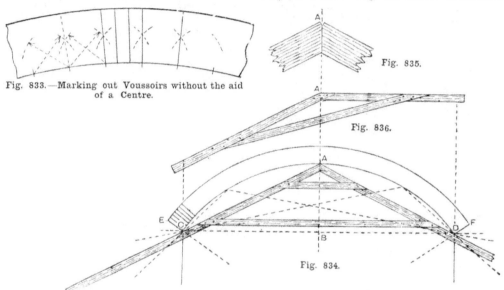

Fig. 833.—Marking out Voussoirs without the aid of a Centre.

Fig. 835.

Fig. 836.

Fig. 834.

Fig. 834.—Drawing Curves of Arch without a Centre by means of a Frame.
Fig. 835.—Enlarged View of Apex of Frame shown in Fig. 834.
Fig. 836.—Alternative Frame to that shown in Fig. 834.

Setting Out Curves with Radius Rod.

Centres for segmental arches of larger dimensions, and others of irregular proportions, should be struck out with a radius rod. After finding the radius by calculation, measure off the distance on the radius rod, and beyond this mark off the depth of the face of the arch. Insert the bradawl in these points, and draw the curves ; cut off a portion of the inner curve so that its chord is equal to the span, and draw the voussoirs or members. This may be done by dividing out either the inner or the outer curve, and bisecting the divisions as shown

of drawing the curves without the aid of a centre. For the purpose of elucidation the figure necessarily shows considerable round-up. The span is marked off on line c d ; the centre line A B is drawn at right angles to it, and the rise is marked off at A. Two French nails are driven in at c and D, and a triangular frame is made as shown, using deal battens about 1 in. thick for a very large arch ; for an arch of moderate size slate battens will do ; the legs touch the nails c D, and at the apex at the point of the rise a small notch is made to accommodate the pencil as shown at A in Fig. 835. The pencil being in position at A, the frame is moved round as indicated by the dotted lines, care being taken that it is always in contact with the nails at C and D, and a true segment of a circle is then struck (Euclid, bk. iii. prop. 21). The frame must be altered, and the nails shifted to E and F to draw the outer curve. The voussoirs or members can be set out as already explained (see Fig. 833). Fig. 836 is a much better appliance, and is arranged to suit the arch shown in Fig. 834; it is not so clumsy to work, and it requires a third nail at D; it is used in the same manner as the board shown in Figs. 837 and 838, and is identical with it.

rise; parallel to the edge of the board a line is then drawn from point A until it intersects with the line of abutment at C, and

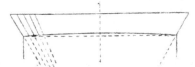

Fig. 839.—Camber or Straight Arch.

the triangular piece from C B is removed. Three nails are inserted at B, C, and D (Fig. 837), and the board (known as a camber

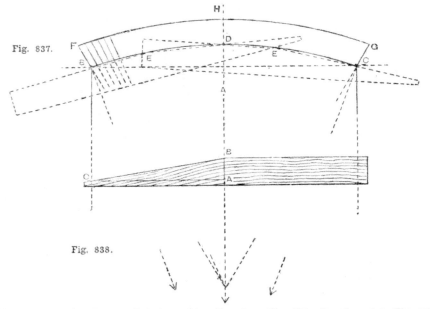

Fig. 837.

Fig. 838.

Fig. 837.—Drawing Curves of Arches with a Board.   Fig. 838.—Board used in Fig. 837.

Setting Out Curve for Camber Arch.

Fig. 837 is an example of a camber arch whose members do not radiate from the centre of curvature, but from some point within it. The centre lying at some considerable distance, the members, if radiating from it, would be so nearly parallel that they would offer no key. In Fig. 837 it is presumed that the rise is less than plank width, namely, 11 in. At the centre of a board (Fig. 838) rather longer than the span required, a line A B is squared equal to the

slip) is moved round (the pencil being held in contact with it at point E), as shown in the dotted lines. As in the former case, the board must be altered and the nails shifted in order to draw the outer curve F, G, H. The camber slip is based on the principle that all angles in a segment of a circle are equal to one another; so by having a long and wide slip tapering both ends, the middle pin marking the rise could be dispensed with; but such a slip is unwieldy, as only half its length is of use in marking a centre.

So-called Straight Arch.

Fig. 839 is an illustration of a so-called straight arch whose intrados is really camber; properly speaking, not less than

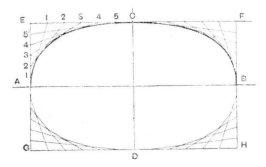

Fig. 840.—Method of Setting Out Ellipse by Intersecting Lines.

$\frac{1}{8}$ in. per foot of span. The method last described is about the only way in which this arch can be expeditiously set out.

Setting Out Curves for Elliptical Arches.

Commonly, an "ellipse" is set out from three centres with compasses, but a three-

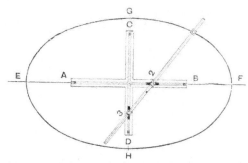

Fig. 841.—Method of Setting Out Ellipse with Trammel.

centred arch is not an elliptical arch—it is only an approximation. The true ellipse is obtained from an oblique section of a cone or cylinder, and no portion of its curve is part of a circle; therefore an ellipse cannot be drawn by compasses or from centres. The following methods are for describing and

setting out true elliptic and oval arches. The first method, illustrated at Fig. 840, is almost universally used by mechanics, as it is easily drawn, and can be adapted to arches of any size. It has also the advantage that no centres are required, the intersection of the lines giving the points through

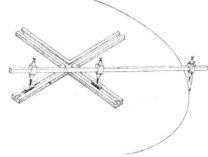

Fig. 842.—General View of Trammel and Cross for Setting Out Elliptical Arches.

which the curve passes. The transverse or major axis A B, and the conjugate or minor axis C D, being given, enclose the space by the parallelogram E F G H. Divide the lines A E, E C, each into any number of equal parts (in this case six), draw the lines 1 1, 2 2, 3 3, 4 4, and 5 5, and the intersection of the lines

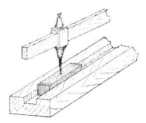

Fig. 843.—Enlarged Detail of End of Trammel and Rod for Setting Out Elliptical Arches.

will give the points in the curve for one quarter of the figure. Repeat the operation for the other three quarters, then bend a thin flexible rod round the points obtained, and draw the curve. It is interesting to note that the granite arches of 50-ft. span crossing the roadway at the Tower Bridge were set out by this method. Although it does not form a perfectly true ellipse, an arch set out in this manner is by no means unpleasing.

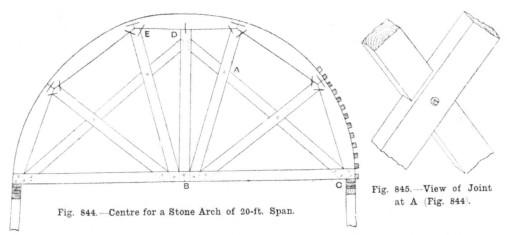

Fig. 844.—Centre for a Stone Arch of 20-ft. Span.

Fig. 845.—View of Joint at A (Fig. 844).

Another Method.—The second method of setting out an ellipse, illustrated by Fig. 841, is probably the best yet devised : it is

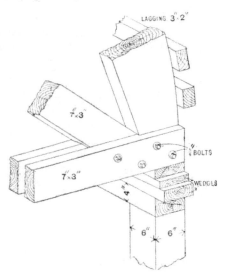

Fig. 846.—View of Joint at C (Fig. 844), also showing Method of supporting Wedges, etc.

done by means of a trammel, which gives a curve by one continuous motion. The trammel consists of a wooden cross A B C D, each arm of which is slotted or grooved. In these grooves two small hardwood sliding bars are carefully fitted, so that they can be moved smoothly to and fro. A rod, 1, 2, 3, provided with three adjustable trammel

heads, is now placed in position as shown in the diagram. This position is obtained by making the distance from 2 to 1 equal to half the shortest diameter of the ellipse G H, and the distance from 1 to 3 equal to half the longest diameter E F. The points

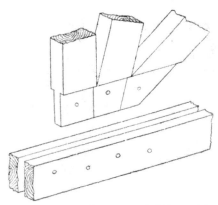

Fig. 847.—Detail of Joint at B (Fig. 844).

of the heads are inserted into the centres of the sliding bars at 2 and 3, and by moving the rod round (the outside head having a pencil fitted in it) the ellipse is described by one continuous line. Fig. 842 shows the trammel in position, while Fig. 843 shows an enlarged detail of one end of the bar. The frame of the trammel is usually made of mahogany or some other hard wood, and the sliding bars are of ebony. The heads are similar to those used for beam compasses, and are adjusted by a screw pressing against the radius rod on which they slide.

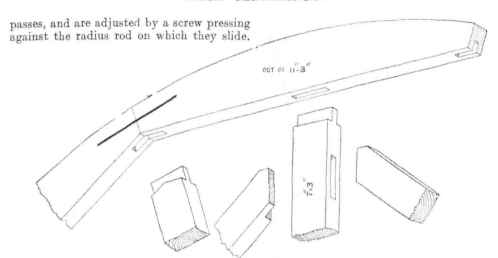

OUT OF 11"-3"

7"×3"

Fig. 848.—Detail of Ribs, and Joints at D and E (Fig. 844).

The frame of the trammel is held together by a small wood screw at each corner. Its size depends, of course, upon the size of the ellipse to be described, but one frame will describe ellipses of various diameters.

Centerings for Semicircular Arches.

Centres for semicircular stone arches, having generally to bear a large amount of weight, are usually built up of plank and batten scantlings, which are roughly framed

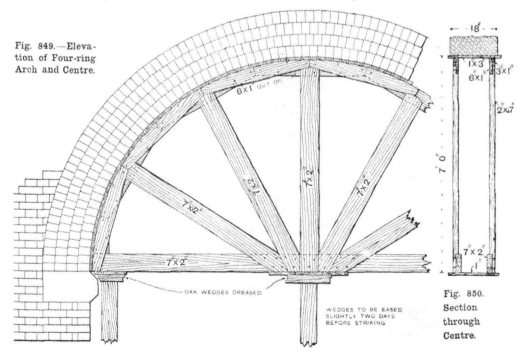

Fig. 849.—Elevation of Four-ring Arch and Centre.

6×1 OUT OF

7×2"

7×2"

7×2

2×1

7×2

OAK WEDGES GREASED

WEDGES TO BE EASED SLIGHTLY TWO DAYS BEFORE STRIKING

18"

1×3
6×1

3×1"

2"×7

7"0"

7×2"

Fig. 850. Section through Centre.

together, and fixed with bolts and dog irons, as indicated in the sketches. Fig. 844 shows a centre braced so that it is only necessary to be supported at each end, leaving a free

wedges are used for raising or lowering the centre slightly, so as to adjust it to its exact position previous to building the arch ; and, secondly, when the arch is finished, to ease

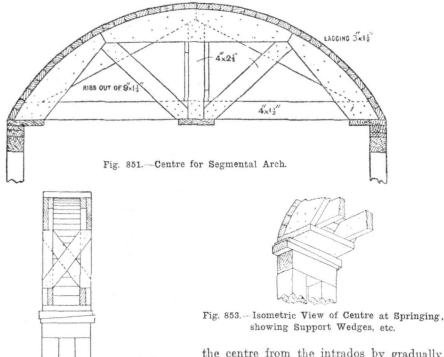

Fig. 851.—Centre for Segmental Arch.

Fig. 853.—Isometric View of Centre at Springing, showing Support Wedges, etc.

Fig. 852.—Section through Figs. 851 and 854.

the centre from the intrados by gradually slackening the wedges. They also allow of the centre being taken down without undue

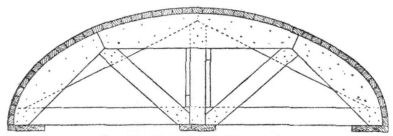

Fig. 854.—Centre for an Elliptical Arch.

passage underneath. The joints used are shown by Figs. 845 to 848. Wedges are placed in pairs directly between the top of the supports and the bottom of the centre, as shown by Fig. 846. In the first place, the

vibration, which would be otherwise caused: Fig. 849 is the part elevation of a four-ring brick arch with centering, the struts of the latter finishing against the ribs as shown in the vertical section (Fig. 850).

Centerings for Segmental Arches.

Fig. 851 is the elevation of a rough centre for a segmental arch 8 ft. wide, 2 ft. 6 in. rise, and 12 in. soffit. Fig. 852 is the section through this centre and also the elliptic one shown later (see Fig. 854). Fig. 853 is an isometric view, showing wedges under end of centre. Close lagging is shown, but

Centering for Elliptical Window.

An elliptical window in brickwork may be built up round a centering of the kind shown in the half internal elevation (Fig. 855) and in the conventional view (Fig. 856). The centering is supported by the struts A, but sometimes, instead of these, a part ring of bricks, laid dry, is used, the bricks resting

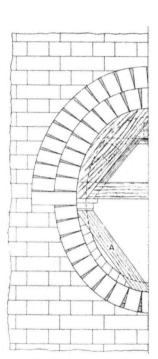

Fig. 855.—Half Internal Elevation of Elliptical Window.

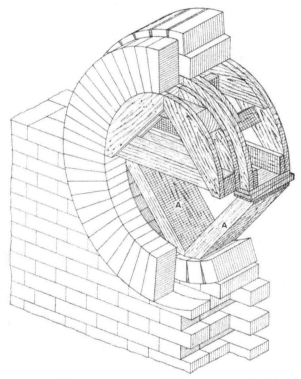

Fig. 856.—Conventional View, showing Centering, Construction of Arch, etc.

frequently it would be formed so that there would be a space between each two strips of wood lagging.

Centerings for Elliptical Arch.

The centering shown by Fig. 854 is to fulfil the conditions laid down for the segmental arch which has been described in the previous paragraph, and the construction of the centering is similar.

on the inverted ring supporting the wedges and centering.

Centre for Circle-on-Circle Arch with Parallel Jambs and Reveals.

Figs. 857 and 858 are elevation and plan of a circle-on-circle window or door opening in which jambs or reveals are parallel and the whole of the soffit of the arch is cylindrical. The elevation of the centre is given at Fig. 859, the plan of the ribs at A, the plan of the laggings at B (Fig. 860), and the development of the soffit at C. The shapes the ribs and soffit as indicated at A and B (Fig. 860). In Fig. 859 the line of ribs is shown on the left, whilst the line of laggings

Fig. 857.—Elevation of Circle-on-Circle Opening, with Soffit parallel at the Springing and level at the Crown.

Fig. 858.—Plan of Soffit (Fig. 857).

of the inner and outer ribs are identical. How to obtain their shapes will now be explained. First draw the elevation of the centre as shown in Fig. 859, and the plans of is shown at the right. In the elevation, divide the arc into an equal number of parts, as a', b', etc., draw ordinates to the springing line, and project down to the plan of the ribs, as a, b, c, d, e, and 1. At right angles to the plan of the ribs, draw the ordinates, making them of the same lengths as those in the elevation; thus a series of points is obtained, as c'' to $1''$, through which the curve for the ribs is drawn. To obtain the soffit mould, divide the line of laggings into equal parts as o' to $1'$, and project down to the plan. Through point o in the plan, draw the horizontal line o to $1''$ and mark

Centre for Circle=on=Circle Arch with Radial Jambs or Reveals.

Figs. 861 and 862 show the elevation and plan of a circle-on-circle window or door opening with radial jambs or reveals. Fig. 863 gives the elevation of the front and back of the centre, at the left and right respectively. The face moulds for the ribs are obtained in exactly similar manner to those

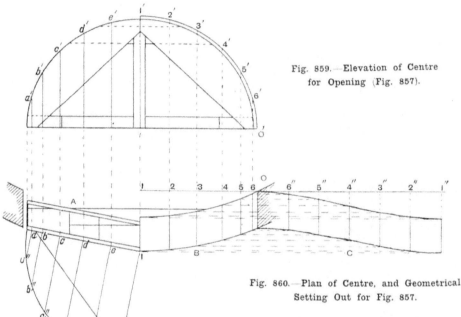

Fig. 859.—Elevation of Centre for Opening (Fig. 857).

Fig. 860.—Plan of Centre, and Geometrical Setting Out for Fig. 857.

off distances on it equal to the divisions o', $6'$, $5'$, etc., in elevation. From these points draw lines at right angles to o $1''$, and project from the corresponding points in the plan. Through the points thus obtained the development of the soffit may be drawn. After the ribs are sawn out, the edge should be planed in such a way that it is level at the top, but gradually bevels until at the springing it is at the angle shown in the plan. The construction of the centre is clearly shown in the illustrations.

in the case illustrated by Figs. 859 and 860. For the development of the soffit mould, a separate drawing must be made as shown at Fig. 865. Set out the elevation and plan of soffit and radial lines meeting at any convenient point a', divide the elevation into a number of equal parts as shown from 1 to 8 (A), and project these points down to the plan, giving corresponding numbered points. From these last points draw lines radiating to a', these being plans of generators of the conoidal surfaces which are shown by the conventional view (Fig. 866). Projectors have not been drawn for the right-hand half of the plan, as this is exactly the same as the left-hand half. From points 1 to 7 in the elevation draw the elevation of the generators meeting the line $a\,8$ in points as shown ; at right angles to $1'\,a'$ draw $a'\,8'$ (c). Mark off

development of the front edge, set a pair of compasses to one of the equal distances of

Fig. 861.—Elevation of Circle-on-Circle Opening, with Soffit converging at the Springing and level at the Crown.

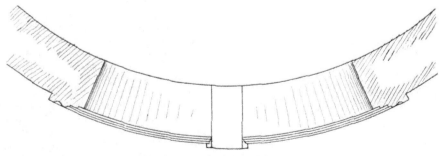

Fig. 862.—Plan of Soffit (Fig. 861).

distances on it exactly equal to those on $8\,a$ (A), which give the points where the generators start from. This will be understood by reference to Fig. 866. For the

the elevation, and another pair or a radius rod to the length of the generators. (Note, these are all one length.) Then using point 1 (B) as centre, describe an arc with the compasses ; then with b as centre in $a'\,8'$ (c) draw an arc with a radius equal to the generators ; where these two arcs intersect is the point 2 (c). Each of the other points in the development is obtained in precisely the same manner, and the curve can be

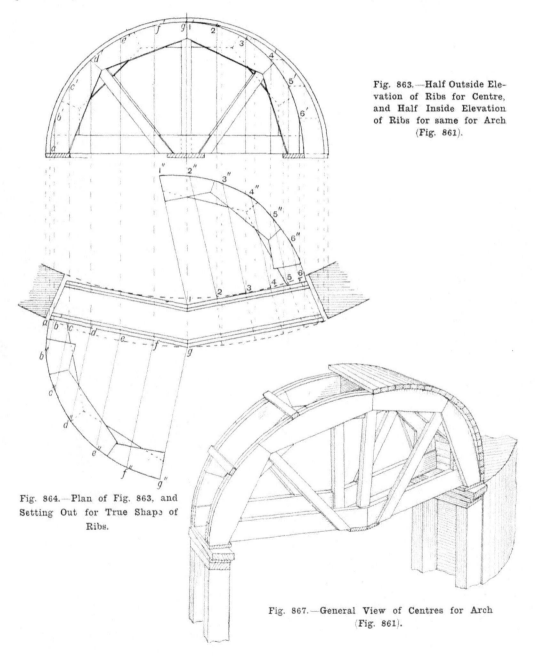

Fig. 863.—Half Outside Elevation of Ribs for Centre, and Half Inside Elevation of Ribs for same for Arch (Fig. 861).

Fig. 864.—Plan of Fig. 863, and Setting Out for True Shape of Ribs.

Fig. 867.—General View of Centres for Arch (Fig. 861).

drawn in as shown. For most practical purposes half would be sufficient, but the whole has been shown. The method of

In a segmental-on-plan centre, supports must be given by sturts fixed at an angle of 45° from the vertical side posts.

Fig. 865. — Setting Out for Development for Soffit of Arch.

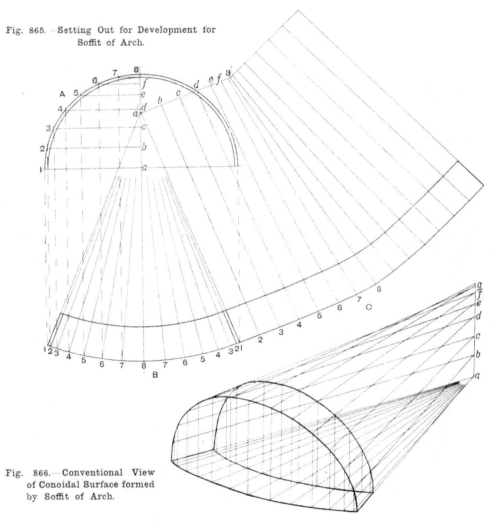

Fig. 866. — Conventional View of Conoidal Surface formed by Soffit of Arch.

building up the centre with the ribs in two thicknesses is fully shown at Figs. 863, 864, and 867. And the last-named figure also shows the general construction of the centre, its support, with wedges, etc. A central support is not shown, but if by any means this can be arranged for it will be found advantageous.

Centre for Opening with Converging Outer Reveals and Parallel Inner Reveals.

At Figs. 868 and 869 are shown in elevation and plan the necessary centering for an opening with converging outer reveals and parallel (cylindrical) inner reveals. The true geometrical working for obtaining the shape of the ribs and the development of half the inner soffit is shown, but not for the

Fig. 868.

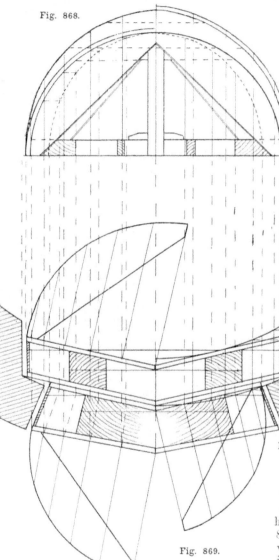

Fig. 869.

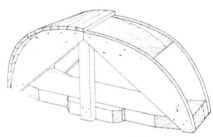

Fig. 870.—Conventional View of Centre for converging Jambs (Fig. 868).

Fig. 871.—General View of Inner Centres for Parallel Reveals (Fig. 868).

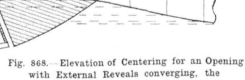

Fig. 868.—Elevation of Centering for an Opening with External Reveals converging, the Internal Reveals being parallel.

Fig. 869.—Plan of Centering, Setting Out for Ribs, etc.

have been so fully described. Figs. 870 and 871, which represent respectively a conventional view of a centre for converging jambs, and a general view of inner centres for converging reveals, will convey a clear idea of the method of constructing these centres when they are for a span of an ordinary sized opening. Fig. 877 shows how a block may be cut to connect the

soffit with the converging reveals; but the careful reader will have little difficulty if he note that this case is a combination of the two previous ones (Figs. 859 to 867), which heads of the ribs, as at A (Fig. 876) ; and the centre may be further strengthened by fixing in a tie, and bracing as indicated by the dotted lines. The close lagging shown is most suitable for brickwork, but in the case

Centering for a Gothic-on-Circle Arch.

The elevation and plan of the soffit of an arch of this description is shown at Figs.

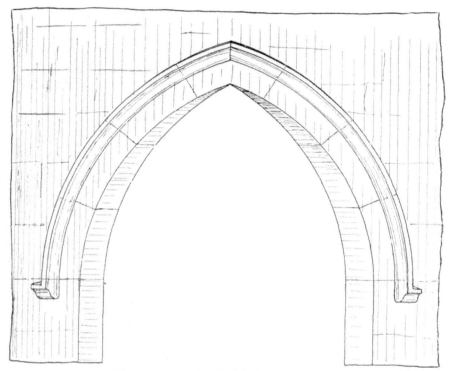

Fig. 872.—Elevation of Gothic-on-Circle Arch.

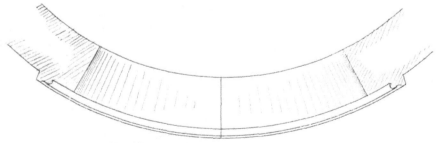

Fig. 873.—Plan of Soffit (Fig. 872).

of a stone arch the laggings are usually of a stouter character, and put at intervals of 2 in. or more apart, so as to support the stonework and tie the ribs together.

872 and 873. The soffit of the arch at the springing converges, but finishes in a level line. Details of the geometrical setting out of the ribs and general construction are

illustrated by Figs. 874 to 877. The geo-metrical construction for obtaining the shape of the ribs would involve the same principles and method of working as given

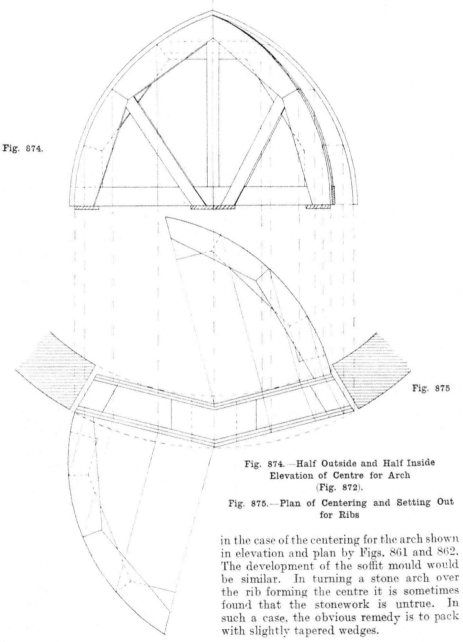

Fig. 874.

Fig. 875

Fig. 874.—Half Outside and Half Inside
Elevation of Centre for Arch
(Fig. 872).

Fig. 875.—Plan of Centering and Setting Out
for Ribs

in the case of the centering for the arch shown in elevation and plan by Figs. 861 and 862. The development of the soffit mould would be similar. In turning a stone arch over the rib forming the centre it is sometimes found that the stonework is untrue. In such a case, the obvious remedy is to pack with slightly tapered wedges.

12

Centering for a Stone Arch and Brick Back Arch.

Figs. 878 and 879 show the arrangement of centering often used when the front of an arch is of stone and the backing is of brick

Centre for Gothic Arch to the Arcade of a Church.

An example of centering of this description is illustrated by Figs. 880 to 885. It will be seen that in this example, where the mould-

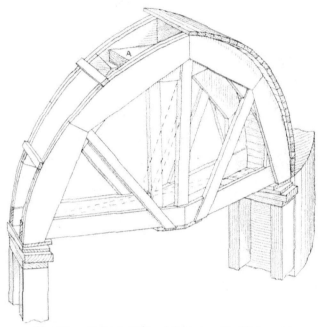

Fig. 876.—General View of Centre, with Wedges, etc.

with a 4½-in. reveal. The centre for the brickwork is larger to allow of the 4½-in. reveal with lagging as shown, whereas a rib centre built up of two thicknesses as shown is sufficient to support the voussoir if the centre of gravity of the stones falls within the ribs. When this is not the case another rib would have to be made large enough to be in contact with some member of the moulding. In the case that is here shown the rib might be adjusted against the flat part of the hollow A or against the square B. The two parts of the centre are connected together by nailing on blocks as shown at C, D, and E (Fig. 879). The block shown at F is nearly behind the head and top of the standards in the same plane, and thus they can be more easily braced.

ings of the arch are for the most part in chamfered planes, the ordinary centre with lagging would be unsuitable, as it would only give direct support to the centre moulding or surface forming the soffit of the arch. As these arches are usually built in at least

Fig. 877.—Block for fixing between Head of Ribs as shown at A (Fig. 876).

two rings of courses, the centre would not directly touch the second ring, therefore the centres have to be constructed so as to give direct support to each ring. This is gener-

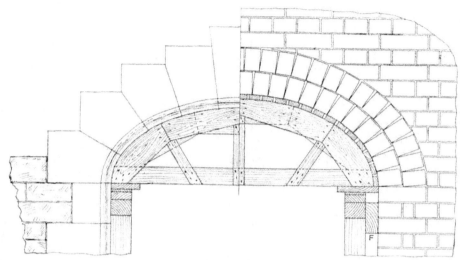

Fig. 878.—Half Elevation showing Centering to a Stone Arch ; also Half Elevation of Centering to Internal Brick Arch.

ally done by supporting each ring upon one or more ribs, the case here dealt with being so treated. By reference to the half-sectional elevation (Fig. 880) and the transverse section (Fig. 882) it will be seen that the middle centre B is smaller. It is formed of two ribs, and gives support to the stones forming the soffit of the first ring of the arch.

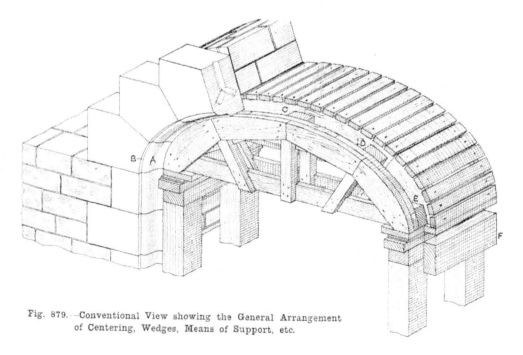

Fig. 879.—Conventional View showing the General Arrangement of Centering, Wedges, Means of Support, etc.

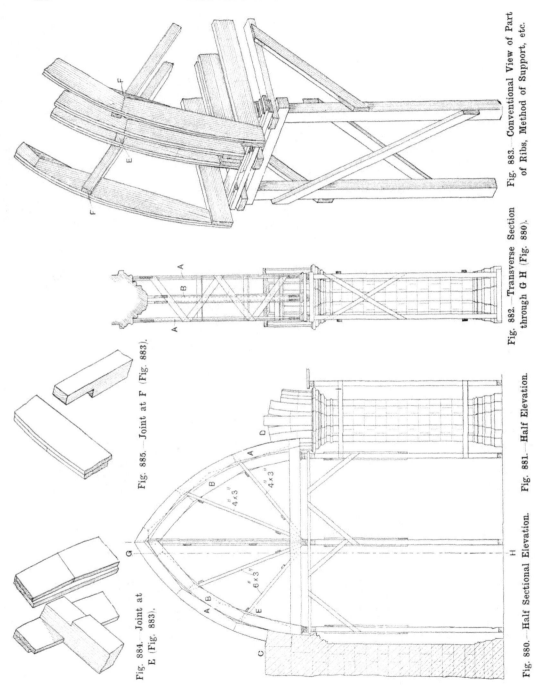

Fig. 883.— Conventional View of Part of Ribs, Method of Support, etc.

Fig. 882.— Transverse Section through G H (Fig. 880).

Fig. 885.— Joint at F (Fig. 883).

Fig. 884.— Joint at E (Fig. 883).

Fig. 881.— Half Elevation.

Fig. 880.— Half Sectional Elevation.

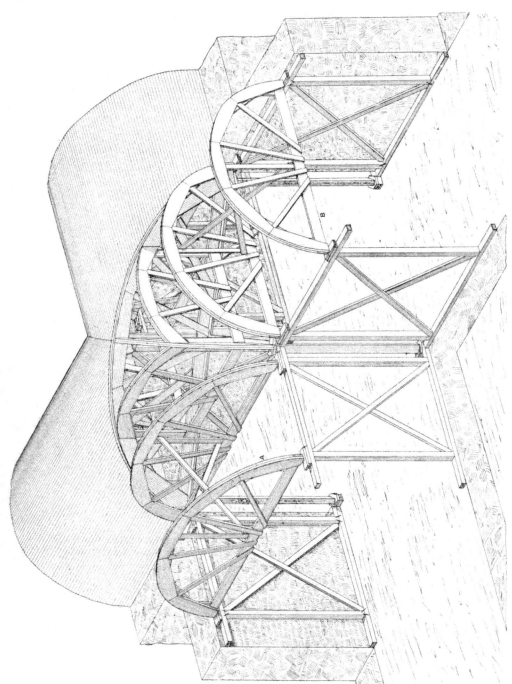

Fig. 886.—Conventional View of Centering, Timbering, etc., for Four Vaults intersecting at Right Angles.

The two outer ribs shown at A (Figs. 880 to 882) are larger so as to support the series of stones forming the second ring. The conventional view (Fig. 883), showing part of the ribs, the wedges, and the method of supporting, will make the construction clear. Each rib is built up of two thicknesses of 11-in.

Centering for Barrel Vaulting.

The isometrical view (Fig. 886) will convey a general idea of the centering and timbering required in the construction of four semicircular barrel vaults, which intersect at right angles in groins as shown. To give a

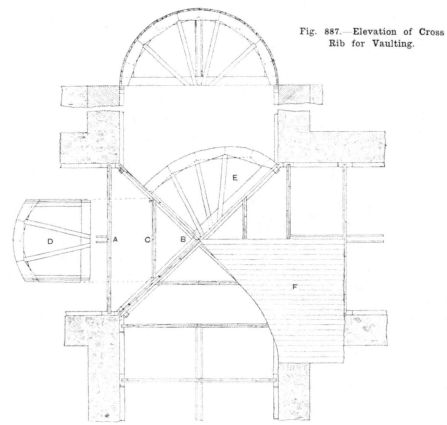

Fig. 887.—Elevation of Cross Rib for Vaulting.

Fig. 888.—Plan of Centering at Intersection of Vaults : C, Plan of Intermediate Principal, or Rib. D, Elevation of Same. E, True Form of Half Diagonal Principal. F, Development of Lagging.

boards from 1 in. to 1½ in. thick, according to the weight to be supported, nailed together in the usual manner. The struts may also be of board stuff, or 4-in. by 3-in. to 6-in. by 3-in. scantling may be used and shouldered to the ribs as shown by Figs. 884 and 885. The ribs should be braced together as indicated in the elevations and section.

better view of the centering connected with the groining of the vaults, the ribs A and B (Fig. 886) have been purposely shown farther apart than they should be in reality, and consequently the standards would be placed much nearer together than shown. All main members have been shown where practicable, but for clearness some of the minor timbers, such as braces, have been omitted. For a similar reason the walls nearest to the spectator have not been shown above the ground level. The strength of the ribs and of the supporting timbers would be varied according to the material they would have to support. Generally, in the case of stonework, the weight to be supported would be much greater than in the case of brickwork, therefore all members of the timbers should be proportionately increased. The laggings at the intersection of the vaults would be supported principally by ribs across the diagonals as shown in the conventional view (Fig. 886), and in the plan (Fig. 888). One principal would be framed up of several thicknesses so as to span one of the diagonals ; this should be well supported by a head, standards, braces, and sill. The other diagonal would be spanned by two half-principals ; these would be fixed to the main principal by straps and bolts, and supported by a head, standards, etc. It will be noticed that these principal ribs are represented as being backed ; that is, the edge of half the thickness of each principal rib is bevelled to keep it in the same cylindrical surface as the lagging it will have to support, so as to afford a firm bearing to the latter. As the space from one of the ordinary ribs (shown in plan at A, Fig. 888) to the intersection at B would be too great for the laggings, part principals would be constructed and fixed to those spanning the diagonals as indicated in plan at C. An elevation of one of these is projected at D. A half-elevation of one of the diagonal principals is shown at E. Also development of portion of the lagging is shown at F. The methods of setting out the curves for the diagonal principals is shown separately at Fig. 889, and that for the development of the laggings at Fig. 890. At G half the transverse section, from the springing to the crown of the vault, is shown from 7 to 0. To set this out, mark off the thickness of the lagging and draw in the quadrant as shown from a' to h' ; then, at any convenient position below, draw lines 9, 11, and 9, 10 ; the former represents the plan of the line of transverse section of the vault and the latter the plan of the line of intersection of two vaults. Divide the

quadrant a' to h' into any number of equal parts, and from these draw ordinates at right angles to 7, 8. Project each of these down to the plan of the diagonal 9, 10. At right angles to this line set up ordinates as shown at H, the length of each of these being, of course, equal to its corresponding member at G. Through the points thus obtained draw the curve. It will be clear to the reader that this is a quadrant of an ellipse, and that in nearly every practical case it would be more convenient to draw in the

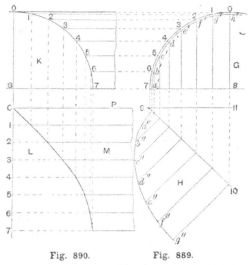

Fig. 890.   Fig. 889.

Fig. 889.—Geometrical Setting Out for obtaining True Form of Edge of Diagonal Principal.

Fig. 890.—Geometrical Construction for obtaining Development of Curve for Intersection of Laggings.

curve by means of an elliptic trammel or other of the practical methods described on pp. 250 to 252. To obtain the development of the laggings on a level with 7, 8 at G, draw 7, 8 at K (Fig. 890) ; then draw in the quadrant 7 to 0, equal to 7 to 0 at G ; project across from points 0 to 7 at G, and obtain corresponding points on the quadrant at K. Produce 0, 8, of course at right angles to 8, 7, and at any point 0 in 0, 7, at L, draw 0 P at right angles. Obtain the stretch out of the quadrant 0, 7, and mark it off from 0 to 7 at L, dividing into the same number of equal parts as the quadrant. Projecting down

from these points in the quadrant, and from
the corresponding numbered points in the
line 0 7 at L, points for the curve are obtained, and this may be drawn in as shown,
by which a portion of the lagging is obtained

turning movement.   In the case of vaulting
spanning a greater distance, thicker material
would be necessary, and in some cases it
would be considered necessary to frame the
principals together out of battens and deals

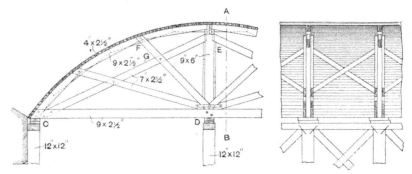

Fig. 891.—Part Elevation of Centering and
Supports.

Fig. 892.—Part Longitudinal Section through A B
(Fig. 891), showing Bracing connecting Principals.

as indicated at M.   The construction of the
ribs or principals for vaulting differs very
much according to the magnitude of the job.

3 in. by 4 in., connecting the joints together
by stub tenons, dogs and straps, as found
most serviceable.

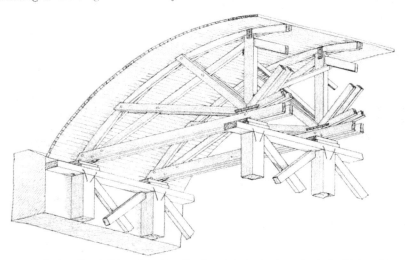

Fig. 893.—Conventional View showing the General Arrangement of the Main Timbers.

For small vaults spanning not more than
8 ft. or 9 ft. they would be made of 1¼-in. or
1½-in. boarding, nailed together in two
thicknesses and supported on standards
6 in. by 4 in. or 6 in. by 6 in., these being
properly braced so as to prevent any

Centre for a Segmental Bridge, 30-ft. Span.

The accompanying Figs. 891 to 897
illustrate the construction of the centre for
a stone and brick segmental bridge of 30-ft.

span and 8-ft. rise, which was actually erected over a shallow river. The sizes of the different members are given. By reference to the conventional view (Fig. 893) and details (Figs. 894 to 897) it will be seen that the ribs were in two thicknesses, out of 9-in. by $2\frac{1}{2}$-in. These were connected to and supported by struts, bolted on each side and meeting at the foot of the king-post as shown.

of from 50 ft. to 60 ft., in which it is desirable for the centering to be supported at each end only so as to leave as much space as possible, such as would be required to allow of navigation in the case of a bridge over water, or over a thoroughfare for vehicular traffic. Fig. 898 also shows part of the elevation of the bridge. The conventional view (Fig. 899) shows clearly the construction of each

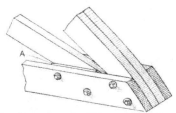

Fig. 894.—Detail of Foot of Principal C
(Fig. 891).

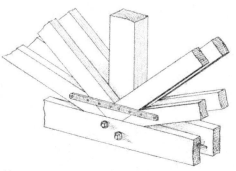

Fig. 895.—Detail of Foot of King-post, etc., D
(Fig. 891).

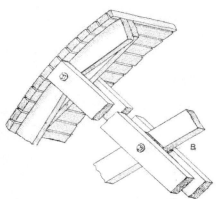

Fig. 897.—Details of Joints at F and G
(Fig. 891).

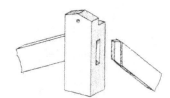

Fig. 896.—Detail of Joints at E (Fig. 891).

All the parts of each principal were connected together by $\frac{3}{4}$-in. bolts and nuts with washers, and straps at the foot of the king-post. Packing pieces were used where necessary, as indicated at A (Fig. 894) and B (Fig. 897). The principals were braced at 6-ft. centres as indicated at Fig. 892.

Design for Centre for an Elliptical Stone Arch for 50=ft. Span, supported at the Ends.

Fig. 898 shows in elevation the centering for the erection of a stone bridge for a span

12*

principal, and also the means of support and bracing. It should be noted that to avoid confusion in this view the bracing connecting the principals is not shown. The leading dimensions are figured on the drawings, which clearly show the construction, and therefore it will only be necessary to refer to a few of the chief points in the centering. The main strut A is connected and strapped to the king-post at its foot; a piece of timber B is bolted to it by 1-in. bolts as indicated, so as to give additional bearing. The ties C are in two thicknesses, one being fixed on each side of the main strut

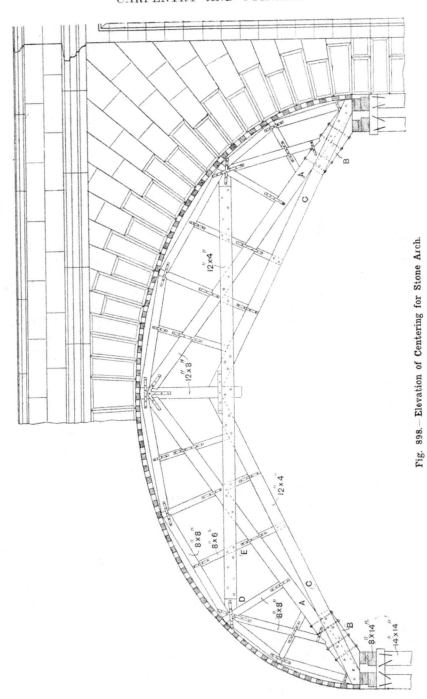

Fig. 898.— Elevation of Centering for Stone Arch.

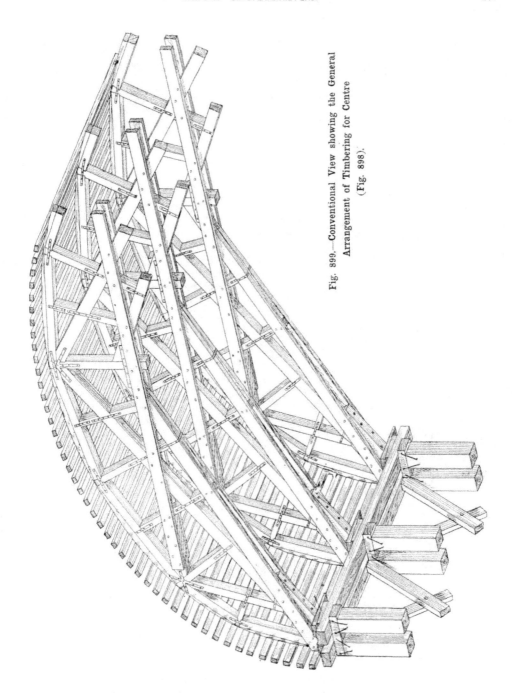

Fig. 899.—Conventional View showing the General Arrangement of Timbering for Centre (Fig. 898).

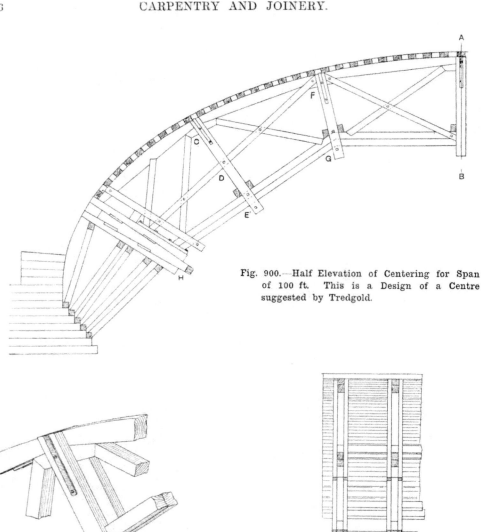

Fig. 900.—Half Elevation of Centering for Span of 100 ft. This is a Design of a Centre suggested by Tredgold.

Fig. 902.—Enlarged Details of Joints, etc., at C, D, and E (Fig. 900).

Fig. 901.—Longitudinal Section through A B (Fig. 900).

A and of the heel piece B by bolts. By reference to Fig. 899 it will be seen that these two thicknesses are gradually brought together so that they become equal to the thickness of the king-post to which they are connected and strapped. This allows an uninterrupted horizontal brace to run through on each side of the principal, so that it can be bolted to each member which it crosses. The ends of these horizontal braces are connected by bolts to a piece of scantling cut between the struts D and E. When the arch is loaded, this allows these braces to serve as struts. Two pairs of wedges are shown under the end of each principal, their purpose being to give a greater bearing surface, to obviate any chance of the wedges being crushed, and also to facilitate the easing or striking of the principals.

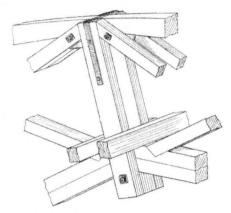

Fig. 903.—Enlarged Details of Joints at F and G (Fig. 900).

Design for Centre suggested by Tredgold.

Figs. 900 to 904 are the elevation, section, and details of the centering for an elliptical stone arch as designed by the well-known authority, Tredgold. This particular construction may be used for any span from 60 ft. to 100 ft., and the centres of similar design, modified to special requirements, have proved successful. In Fig. 900 a special form of wedging for adjusting the principals and striking them is shown in elevation at H. The particular shape of each piece will be understood by reference to Fig. 904. The pairs of wedges K are for insertion in the four spaces shown at H (Fig. 900). These wedges are used for adjusting the principals. When it is desired to ease the principals, to prevent any slipping of the main wedges, these smaller wedges are loosened, and the centre wedge is struck at the end H. Or, upon the centres being struck, the smaller wedges would be taken right out and the centre wedge H driven back to its full extent. Sometimes the end H is shod with iron to prevent splitting whilst being driven.

Centering for a Segmental Stone Arch, 70-ft. Span, resting on Five Supports.

The design for the centering for a segmental stone arch for a 70-ft. span and 30-ft. rise

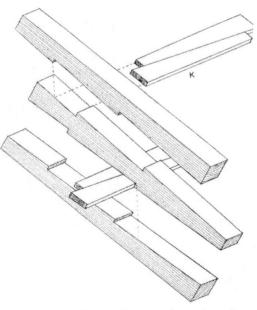

Fig. 904.—Enlarged Detail of the System of Wedging for Striking Centre at H (Fig. 900).

is shown by Figs. 905 and 906. This form is applicable where intermediate supports

The right-hand half shows an alternative method where the head beam A is supported

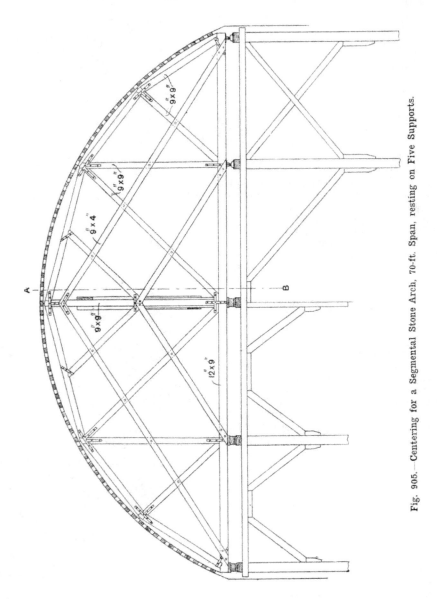

Fig. 905.—Centering for a Segmental Stone Arch, 70-ft. Span, resting on Five Supports.

are permissible. The left-hand half shows how the timbering would be arranged if supported on five rows of piles or standards.

at five positions, but only four rows of piles or standards are needed. This arrangement would provide for a space of about

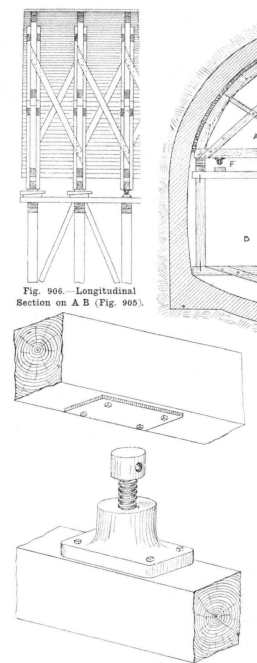

Fig. 906.—Longitudinal Section on A B (Fig. 905).

Fig. 907.—General View of Screw Jack for easing Principals.

Fig. 908.—Transverse Section through a Tunnel, showing Alternative Methods of Centering.

32 ft. for navigation or traffic. The left-hand half shows ordinary pairs of wedges for easing and striking, whilst on the right simple forms of screw jacks are shown. The bodies of these jacks are castings fixed on the transom beams, the heads of the screws bearing against malleable cast plates bolted to the under edge of the tie beam. An enlarged detail of one of these jacks is shown at Fig. 907.

Centering for a Tunnel.

A transverse section through a tunnel is given at Fig. 908, where it will be seen that the trusses for the ribs have queen-posts; this principle of construction gives great strength, and is in general favour for this class of work. Alternative designs for trusses are given at A and B; as also for the supporting timbering at D and E. When a section of the arch has been completed, arrangement is made to lower the centering a little from the soffit, and then to push the centering forward and raise it to its

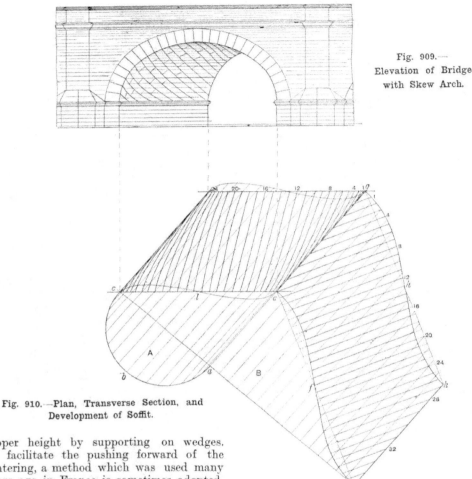

Fig. 909.—
Elevation of Bridge
with Skew Arch.

Fig. 910.—Plan, Transverse Section, and
Development of Soffit.

proper height by supporting on wedges.
To facilitate the pushing forward of the
centering, a method which was used many
years ago in France is sometimes adopted,
by fixing strong axle rollers as indicated at
F. Thus when the wedges are taken out
the rollers are received by planks as shown,
and the centering is easily levered forward.

Centering for a Skew Arch Bridge.

The elevation of a bridge with a skew arch
is shown by Fig. 909, where it will be noted
that the face of the arch is elliptical; but
by reference to Fig. 910 it will be seen that
the transverse section A of the arch is a semicircle. At Fig. 911 is given an elevation
of one elliptical rib and the plan of seven
ribs. Sometimes the ribs for skew arches
are made to fit the transverse section, and

placed as indicated in the plan (Fig. 912);
but it will be seen that for the present case
(shown by A, B, C, D) this arrangement would
be unsuitable, because of the large proportion of some of the ribs at each end, which
is not required to support the arch; also the
thrust of the masonry at the loaded ends
would necessitate the other ends being
strongly shored to prevent movement.
When the axis of the arch is less oblique to
the face, as indicated by the dotted lines
E F and G H, square centering is permissible,
and often an advantage. The development

of the soffit of the arch is shown at B (Fig. 910), the line *a d* being equal in length to the semicircle *a b c*, and each one divided into an equal number of parts, giving points from which generators are drawn as shown. By projecting from where the plan of each generator cuts the face line *c e* to its respective position in the development B, a number of points of intersection are obtained through which the curve of the front edge of the soffit can be drawn as shown by the line

of the voussoirs of the arch, as indicated at Fig. 918. This centre is supported on the same timbering, but is independent of the centering for the soffit of the dome. The front rib of the centering for the soffit is built up of two thicknesses of 1¼-in. stuff, with 4-in. by 2½-in. stuff for struts. The transverse ribs are made of 1-in. stuff, and constructed as indicated at Figs. 920 to 923. It will be observed that these ribs have their curved edges bevelled so as to fit the

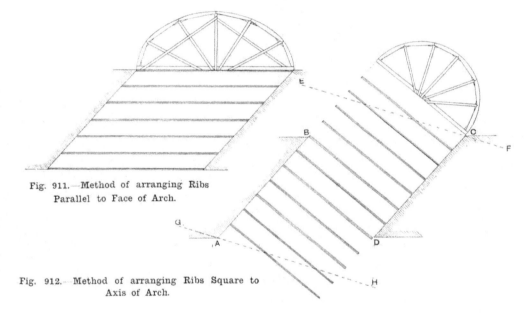

Fig. 911.—Method of arranging Ribs Parallel to Face of Arch.

Fig. 912.—Method of arranging Ribs Square to Axis of Arch.

e f d ; g h and *k* are obtained in a similar manner. The coursing joints are shown on the development, and from this they have been projected to the plan, and from the plan to the elevation.

Centering for a Large Elliptical Niche or Semi=Dome.

Fig. 913 is half the sectional elevation of an elliptical stone niche or semi-dome on line A B (Fig. 914). It is 24 ft. span, rise 8 ft. 6 in., and depth 8 ft. 6 in. The centre for the arch in the front of the dome is formed of two ribs, upon which lagging pieces are fixed ; these are sufficiently long to give support to the soffit

soffit. The amount of this bevelling, and also the size of each rib, has been projected from the plan (Fig. 916) to the end view (Fig. 917). The transverse ribs are fixed to the front rib in the manner clearly shown at Fig. 919. The whole would be supported on timber staging as shown, with wedges inserted under the centering for striking purposes. If the stones were of a large size, little or no lagging would be required, provided the ribs were sufficiently near together.

In the event of constructing centering of this character for a dome to be built in brick-work, it would be necessary to close-board the top of the centre ; and in this case the boards would have to be shaped somewhat

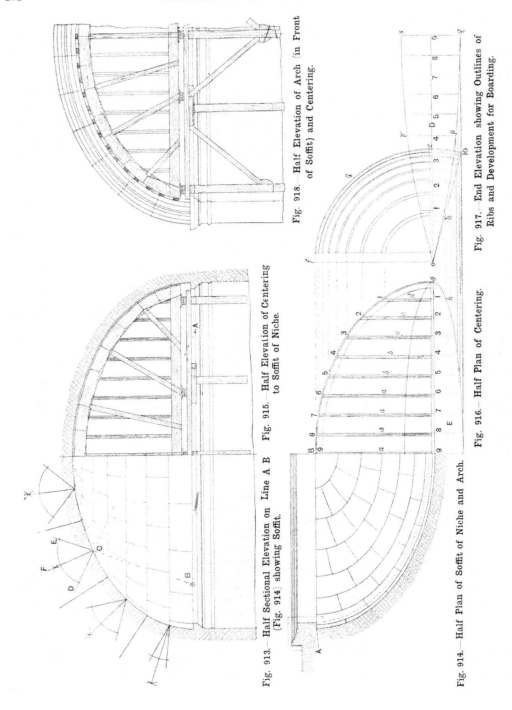

Fig. 918.—Half Elevation of Arch (in Front of Soffit) and Centering.

Fig. 917.—End Elevation showing Outlines of Ribs and Development for Boarding.

Fig. 915.—Half Elevation of Centering to Soffit of Niche.

Fig. 913.—Half Sectional Elevation on Line A B (Fig. 914) showing Soffit.

Fig. 916.—Half Plan of Centering.

Fig. 914.—Half Plan of Soffit of Niche and Arch.

as shown at D (Fig. 917). The geometrical method of doing this is as follows :—From any convenient points in the plan of the face of the rib, as 0 to 9, project up to the front arris of the soffit in elevation. But as

to half the breadth of the board in its widest part, join 0 10, and continue the other arcs to touch 0 10 as shown by the dotted lines (Fig. 917). Now draw the plans of these as shown by 1 to 9 (Fig. 916). Then

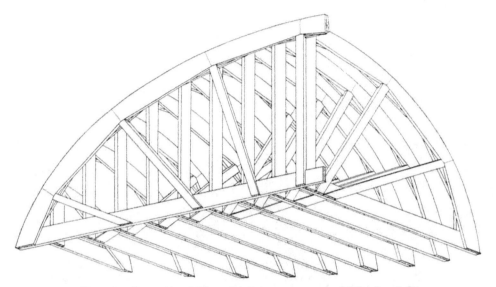

Fig. 919.—Conventional View showing arrangement of Ribs for Soffit.

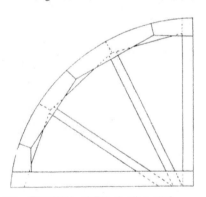

Fig. 920.—Elevation of Rib A (Fig. 916).

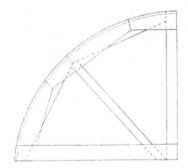

Fig. 921.—Elevation of Rib B (Fig. 916).

in this case the ellipse in plan is the same as in elevation, points 0 to B may be used as shown. Draw 0 9 (Fig. 917) in the same line as 0 9 (Fig. 916). Along 0 9 (Fig. 917) mark off distances 0 to 9 equal to 0 to 9 along the elliptic curve 0 to 9 (Fig. 916), continue the arc ge as e 10, make e 10 equal

from where each of the dotted arcs cuts line 0 10 (Fig. 917) project parallel to 0 9 to intersect with their plans at Fig. 916. One of the projectors is lettered hh (Figs. 916 and 917). Producing these projectors to the right, until they cut the ordinates 0 to 9 (Fig. 917), a series of points are obtained through which the curve 0 pq can be described as shown. The other half 0 vs is the same shape. To keep the working on the illustrations as clear as possible, the spheroidal surface has been continued in front as indicated at E (Fig. 916). The

Centering for Groin Vaulting.

Fig. 924 is a sketch of groin vaulting over part of an octagonal space as shown by the plan (Fig. 925), which also shows the plans of the ribs for the centering. The

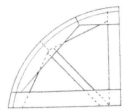

Fig. 922.—Elevation of Rib C (Fig. 916).

Fig. 923.—Elevation of Rib D (Fig. 916).

Fig. 924.—Sectional Elevation of Vaulting taken on Line A B (Fig. 925).

method of drawing normals, or rather joint lines, of the voussoirs of the arch is shown at Fig. 913. First find the foci points of the ellipse (A B), then, at the points where the joint commences, draw lines from each focus point and continue it as shown by D and E; bisect the angle by C F, which gives the direction of the joint line required. The other lines that are required are obtained in the same manner.

left half illustrates only the elevation of the centres carrying the main ribs of the vaulting shown in plan by A and L (Fig. 925), but on the right all the ribs of the centering are shown. It will be observed that on each side of the centres carrying the main ribs of the vaulting there is provided a rib made of two thicknesses to help to carry the stones of the panels; these ribs are lettered G H, D E, etc. Under the intersection of the panels

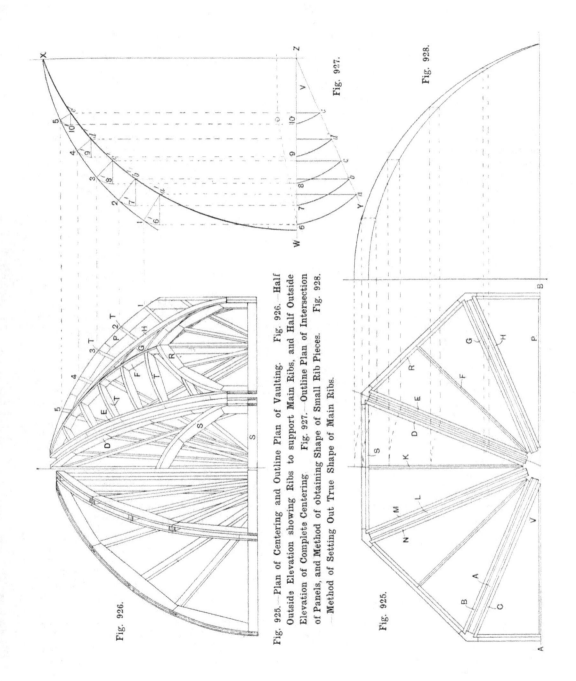

Fig. 927.

Fig. 928.

Fig. 926.

Fig. 925.

Fig. 925.—Plan of Centering and Outline Plan of Vaulting.   Fig. 926.—Half
Outside Elevation showing Ribs to support Main Ribs, and Half Outside
Elevation of Complete Centering   Fig. 927.—Outline Plan of Intersection
of Panels, and Method of obtaining Shape of Small Rib Pieces.   Fig. 928.
—Method of Setting Out True Shape of Main Ribs.

a rib is shown at F and P, and at the wall end these ribs are shown by R and S. Small rib pieces marked V are shown, connected to the ribs E, F P, etc. These are cut to the form that is desired for the soffits of the panels. A method of obtaining the shape of these is shown at Fig. 927. The line 1 to X is the true shape of the intersection between the panels ; 6 a' X is the elevation of the line of intersection as indicated by 1 to 5 (Fig. 927) ; radial lines are drawn to this curve, meeting the intersection 6 a' X in a' to e'.

By projecting horizontally from points 1 to 5 and a' to e' (Fig. 927) the positions of the rib pieces have been determined at 1 to 5, etc., in the elevation (Fig. 926). A method of obtaining the true curvature of the outline for the centre for the main ribs is shown at Fig. 928, which is projected from Fig. 925.

Centering for a Hemispherical Dome.

The illustrations (Figs. 929 to 935) show the necessary centering and timber support

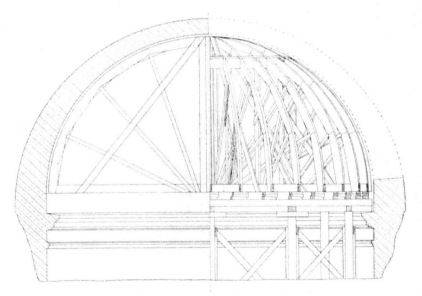

Fig. 929.—Half Sectional Elevation of Dome, and Elevation of a Main Rib.

Fig. 930.—Half Elevation of Centering, and Timber Stage for supporting it.

Horizontally from these, points are projected as shown ; also with these points as centres, arcs are drawn meeting the horizontals in points 6′ to 10′ ; then project down from these points and obtain points 6 to 10 in W Z. From points a' to e' project down to Y Z, obtaining points a to e. Now by drawing arcs joining a 6, b 7, etc., the shape of the edge of each rib-piece is obtained as shown. The curvature of these arcs can be varied from almost a straight line to an amount which would make the intersection of the panels disappear by taking the centres on the plan of it (W Z).

for the erection of a hemispherical dome, 30 ft. to 40 ft. diameter. Fig. 929 is a half-sectional elevation of the dome, and also shows the construction of one of the main ribs. At Fig. 930 are shown a half elevation of the main and secondary ribs with wedges, and also the method of timber supports under. The plan of all the ribs and main timbering is clearly shown at Fig. 931. A quarter plan, looking up, and a quarter plan of the upper side of the supporting stage, are given at Fig. 932. The conventional view (Fig. 933) will convey a fair idea of this supporting timber work. It also shows a central post, or mast, to which some of the main members are attached. The necessary

at A (Fig. 934), which also shows a form of lagging very convenient for work of this

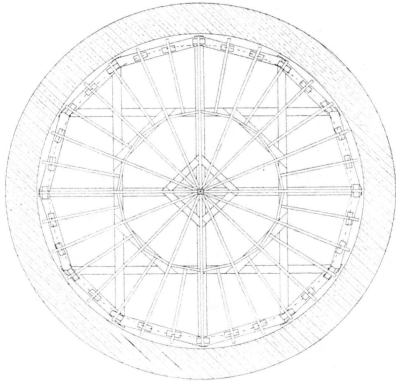

Fig. 931.—Plan of Ribs and
Main Timbering.

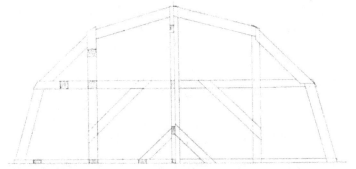

Fig. 932.—Quarter Plan, looking up, of Timber Stage, and Quarter Plan of Top of Stage

wedges, etc., are also illustrated. The method of connecting the secondary ribs to the main ones by a trimming piece is shown

class, made by pieces of scantling having an edge cut to the curvature of the soffit of the dome; their edges are all struck from the

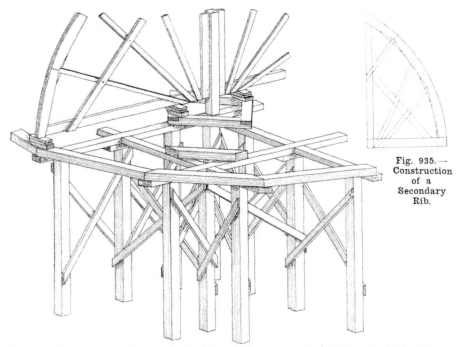

Fig. 935.—
Construction
of a
Secondary
Rib.

Fig. 933.—Conventional View of Timbering and Staging, and Part View of a Main Rib.

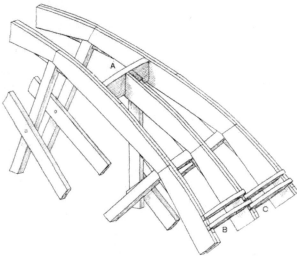

Fig. 934.—Conventional View showing Connection of Secondary Rib with Main Ribs, and Method
of Lagging Centering.

same centre, thus they can be cut from one
mould by a bandsaw, then notched down
on the ribs so that they all project up the
same distance, as indicated at B and C (Fig.
934). The construction of a secondary rib
is shown by Fig. 935.

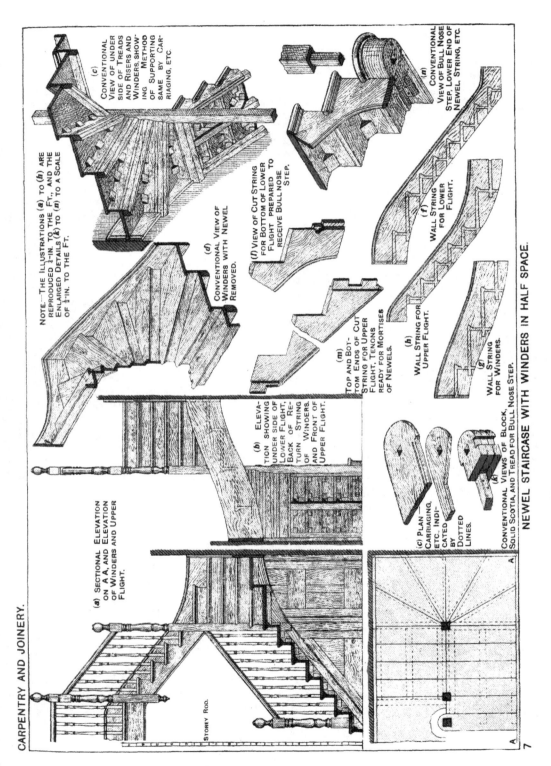

CARPENTRY AND JOINERY.

NEWEL STAIRCASE WITH WINDERS IN HALF SPACE.

NOTE.—THE ILLUSTRATIONS (a) TO (h) ARE REPRODUCED ¾-IN. TO THE FT., AND THE ENLARGED DETAILS (k) TO (m) TO A SCALE OF ½-IN. TO THE FT.

(a) SECTIONAL ELEVATION ON A A, AND ELEVATION OF WINDERS AND UPPER FLIGHT.

STOREY ROD.

(b) ELEVATION SHOWING UNDER SIDE OF LOWER FLIGHT, BACK OF RETURN STRING OF WINDERS, AND FRONT OF UPPER FLIGHT.

(c) CONVENTIONAL VIEW OF UNDER SIDE OF TREADS AND RISERS AND WINDERS, SHOWING METHOD OF SUPPORTING SAME BY CARRIAGING, ETC.

(d) CONVENTIONAL VIEW OF WINDERS WITH NEWEL REMOVED.

(l) VIEW OF CUT STRING FOR BOTTOM OF LOWER FLIGHT PREPARED TO RECEIVE BULL NOSE STEP.

(m) CONVENTIONAL VIEW OF BULL NOSE STEP, LOWER END OF NEWEL, STRING, ETC.

(f) WALL STRING FOR LOWER FLIGHT.

(m) TOP AND BOTTOM ENDS OF CUT STRING FOR UPPER FLIGHT, TENONS READY FOR MORTISES OF NEWELS.

(h) WALL STRING FOR UPPER FLIGHT.

(g) WALL STRING FOR WINDERS.

(c) PLAN—CARRIAGING, ETC. INDICATED BY DOTTED LINES.

(d) CONVENTIONAL VIEWS OF BLOCK, SOLID SCOTIA, AND TREAD FOR BULL NOSE STEP.

A

7

To download a PDF of the original plates in color, please visit
http://brooklyntoolandcraft.com/hasluckplates.

JOINERS' RODS.

Introduction.—A rod stands in the same relation to a craftsman as a scaled drawing does to a designer. In most shops the work is done by a setter-out, who makes his drawings from full-sized details prepared by the architect. The most convenient size of rod for general use is about 10 ft. by 11 in. by $\frac{5}{8}$ in., but a varied stock should be kept. Rods should be of pine, free from shakes and loose knots. Pine is chosen on account of its softness and evenness of grain, which enables lines of equal firmness, and not easily erasable, to be drawn. The boards should be nicely smoothed, whitened over, and rubbed with fine glass-paper to produce an even surface. The edges should be kept square. For making drawings on the rods, squares with 6-in., 12-in., and 36-in. blades, a trammel, dividers, pencil compasses, and a five-foot rule, will be found most useful.

Rods for Ledged and Beaded Door and Frame.

Fig. 936, Rod 1 (scale = $\frac{3}{4}$ in. to 1 ft.), shows the plan of a ledged and beaded door, in a $4\frac{1}{2}$-in. by 3-in. rebated and beaded frame, fixed in a $4\frac{1}{2}$-in. wall. (Fig. 937 shows a section of the door and frame, and will be referred to later.) This kind of door is generally used for outhouses. First lay the rod on the bench and draw a line parallel to the front about 1 in. from the edge, which will represent the face of the wall. At a distance of $4\frac{1}{2}$ in. from it draw a parallel line to represent the thickness of the wall. As plaster is not required, the framework will be of the same thickness, an opening 3 ft.

wide being made in the brickwork to receive it. The lines meet at A A. The two posts B B are next filled in, the outside of the frames being marked off $\frac{1}{2}$ in. less than the opening, which saves scribing, as the brickwork is always more or less rough. Fig. 938 shows a mould of a wood jamb or post used in Fig. 936. If several patterns of moulds are kept in stock, much time will be saved in setting out. A space of 2 ft. 6 in. is required in the clear of the frame, the posts being $2\frac{3}{4}$ in. when planed, and a 3-in. jamb is allowed for. A line is next drawn joining the rebate at each end, as shown, the thickness of the door and a depth of $\frac{1}{2}$ in. being included. The space between the two jambs is divided into five equal parts, the two outer boards which fit into the rebates being $\frac{1}{2}$ in. wider than the others, so that they show equal on the face. On each side of the tongued edge at the boarding joints a tongue and bead is filled in. The projection marked c (Fig. 936) represents the ledge on the back, forming a rail to which the boards are fixed. It is made $1\frac{1}{2}$ in. thick, with a $\frac{3}{8}$-in. chamfer. Where possible, an elevation is drawn at one end of the plan rod, as shown at Fig. 936 (enlarged at Fig. 939), for general guidance. The rod is turned over and the height drawn in section (see Fig. 937). Parallel lines $4\frac{1}{2}$ in. apart are drawn as before. A A is squared across, and the head filled in with the mould shown in Fig. 938. From A A set down 6 ft. 9 in., the height required, which gives a 6-ft. $6\frac{1}{2}$-in. door. The ledges or rails are filled in as shown, the middle one being 3 ft. from the bottom, the lower one 1 ft. from

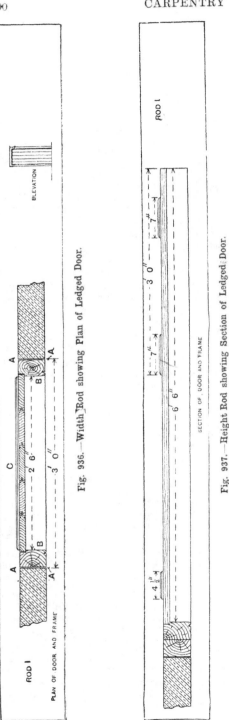

Fig. 936.—Width Rod showing Plan of Ledged Door.

Fig. 937.—Height Rod showing Section of Ledged Door.

the bottom, and the top one 9 in. down to the under side. The top rail is $4\frac{1}{2}$ in. wide

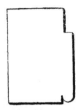

Fig. 938.—Mould of Jamb used in Fig. 936.

and chamfered on the top and bottom edges, the lower ones 7 in.

Rods for Four-panelled Moulded and Square Door and Frame.

Rod 2 (Fig. 940) (scale = $\frac{3}{4}$ in. to 1 ft.) represents the plan of a four-panelled moulded and square door (that is, with a mould on one side only), having double rebated casings, set in a 9-in. brick wall

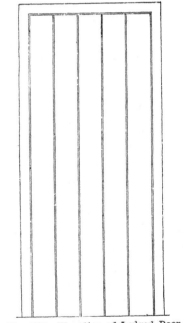

Fig. 939.—Elevation of Ledged Door.

which is plastered on both sides. (Fig. 941 is the height rod showing section of the door.) The door casing has deal splayed grounds and moulded architraves. First draw parallel lines $1\frac{1}{2}$ in. from the edge for the face of the plaster. The rod being only 11 in. wide, there will not be sufficient space to show the width with the architrave in full. It has therefore to be set out with a broken line, the out-to-out widths being indicated by figures. At a distance of 8 in. from the last, a second parallel line is drawn, representing the face of the plaster on the opposite side, $10\frac{1}{2}$ in. being figured in; that is, 9 in. for the brickwork and $\frac{3}{4}$ in. on each side for the plaster. As a door 6 ft. 8 in. by 2 ft. 8 in. by 2 in. is required in this case, an opening of 3 ft. is made, which will allow for $1\frac{1}{2}$-in. casings with proper backings. Draw two lines 3 ft. apart, meeting the parallel lines at D D, and draw another line, at a distance

Fig 942.—Mould of Architrave used in Fig. 940.

of $\frac{1}{2}$ in., to form the back side of the casing. Another line, drawn at a distance of $1\frac{1}{8}$ in. from the latter, will form the face of the skeleton frame and the door rebate. The width of the stiles is filled in at E. They should always be made at least 1 in. wider than the thickness of the door, which in this case is 2 in. The stiles should therefore not be less than 3 in., so that sufficient width is allowed for the proper fixing of the stop. The finished thickness of a 2-in. door being $1\frac{7}{8}$ in., the stop will be $3\frac{3}{4}$ in. narrower than the entire width of the framed lining, and as the latter is $10\frac{1}{2}$ in., the stop will be $6\frac{3}{4}$ in. wide and $\frac{1}{2}$ in. thick. Jamb linings are frequently made in the solid, and, if so required, the setting out will be as shown on the left-hand side of the opening F. A 2-in. by $\frac{3}{4}$-in. splayed deal ground G is provided at the back of the linings, and a 3-in. moulded architrave as seen in Fig. 942 filled in. This will cover the joint where the plaster meets the ground. Next fill in the door, and draw a parallel line to meet the rebate at H H on each side. This gives the thickness, the stiles being $4\frac{1}{4}$ in. wide. Take a mould

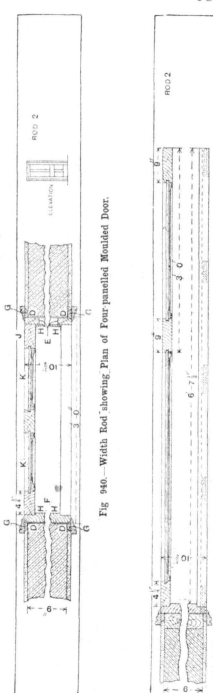

Fig. 940.—Width Rod showing Plan of Four-panelled Moulded Door.

Fig. 941.—Height Rod showing Section of Four-panelled Moulded Door.

similar to that shown at Fig. 943, and mark its outline on the rod at each side in the position indicated by J. Fill in the muntin in the centre of the same width, with a panel 10½ in. by 9⅝ in., sight size. Draw the two parallel lines K K representing the panels, the moulding on the face of the latter being filled in with the mould shown in Fig. 944. An elevation is given at the end of the rod (see Fig. 940, and enlargement at Fig. 945). Turn over the rod and fill in the height, as at Fig. 941, following the same rules as before. The door is shown on one side 6 ft. 8 in. high, with bottom and middle rails each 9 in. wide, and a top rail 4¼ in. wide. The distance from the top edge of the middle rail to the bottom of the door is 3 ft. In the opposite rebate the rails of the skeleton

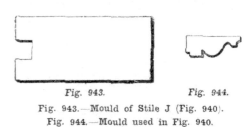

Fig. 943.　　　　　Fig. 944.
Fig. 943.—Mould of Stile J (Fig. 940).
Fig. 944.—Mould used in Fig. 940.

frame may be shown as seen in Fig. 941, or left blank if a solid lining is to be provided.

Rods for 2¼-in. Deal Door, Bead Butt and Square.

On Rod 3, shown by Fig. 946 (scale = ¾ in. to 1 ft.), set out a 4½-in. by 3-in. solid rebated, beaded, and staff-beaded frame fixed in a 14-in. wall. The door is a 2¼-in. deal one, bead butt and square, and is provided with a left-hand mortice lock. (Fig. 947 shows the height rod, giving section of door and frame.) The frame has 1-in. side linings and soffit, with splayed grounds, and moulded architraves on the inside. The total width of the frame and linings is 10¼ in., being a 4½-in. brick reveal. The reveal is shown by two square lines 3 ft. apart, and 1 in. from the edge of the rod. At each end draw a parallel line 3 in. long to form a rebate for the deal frame, which is 3 in. thick. Mark the mould shown in Fig. 948

on the rod as before, leaving ¼ in. over the quirk of the bead (see L, Fig. 948). Draw a line parallel to the back of the frame (the inside edge), and one to meet the rebate for the door. Fill in the stiles and the muntin with the same mould as before, the panels being bead flush—that is, beaded all round, and level with the face of the door. Fill in with the mould shown in Fig. 949, a tongue and bead being formed. Square across from the groove on the inside edge

Fig. 945.—Elevation of Four-panelled Moulded Door.

of the frame to meet a line 10¼ in. from the face of the frame. This will form the face of the lining. The thickness is given by setting back 1 in., and squaring across as before. The plan will be completed when the 2-in. splayed ground and moulded architrave have been filled in, and a sketch elevation has been drawn as before (see Figs. 946 and 950). On the other side of the rod set out the height, with rails of the same depth, and complete the rod by filling in the head and soffit lining, the ground and the architrave.

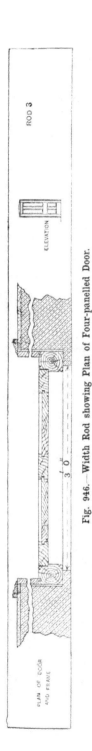

Fig. 946.—Width Rod showing Plan of Four-panelled Door.

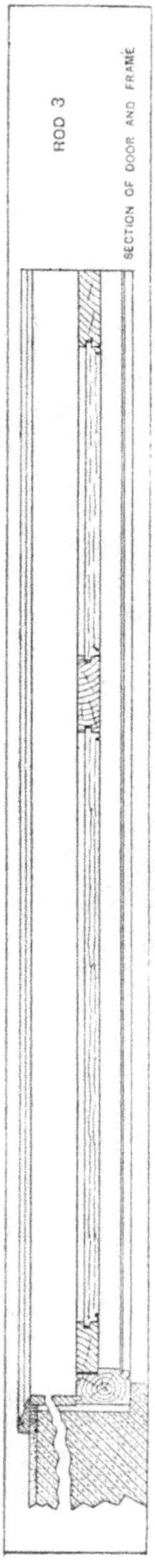

ROD 3

PLAN OF DOOR AND FRAME

ELEVATION

3′ 0″

Fig. 947.—Height Rod showing Section of Four-panelled Door.

ROD 3

SECTION OF DOOR AND FRAME

Rods for Boxed Sash Frame.

It is desirable that not only the joiner's work shall be shown on the rod, but the

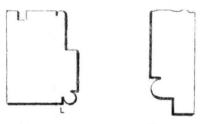

Fig. 948.—
Mould of Frame used in Fig. 946.

Fig. 949.—
Mould of Panel used in Fig. 946.

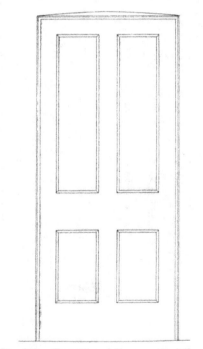

Fig. 950.—Elevation of Four-panelled Door.

brickwork in which it is placed, and the method of obtaining the interior finishings. The setting out a rod for an ordinary boxed sash frame, of which Fig. 951 is a plan, at the window-board level, and a horizontal section through the frame and brickwork, will be first considered. (Fig. 952 is the vertical section of the sash.) The frame is 6 ft. high by 3 ft. wide—that is, from the top of the stone sill to the springing of the

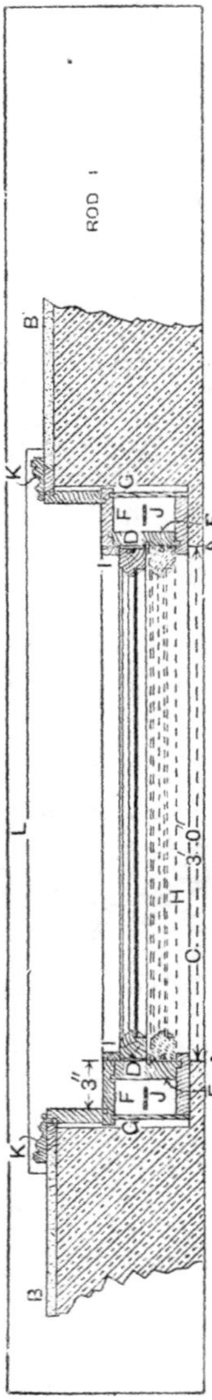

Fig. 951.—Width Rod showing Plan of Sash Frame.

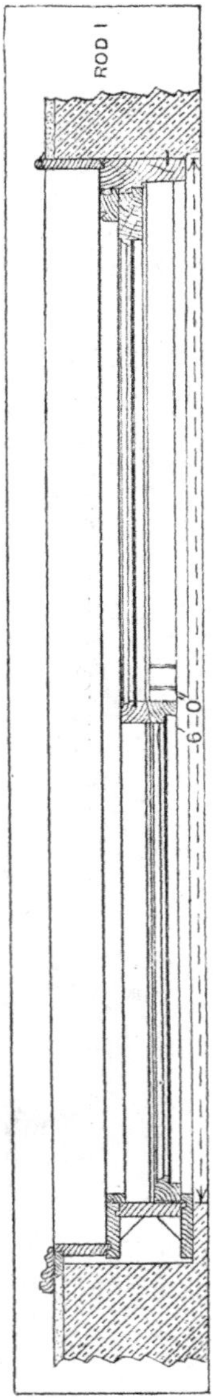

Fig. 952.—Height Rod showing Section of Sash Frame.

arch, and between the brick reveals, which is the general method for giving the sizes of frames. Take a rod 8 ft. long and 12 in. wide, and prepare it as before. Draw a line

1 in. from the edge 3 ft. 9 in. long, parallel with it, and from each extremity set in 4½ in., leaving the 3-ft. reveal A A. It is immaterial what portion of the brick reveal is shown ; the rod being too narrow to show the usual 4½ in., an inch is sufficient for the purpose. Square across from each extremity a line 9¾ in. long, and from each *point of the last lines draw another parallel* to B B. These lines represent the face of the plaster, and a line ¾ in. from the line B B will give the thickness of the plaster. These parts, from A to B on each side, should be marked with red crayon to indicate that it is solid brickwork, blue crayon being put on

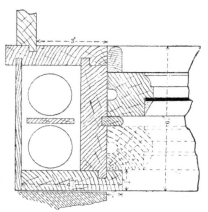

Fig. 953.—Enlarged Section through Side
Boxings of Frame.

for the plaster. Next proceed to fill in the frame and finishings. In this instance the sashes are 2 in. thick with 1 in. outer and inner linings, making the total thickness 6¼ in. as shown in the enlarged detail (Fig. 953). Draw a line 6¼ in. from the face line C to represent the thickness of the frame ; at a distance of ⅞ in. from the same line draw another to represent the thickness of the outer lining ; fill in the lines at D D to form the pulley stiles, and draw the line E 1⅛ in. from the line D D for the thickness of the pulley stile. From the inner line of the frame fill in the inside lining ⅝ in. thick at F F ; also add tongues to the pulley stiles. The width of the inner and outer linings is the same, being 4½ in., the outer one projecting ⅝ in. to form the stop for the sash and a margin round the brick reveal. The inner one is level with the face of the pulley stile, and projects ⅝ in. at each side, allowing sufficient room to enable a plough groove to be made to receive the back linings at G G. Care should be taken to fill in all details. The sashes being cut out of 2-in. stuff will finish 1⅞ in. Fill in a dotted line 1¹⁵⁄₁₆ in. from the dotted line H to represent the upper sash. This is shown dotted because the section line does not cut through it. At a distance of ⅜ in. from the last line draw a line for the parting bead, and another for the thickness of the sash, leaving 1⅛ in. for the width of the stop bead 1 and the draught bead which is tongued to sill. In setting out the thickness of a frame, take care to allow the stop bead to cover the joint between the pulley stile and the inside lining. Fill in the parting bead, allowing it

Fig. 954. Fig. 956. Fig. 955.

Fig. 954.—Mould for Bead.
Fig. 955.—Mould for Sash Stile.
Fig. 956.—Mould for Architrave.

to pass into a groove formed in the pulley stile ¼ in. deep. This will then project ⅝ in. beyond the face of the pulley stile. Take the mould (Fig. 954), and fill in the bead I; take also the mould (Fig. 955), and fill in the stiles of the sashes. Put in the parting slips J which separate the weights. From the inside face of the pulley stile mark on 3 in., and allow it to form a margin round the frame, as shown; this will show the position of the face of the side linings and soffit, which are tongued with the inside lining of the frame, which is grooved to receive the tongue. This lining will finish, in width, level with the face of the plaster, and will be an inch thick. Fill in the architrave K with the mould (Fig. 956) as shown. Draw a line L ½ in. from the face of the architrave to represent the front edge of the window-board. Square across

a line at each end in the same position as the architrave to form the return ends of the window-board. The rough grounds behind the architraves, to which the latter are

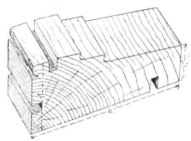

Fig. 957.—Mould for Oak Sill.

fixed, provide a good base for fixing the linings, and a key for the plaster.

Height Rod.—Turn the rod over and set out the vertical section or height as at Fig. 952. The height of the frame is 6 ft. Draw a line as before 1 in. from the edge of the rod, stopping at the line forming the top of the stone sill, and running 4½ in. beyond the 6 ft. height. Fill in the sill with the mould shown in Fig. 957, and continue the 6¼-in. line (the thickness of frame) up to the head. Square across the line representing the soffit of the frame, and draw the lines for inner and outer linings as before. Fill in the draught bead with the mould (Fig. 958). The inside face of the bottom rail of sash will be slightly bevelled to allow the sash to pass the wide draught bead when closing. With the mould (Fig. 955) fill in the moulding of the bottom rail of the sash 4¼ in. wide to the splayed edge, also fill in the top rail of the upper sash. Divide the space between the sight of each rail into two equal parts, the centre thus obtained being the centre line of the meeting rails. The latter

Fig. 958.—Mould for Draught Bead.

are 1½ in. thick. Fill in the rails as shown, the size of the glass being the same in both sashes. Fill in the upper part of the frame in sectional plan as before. Write th

number of the job on the rod, and sketch in
the elevation as shown at Fig. 959. Figs.
951 and 952 are drawn to a scale of $\frac{3}{4}$ in. to
1 ft., Fig. 959 to a scale of $\frac{1}{2}$ in. to 1 ft.,
Fig. 954 half full size, and the remaining
figures one-third full size.

Rods for Solid Casement Frame.

Figs. 960 to 962 show a plan, section, and
elevation of a solid casement frame. Take
a rod of the same size as the last and set

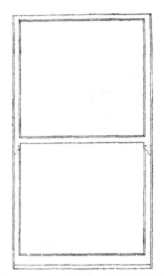

Fig. 959.—Elevation of Sash Frame.

out the plan as at Fig. 960. The frame is
set in a 9-in. wall with a 3-ft. 6-in. reveal.
Draw a line A of any length an inch from
the edge of the rod, and square lines across
at B B $4\frac{1}{2}$ in. long, another line C being drawn
parallel with A and projecting $3\frac{1}{2}$ in. at each
side beyond the 3-ft. 6-in. reveal. The
thickness of the wall is given by drawing a
line 9 in. from the line A. With the mould
shown in Fig. 963, fill in the jamb of the
frame at each side, projecting $\frac{5}{8}$ in. from
brickwork as shown (see line D, Fig. 963).
Fill in the sash stiles with the mould shown
in Fig. 955 (p. 295), allowing sufficient width
in the meeting stiles to form the hook
joint, and sufficient width on the outer stiles
to form the circular tongue. The sight
size—that is, the clear size between the

square of the moulding—is the same in both
lights. To obtain the exact positions of the
centre or meeting stiles, first fill in the

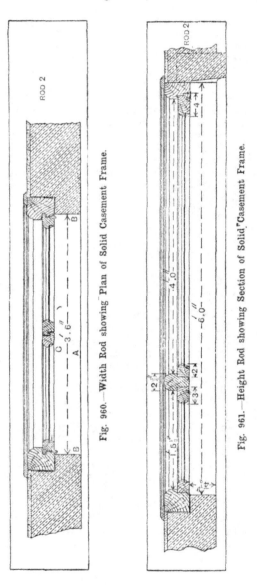

Fig. 960.—Width Rod showing Plan of Solid Casement Frame.

Fig. 961.—Height Rod showing Section of Solid Casement Frame.

hanging stiles and divide the space into
two equal parts, working from the centre
point outwards for distance. The method
of working from a centre will be found far

preferable to any other, as it enables the work to be done more exactly. After filling in the sashes, continue all lines as shown, and fill in the architraves with the mould seen in Fig. 964, leaving a proper margin. The brickwork and plaster are filled in from the moulding as before. Turn the rod

of it set up 1 ft. 5½ in. to the under side of the head. The head is filled in with the mould shown in Fig. 967. With the mould shown at Fig. 964 fill in the architrave, continuing all lines as shown. The mouldings of the sash rails are filled in with the mould shown in Fig. 955 (p. 295), the widths being those usual,

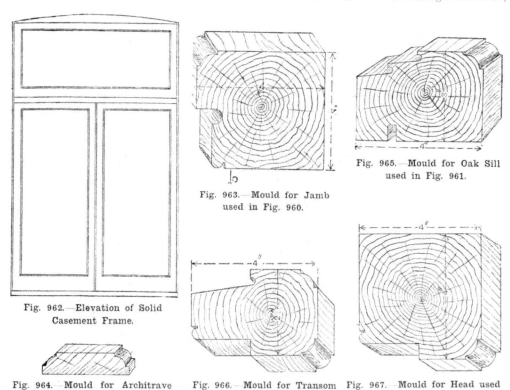

Fig. 962.—Elevation of Solid Casement Frame.

Fig. 963.—Mould for Jamb used in Fig. 960.

Fig. 965.—Mould for Oak Sill used in Fig. 961.

Fig. 964.—Mould for Architrave used in Figs. 960 and 961.

Fig. 966.—Mould for Transom used in Fig. 961.

Fig. 967.—Mould for Head used in Fig. 961.

round and set out the section shown in Fig. 961. The distance from the top of the stone sill to the under side of the arch at springing is 6 ft. From the front edge of the rod, square a line across to represent the top of the sill, and another 6 ft. from it for the springing. Show the reveal as before 4½ in. from the face of the frame to the face of the wall. With the mould shown in Fig. 965 fill in the oak sill, and from the top of the latter set up 4 ft. to the under side of the transom. With the mould shown by Fig. 966 fill in the transom, and from the top

as shown. A mould for each width should be kept for general use. Figs. 960 and 961 are drawn to a scale of ¾ in. to 1 ft., Fig. 962 to a scale of ½ in. to 1 ft., and the remaining figures one-third full size.

Rods for Square Bay Window.

Take a rod 8 ft. long by 3 ft. wide, and set out a square bay window, with solid frame, 6 ft. by 2 ft. The width rod is represented by Fig. 968, and the height rod by Fig. 969. Draw a line for the face of the wall (see Fig. 968), and from it draw two

13*

lines 6 ft. apart and each 2 ft. long. Join
the two points forming the outside of the
bay, fill in the angle posts with the mould

angle posts and the mullions being 1 ft. 7 in.
Mark on the section of the mullion with the
mould shown in Fig. 972, the hollows in the

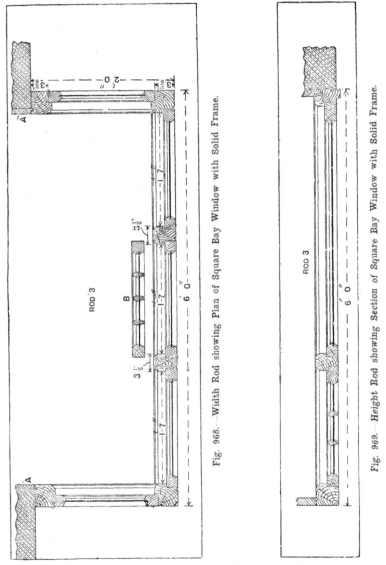

Fig. 968.—Width Rod showing Plan of Square Bay Window with Solid Frame.

Fig. 969.—Height Rod showing Section of Square Bay Window with Solid Frame.

seen in Fig. 970, and with the mould seen
in Fig. 971 fill in the wall posts as shown.
Divide the distance between the angle posts
into three parts, allowing 7 in. for the thickness of the mullions, the distance between the

rebating being filled in as required. With
the mould seen in Fig. 955 (p. 295), fill in
the sash stiles, and continue all the lines as
before, carrying the width of the casement
into the space as shown. Fill in the stiles,

divide the space between the sight lines into four parts, deducting 3 in. for bars, and the plan of the fanlight will be complete (see B, Fig. 968). In the inner edge of the

and fill in the head with the mould seen in Fig. 974, and the transom with the mould seen in Fig. 966 (p. 297). The rails are filled in as described in the previous

Fig. 970.—Mould for Angle Post used in Fig. 968.

Fig. 971.—Mould for Wall Post used in Fig. 968.

Fig. 972.—Mould for Mullion used in Fig. 968.

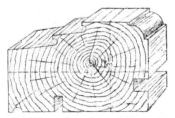

Fig. 973.—Mould for Sill used in Fig. 969.

Fig. 975.—Elevation for Square Bay Window.

Fig. 974.—Mould for Head used in Fig. 969.

wall posts plough a groove to receive the tongue on the lining A, and fill in the lining and bead to the broken line. Turn over the rod, and set out the section or height shown in Fig. 969. This is 6 ft. from the top of the stone sill to the top of the head of the frame. Commence as before, and fill in the sill with the mould seen in Fig. 973,

paragraph, and the upper sash divided into squares as shown in Figs. 969 and 975, the latter figure also showing the stops on the moulding of the angle post. Figs. 968 and 969 are drawn to a scale of $\frac{3}{4}$ in. to the foot, Fig. 975 to a scale of $\frac{1}{2}$ in. to the foot, and the remaining figures are one-third full size.

Rods for Canted Bay Window.

Figs. 976 and 977 represent respectively the width rod and height rod for a canted

the parallel lines are made with the aid of a straightedge and the perpendicular with a set-square. If the bay window is required to be made in wood without stone mullions,

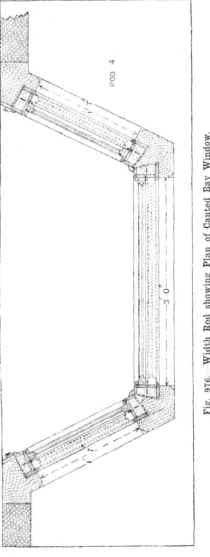

Fig. 976.—Width Rod showing Plan of Canted Bay Window.

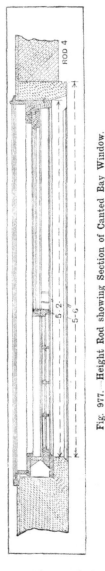

Fig. 977.—Height Rod showing Section of Canted Bay Window.

Fig. 978.—Mould for Horn used in Fig. 977.

bay window set in stone mullions and jambs. These are set out in the same manner as the frame described on Rod 1 (Figs. 936 and 937, p. 290), except that at the sides or cants

etc., the work is similarly set out, but the boxings for the weights are kept as small as possible, and the outer linings mitred at the angles, a moulding being generally intro-

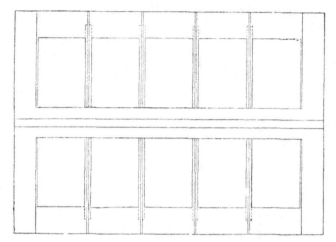

Fig. 979.—Joiners' Rods for Skylight.

Fig. 980.—Section through Side of Skylight.

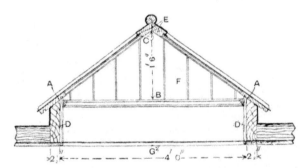

Fig. 981.—Transverse Section through Skylight, etc.

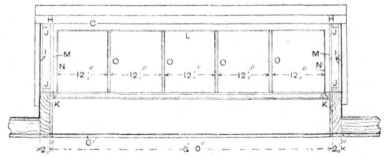

Fig. 982.—Longitudinal Section through Skylight.

duced to form a margin round each frame. Fig. 978 shows the mould for the horn or projecting end of the upper sash at the meeting rail. All the other moulds for this rod can be made by modifying those previously described. Figs. 976 and 977 are drawn to a scale of $\frac{3}{4}$ in. to the foot, and Fig. 978 is one-third full size.

Rods for Skylight.

To set out the skylight shown in Figs. 979 to 982 it will be first necessary to set out Fig. 981, which, being the section, shows

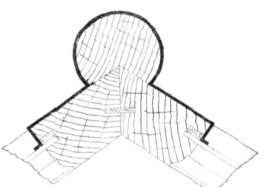

Fig. 983.—Detail at Apex of Skylight, showing Ridge Roll.

all finishings. The curb forms the trimmer joists of the flat. The two long sides are splayed to the pitch of the roof or light. The rough curb is 6 ft. by 4 ft. in clear, the roof rising 1 ft. 6 in. ; the apron lining which covers the rough curb is of 1-in. stuff beaded on the bottom edge and grooved for the plastered ceiling on the back side ; the moulding running round along the top of the apron is hollow on the top to form a condensation gutter. The light itself is 2 in. thick, with stiles, top rails, and bars moulded. The bottom rail is left square on the top edge, and grooved on the under side at the bottom to form a drip for rainwater ; the top rail and ridge roll are covered with 5-lb. lead, dressed so as to cover the putty in the groove of the top rail. The ends are formed by fixing a fillet on the upper edge of the curb ; also on the under side of the stiles of the light 1-in. rough boarding is cut spandril

shape and fixed on the outside ; $\frac{3}{4}$-in. boarding on the inside finishing flush with the face of the apron lining, and the ends butting on the top edge of the latter lining. Get a rod 8 ft. long by 3 ft. wide ; from the edge draw a parallel line 1 in. from it ; set up a perpendicular for the centre line, the rough curb being 4 ft. in the clear ; measure off 2 ft. on each side from the centre line, and set up lines $11\frac{3}{4}$ in. high, representing trimmers and plaster. Now draw a line

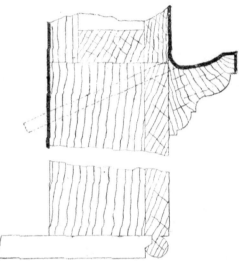

Fig. 984.—Section through Curb of Skylight, showing Apron Lining and Moulding.

parallel with the first, meeting at the two points A A (Fig. 981). The latter line will be the level of the upper edge of the curb. From the point of intersection of the latter line with the centre at B, set up 1 ft. 6 in., this being the rise of the roof on the inside face of the framing at C. Now draw lines cutting C A on each side, allowing the lines to project to beyond A. Measure off 2 in. from this line, and fill in the thickness of the skylight. Add also the ridge roll (see Fig. 983). Now fill in lines representing the apron lining at D. Seven-eighths of an inch from the curb, form a $\frac{7}{8}$-in. bead on the bottom edge as shown. The upper edge will be splayed the same as the curb ; fill in the thickness of the curb $2\frac{1}{2}$ in. ; form a slight rebate on the top edge, outside, to receive the lead dressing ; and form the moulding on the top rail at E, allowing $3\frac{1}{2}$ in. for the width. Note that the top rail is grooved for the glass instead of being rebated (as is usual) at the sides. The bottom rail includes the thickness of the rough boarding at I. Fill in the fillets J J at the top and bottom. Now from the two lines, first drawn on the inner side, measure $\frac{7}{8}$ in. for the thickness of the apron lining and matched boarding ; continue the line from

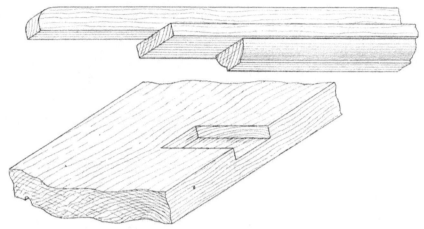

Fig. 985.—Detail showing Method of Fixing Bar to Bottom Rail.

should carry about the same line of sight as the moulding vertically ; it will therefore project 2 in. on the horizontal line. Indicate roughly the depth of the joist, and show boarding and lead to flat. Divide the space at the spandril end, and fill in the lines to represent the beaded joint of the matched

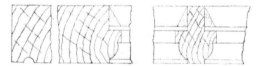

Fig. 986.—Detail Section through Stile and Side of Skylight.

boarding F. Turn the rod over, and set out Fig. 982 ; draw a line 1 in. from the edge as before, and set up two perpendiculars 6 ft. apart. These lines must be the same length as from C to G (Fig. 981), being the height to the under side of the framing. Draw a parallel line cutting the two points at H H, allowing $4\frac{1}{2}$ in. longer at each end. Measure off $2\frac{1}{2}$ in. at each side for the thickness of the curb and set up another line to C. This

c to G ; form a bead on the bottom edge of the apron as shown in Fig. 981 ; also the moulding at K. Draw line L parallel with C and 2 in. from it, representing the sight line on the top rail. From face line of boarding M measure off 2 in., and set up perpendicular N, meeting at line L. Now divide the distance between these two lines into five parts, allowing $1\frac{1}{2}$ in. each for the four bars O, making the spaces $12\frac{1}{2}$ in. each ; the lines forming these openings are sight lines. Fill in the two lines above, and parallel with C, representing the ridge roll. Fig. 980, which is a sectional plan through the framing, may be set out on an ordinary rod 8 ft. long by 11 in., and to figured dimensions ; it is unnecessary to set out Fig. 979 full size on a rod, the figure being intended simply for illustration and as a guide. Figs. 983 to 986 are enlarged details of the several parts.

Rods for Recess Cupboard Front in Two Heights.

To set out a cupboard front for a recess at the side of a chimney breast, take a rod

8 ft. long by 11 in. wide, and proceed with the setting out shown at Figs. 987 and 988. The cupboard front will be 7 ft. high from

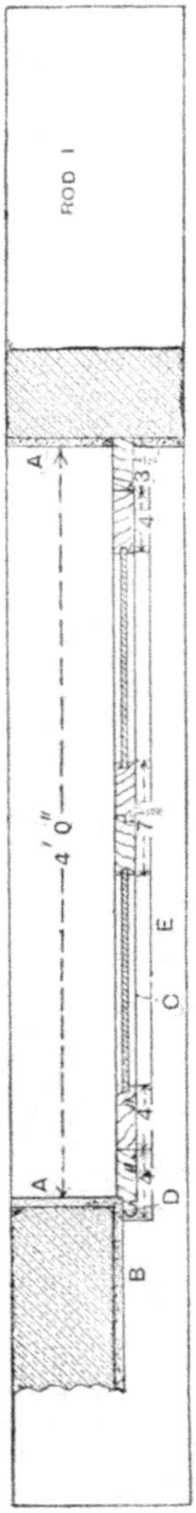

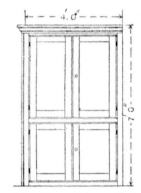

Fig. 987.—Width Rod, showing Plan of Recess Cupboard Front.

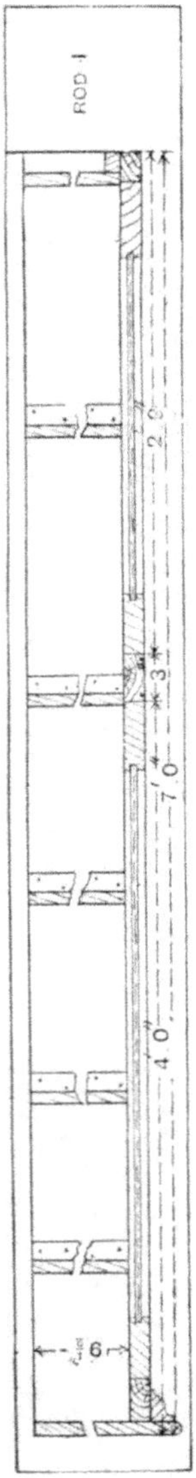

Fig. 988.—Height Rod, showing Vertical Section of Cupboard Front.

1½ in., allowing ¾ in. on each face for the plaster A A. The brickwork on the right-hand side is part of the side wall of the room, while that on the left hand is part of the side of the chimney breast, which is

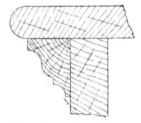

Fig. 989.—Detail of Cornice of Recess Cupboard.

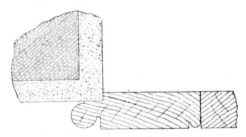

Fig. 990.—Elevation of Recess Cupboard Front.

Fig. 991.—Detail of Angle Stile to Chimney Breast.

the floor level to the top of the cornice, 4 ft. 1½ in. wide between the brickwork of the recess, and 9 in. deep between the faces of the brickwork reveal. Begin as before with the rod by setting out the brickwork 4 ft.

9 in. deep. Fill in a line ¾ in. from the chimney breast to represent the face of the plaster at B. Next draw a line parallel with the edge of the rod, and ⅞ in. from the face of the plaster B, to represent the face of the cupboard framing c. Draw a parallel line

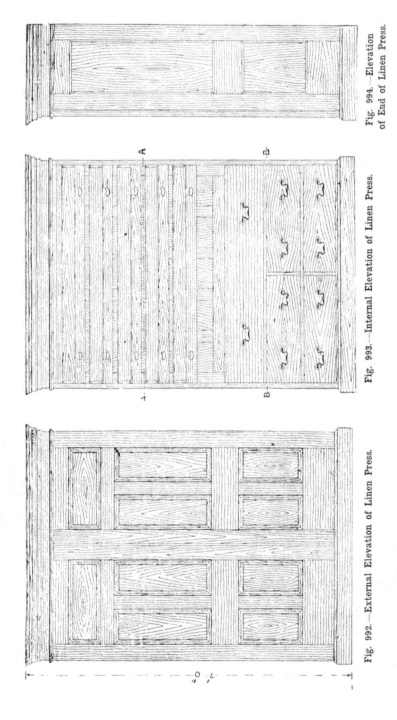

Fig. 994.—Elevation of End of Linen Press.

Fig. 993.—Internal Elevation of Linen Press.

Fig. 992.—External Elevation of Linen Press.

$1\frac{1}{8}$ in. from the last, to represent $1\frac{1}{4}$ in. front, finishing the line on right-hand side at the brick-work, and on the left-hand side at the face of the plaster. Fill in the stile and angle bead D with the mould seen at Fig. 991. The stiles must be set out to the figured dimensions. The doors are hung folding in two widths, the centre of the rebate being equidistant from the outer hanging stiles. Square across the sight lines of the framing from the edge of the rod, and fill in the panels $\frac{1}{2}$ in. thick and $\frac{3}{8}$ in. from the face of the door framing. If moulding is required, the panel will need to be set back accordingly. Turn the rod over, and set out the height as shown in the section (Fig. 988). The height being 7 ft., square a line across the rod at each end to represent it, and draw the face and thickness lines parallel as before. Working from the dimensions given on the width rod (Fig. 987), the height from the floor to the top of the lower doors will be 2 ft. 9 in., these being surmounted by a 3-in. dividing rail. Fill in the beads on the edges of the horizontal rails, and project the top of the cupboard sufficiently to take the moulding shown at Fig. 989, which forms a cornice, and returns into and stops on the face of the

driven into the joints of the brickwork. When fixed they form a stop for the doors. The left-hand doors are secured by necked bolts fixed on the inside and shot into the

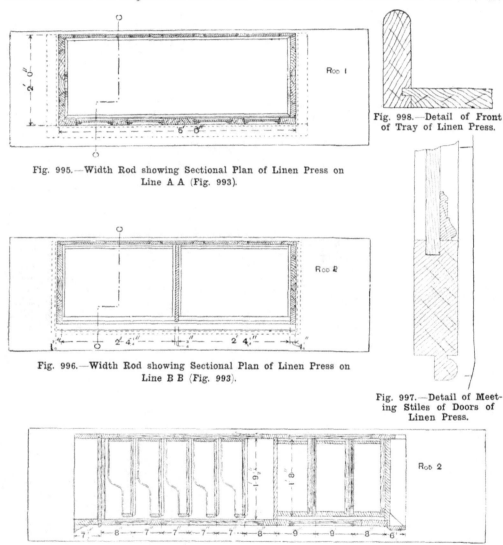

Fig. 995.—Width Rod showing Sectional Plan of Linen Press on Line A A (Fig. 993).

Fig. 998.—Detail of Front of Tray of Linen Press.

Fig. 996.—Width Rod showing Sectional Plan of Linen Press on Line B B (Fig. 993).

Fig. 997.—Detail of Meeting Stiles of Doors of Linen Press.

Fig. 999.—Height Rod, showing Vertical Section of Linen Press on Line C C (Fig. 995).

plaster as seen at Fig. 990, or on the face of the top rail. The shelves are conveniently distanced apart, and are generally fixed on fillets or bearers, secured to plugs

upper face of the shelf. Figs. 987 and 988 are drawn to a scale of ¾ in. to the foot, Fig. 990 to a scale of ¼ in. to the foot, the remaining figures being one-third full size.

Rods for Linen Press.

Figs. 992, 993, and 994 show respectively an external elevation, an internal elevation, and an end elevation of a linen press. The interior, as will be seen, is fitted up with

in the position of the door stiles ; add the panels and moulding. The ends are framed together with solid panels, the bead being flush on the outside. Fill in the stiles and the tongues and beads on the panels, and allow $\frac{7}{8}$ in. for the panelled framing at the

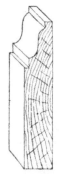

Fig. 1002.—Detail of Plinth of Linen Press.

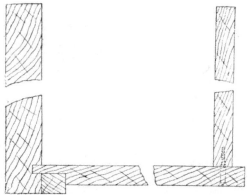

Fig. 1000.—Detail of Section through Drawer of Linen Press.

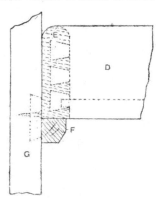

Fig. 1003.—Detail showing Dovetail Runner with Dovetailed Front and End to Tray.

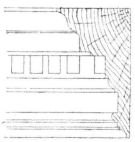

Fig. 1001.—Detail of Cornice of Linen Press.

drawers and trays, the former being arranged at the bottom and the latter at the top. The press is 7 ft. high, 5 ft. wide, and 2 ft. deep, and the front is formed by a pair of five-panelled moulded doors. The cornice is made separate, and drops on the top of the carcase. Take a rod 8 ft. long and 2 ft. 6 in. wide, and proceed to set out the plans, shown at Figs. 995 and 996. Fig. 995 is a horizontal sectional plan at A A (Fig. 993), and Fig. 996 is a similar plan at B B (Fig. 993), showing the drawers. Draw the outer lines in the usual manner, and take a mould, similar to that shown by Fig. 997, and fill

back. Divide the press up as shown in Fig. 996, making the panels flush on the inside. The double lines shown on the inside of the framing at Fig. 995 represent the thickness of the sides of the tray, which is $\frac{7}{8}$ in. (see detail Fig. 998). These trays are made of mahogany, and have holes cut in the front for the hand, as shown at Fig. 993. Turn over the rod and set out the plan of the drawers. The length is 2 ft. $4\frac{1}{4}$ in. between the standard and ends. Let the drawers stand back 1 in. from the face edge of the end to give room for the drop handle. On another rod (2) set out the vertical section, shown at Fig. 999. Divide up the spaces as figured in. A detail of the drawer in section is given at Fig. 1000 ; a similar door, the square on the upper edge of the plinth forming a margin along the front and ends. Fig. 1003 shows the front D and the

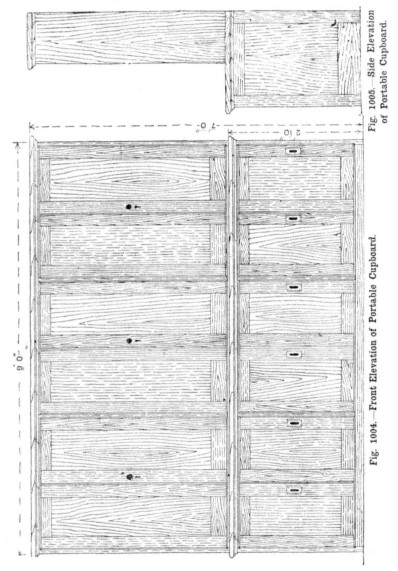

Fig. 1005.—Side Elevation of Portable Cupboard.

Fig. 1004.—Front Elevation of Portable Cupboard.

detail of the tray is shown at Fig. 998 ; a detail of the cornice at Fig. 1001 ; and the plinth at Fig. 1002. It will be noticed that the bottom projects sufficiently from the face of the ends to take the thickness of the end E of a tray, dotted lines representing the dovetails. The top edge of the rim is mitred at the angles, and the hardwood runner F is dovetailed into the end G and stopped 2 in. from the face edge of the end

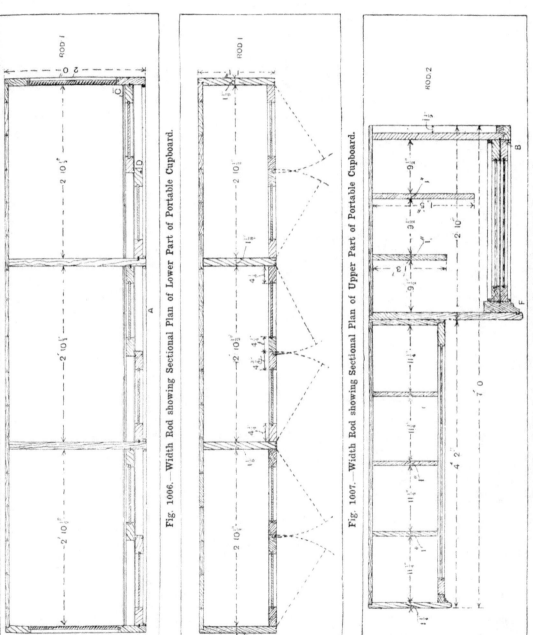

Fig. 1006.—Width Rod showing Sectional Plan of Lower Part of Portable Cupboard.

Fig. 1007.—Width Rod showing Sectional Plan of Upper Part of Portable Cupboard.

Fig. 1008.—Height Rod showing Vertical Section of Portable Cupboard.

of the carcase. Figs. 992 to 996 and Fig. 999 are drawn to a scale of $\frac{1}{2}$ in. to the foot; the remaining figures are one-third full size.

and each compartment is 2 ft. 10½ in. in the clear, with 1⅛-in. ends and division standards, which stand flush with the face of the doors

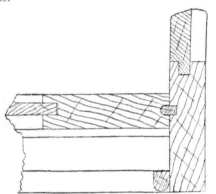

Fig. 1009.—Detail of Portable Cupboard at C (Fig. 1006).

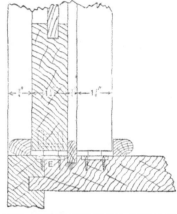

Fig. 1011.—Detail of Portable Cupboard at B (Fig. 1008).

Rods for Portable Cupboard.

Figs. 1004 and 1005 show a portable cupboard adapted for glass and china. Two heights are given for this cupboard, but

in each case. The width from back to front of the lower part, exclusive of the projection of the top, is 2 ft.; the upper one is 13 in.

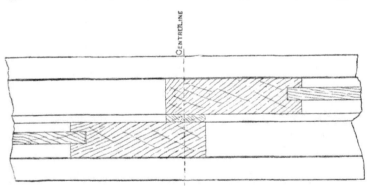

Fig. 1010.—Detail of Portable Cupboard at D (Fig. 1006).

each is complete in itself, and each is divided into three compartments, with folding doors to the upper part and sliding doors to the lower. The back of the cupboard is composed of ¾-in. V-jointed matching in narrow widths, the top, bottom, and ends being rebated to receive it. The ends of the lower cupboard are panelled; those of the upper one are solid. The extreme width of the cupboard is 9 ft., exclusive of the projection,

The height from floor level to the top of the lower part is 2 ft. 10 in.; the upper part from top to top, 4 ft. 2 in., making in all 7 ft. The shelves are divided equally, 9½ in. and 11¼ in. respectively. Prepare moulds for each separate part to details, and retain for future use. Take a rod 2 ft. 4 in. wide and 10 ft. long, and set out the plan of the upper and lower cupboards as in Figs. 1006 and 1007, one on each side. Begin by drawing all the outer lines, and work inwards. Prepare a slip $\frac{1}{8}$ in. thick and $1\frac{1}{8}$ in. wide, the finished thicknesses of the outer standards or ends and divisions. Divide the plan into three equal parts, as shown, 2 ft. $10\frac{1}{2}$ in. in the clear. From the face line A (Fig. 1006) draw lines at distances figured in detail in Fig.

Fig. 1012.—Sketch of Roller for Portable Cupboard.

Fig. 1013.—Detail of Portable Cupboard at F (Fig. 1008).

1008—$\frac{3}{4}$ in. for bead, $1\frac{1}{8}$ in. for front door, $\frac{1}{4}$ in. for parting bead, $1\frac{1}{8}$ in. for back door —form the tongue, and groove in the stiles as shown in detail in Fig. 1009. Fill in the stiles and panels as in Fig. 1010. Mark along the back the spaces indicating the widths of the matched lining. Turn the rod over and set out Fig. 1007 in the same way. Take another rod the same length, 2 ft. 4 in. wide, and set out the section or height, following the same rules as before. This is shown on rod 2 (Fig. 1008). Arrange the lower sliding doors as shown in Fig. 1011. The roller (Fig. 1012) is fitted into the bottom

edge of the door; two are fixed to each door, the face of the roller being flush with the edge. The brass strip E, fixed with screws, is let in flush with the bottom or pot board, and forms a runner. The beads must be

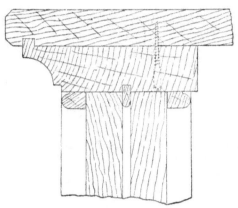

Fig. 1014.—Mould for Scotia of Portable Cupboard.

shown slack with the door, because if the latter is fixed tight it will not work freely. The thicknessing piece F (Fig. 1008), shown in detail in Fig. 1013, is continuous, and is intended to strengthen the top, so that any extra weight put upon it may not interfere with the easy working of the doors. Fig. 1014 shows the mould for the scotia. Figs. 1004 and 1005 are drawn to a scale of $\frac{1}{2}$ in. to the foot, Figs. 1006 to 1008 to a scale of $\frac{3}{4}$ in. to the foot, and the remaining figures one-third full size.

DOORS AND DOOR FRAMES.

Varieties of Common Doors.

COMMON doors are constructed in a number of styles. The principal three are the following, which are presented in the order of their cost and strength : the ledged door

for doors are given in the section on joiners' rods (see pp. 289 to 311), and these instructions will form a basis from which to prepare rods for any of the doors here mentioned.

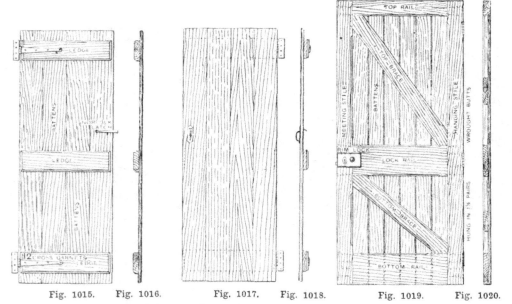

Fig. 1015.   Fig. 1016.   Fig. 1017.   Fig. 1018.   Fig. 1019.   Fig. 1020.

Fig. 1015.—Back of Ledged Door.   Fig. 1016.—Back Edge of Ledged Door.   Fig. 1017.—Front of Ledged Door.   Fig. 1018.—Front Edge of Ledged Door.   Fig. 1019.—Back of Ledged and Braced Door.   Fig. 1020.—Section of Ledged and Braced Door.

(Figs. 1015 to 1018); the ledged and braced door (Figs. 1019 to 1022); and the framed and braced narrow batten door (Figs. 1023 to 1026). With regard to the setting out of these, instructions on preparing joiners' rods

Ledged Doors and Frame.

The construction of one of the simplest and commonest forms of ledged door and frame is illustrated by inside and outside

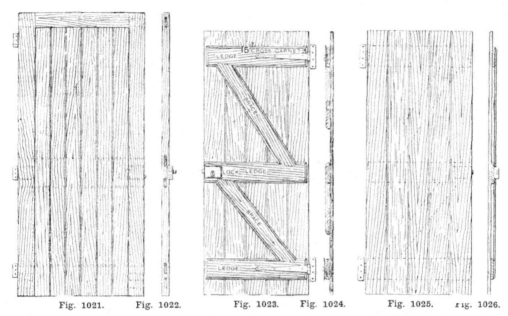

Fig. 1021.   Fig. 1022.   Fig. 1023.   Fig. 1024.   Fig. 1025.   Fig. 1026.

Fig. 1021.—Front of Framed and Braced Door.   Fig. 1022.—Section through Stile.   Fig. 1023.—Back of Ledged and Braced Door.   Fig. 1024.—Vertical Section through Door.   Fig. 1025.—Front of Ledged and Braced Door   Fig. 1026.—Elevation of Shutting Edge.

elevations and edge view at Figs. 1027 to 1029. The frame is quite square without any beads and stops, the door overhanging and shutting against the outer faces of the jambs. Only a few chief points in the preparing, making, and fixing of the frame will here be explained, because the general processes involved in the making are somewhat similar, although perhaps not requiring the same degree of accuracy as for more important doors and frames. The leading operations in the making of these will be described fully in the examples that will follow. The head and jambs of the frame are generally made of scantling 3 in. by 2 in. or more. This should be of good red deal for external work, and when there is a wood sill it should be of English oak. The three or four pieces composing the frame are planed up, the jambs or posts are then set out from the rod marking the shoulders and tenons at the top end, and a scribe line is drawn at the bottom for fitting to the stone sill ; or if the sill is of wood the posts are marked for tenons. The head is set

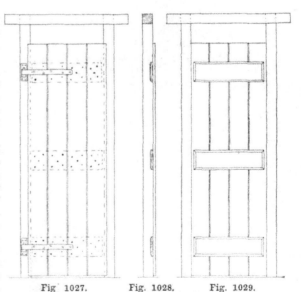

Fig. 1027.   Fig. 1028.   Fig. 1029.

Fig. 1027.—External Elevation of Ledged Door and Frame.   Fig. 1028.—Vertical Section.   Fig. 1029.—Inside Elevation.

14

out for either a close or a slot mortice ; the latter is the one shown in this example (Fig. 1030). As there are no stops or beads to work on the frame, the setting out is

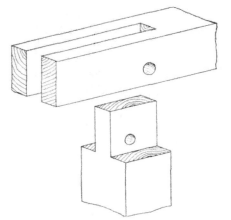

Fig. 1030.—Slot Mortice and Tenon Joint between Head and Jamb.

very simple. After the jambs are made and fitted, they should be draw-bored for draw-pinning ; the process is often termed draw-boring. This is done by boring through the cheeks of the mortice as indicated at Fig. 1031, using a $\frac{1}{2}$-in. or $\frac{5}{8}$-in. auger. The shoulders of the joint are then put together,

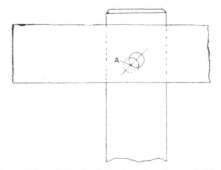

Fig. 1031.—Method of Boring for Draw-pinning.

and a marking awl (or pricker) is used in the hole of the cheek to make a mark in the tenon as indicated at A (Fig. 1031), where it will be seen that this mark is made against the

side of the hole of the cheek and on a line about 45 degrees from the centre. This mark A shows the centre for the hole to be bored in the tenon. If the holes in the cheeks and the tenon are bored thus, the pin, when driven in, tends to draw together the shoulders of the joint and also the side of the tenon against the end of the mortice. Another method largely adopted, but not so good, is to cramp the joints together ; then at one operation bore through cheeks and tenon, and drive in the prepared pin. In very common door frames a couple of 3-in. or 4-in. nails are driven obliquely into the top of the frame-head, the nails passing into the tenon and shoulders. Usually a piece of wood is nailed across the lower part of the jambs so as to keep them parallel until the frame is fixed across in its place. Before being put together, the parts

Fig. 1032.—Detail of Hook and Plate Hinge.

of the joints are generally painted, which, for exposure to the weather, is considered more durable than gluing.

Fixing the Door Frame.

This class of door frame is largely used for outbuildings, and when these are built of brick or stone, the frame is usually placed in its position and held by one or two raking struts so as to keep it plumb and firm until the brickwork or masonry is built around it. Usually the head-piece is made to project a few inches beyond the posts, as indicated in the illustrations. These projections are called " horns " and are useful for bonding into the brickwork. When there is a wood sill it is similarly shaped. As the brickwork is built up, two or three wood bricks are built in as a joint between two courses. These wood bricks may be peices of 3 in. by 4 in. by 6 in. or 9 in. long, or pieces about $\frac{1}{2}$ in. thick, 4 in. wide, and

4 in. to 9 in. long. These are built in the brickwork against the frame, which is nailed to them.

Preparing the Ledged Door.

The vertical boards for the door vary from 3 in. to 7 in. in width, the narrower then beaded, or the meeting edges are chamfered to form a **V** joint. (See page 62, Fig. 263.) The ledges are next prepared and chamfered as shown in Figs. 1028 and 1029. Two pieces of quartering are laid across the bench, and the boards are placed face downwards on these. The cramp is applied,

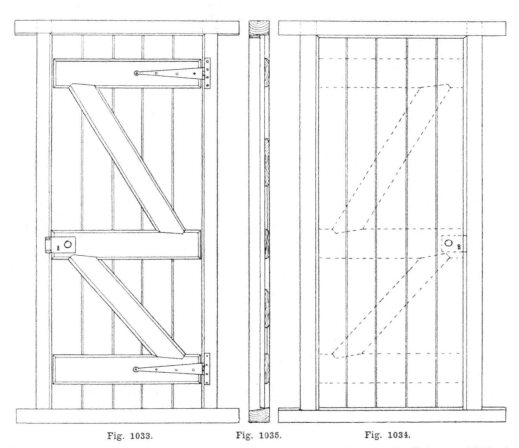

Fig. 1033. Fig. 1035. Fig. 1034.

Fig. 1033.—Inside Elevation of Ledged and Braced Door. Fig. 1034.—Outside Elevation of Ledged and Braced Door. Fig. 1035.—Vertical Section of Ledged and Braced Door.

being better, as when shrinkage occurs there is less space between the joints with the narrower boards. The thickness varies from $\frac{5}{8}$ in. to $1\frac{1}{4}$ in. The boards are faced up, thicknessed, and jointed ; then ploughed and tongued, or more frequently grooved and tongued. One edge of each board is and then lines are squared across to show the position of the ledges, the cramp being applied near each ledge, so as to keep the joints of the boards close, and these are secured by a few nails or preferably screws. The door is now turned over face side up and lined out for nailing. Care is taken to space the nails in diagonal lines so as not to split either boarding or ledges. A piece of waste wood is placed under each ledge in turn during nailing. Wrought-iron nails

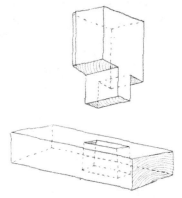

Fig. 1036.—Joint between Sill and Post.

are often used sufficiently long to project beyond the face of the ledges after being punched in from the face. The points of these nails penetrate into the piece of waste wood so as to prevent splitting pieces out of the ledges. The door is again turned

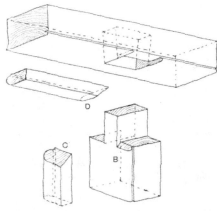

Fig. 1037.—Joint between Head and Post.

over, and the points of the nails are bent over by means of the nail punch and hammer, and in the form of a hook. They are then driven slightly below the surface by

the hammer and punch ; thus the nails are clenched. In doing this the punch must be held slanting so as not to drive the nails back. The top and bottom of the door should be sawn square and planed.

Hanging the Door.

This door is cut off level at the bottom, and it does not require fitting between the

Fig. 1038.—Alternative Method of Joining Head and Post.

jambs. The hinges shown are of the hook and plate pattern, a detail of which is given at Fig. 1032. These are screwed on in the positions illustrated, care being taken to keep an equal margin of the lap of the door over the frame.

Fig. 1039.—Joint between Brace and Ledge.

Ledged and Braced Door and Frame.

A door and frame of this description is shown by Figs. 1033 to 1039. The chief points in the preparing and fixing of this door and frame will now be explained. It will be seen from the figures that the frame is beaded round the inner edge, and that the beaded stops are nailed on.

Preparing the Door Frame.

All the stuff should be carefully planed up out of winding and square, and the oak sill is slightly splayed. Place the two posts face sides together and face edges outwards, as indicated at Fig. 1040. Then setting out for the shoulder as well as the gauging should be completed as shown at Fig. 1041. The rod should be applied and the frame head set out for mortising as shown at Figs. 1042 and 1043. The former shows the setting out on the soffit of the head and the latter the top side of the head,

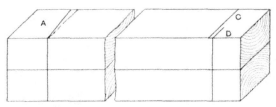

Fig. 1040.—Setting Out Posts for Shoulders.

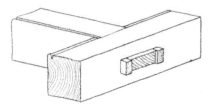

Fig. 1045.—Mortice and Tenon Joint Wedged.

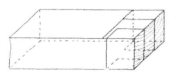

Fig. 1041.—Setting Out Head for Mortices.

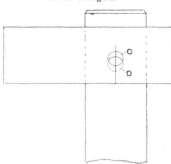

Fig. 1046.—Draw-boring Mortice and Tenon.

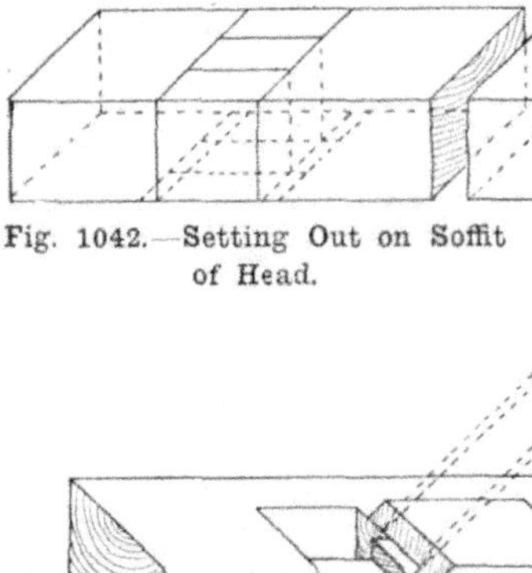

Fig. 1042.—Setting Out on Soffit of Head. 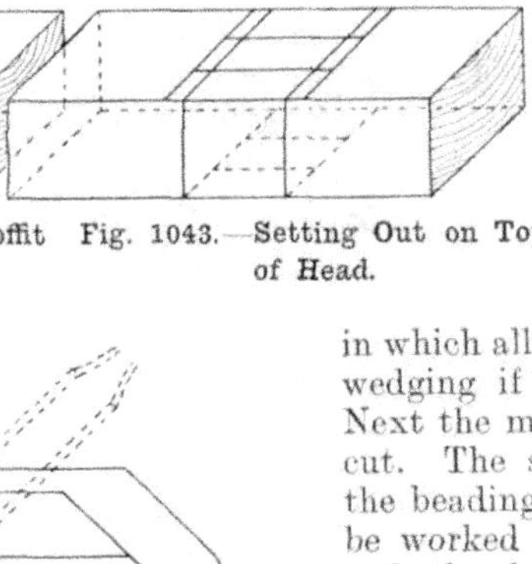   Fig. 1043.—Setting Out on Top of Head.

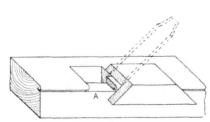

Fig. 1044.—Mitering Bead.

apply the rod and mark off for the shoulders at top and bottom. The splay for the sill (Fig. 1036) must be allowed for as shown at A (Fig. 1040). The inner shoulder D must be carried beyond the outer one C, so as to allow for the bead on the inner edges of the frame (Figs. 1037 and 1038). The

in which allowance is made for the necessary wedging if the mortices are to be closed. Next the mortices are made and the tenons cut. The shoulder must not be cut until the beading is done. The bead should now be worked on the inner edge of the posts and the head. The shoulders should be cut, and the part between the head and mortice sawn out and pared as shown at A (Fig. 1044); then with a mitre template and chisel form the mitre as indicated. The beads at the top of the post should also be mitred as shown at B (Fig. 1037). The joints should be fitted, painted, cramped together, and wedged as shown at Fig. 1045. If the joint is draw-bored and pinned, the holes should be made on the cheeks of the mortice as indicated by the circle C (Fig.

1046), and the hole in the tenon as indicated by the circle D, partly shown by dotted line. It will be noticed that the end of the tenon (Fig. 1046) projects slightly and is also chamfered off. This allows the tenon to enter the mortice easily, and the projection allows a little extra strength behind the pin. The projecting part is generally sawn off at the time of fixing.

Forming the Stops.

The stops are prepared by facing up a board, shooting the edge, setting a gauge

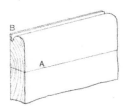

Fig. 1047.—Preparing Beaded Stop.

to the breadth of the stop, and gauging as shown at A (Fig. 1047). A $\frac{1}{2}$-in. or $\frac{5}{8}$-in. bead is next planed along the edge, forming as shown. A saw-cut is now made just outside the gauge line (Fig. 1047), then the edge is planed just down to the line. Any projection left beyond the quirk, as indicated at B (Fig. 1047), is removed by planing, and thus the stop is brought to thickness as shown at Fig. 1048. The stops are next mitred at each corner of the head, as indicated at C and D (Fig. 1037). These may be painted at the back and nailed on at the

Fig. 1048.—Preparing Beaded Stop.

bench, or just tacked on temporarily and finally fixed at the time of hinging the door. These door frames are usually built in the walls as explained on p. 314 in describing the previous example. The making of this door would be a somewhat similar process to making the ledged door, except that braces are introduced to prevent the door dropping out of square. The method of

joining the braces to the ledges is shown clearly at Fig. 1039. Of course, the braces are nailed to the boards. This door is hung with cross-garnet hinges, also known as T hinges. A rim lock and staple are shown. The methods of fitting and fixing these will be explained in a subsequent example.

Framed and Braced Door and Frame.

The general details of a door and frame of this description are shown at Figs. 1049 to 1061. All the chief measurements are figured on the illustrations. The chief points to notice in preparing the frame are that it is rebated out of the solid, and beaded inside and outside, this involving more care in setting out. At A in Fig. 1057 part of the rod is represented, and at B the top part of a post is shown raised above, and the projectors show the relation of the rod to the setting out on the post. It will be noticed that the shoulder lines are not in the same plane, and each is marked long enough to fit against the quirks of the beads in the heads of the frame. Fig. 1058 shows the top part of the post gauged for the rebate, and with the tenon cut. The depth of the rebate is usually $\frac{1}{2}$ in. to $\frac{5}{8}$ in. The head should next be set out from the rod. This is shown at Fig. 1060 ; a and b indicating the mitre lines for the beads.

Rebating the Frame.

The next operation would be the rebating, which may be done in several ways. Only one will be explained here, and other methods will be treated of in other examples. With a plough fitted with a $\frac{3}{8}$-in. or $\frac{1}{2}$-in. iron, make a groove on the face of the post, as indicated at A (Fig. 1059). Next chisel away the greater part of the waste, using a mallet ; then the rebate may be finished with a rebate plane and a trying and smoothing plane. But a quicker and better result can be obtained by using a panel plane, if one is available, than by using a rebate plane and trying plane. The beads should next be stuck. Usual sizes are : for the inside, $\frac{1}{4}$ in. to $\frac{3}{8}$ in. ; for the outside, $\frac{1}{2}$ in. to $\frac{5}{8}$ in.

In fitting the tenon it will be seen (Fig. 1056) that this projects both into the rebate and stop part of the head, and therefore to save making two tenons (which will be shown in another example) it is pared back flush with the rebate as shown at A (Fig. 1056). In this case the beads are mitred only ½ in.

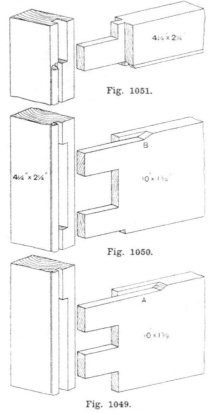

Fig. 1051.

Fig. 1050.

Fig. 1049.

Fig. 1049.—Joint at A (Fig. 1062), showing Barefaced Double Tenons. Fig. 1050.—Joint at B (Fig. 1062) showing Barefaced Double Tenons. Fig. 1051.—Joint at C (Fig. 1062).

or so in, and then the parts C and D are cut so as to butt against the corresponding parts on the head, as shown at C and D (Fig. 1056). The wedging up, finishing, and fixing of this door frame would be done in the same way as explained for previous examples. The bottom of the post is shown at Fig. 1055 with a square metal dowel partly inserted, the other portion projecting to be received by a hole cut in the stone sill.

Making the Door.

The chief features to be noticed in the construction of the door will now be given. Having planed up the stuff out of winding, and to a thickness and breadth for the framing of the door, the stiles would be set out from the rod for the mortices. These would have to correspond with the tenons as shown at Figs. 1049 to 1051, where they are shown notched at A and B, because of the small rebates made to receive the short tenons of the braces. If the braces were cut in nearly square, this would not be necessary. The stiles and top rails are ploughed

Fig. 1052.—Joints of Boarding Grooved and Tongued and Beaded.

Fig. 1053.—Joints of Boarding when Rebated and Beaded.

Fig. 1054.—Joints of Boarding when Ploughed and Tongued, and showing a **V**-joint on each side.

to receive tongues on the boarding as shown. The bottom and middle rails have double tenons with shoulders on one side only, known as barefaced tenons. This is to allow the boarding to pass down in front of them, and also to be nailed to them. The inside of the framing is stop-chamfered as shown at Fig. 1064. The boarding may be of any of the forms shown by Figs. 1052, 1053, or 1054, or a **V**-joint may replace the beads shown at Fig. 1052. The boarding should be fitted in accurately before wedging up the framing. After this it should be nailed to the bottom and middle rails. Other particulars of preparing the mortices, fitting, wedging up, finishing, etc., being common to many examples of joiners' work, will be fully treated in the cases which follow.

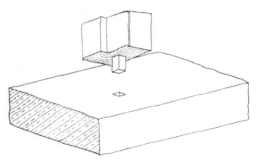

Fig. 1055.—Bottom of Door Post with Metal Dowel for Fixing to Stone Sill.

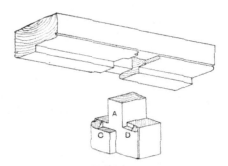

Fig. 1056.—Joint between Post and Head of Frame.

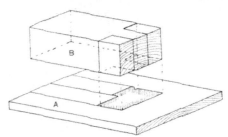

Fig. 1057.—Showing Top of Post Set Out from Rod.

Hanging and Fastening.

This form of door would be hung with 4-in. to 5-in. wrought butt hinges, the lower one being fixed about 11 in. from the bottom, so that the screws should not be too near a tenon, the top butt being 6 in. or 7 in. down. Doors of this kind are often hung with three butts, the third

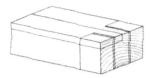

Fig. 1058.—Post Gauged for Rebating and Tenons Cut.

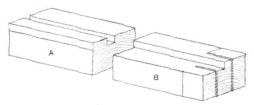

Fig. 1059.—Post Ploughed for Rebate shown at A ; Rebate completed shown at B.

Fig. 1060.—One End of Head of Frame Set Out.

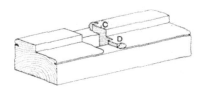

Fig. 1061.—One End of Head of Frame completed.

being central. Heavier doors of this class are usually hung with some form of hook and strap hinges, of which there are various kinds. Fastenings for this door are : A Norfolk latch, and a dead lock, which is a lock without a spindle and handles.

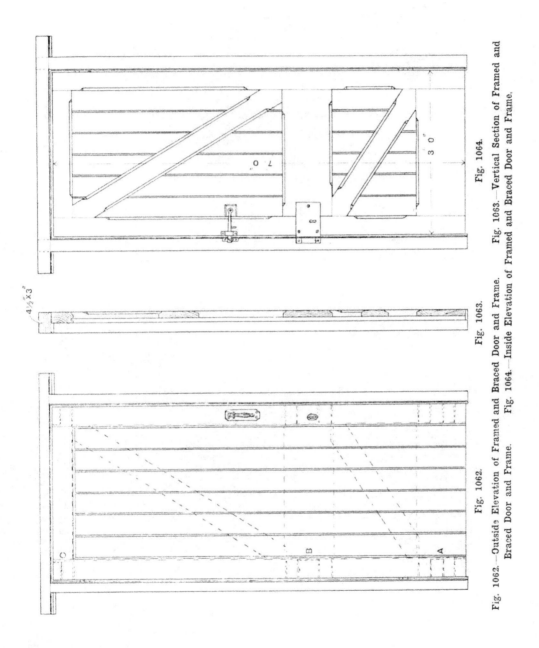

Fig. 1064.

Fig. 1063.

Fig. 1062.

Fig. 1062.—Outside Elevation of Framed and Braced Door and Frame.   Fig. 1063.—Vertical Section of Framed and Braced Door and Frame.   Fig. 1064.—Inside Elevation of Framed and Braced Door and Frame.

Fig. 1066.—Inside Elevation of Large Framed and Braced Door.

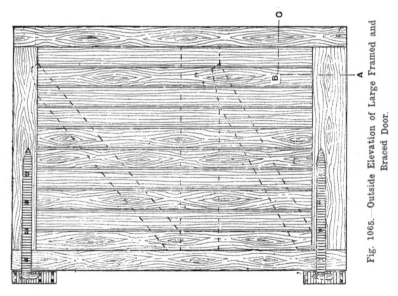

Fig. 1065.—Outside Elevation of Large Framed and Braced Door.

Fig. 1067.—Horizontal Section of Large Framed and Braced Door.

Large Framed and Braced Door.

The construction of a single framed and braced door, such as is often used for stables, archway entrances, etc., is shown in Figs. 1065 to 1067. The finished dimensions are: Door, 8 ft. high and 6 ft. wide; stiles, 6 in. by $2\frac{1}{2}$ in.; top rail, $6\frac{3}{4}$ in. by $2\frac{1}{2}$ in.; bottom rail, $10\frac{3}{4}$ in. by $2\frac{1}{2}$ in.; middle rail, $10\frac{3}{4}$ in. by $1\frac{3}{4}$ in.; boarding, $1\frac{1}{8}$ in.; braces, 6 in. by $1\frac{3}{4}$ in. The joints connecting the stiles and rails are shown by Figs. 1068 and 1069. At the connection of the middle rail and stile barefaced tenons are shown. The bottom and top rails finish flush with the outside of the stiles, and the joint between these and the boards forming the panel is broken by means of a bead, as shown by

Fig. 1068.—Joints between Rails and Stiles.

Fig. 1070. The plough groove is so arranged as to have one side of it in the same plane as the mortices. The braces butt against the rail and stile at their lower ends, and their upper

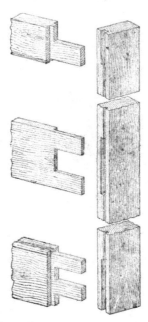

Fig. 1069.—Joint between Brace and Rail.

ends are lapped and joggled in as illustrated at the bottom of Fig. 1069. It is frequently considered sufficient merely to butt the ends of these braces as described, but sometimes

it is preferred to form short barefaced stub tenons on the lower ends of the braces, and let these fit into corresponding mortices in the rails and stiles. Strong hinges are necessary

Fig. 1070.—Conventional Detail of Corner of Door at A, B, C (Fig. 1065).

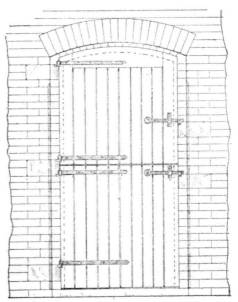

Fig. 1071.—Outside Elevation of Stable Doors.

Fig. 1072.—Horizontal Section of Stable Doors.

for this door. These may be specially made by a good smith, but excellent hinges for the purpose are Collinge's patent.

Frameless Stable Doors.

In many country districts—particularly in certain parts of Shropshire—wood frames to stable doors are often dispensed with, it being urged against them that they are apt to become loose, and that rot is apt to

the doors. The exterior and interior jambs have rounded bull-nosed bricks, to prevent injury to the horses passing in and out of the doorway. Fig. 1078 shows the method used for door jambs where 9-in. walls are erected, the 14-in. jambs being formed as a pier, and then reduced to the 9 in. The doors should

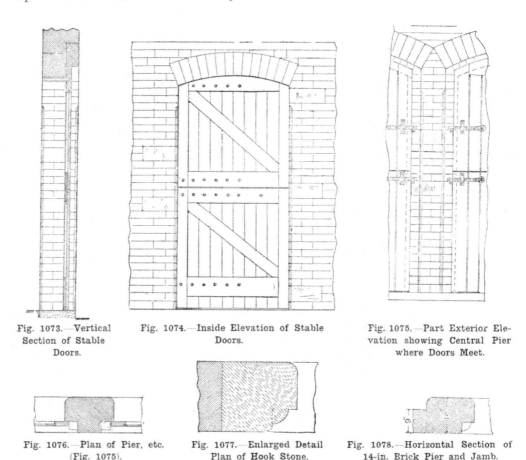

Fig. 1073.—Vertical Section of Stable Doors.

Fig. 1074.—Inside Elevation of Stable Doors.

Fig. 1075.—Part Exterior Elevation showing Central Pier where Doors Meet.

Fig. 1076.—Plan of Pier, etc. (Fig. 1075).

Fig. 1077.—Enlarged Detail Plan of Hook Stone.

Fig. 1078.—Horizontal Section of 14-in. Brick Pier and Jamb.

set in towards the bottom when the frame is let into the step. Figs. 1071 to 1078 show frameless doors, Fig. 1071 being the front elevation of a door in two leaves. The hinge-hooks and latch-hooks are leaded in the stone and built in the brick jambs as the work proceeds. Figs. 1072 and 1073 show the brick jambs rebated to receive

be of good sound and dry red deal, framed, ledged, and braced as shown. The two stiles and the top rail are of equal thickness—namely, $2\frac{1}{4}$ in. ; the other rails and braces are only $1\frac{1}{2}$ in., and are flush on the rear side of the framing, and so arranged that the $\frac{3}{4}$-in. sheeting when nailed to them is flush on the face side. The upper ends and

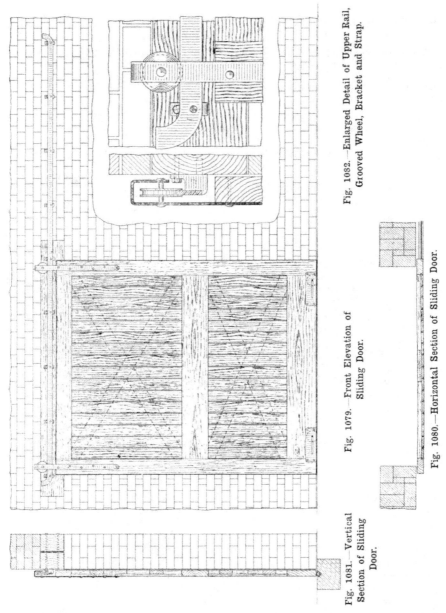

Fig. 1082.—Enlarged Detail of Upper Rail, Grooved Wheel, Bracket and Strap.

Fig. 1079.—Front Elevation of Sliding Door.

Fig. 1080.—Horizontal Section of Sliding Door.

Fig. 1081.—Vertical Section of Sliding Door.

outside edge of the sheeting are tongued into a groove running round the framing. All joints in these doors should be painted, and not glued ; and this remark applies also to all the edges and tongues in the joints of the sheeting. The wrought-iron straps and latches should be strong, and secured with bolts and nuts. Take care that the hinges and hooks are so fixed that the doors will open and lie back against the face of the wall.

Large Framed and Braced Sliding Door.

Sliding doors and gates of this description are often used for coach-houses, entrances to factory yards, and similar purposes. They have no frame, the opening being formed by the brickwork and a wooden lintel, which is often flitched as shown (Fig. 1081). The general construction would be similar to the last example (Fig. 1065) except, as will be noticed, the rails are of the

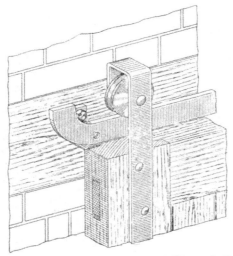

Fig. 1083.—Conventional View of Upper Rail, Wheel, and Strap (Fig. 1082).

same thickness as the stiles, although sometimes the bottom rail may be thinner, so as to allow the boarding to extend down; also the middle rail may be treated in the same way. To prevent any chance of racking, four braces are introduced. The important parts to notice are : To facilitate sliding, a rail shown in section by Fig. 1084 has screws inserted at intervals in the under side. A groove is cut into the stone floor to the shape shown, the rail is then placed in position, and molten lead is run in. Reference to Figs. 1079, 1081, and 1084 shows that the door does not touch the ground; therefore to enable the rail to take some of the weight a couple of grooved wheels are fixed to the door, the axles are cast on the solid, and their

ends work in slots cut in the iron boxes. Fig. 1085 shows a sectional view of one of these boxes and wheels. To fix them, sockets should be cut to the depth of a box ; four stout screws will hold each in position.

Preparing Joiners' Work principally by Hand.

The following particulars treat of some of the most usual methods adopted by joiners in the cutting out, planing up, setting out, and finishing at the bench, doors, panelled framing, and bench work generally. After setting out the rod, careful note

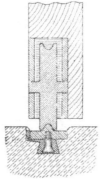

Fig. 1085.—Conventional Sectional View of Box and Wheel.

Fig. 1084.—Vertical Section through Lower Rail, Wheel and Box.

must be taken of the number of pieces required, together with their sizes. When dealing with a large piece of framing, it is well to write down these particulars. The stuff should next be carefully selected and lined out for the various pieces. Although this may seem a simple matter, it often requires a great deal of judgment. So much is this a fact that, in large shops, a leading hand or deputy foreman is appointed specially for this work, an arrangement that is undoubtedly advantageous to the firm. The stuff should next be cut down the grain with a rip saw, and across the grain with a panel or hand saw.

Trying Up.—Most joiners, before putting a plane on timber, brush the latter with a little wire brush kept on the bench for the purpose. This removes all the grit from the beard of the wood, and prevents the plane-iron getting notched. Before planing, if there is a black end, cut it off or remove with a chisel the arris or edge from the end at which a start will be made, as indicated at Fig. 1086 ; this will obviate notching the irons with dirt brought, perhaps, from the timber perch. In executing all framed work the trying up is the most important process. Having brushed away the grit, test for winding. Use a pair of boning or winding strips ; these are two pieces of stuff about 12 in. long, 2 in. wide, and $\frac{1}{2}$ in. thick ; place one on each end of the stile as it lies on the bench. Look along the top edges of these as indicated at Fig. 1086, and thus discover

continuing the tried-up mark on this also, as shown on Fig. 1087. Note that these tried-up edges and faces must always be used to gauge and to square from. To ascertain whether the edge is shot straight, use a long straight-edge, or lift the end of the stile nearest the worker to the level of one of the eyes, closing the other, when any inequality will be seen along the edge and must be planed off accordingly. Now gauge the stuff to breadth along its whole length, plane off down to the gauge mark (which must be left on), and put aside. Serve all the rest of the pieces in the same way, altering the gauge, of course, for the wider or narrower pieces. The stuff can now be gauged and

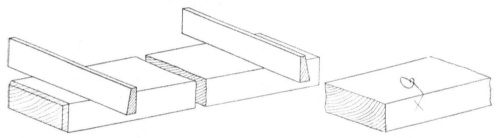

Fig. 1086.—Application of Winding Strips.
Fig. 1087.—Face and Face Edge Marks.

how much the timber twists or is in winding. A practical joiner can test the stuff out of winding by the eye alone, or, if the stuff is very long, by laying his jack- and trying-plane at either end and boning from them. Take off the highest parts with the jack-plane, but do not scoop thick shavings off, or the framing will be too thin. Always leave from the jack-plane what a joiner calls a witness mark—that is, the faintest impression of the weather stain or saw—before using the trying-plane. It is sometimes well to leave these marks showing after the trying-plane, to indicate that the stuff has not been planed too much. The trying-plane must be used until both the top edges of the winding strips are parallel with each other ; otherwise the framing will twist when wedged up.

Completing the Trying-up.—When the face is true, put on the tried-up mark (see Fig. 1087), jack and shoot perfectly straight, and square the edge with the trying-plane,

planed to thickness, or the back faces may be simply jacked off for the present.

Setting Out Stiles.—One of the stiles should be placed on the rod, and the proper positions of the rails should be pricked off on the face edge, also marking off spaces to be allowed for in the cases of plough grooves, rebates, beads, mouldings, etc. The relation between the rod and marking off for the mortices on a stile for a top rail is clearly indicated by the projections at A and B (Fig. 1089). Now place the stiles in pairs with their face sides together, and face edges outwards, the ends of the lower one resting on two planes on a couple of blocks as shown at Fig. 1088. This facilitates the use of the try square, as indicated. Through the marks pricked off from the rod, square down the edges as shown at Fig. 1088. When there are muntins they may be placed on the stiles and squared down for their shoulders when marking on the edges for the stiles, where they will meet the edges of the rails ; this is clearly shown at Fig. 1088. The muntins can now be removed

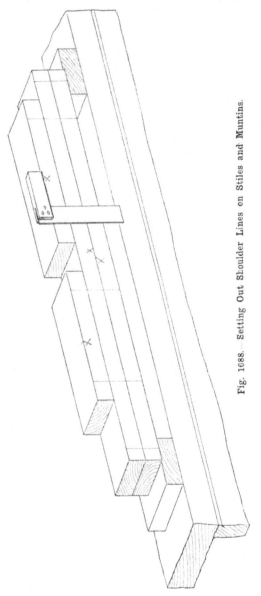

Fig. 1088.—Setting Out Shoulder Lines on Stiles and Muntins.

and scribed all round for the shoulders. The setting out on the edges for the mortices should be complete as shown at Fig. 1089 ;

then by using the square and ticking off on the back arris, as indicated by the dotted lines, and also marking off about ¼ in. or ⅜ in. (according to the size of the work) for wedging. the back edges should be marked for the mortices as indicated at Fig. 1090.

Setting Out Rails.—When there is a bottom rail, a middle rail, and a top rail, they usually are in panel framing, all of one length, and therefore the distances for shoulders, positions for mortices and muntins, should be pricked off from the rod. The three rails can now be placed with their face edges outward, and then squared down as shown at Fig. 1091 at one operation. The lines b and g are for shoulders, a and h for haunchings, d and e for mortices for the muntins. and c and f positions where edges of muntins meet the rails, the distances between each pair of lines being $\frac{1}{16}$ in. less than the depth of the ploughed groove, which usually ranges from ⅜ in. to ⅝ in. The setting out (without gauging) of the top rail is shown by Fig. 1092, of the middle rail by Fig. 1093, and of the bottom rail by Fig. 1094 respectively. The breadth of the tenons is next set out, and the waste parts removed as shown at Figs. 1095, 1096, and 1097, the cuts along the haunch lines a and b (Figs. 1095 and 1096) being made with a bow saw. Some joiners prefer to cut through the whole breadth of the rail, and use the parts between the tenons for wedges, as indicated at Fig. 1099. Select a mortice chisel about one-third the thickness of the stuff, and set the teeth of a mortice gauge, so that the chisel just sinks between the points (see Fig. 1098). Adjust the gauge stocks so that the teeth will be in a position to scribe the mortice in the desired place, which usually is in the centre of the stuff ; but there are, of course, exceptions to this rule. All the parts for mortices and tenons should now be gauged, taking care that the stock of the gauge is used always against the face side of the stuff.

Making Mortice and Tenon Joints.—The next thing is to mortise the stiles with a mortice chisel and mallet. The mortices should be made halfway through from the back edges ; then the stiles should be turned over and mortised through from the face

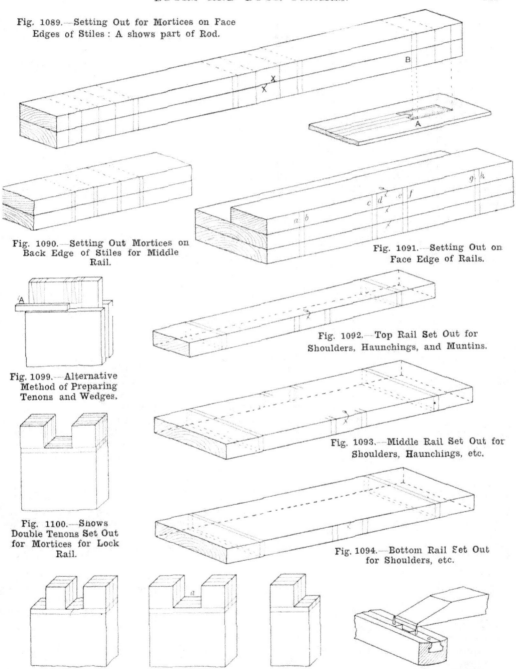

Fig. 1089.—Setting Out for Mortices on Face Edges of Stiles : A shows part of Rod.

Fig. 1090.—Setting Out Mortices on Back Edge of Stiles for Middle Rail.

Fig. 1091.—Setting Out on Face Edge of Rails.

Fig. 1099.—Alternative Method of Preparing Tenons and Wedges.

Fig. 1092.—Top Rail Set Out for Shoulders, Haunchings, and Muntins.

Fig. 1093.—Middle Rail Set Out for Shoulders, Haunchings, etc.

Fig. 1100.—Shows Double Tenons Set Out for Mortices for Lock Rail.

Fig. 1094.—Bottom Rail Set Out for Shoulders, etc.

Figs. 1095, 1096, and 1097.—One End of each Rail with Waste removed, and completely Set Out ready for Tenon Cutting.

Fig. 1098.—Method of Setting Mortice Gauge to Chisel.

edges. The wedging should be done carefully from the back edges. Clean out these
mortices by driving the core through from
front to back with a slip of hard wood called
a core-driver. Clean the mortices sparingly
with a paring chisel, and the stiles are ready
for ploughing. Now mortise the top, lock,
and bottom rails. These mortices should go
in only about 2 in., or 2½ in., as the tenons
of the muntins are only stumped in. Clean
these mortices out, and the rails will then be
ready for ploughing. For cutting tenons,
put one of the rails in the bench screw, tilted
as shown in Fig. 1101, and with a rip saw
cut down the tenon just by the side of
the gauge-mark, leaving half of it visible.
Do not force the saw, but work it freely,
and keep it parallel with the gauge-mark,
both down the side and across the end.
When the saw is down about 3 in., take out
the rail and serve the opposite side of the
tenon in the same way for about the same
depth ; then, screwing the rail perfectly
upright in the screw, connect the two saw
kerfs by nice easy strokes, and cut down
to the shoulder ; serve all the rails and
muntins like this. The two outside portions
are called the tenon cheeks, and the inner
portion the tenon. The tenon cheeks must
not be cut off until later on. An experienced man can cut his tenons down straight
away. With a plough-iron of the proper
size, plough from the tried-up face the tried-
up edges of the stiles from end to end, also
the bottom edge of top rail, both edges of
the middle rail, top edge of bottom rail, and
both edges of top and bottom muntins. If
the plough is set to the right depth, it is impossible to go any deeper. The shoulders
can now be cut with a tenon saw, leaving
half the scribe line as when cutting the
tenons, also under-cutting them the least
bit so that the joints will come together close
on the face. Having cut off all the tenon
cheeks, if the rails have been cut through
their whole breadth, as in Fig. 1099, prepare
a strip of wood about 9 in. long and about
$\frac{1}{16}$ in. narrower than the depth of the
plough groove ; lay this on the shoulders
just cut, and mark in lead pencil across
the tenons A (Fig. 1099). The portions thus
marked will form the haunchings to fit in the
grooves (see Fig. 1099). Serve all the rails

only like this, the muntin tenons being left
from the plough. Next mark out the tenons
as shown in Fig. 1099 ; but before cutting out
the portions indicated by the dimensions,
mark with lead pencil as many wedges as
these portions will allow, as shown, and cut
them down ; then, in cutting along the
haunching lines with a bow saw, these
wedges will fall out, and, when trimmed
and pointed, can be utilised for wedging up
the tenons, and will be the exact thickness
necessary to fill the mortices. Just nip the
extreme corner off each tenon with a chisel
to give a start in the mortices, and put the
framing temporarily together, and let it
stand while the panels are prepared. It might

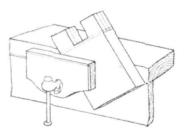

Fig 1101.—Method of Fixing Rails in Screw for
starting Saw Kerf of Tenons.

be left for several months with advantage,
and, if kept dry, all shrinkage would take
place while all the shoulders were free. On
wedging up, it would be thoroughly seasoned,
and would remain a good job throughout its
ordinary life as a door.

Panels.—For ordinary work panels should
be made from sound yellow pine, free from
knots and shakes. Their dimensions can be
measured from the framing. Cut them out
in the rough, jack them over, and bend
each one on the edge of the bench. If there
should be a shake it will betray itself, and
for a first-class job the panel would be rejected ; for commoner work, however,
these shakes should have a little whiting
and glue rubbed in with a hammer and allowed to dry. Now try up the face and shoot
one edge, not forgetting to put on the tried-
up marks. With the panel gauge, scribe the
finished width, which, to allow for swelling,
must be about ⅛ in. less on each side than the
actual width when finally driven home into

the plough grooves. Square and cut off both ends, allowing the $\frac{1}{8}$-in. play also. Make a mullet (Fig. 1102) from any odd bit of stuff ploughed with the iron used for the framing. Slide this round the edge of each panel as it lies projecting a little over the edge of the bench, and it will indicate at once any place that will be tight when driven home, and which must be eased accordingly. The panels must fit the mullet without binding. With a sharp smoothing plane, set very fine, smooth up both sides, and, with some fine glasspaper folded on a cork pad, rub both sides across the grain until a fine level surface is acquired. The panels may be put in by removing one stile at a time and gently driving them home with the hand.

Gluing and Wedging-up Doors and

squaring rod mentioned below is not so necessary when gluing up four-panel doors as for skeleton frames and doors containing very narrow rails. Carefully note that the rails are in their correct position, which can be seen by the marks on the stiles; the rails can be regulated to these by a blow or two on their outer edge. This knocking may also be necessary to close the shoulders between the rails and muntins. Then, dipping the trimmed and pointed wedges in the glue, insert them just tight. It is then usual to drive the outer wedge, c or d (Fig. 1103), most at first, so as to ensure the joints of the muntins being closed; both wedges c and a are afterwards driven to hold the tenon and mortice. It may be necessary to give the muntins a knock or two to get

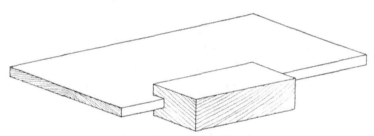

Fig. 1102.—Mulleting Panels.

Framing.—This is a two-handed job, and to carry it through the services of a mate on the other side of the bench will be necessary. Clear away all tools, etc., lay some scantling across the bench, place the door on these (see Fig. 1103, which shows the wedging up of a two-panel door), knock all shoulders about 3 in. apart, and then, with some thin hot glue, rub the haunchings, shoulders, and tenons with the brush. Also brush the glue in the mortices from the back edge. Turn the door quickly and serve the other side the same way. Knock the stiles up and put on a cramp, screwing up tight until all shoulders are up. The cramp should be placed in the centre of the middle rail, or, better still, two cramps should be in use one on either side of the rail. The shoulders of the rails having been cut quite square, the door may be wedged up so that the shoulders fit. The

them in their exact positions, but in doing this interpose a piece of wood between work and hammer to avoid bruises—a precaution which will also apply in knocking up the stiles and rails. Also have waste pieces for the cheeks of the cramp to screw against. Having finished wedging up, take off the cramp, oil the tenon saw with olive oil (not linseed), and cut off the projecting ends of the tenons and wedges. When the door or framing has stood for a day or two it will be ready for cleaning off in the manner described below.

Cleaning Off.—When ready to clean off, lay the door or framing on the bench, and cut and nail two pieces of stuff between the horns at each end, so as to keep the door solid for planing. Clean the superfluous glue from the joints with a chisel, and try, smooth, and glasspaper the face side first. Now set the gauge to the thickness the door or framing is to be, which is the finished thickness, and run it down each edge, and clean off the

some kinds of doors and framing have to be treated in a different way as regards the

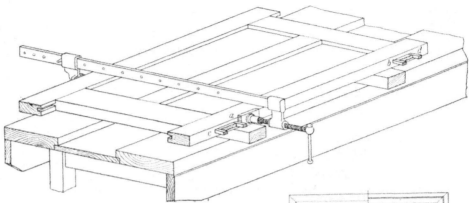

Fig. 1103.—Wedging up a Two-panel Door.

other side, down to this gauge mark, in the same way. Of course, this last process of gauging is unnecessary if all the stuff is thickened as planed ; but for commoner work the method described is often adopted. Do not shoot the edges or cut the horns off the ends. This is done when the door is fitted and hung or the framing fixed.

Planting Door Mouldings.—To plant the mouldings, get a length of ordinary ogee mould, see that it is quite clean ; if not, get some glasspaper and take out the rough parts ; then cut a mitre at one end. push it gently up into place, mark the opposite end, and cut that mitre to it. Lay this on the framing by the side where it is to go, and proceed to cut the rest in the same way. Note that these pieces will all be the dead length from shoulder to shoulder. Now place the end pieces in first, and then, placing one of the side-piece mitres in position, bend the moulding over the fingers and spring it into position against the bottom one. When this is pressed and bradded down, both mitres will press home, and good mitres will be the result. Serve all the mouldings in the same way, and then brad in, taking care to keep the brads well bevelled from the worker, or the panel will be split and choked. Now drive in the brads with a small steel punch, and the door or framing is completed. As already stated,

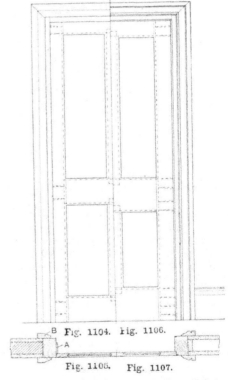

B Fig. 1104. Fig. 1106.

Fig. 1105. Fig. 1107.

Fig. 1104.—Half Elevation of a Battened Framed Door and Solid Frame in a Partition Wall. Fig. 1105.—Half Horizontal Section (Fig. 1104). Fig. 1106.—Half Elevation of a Door, Framed Square out of Deal Stuff, with Solid Frame in a Partition Wall. Fig. 1107. — Half Horizontal Section (Fig. 1106).

panels and moulds, but they are all constructed in the same way as regards the framing.

Four=panelled Doors and Solid Frames in Partitions.

Fig. 1104 is a half elevation and Fig. 1105 is a half-horizontal section of a battened 5 in. by 3 in., tried up, and squared on three sides. The posts and head are mortised and tenoned together as represented at Fig. 1108. The upper part of the post A, above the head, is reduced on each side by about $\frac{7}{8}$ in., as shown, thus making it the same thickness as the bricknogging or studding of the partition. Each side is

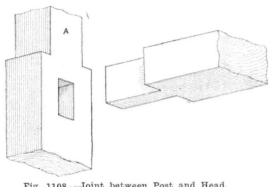

Fig. 1108.—Joint between Post and Head.

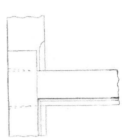

Fig. 1110.

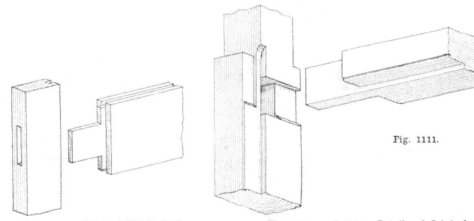

Fig. 1109.—Joint of Middle Rail and Stile.

Fig. 1111.

Figs. 1110 and 1111.—Details of Joints between Post and Head of Frame.

square-framed door, 6 ft. 6 in. high by 2 ft. 6 in. wide, with solid wrought frame and stops A nailed on, and 3-in. by 1⅛-in. square and splayed architrave. Doors and frames of this description are often used for attics and small houses, and for openings in studded or bricknogged partitions.

The Frame.—This usually forms a part of the partition, and is made of stuff about

generally finished by fixing some form of plain architrave, as shown at B (Fig. 1105), which projects over the plastering as indicated. The stop A is square-edged and nailed on.

The Door.—This is frequently made out of battens 7 in. by 1½ in., used full width at the bottom and middle rails, and sawn down the centre to make the top rails, muntins,

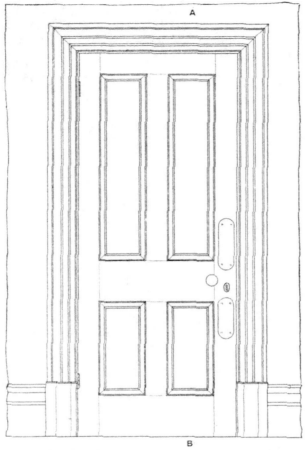

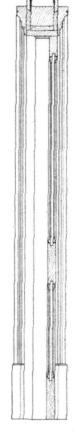

Fig. 1114.—Elevation of Four-panel Framed and Moulded Door with Linings, etc., in Partition.

Fig. 1116.—Vertical Section through A B (Fig. 1114).

Fig. 1115.—Horizontal Section of Fig. 1114.

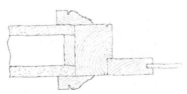

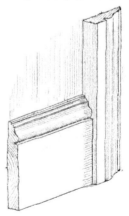

Fig. 1112.—Enlarged Detail through Jamb, Door Stile, Architraves, etc. (Fig. 1106).

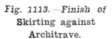

Fig. 1113.—Finish of Skirting against Architrave.

and stiles. The finished sizes thus usually come : Bottom and middle rails, $6\frac{3}{4}$ in. by $1\frac{3}{8}$ in. ; stiles, top rails, and muntins, $3\frac{1}{4}$ in. by $1\frac{3}{8}$ in., the panels finishing about $\frac{3}{8}$ in. or $\frac{7}{16}$ in. thick. The bottom and middle rails being narrow, it is usual to have only one tenon ; if this is 3 in. or $3\frac{1}{2}$ in. wide with haunchings on each side, as shown at Figs. 1104 and 1109, it will, as a rule, be found sufficient.

Framed Door.

A 6-ft. 8-in. by 2-ft. 8-in. door, framed up out of deal stuff, namely, 9 in. by $1\frac{3}{4}$ in., is shown at Figs. 1106 and 1107 in half elevation and horizontal section.

The Frame.—This, as in the last example, is solid ; but it is rebated and beaded. It is intended that the frame forms a direct portion of the partition without jamb linings, etc. The enlarged details at Figs. 1110 and 1111 will make the construction of the joints, beading, etc., of the frame quite clear. The architraves are shown by the detail Fig. 1112, as fixing on to the frame and covering its joint with the plastering. Fig. 1113 shows the skirting finishing against the architrave. The door being framed of deal stuff, the finished sizes of the members will be : Bottom and middle rails, $8\frac{3}{4}$ in. by $1\frac{3}{4}$ in.; muntins, top rail, and stiles, $4\frac{1}{4}$ in. by $1\frac{3}{4}$ in. (of course, the thickness might vary up to 2 in. or even more). The tenons, wedging, plough-grooves, etc., are indicated by the dotted lines.

Four-panelled Moulded Door, with Jamb Linings, etc., in a $4\frac{1}{2}$-in. Wall or Studded Partition.

A good ordinary door, with its fitments, is shown in elevation, plan, and section at Figs. 1114 to 1116. The construction of the joints will be clearly understood by reference to Fig. 1117, and sections of the mouldings in the panels are given at A (Fig. 1118).

Jambs, Grounds, etc.—Fig. 1119 illustrates an ordinary form of plain jamb linings, which are grooved and tongued together as represented. They are made out of 1-in. to $1\frac{1}{2}$-in. stuff, sufficiently wide to project on each side of the post about $\frac{7}{8}$ in., the amount required for lathing and plaster-

ing, or for plastering only in the case of brickwork. The jamb linings are fixed plumb and straight by packing pieces, or

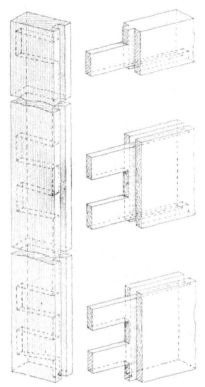

Fig. 1117.—Conventional View of Mortice and Tenon Joints of Door (Fig. 1114).

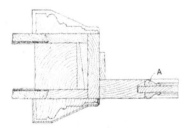

Fig. 1118.—Enlarged Detail of C at Fig. 1115.

wedges, which are first placed between the back of the lining and the posts as represented at B (Fig. 1119), and then nailing through, as also shown. The head is fixed in a similar manner. The advantage of using wedges is that the jambs can be adjusted more accurately for straightness. For fixing the architraves, grounds are fixed

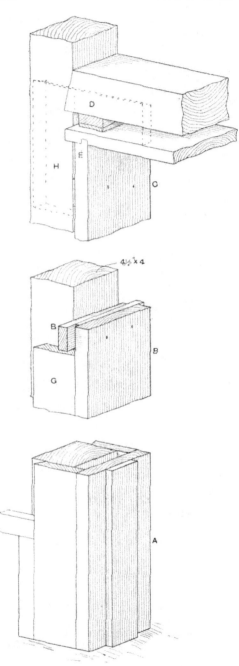

Fig. 1119.—Conventional View showing Method of Fixing Jamb Linings, Grounds, etc.

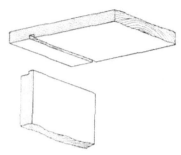

Fig. 1120.—Detail of Groove and Tongue Join at E (Fig. 1119).

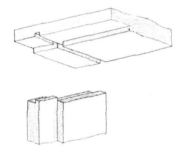

Fig. 1121.—Showing a Corner of Jambs with Single Rebate out of the Solid.

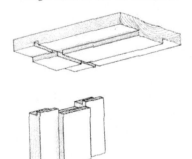

Fig. 1122.—Showing an Angle of Jambs with Solid Stops.

to the posts and head, so that their faces are flush with the edges of the jamb, as indicated at Fig. 1119. These grounds are generally splayed at the back edges as illustrated, to form a key for the plastering. In

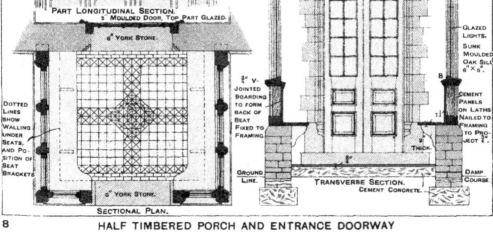

DETAIL SECTIONAL PLAN THROUGH ANGLE POST AND WINDOW MULLION.

BROSELEY TILES

NOTE. — THE GENERAL VIEWS ARE REPRODUCED ⅛-IN. TO THE FT., AND THE ENLARGED DETAILS TO A SCALE OF ½ IN. TO THE FT.

CEMENT PANELS.

2½"

FRONT ELEVATION.

SIDE ELEVATION.

DETAIL SECTION THROUGH DOOR-HEAD AT A.

11" × 1"

4½" × 3½"

DETAIL SECTION THROUGH BARGE-BOARD.

1" DEAL REBATED BOARDING.

4½" × 3"

DETAIL SECTION THROUGH SILL AT B.

¾" TONGUED AND GROOVED V-JOINTED BOARDING IN NARROW WIDTHS MOULDED ANGLE STRIPS 3" × 2"

SPROCKET PIECE.

6" × 3"

PART LONGITUDINAL SECTION.
2" MOULDED DOOR, TOP PART GLAZED.

6" YORK STONE.

GLAZED LIGHTS.
SUNK MOULDED OAK SILL 8" × 5".

¾" V-JOINTED BOARDING TO FORM BACK OF SEAT FIXED TO FRAMING.

B

CEMENT PANELS ON LATHS NAILED TO FRAMING TO PROJECT ¾".

1½"

DOTTED LINES SHOW WALLING UNDER SEATS, AND POSITION OF SEAT BRACKETS

THICK.

GROUND LINE.

¾"

DAMP COURSE.

6" YORK STONE.

TRANSVERSE SECTION.
CEMENT CONCRETE.

SECTIONAL PLAN.

HALF TIMBERED PORCH AND ENTRANCE DOORWAY

the case of commoner work, instead of preparing and fixing grounds, pieces of board, 2 in. or 3 in. wide and 1 ft. or so long, are fixed at intervals of 12 in. or 15 in., as indicated at H and D (Fig. 1119). This example shows the stops nailed on. Two other kinds

Fig. 1123.—Taking Width of Door with Two Rods.

of jamb linings are represented at Figs. 1121 and 1122. The form at Fig. 1121 has a single rebate made out of the solid, the jamb and head being grooved and tongued together as shown. A rebate on each side is shown at Fig. 1122; thus a solid stop is

door should this happen to be a little out of truth; and should the jamb-linings be fixed out of truth—especially when the jambs wind one way and the door the other way—the difference is intensified, and causes considerable trouble to the workman. For the sake of example, suppose a door 6 ft. 8 in. by 2 ft. 8 in. by 2 in. thick, prepared for a mortice lock, has to be hung This means that one end of the middle rail has four tenons instead of two, this provision being necessary in order that when the mortice for the lock has been cut, the wedging between the tenons shall not be cut away; were they cut away, there would be nothing to hold the stile to the rail, and very soon the stile would come away from the shoulder. If it has not previously been arbitrarily decided to which jamb the door has to be hung, it should be so arranged that the door, when slightly open, will hide most of the interior of the room. When a doorway is arranged near the centre of one side of a room, it is not

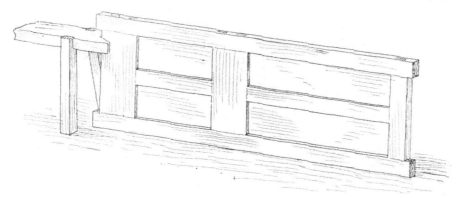

Fig. 1124.—Door in Position for Shooting Edge.

formed as illustrated. The grooving and tongueing of the jambs and head are of a more complicated character.

Hanging Ordinary Four-panel Door.

The method of hanging an ordinary four-panel door will now be described. The frames or jambs to which doors are hung are either rebated out of the solid or have stops nailed on. For inside work, if the stops are planted on after the door is hung, the carpenter is able to make them fit the

so very important which side the door is hung, although it is more usual for a door to open against the fireplace; but when, as is generally the case, the doorway is near a corner, the door should be hung to the jamb farthest from the corner. The side of the door which has double tenons is the side for the lock, so the other must be the one for the hinges. Mark on the muntin of the door to indicate which is the inside face, and stand it against the wall. Then take two short strips of wood, and,

15

holding them together in the middle, carefully take the width of the opening about a foot above the floor, as indicated at Fig. 1123. Mark this measurement on the door at about the same height, allowing for taking as much off one edge as the other, so as not to make one stile appear narrower than the other; also allow a good $\frac{1}{16}$ in. each side for the joint. Follow the

way, put a wedge, about 1 ft. long and 1 in. thick, the thin edge under the low stile, and force it in until the door stands square with the jambs, and shows an equal joint from top to bottom. Observe how each stile fits each jamb, and also whether the joints are even. If not quite satisfactory, mark the parts, take down the door, and plane the stiles where necessary. Replace in position and keep one stile close to a jamb. Now push the blade of a square under the head of the door jamb, in the joint of the door, and mark across the edge (Fig. 1125); then

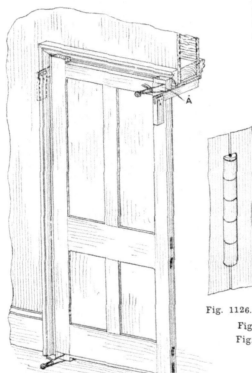

Fig. 1125.—Scribing Door for Height.

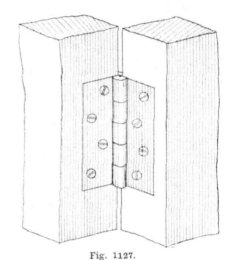

Fig. 1126.   Fig. 1127.

Fig. 1126.—Projection of Knuckle of Butt.
Fig. 1127.—Joint between Door and Frame.

same procedure at the upper part of the door about a foot below the head. Test each jamb with a straightedge so as to note whether to plane the stiles straight or to allow for any inequalities. Next hold the door on its edge as shown at Fig. 1124 and plane the edges down to the marks, taking care to leave them a little out of square in favour of the outside of the door. Stand the door up in its place, and, if it leans either

square this mark across the stile of the door, and do the same on the opposite stile. Take the compasses, and, setting them to the narrowest part of the top rail left above the marks just made (see A, Fig. 1125), prick off two marks above those made with the square on the stile. This will give the actual shape of the door head, irrespective of its squareness; and if, without shifting the compasses, the floor line is scribed across both stiles and the bottom rail as shown at Fig. 1125, the exact height of the door, without allowing anything for joint, will be obtained. Lay the door on two stools, mark across from the marks at the top made by the compasses, and also at the bottom, after allowing $\frac{3}{8}$ in. for the joints at top and bottom. Saw off the surplus stuff to these marks, take off the rough arrises with the jack plane, and the door is fitted. Stand the door on one side for a few minutes, and fit the door stops. Cut the head to length first, and tap it in; then square off the side stops a little shorter than the height of the opening, mark them to exact length, and spring them into their places. Try the door in position, and ease it if necessary; if not, then stand it on one

and the door is left slightly open, people can peep in and see all over the room. No doubt there is truth in this; but, on the other hand, if this course is followed, the joint of the door comes so close that it will not open much more than square before it binds on the mouldings. The first man coming in with the furniture pushes the door right open, as he thinks; but as

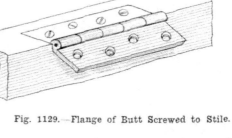

Fig. 1129.—Flange of Butt Screwed to Stile.

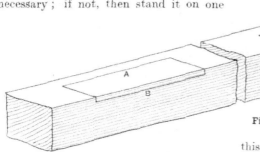

Fig. 1128.—Setting Out on Edge for Flange of Butt.

this cannot be done, the result is that the door is partly torn from its hinges. It is therefore preferable to keep the top hinge out nearly $\frac{1}{4}$ in., and the bottom one $\frac{1}{8}$ in.

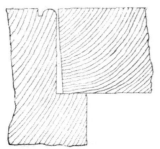

Fig. 1131.—Showing Stile of Door Planed to Allow for Clearance in Opening.

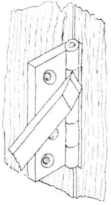

Fig. 1130.—Marking along Knuckle of Butt for Letting into Jamb.

side and let in the hinges, the top one about 6 in. or 7 in. down, and the bottom one just above the bottom rail. There is no particular rule to go by as to height, and half an inch more or less makes no difference, as long as the butts are never let into the stiles at the ends of the tenons of the rails.

Putting on the Hinges.—Some joiners insist that the hinges should be let in so that their centres come to the centre of the joint of the door (Figs. 1126 and 1127); the reason given is that if this is not done,

more. Set out the hinges on the stile as shown at Fig. 1128, the lines A and B being made with a marking gauge. Saw and pare out and then screw the flange of the butt in position as shown at Fig. 1129; its surface should be just flush with the edge of the door. Next offer the door in position, pushing the wedge underneath until a joint about the thickness of a penny is obtained at each side and at the top. Mark the position of the hinges both at top and bottom with a chisel; then with the chisel mark the thickness of the knuckle on the edge of the jamb, as shown at Fig. 1130. This will give the depth to which to let the hinge in. The other edge of the hinge should not be let in more than its own

Repairing and Replacing Door Panels.

Cause of Panels Splitting.—One of the most general causes of door panels splitting up the middle is the improper fixing of the mouldings: the nails being inserted so that they pass through a portion of the panel into the framing as indicated at A (Fig. 1132), whereas the moulding should be secured to the framing only, as indicated at B. It will

Fig. 1132.—Part Horizontal Section through Split Door Panel, etc.

Fig. 1133.—Part Elevation of Split Door Panel, etc.

thickness, and care should be taken to drive the screws in square with the hinge.

Completing the Hanging.—Get the position of the stops, close the door, and knock it gently until it is flush with the jamb at the spot where the lock will come; then nail on the stops so that they fit close on the lock stile, and allow about $\frac{1}{16}$-in. joint on the head and hanging stile to allow for paint. Drive in the nails about 1 ft. apart, and on alternate edges, so that the stops shall not curl away from the jamb. The planing under of door stiles—that is, planing them out of square so that they do not bind on the inner edge when closed—is shown in the section (Fig. 1131).

be seen that in the former case the panel has no chance of shrinking a little in the plough groove; hence the splitting. (Figs. 1132 and 1133 are part section and part elevation respectively of a split pane in a door.) Another cause of splitting is the fitting of the panels too tightly into the plough grooves.

Repairing Split Panel.—To repair a split panel in which the split is not of long standing, and in which, when the parts are forced together, a fair joint will result, first carefully take out the mouldings from each side, then make six or eight blocks and wedges similar to those shown in Fig. 1134; screw the blocks to the panel as indicated, taking care to keep the screw-holes in the panel so that they will be covered by the mouldings : then by carefully levering with a chisel, and lightly striking the wedges, the parts of the panel can be forced together. If the result

Inserting Strip in Split Panel.—Another method, which, in certain circumstances, would be the only satisfactory one, of repairing a split panel is as follows :—Take out the moulding at the top and bottom of the panel (this being necessary on one side only), and set out as shown in Figs. 1132 and 1133 in which B and B′ indicate the crack. On one side make the distance, indicated by the solid

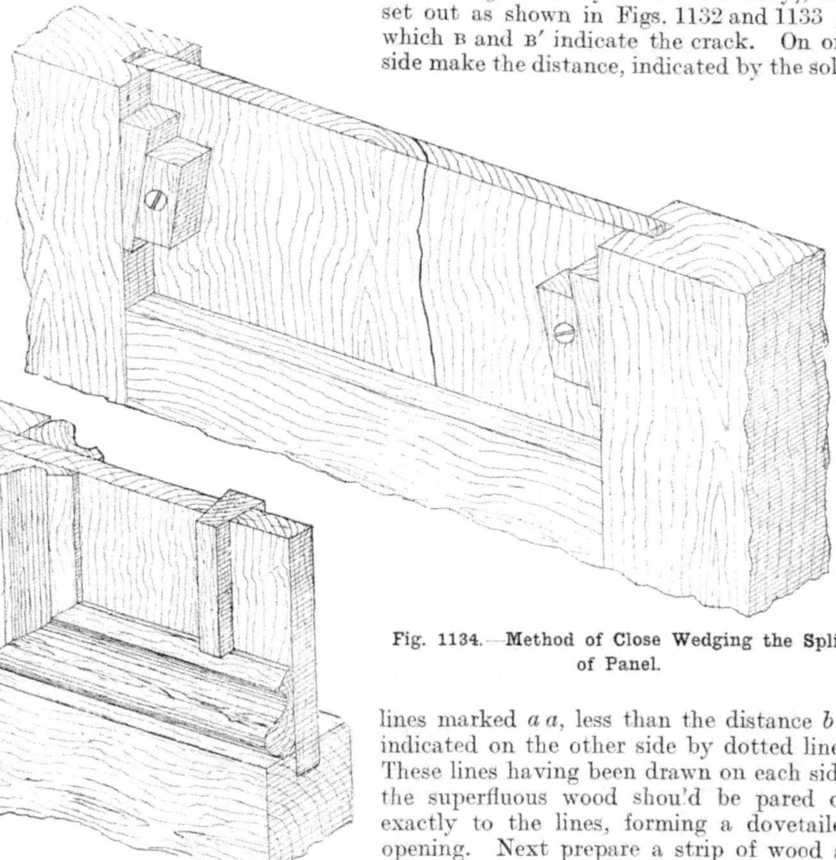

Fig. 1134.—Method of Close Wedging the Split of Panel.

Fig. 1135.—Conventional Sectional View showing Piece Inserted in Panel.

of the trial is found to be satisfactory, the wedges can be released, the crack opened, and some good glue run in. Then the wedges are tightened, and the parts forced as close as possible, any superfluous glue being carefully washed off. After the glue is dry, any projecting parts may be removed by means of a sharp scraper and glasspaper ; then the mouldings can be replaced.

lines marked $a\,a$, less than the distance $b\,b$, indicated on the other side by dotted lines. These lines having been drawn on each side, the superfluous wood should be pared off exactly to the lines, forming a dovetailed opening. Next prepare a strip of wood so that it fits in as indicated in Fig. 1135. It will be noticed that it is necessary to notch each end out just between the mouldings at the top and bottom on that side where the mouldings have not been taken out. The arrangement being satisfactory, secure the strip with glue, taking care to keep the two parts of each surface of the panel in the same plane. The whole should then stand for a time, to allow of the glue setting thoroughly. After this the strip may be cleaned off on each side flush with the panel, a small iron plane being extremely useful for this purpose ; and after being finished off with fine glasspaper, the pieces of moulding may be re-inserted.

Replacing Panel in Door.—Sometimes a panel may be so much damaged that it must be replaced, and in some cases, especially with good doors, it is objectionable to take off the stiles because of the liability to spoil the latter and the rails, particularly where the joints have been well glued and wedged together. The method about to be described will obviate these objections and produce a good sound job :—First take out the mouldings on each side of the panel, and cut out the panel. This may be done by making a hole with a brace and bit and sawing down a short distance with a pad saw, the remainder being cut with a panel saw. The main portion having been taken out, the pieces can be removed from the

plane. Of course, little of this will be necessary if the parts have been carefully fitted. The mouldings may then be re-inserted, and the job thus completed as far as the joiner is concerned.

Four-panelled Moulded Door, with Plain Framed Jamb Linings in an 18-in. Wall.

The Door.—A door and jamb linings which will usually be found in a larger building than in the preceding case are shown at Figs. 1137 to 1142. This door is represented as being 7 ft. by 3 ft. and 2 in. to 2¼ in. thick, panelled and moulded as shown. The framing of doors of this kind, when of deal or pine, is generally made from stuff 11 in. wide ; therefore the finished sizes are usually about as follows : Bottom rail and middle rail 10½ in. wide, and stile, top

Fig. 1136.—Section of Panel with Rebated Fillets.

plough grooves with a chisel. Now make a new panel in the following way. Prepare two strips about ¾ in. wide, and the same thickness as the panel. Next prepare the panel, and rebate this and the strips together as shown at Fig. 1136, so that when they are put together their combined width will be exactly the same as the distance between the plough grooves of the stile and muntin. Next fit the rebated fillets into the plough grooves, and cut off the panel to length. It is not possible to cut it off long enough to go the full distance into the top and bottom plough grooves ; but if it is cut off the length between the rails plus the depth of one plough groove it will, when put in position, be of sufficient length to extend halfway into each plough groove. When found to fit satisfactorily, the panel may be slipped out, and, its rebated edges and also those of the fillets being glued, it may be pushed back into its proper position, care being taken that at the top and bottom it extends into the plough grooves. Additional security may be obtained by inserting a few fine screws diagonally as indicated in Fig. 1136. When the glue is dry, the joints may be cleaned off with a small rebate

rail, and muntins 5¼ in. wide. The upper part of the door is divided into two panels by the horizontal frieze rail. The construction of the different joints would be very similar to that shown and explained in the example on p. 329, except for the tenons on the middle rail where the mortice lock is provided for. In the present example double twin tenons would be made and fitted into corresponding mortices, cut in the stiles as represented at Fig. 1141. These double tenons are provided with the object that when the mortice is cut through the stile for the mortice lock it does not interfere with or weaken the wood in the same vertical planes as the tenons.

Framed Jamb Linings, Grounds, etc.—These are fitted in an 18-in. wall, which means that, with the plastering, the jamb linings will have to be 20 in. wide. It is usual to frame the jamb linings out of stuff 2¾ in. to 3¾ in. wide and about 1⅛ in. to 1¾ in. thick. The rails and stiles of these linings are mortised and tenoned together, wedged up, and cleaned off in the usual manner, as clearly shown at Fig. 1142. The jambs and head stiles are also grooved and tongued together, as

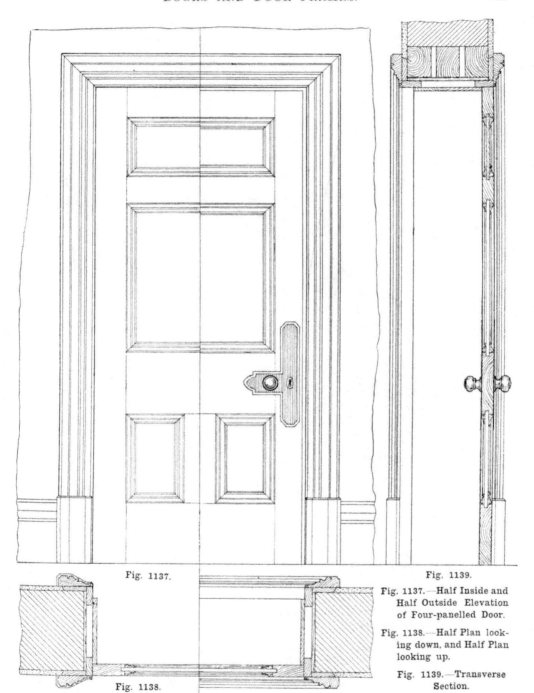

Fig. 1137.

Fig. 1138.

Fig. 1139.

Fig. 1137.—Half Inside and
Half Outside Elevation
of Four-panelled Door.

Fig. 1138.—Half Plan looking down, and Half Plan
looking up.

Fig. 1139.—Transverse
Section.

shown. The jambs are placed plumb and out of winding, and fastened to wooden

driven in. Any necessary firring, either in the form of wedges or strips, would be placed

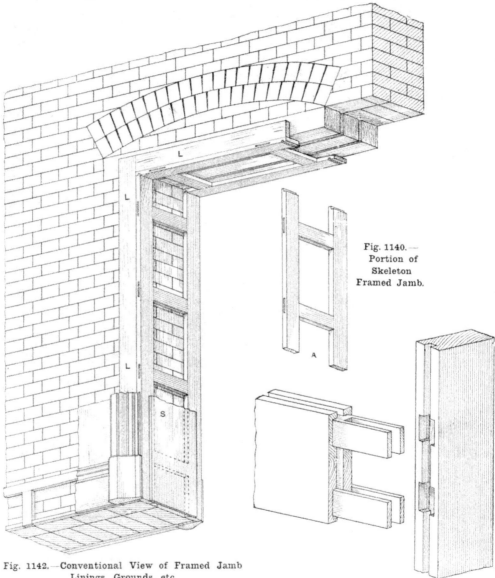

Fig. 1140.—
Portion of
Skeleton
Framed Jamb.

Fig. 1142.—Conventional View of Framed Jamb
Linings, Grounds, etc.

Fig. 1141.—Conventional View of Double Twin
Tenons for Lock Rail.

bricks or to breeze bricks built in the sides of the opening; or if these are not provided, some of the mortar joints between the brick-work would be cut out and wooden plugs

between the wood bricks and the back of the linings, so that the latter might be fixed

Fig. 1143.

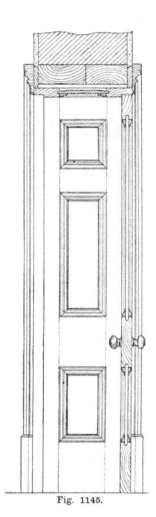

Fig. 1145.

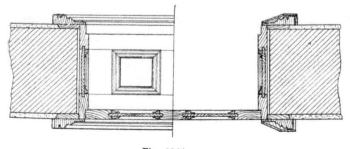

Fig. 1144.

Fig. 1143.—Elevation of Six-
panelled Door

Fig. 1144.—Half Plan looking up,
and Half Plan looking down.

Fig. 1145.—Transverse Section
through Door Head and
Elevation of Jamb.

15*

straight, plumb, and out of winding. Where wood lintels are used as shown, there is very little difficulty in fixing the head. Grounds to form a base for the architraves are now fixed round each side of the opening, flush *with the edges of the jamb linings as shown* at L (Fig. 1142). A board of the necessary width, which in this case is 15½ in., by ½ in. to ⅝ in. in thickness, would be fixed to the framing round the jamb linings, as indicated by the portion shown at s (Fig. 1142), thus forming on each side a rebate equal to the *thickness of the door.*

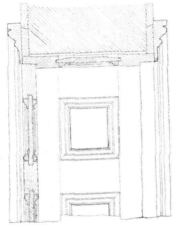

Fig. 1146.—Enlarged Detail through Head of Linings, Architraves, etc.

Six-panelled Door with Framed and Panelled Linings in an 18-in. Wall.

A door and jamb linings of rather a more important character are illustrated by *Figs. 1143 to 1146. The general construc-*tion of the door would be similar to that in the previous case. The jamb linings in this example are formed of panelled and moulded framing, so as to correspond with the door. It will be noticed that the stiles are rebated out of the solid, forming a recess and stop for *the door on one side and* simply a rebate on the other. The fixing of the linings and grounds would be very similar to that in the last example, except that in the last example the fixing of the linings is mostly hidden by the stop.

Outer Door with Bead Butt Panels and Frame with Fanlight.

A four-panelled outer door, the inside being moulded, and the outside having bead *butt panels, is illustrated in elevation, plan,* and section at Figs. 1147 to 1149. The frame is fixed in an 18-in. brick wall with 4½-in. reveals; the finish to the opening on the inside is by splayed linings, as shown. The sizes are figured on the drawings.

Door Frame.—The chief points to notice *in this are the forms of the joints between* the head and jambs, as represented at Fig. 1150, and that between the jambs and transom shown at Fig. 1151. Another view of the transom is given at Fig. 1152 so as to show the construction more clearly. It will be seen that the frame is rebated out *of the solid and beaded on the inside; it* is also ploughed to receive the tongues of the jambs, the splayed linings, and the soffit of same; the outside edge of the frame has an ovolo worked on. The mitering and intersection of the beads between the head post and transom are shown at Figs. 1150 *to 1152. This frame would be built in be-*tween the brickwork as described for previous examples, or it would be fixed to wood bricks or plugs. The rebate of the head of the frame is splayed as shown, to facilitate the opening and closing of the fanlight.

Splayed Linings.—These would be tongued *and grooved together in a somewhat similar* manner as the jamb linings (Fig. 1042, p. 344). They should also have tongues formed on their inner edges so as to fit into the corresponding grooves in the frame as indicated at Figs. 1153 and 1154. These linings would be sufficiently wide to project *about ⅞ in. beyond the brickwork, which is* the thickness of the plaster. The linings would be fixed to wood bricks or plugs provided in the brickwork, with necessary packing pieces or wedges, so that they are straight and out of winding. The grounds vary from 3 in. to 5 in. wide, according to *the breadth of the architrave which is to be* fixed to them. These should next be fixed, so that their faces are flush with the face of the splayed linings round the edge which has to be next to the plastering, this being splayed so as to form a key.

Fig. 1147.

Fig. 1148

Fig. 1149.

Fig. 1147.—Half Inside and Half Outside Elevation. Fig. 1148.— Horizontal Section. Fig. 1149. — Vertical Section.

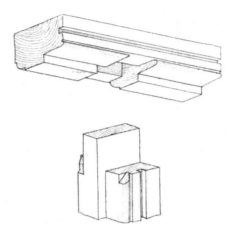

Fig. 1150.—Joint between Post and Head.

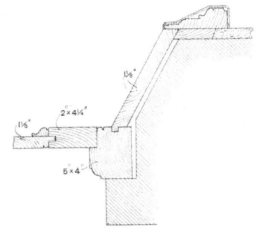

Fig. 1154.—Enlarged Detail through Stile, Post,
Linings, etc.

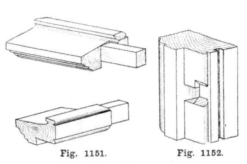

Fig. 1151.   Fig. 1152.

Figs. 1151 and 1152.—Joint between Post and
Transom.

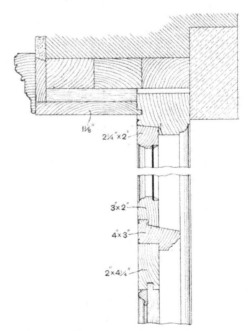

Fig. 1155.—Conventional View showing Portion
of Head and Butt Panels and Framing.

Fig. 1153.—Enlarged Detail through Head and
Transom of Frame.

The Door.—In making this, the points to notice beyond those already treated on are that the face of the panels on the outside usually finish flush with the framing. The joiners prefer making the shoulders of the panels and muntins all exactly one length, but perhaps a better plan is to have each muntin a shade (say $\frac{1}{16}$ in.) long; then at

Fig. 1158.—Half Inside Elevation and Half Outside Elevation with Shutter Removed.

panels are rebated on the outside to fit into the ploughed groove of the framing. A $\frac{1}{2}$-in., $\frac{5}{8}$-in., or $\frac{3}{4}$-in. bead is worked on the vertical edges of the panels, but the ends of the panels fit square to the edges of the rails, as illustrated at Figs. 1147 and 1155. Some the time of wedging up to cramp from the top and bottom rails, so as to bring all up close. The fanlight of this door, being sash-work, will be dealt with in that section. In Figs. 1147 to 1149 a well for a mat is partly shown.

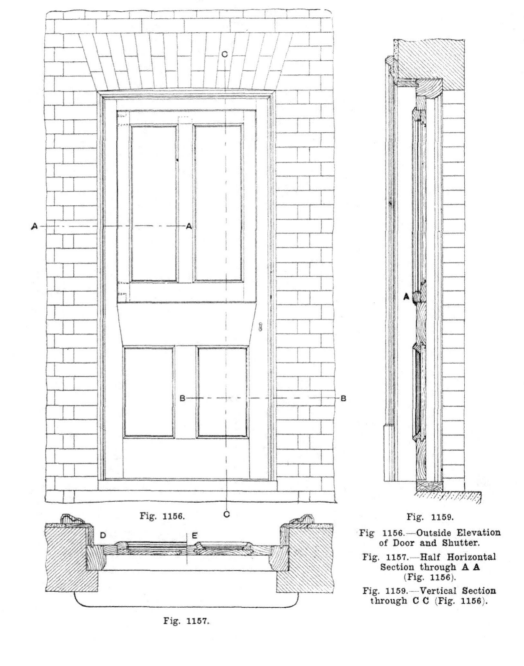

Fig. 1156.

Fig. 1157.

Fig. 1159.

Fig 1156.—Outside Elevation of Door and Shutter.

Fig. 1157.—Half Horizontal Section through **A A** (Fig. 1156).

Fig. 1159.—Vertical Section through C C (Fig. 1156).

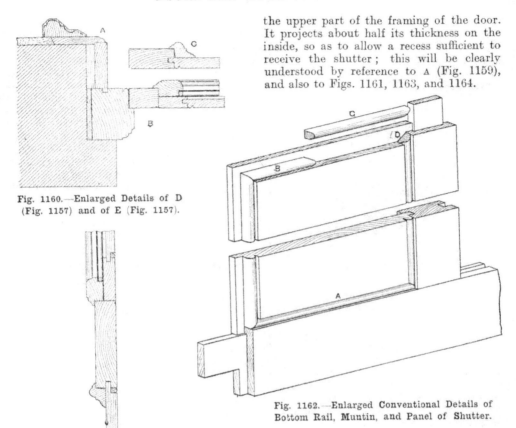

Fig. 1160.—Enlarged Details of D (Fig. 1157) and of E (Fig. 1157).

Fig. 1161.—Enlarged Detail through Lower Panel, Middle Rail, and Bottom Rails of Sash and Shutter.

the upper part of the framing of the door. It projects about half its thickness on the inside, so as to allow a recess sufficient to receive the shutter; this will be clearly understood by reference to A (Fig. 1159), and also to Figs. 1161, 1163, and 1164.

Fig. 1162.—Enlarged Conventional Details of Bottom Rail, Muntin, and Panel of Shutter.

Outer Door and Frame, Lower Panels Bead and Flush. Upper Part of Door prepared for Sash and Lifting Shutter.

This example is illustrated by Figs. 1156 to 1180. Reference to Fig. 1156, which is an outside elevation, will show that the bottom panels have a continuous bead round them, and this kind of panel is known as bead flush. The upper part has a movable shutter, which also has bead flush panels : a conventional view of one of these is shown at Fig. 1162. A half inside elevation with the shutter removed is given at Fig. 1158, which brings into view the sash, fitted into

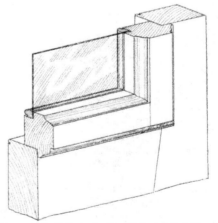

Fig. 1163.—Enlarged Conventional Details of Corner of Sash, showing it fitted to Framing of Door.

Frame, Linings, Grounds, etc.—The construction of the frame being similar to previous examples, it will not be necessary to enter into a lengthy description of it. It is ploughed to receive the tongues of the square linings, as shown in plan and section

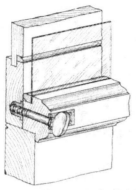

Fig. 1164.—Conventional Detail showing Thumb Screw, etc., for Fastening Bottom of Shutter.

(Figs. 1157 and 1159), and in an enlarged detail (Fig. 1160). In this case the grounds have ovolo-moulded edges, and are slightly rebated so as to fix to the outer edges of the linings, as shown at A (Fig. 1160). These grounds also serve as part of the architrave, and thus allow a narrower architrave moulding being used round the frame, which at

Fig. 1165.—Stub and Plate for Securing Top of Shutter.

the same time produces an effect equal to a broader architrave moulding.

Door.—The construction of this door is similar in some respects to those already described, and therefore these points need not be recapitulated. However, as in the

setting out and making new features are introduced, it will be necessary to explain them. The stuff for cutting out the pair of stiles can usually be lined out on a board as represented at Fig. 1167. Here the upper part of one stile is shown adjacent to the lower part of the other stile. This method leaves a spare strip between the stiles, which may be taken out in two pieces. Perhaps a better method is to cut off from one edge of the board a long strip, as shown at A (Fig. 1167). The more general method in trying up the stiles is to first true up the face sides ; then the back edges are shot straight and also square to these sides, and then the lower parts of the stiles are gauged and planed to a breadth, and, of course, made square to the face side as low as convenient, which is usually a few inches off the exact distance. The sash parts of

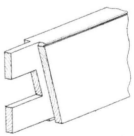

Fig. 1166.—Completed End of Middle Rail.

the stiles are sometimes gauged and worked, but perhaps the better plan is to leave this operation until after the stiles have been set out and mortised. When all the stuff for the framework of the door has been trued up, the setting out would be proceeded with. The rod for the length is represented at A (Fig. 1168). From this the sash and shutter have been purposely omitted for the sake of clearness. One of the stiles should be placed on the rod, and the positions for the rails marked off ; then the sizes of the tenons and allowance for haunchings should also be pricked off. The face sides of this pair of stiles should next be placed together with the face edges outwards ; then squared down for the mortices, etc., as represented at B (Fig. 1168). The positions for the mortices can then be transferred to the back edges of the stiles, as indicated by the dotted lines.

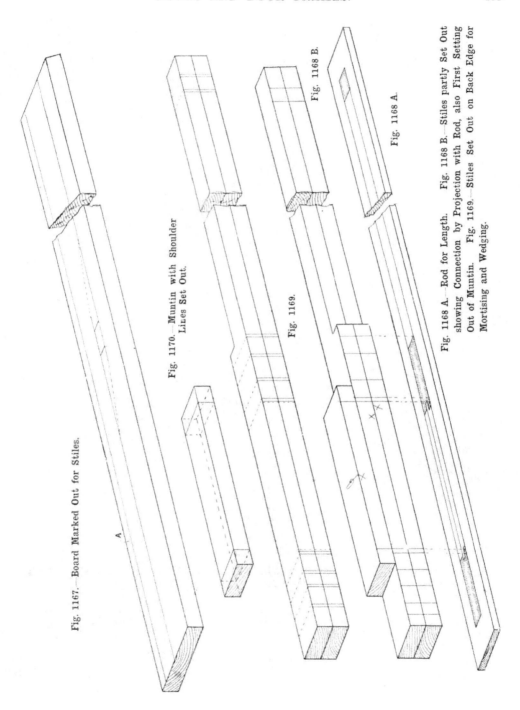

Fig. 1167.—Board Marked Out for Stiles.

Fig. 1170.—Muntin with Shoulder Lines Set Out.

Fig. 1169.

Fig. 1168 B.

Fig. 1168 A.

Fig. 1168 A.—Rod for Length. Fig. 1168 B.—Stiles partly Set Out showing Connection by Projection with Rod, also First Setting Out of Muntin. Fig. 1169.—Stiles Set Out on Back Edge for Mortising and Wedging.

The relation between the rod and the setting out of the stiles is clearly shown by the projectors. The setting out for the mortices and wedging can be completed on part of the setting out of the three rails from this is clearly shown projected above at D. The setting out for the shoulders, haunchings, and mortices for muntins for

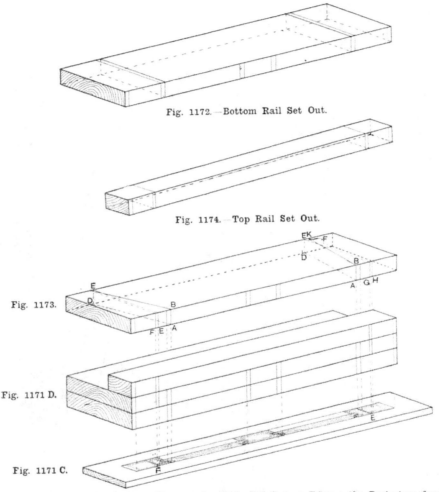

Fig. 1172.—Bottom Rail Set Out.

Fig. 1174.—Top Rail Set Out.

Fig. 1173.

Fig. 1171 D.

Fig. 1171 C.

Fig. 1171 C.—Width Rod.   Fig. 1171 D.—Rails Set Out on Edges; the Projectors from the Rod show the Connection of the Setting Out with it.   Fig. 1173.—Middle Rail Set Out.

the back edges, as represented at Fig. 1169. The muntins can also be placed on the stiles and marked for the shoulders, as represented; then these can be taken off, and squared round for the shoulders as shown at Fig. 1170. The rod for the width of the door is shown at c (Fig. 1171), and the first the bottom rail is shown at Fig. 1172. The complete setting out for the shoulders, etc., of the middle rail is shown at Fig. 1173. By referring to Fig. 1171 c at E and F, it will be seen that the front and back shoulders of the middle rail are not in one plane, owing to the small bead worked round the outside of the framing to break joint with the shutter. The beads start from the same square line at the bottom edge as A B (Fig. 1173), but the one finishes the breadth of the bead in front of the other, as indicated at D and E, the exact amount being the bead and quirk, the stile diminishing the distance A H on the outside and A G on the inside. Squaring the line across from G, as represented by the dotted line, and then

Fig. 1175.—Setting Out of Shoulders on Face Side of Stiles.

Fig. 1176.—Setting Out of Shoulders on Back Side of Stiles.

down the edge, the point D is found, and joining D to A, the inside shoulder is shown. Now squaring across from H as shown by the second dotted line, the point K is obtained; from the arris measuring the thickness of the bead, the point F is obtained; joining F to B gives the outside shoulder. The

M (Fig. 1175), and then marking the thickness of the bead from the gauge line N, draw the short line parallel to this as shown at

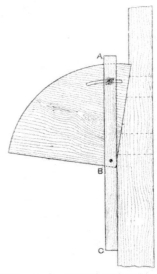

Fig. 1177.—Application of Adjustable Square for Setting Out Shoulders on Stiles.

O. Thus point F is obtained, and joining F to B gives the shoulder line. On the inside of the stile (Fig. 1176) square across from the line L, and where this intersects

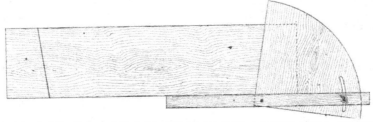

Fig. 1178.—Application of Adjustable Square for Setting Out Shoulders on Middle Rail.

setting out for the top rail is shown at Fig. 1174, the difference between the front and back shoulders being equal to the thickness of the bead. The shoulder lines can now be set out on the stiles; Fig. 1175 represents the outside portion of the stile, whereas Fig. 1176 represents the inside. Remembering what has been stated about setting out the middle rail, first squaring on the side

with the gauge line P, it gives the point D, and joining D to A gives the shoulder line.

An adjustable square will be found very useful in setting out the shoulders both on the stiles and rails. One is illustrated at Figs. 1177 and 1178. Usually the stock is the length shown from A to B, but by having it longer, as shown, the inner edge of the lower part of the stile can be worked from instead of the back edge, and this is an advantage. The longer stock is also an advantage, as there is more of it to adjust to the rail (see Fig. 1178). When the square is true and properly adjusted, then obviously, if the stuff is planed up true, the shoulders can be accurately marked out by the aid of this square.

The work can now be gauged, the mortices made, and the tenons cut. Perhaps the best method of mortising is to do all the mortices of the stiles before the splay shoulders are cut; especially is this more

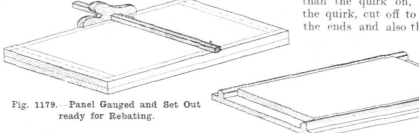

Fig. 1179.—Panel Gauged and Set Out ready for Rebating.

convenient when the mortising is to be done by a mortising machine. When having to mortise entirely by hand, many joiners prefer, before mortising, to trim the splayed shoulders near to within the lines, considering it saving in labour.

The stiles and rails should be ploughed, the beads worked on, and the splayed shoulders of the stiles accurately formed to the lines, a bullnose plane sometimes being found useful for this kind of shoulder. The shoulders of the rails can now be cut and the framing fitted together in the usual manner. A conventional view of a completed end of the middle rail is shown at Fig. 1166. If it is desired to make provision for a mortice lock, double twin tenons, as shown at Fig. 1141, p. 344, would take the place of those shown at Fig. 1166.

Bead and Flush Panels.—In making the bead and flush panels, the special points are: being faced up and thicknessed, they should next be gauged for width and set out for length ⅛ in. less each way than the distance between the plough grooves, then sawn and shot to these lines. They should next be gauged for rebating, and set out for length. The length here referred to is the distance

between what will form the quirks for the top and bottom beads, as indicated by solid lines in Fig. 1179, in which the dotted lines represent the full length of the panel after the insertion of the top and bottom bead. Next rebate and bead the edges, and then, with a short thin chisel and a mallet, cut out the wider rebates at each end, as shown at Fig. 1180. Take a piece of board equal to about the thickness of the panel less the tongue which goes into the plough groove, and on this stick sufficient beading for the ends of the panels, saw this down, leaving a little more than the quirk on, plane down to the quirk, cut off to lengths, mitre the ends and also the ends of the

Fig. 1180.—Panel Rebated, Side Beads Stuck ready for Mitering for Ends of Beads.

bead stuck on the side of the panels; then secure these beads in their position by a little glue and a few brads. In Fig. 1162 at A the bead is shown fixed in position. At B part of the bead is in position, and at C the other part is shown projected up so as to give a view of the mitering at D. When the door is fitted together, the combined breadth of the panels and muntin should be from $\frac{1}{16}$ in. to $\frac{1}{8}$ in. less the length of the middle and bottom rails measured between the shoulders so as to allow for closely cramping up the shoulders of the stiles and rails. The making of the sash will be described in the section on sash-making.

Making the Shutter.—This should present no difficulty, its construction being similar to that of the lower part of the door, as shown in the illustrations. After being wedged and cleaned off, it should be fitted in the recess in the framing, so as to leave a good $\frac{1}{16}$ in. all round the joint to allow for paint and a slight clearance. A thumb-screw and plates suitable for securing the bottom

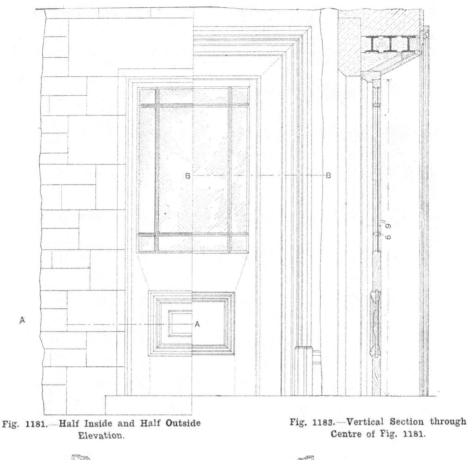

Fig. 1181.—Half Inside and Half Outside
Elevation.

Fig. 1183.—Vertical Section through
Centre of Fig. 1181.

Fig. 1182.—Half Horizontal Section through A A, and Half Horizontal Section through B B.

of the shutter from the inside of the sash is illustrated at Fig. 1164; the nut plate being let in and screwed to the inside of the shutter, and a second plate fixed to the sash receives the shoulder of the thumb-screw. A stub and plate for securing the top of the shutter is shown at Fig. 1165.

External Two-panelled Door, Lower Panel Raised and Bolection Moulded; Upper Part of Framing with Marginal Lights prepared for Glass.

The door, frame, linings, etc., illustrated by Figs. 1181 to 1201, are of a kind that is commonly used in houses of the suburban villa class. The door has a lower panel plain-faced on the inside, with moulding round the framing. The outside of the panel has a sunk margin with a moulding worked on the edge of the raising, and

finished off with bolection moulding. The stiles are known as diminished or gunstock stiles, in which the upper part, being narrower, forms a larger opening for glazing. Four bars are provided, forming what is known as marginal lights. The frame is

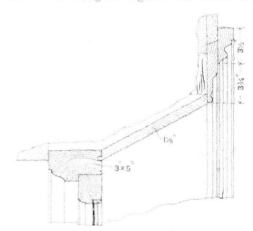

Fig. 1186.—Enlarged Details of Head of Frame, Door, Soffit, etc.

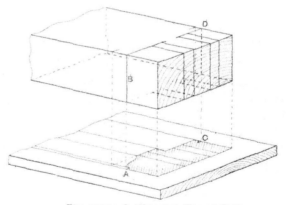

Fig. 1187.—Setting Out Top of Post.

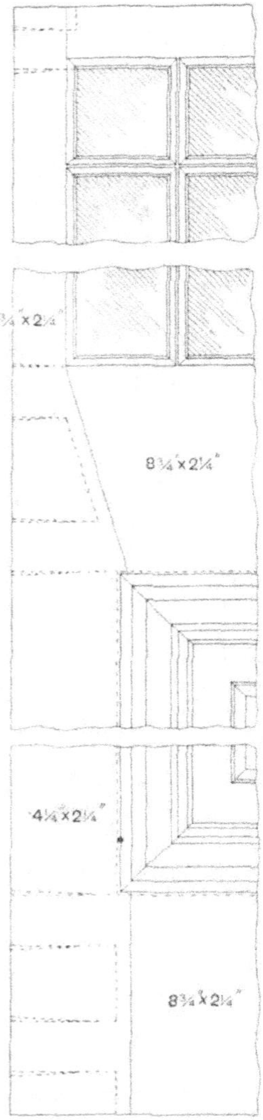

Fig. 1184.   Fig. 1185.

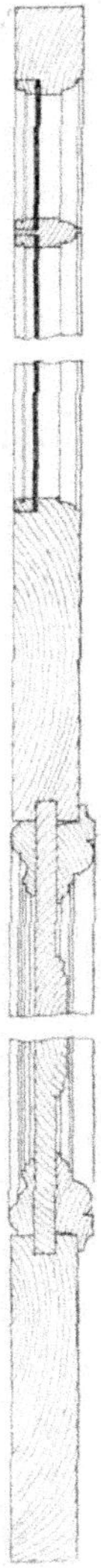

Fig. 1184.—Enlarged Details of Outside Elevation (Fig. 1181).

Fig. 1185.—Enlarged Details of Vertical Section.

beaded on the inside and moulded on the outside. The frame is inserted in a 20-in. stone wall. Splayed linings, which also have a splayed soffit, are tongued into the frame as shown. In the making and fixing of this door, frame, linings, etc., the points which are common to other examples, and which have been already treated, will not be recapitulated ; only the principal new features will be explained and illustrated.

Frame.—A portion of the rod with the head set out is shown at Fig. 1187; projected above this, the top end of the post

is less complicated, whilst the amount of labour involved is no more by having double mortices and tenons each of 1 in., as shown, and the shoulders can be kept up better than with one large tenon. Next set out

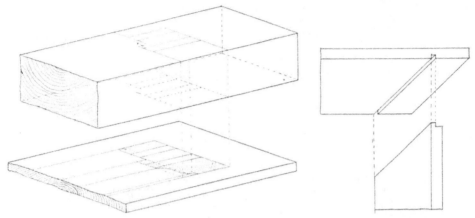

Fig. 1188.—Setting Out for Mortices in Head of Frame.

Fig. 1190.—Showing Grooved and Tongued Joint between Jamb and Soffit of Linings.

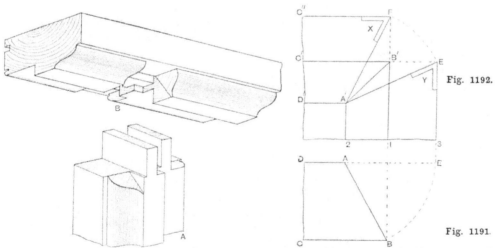

Fig. 1189.—Completed Joint between Head and Post of Frame.

Fig. 1192.

Fig. 1191.

Figs. 1191 and 1192.—Setting Out of Bevels for Intersection at Head of Linings.

is shown set out. The shoulder at B has been set out from the square of the moulding at A, and the inner shoulder is set out from the quirk of the bead C, indicated by the dotted line D. When moulded and beaded as here illustrated, the setting out of a frame

the head. Part of the rod for this is shown at Fig. 1188, and the setting out of the head is projected over it. The mortice adjacent to the rebate is narrow, and the other mortice is equal to the whole thickness of the stuff. The mortices and tenons can now be marked, gauging from the face edge only for both of them. Having cut the mortices and the tenons, the ploughing for the tongues of the linings and rebating should be done, after which the bead may be stuck and then

now be cut, and the mitering of the moulding and bead done to the post as shown at A (Fig. 1189), and that for the head as represented at B (Fig. 1189). In these large mouldings sometimes that on the jamb is

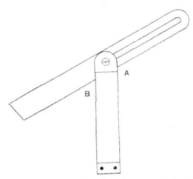

Fig. 1193.—Bevel Set Out for preparing Edges of Linings.

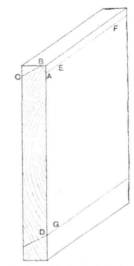

Fig. 1194.—Lining Marked Out and Gauged for Bevelling of Edges.

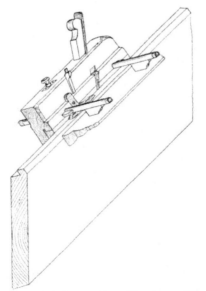

Fig. 1195.—Rebating to form Tongue on Edge of Lining.

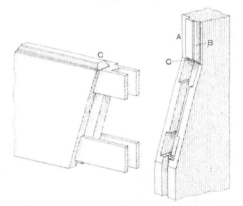

Fig. 1196.—Joint between Stile and Middle Rail prepared for Mortice Lock.

the ogee moulding should be worked as will be explained in a later section. Before sticking a large moulding, it is as well to run gauge lines, one for working the distance on, and the other for working the distance down to. The shoulders of the post can

scribed over that on the head instead of mitering, and this method will be illustrated in a future example. The frame, after fitting, would be wedged up, and a stretcher nailed across the bottom in the usual way.

Splayed Linings.—In these the only new

feature is that as the soffit is splayed as well as the jambs, this involves a little more geometrical construction, which is illustrated in Figs. 1191 and 1192. Let D E represent the line plan of the inner edge of linings, C B the outer edge, and A B the face of the

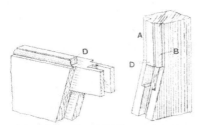

Fig. 1197.—Portions of Stile and Middle Rail, showing Mouldings Scribed Together.

linings, and it will also represent the plan of the intersection between the jamb and the soffit. Projecting up, obtain the portion of the elevation shown by 1 B' C', 2 A' D', and then A' B' is the elevation of the intersection between the two linings. Now with A as centre and B as radius draw the arc B E; project up from E, drawing the line 3 E'. Projecting horizontally from B' we determine the point E'; join this to A', and thus the bevel at X is that required for the jamb. If the face of the soffit marks the same angle with the plan of the door frame, the same bevel will do for both. Where this is not the case, project vertically from B', then with A' as centre and E' as radius draw an arc which intersects with the line projected from B', giving point F; draw F C'' parallel to B C'. Then join F to A', giving the bevel Y for application to the soffit. Fig. 1190 shows one angle of the linings grooved and tongued. A few hints on preparing splayed linings will here be given. The stuff is faced up, and the edge to be tongued is shot straight. The bevel (Fig. 1193) is set to the splay that the lining makes with the frame, and is applied so that the line A B (Fig. 1194), equal to the depth of the tongue, is drawn with the angle A of the bevel, this latter being applied to the edge of the lining. Then with the angle B (Fig. 1193), the line B C (Fig. 1194) is drawn. From A the line E F is gauged. From the rod obtain the breadth E to G, and draw

the gauge line through the latter; with the angle A of the bevel draw the angle shown at D. Plane off these edges, using the bevel in the same way as the try square. Next place the wood edgewise up, holding it in some convenient manner, and with the fillister rebate the back, thus forming the tongue as illustrated at Fig. 1195. Reference to Fig. 1183 shows that the head is cradled out to admit the soffit lining.

Setting Out the Door.—In setting out the stiles, the mortices must be marked off from the rod for the cross bars. The vertical bars may be placed on the stiles and set out both for shoulders and where they intersect each other. The rails can next be set out, and also the cross bars with them. The setting out of the splayed shoulders, both in the middle rail and in the stiles, will be rather simpler than in the previous example, because the sticking down of the ovolo and the depth of the rebate being equal, the shoulders on each side will be in one plane. The setting out of one end of a vertical bar and one end of a cross bar, and also for their intersection, is clearly shown at Fig. 1199. The dotted lines indicate where the square of each member intersects with that adjacent to it, and the space between these and the solid lines shows the amount that must be allowed for moulding or rebating.

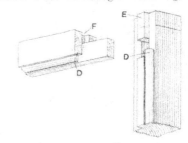

Fig. 1198.—Joint between Top Rail and Stile of Door

Tenons of Bars.—The tenons of the bars should next be sawn down to the shoulder lines as shown at Fig. 1199. The shoulders should be cut in with a fine saw about $\frac{1}{4}$ in. The sawing of the shoulders must not be completed until after the moulding and rebating are finished.

16

Rebating and Moulding.—For a post (shown at A, Fig. 1196), as the sash fillister cannot be used up to the shoulders of the stiles, sufficient of the rebate should first be rebated and stuck with the ovolo planes. A sticking board will, of course, be necessary for this purpose. Particulars of this will be found in the section on sash-making.

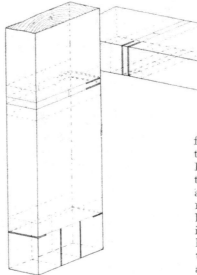

Fig. 1199.—**Bars Set Out, Tenons Cut, and Shoulders Entered.**

made by gauging, and paring out with a chisel a portion as shown at A (Fig. 1196); a bull-nose plane is useful for this purpose. The same remark applies to the moulding. A portion B (Fig. 1196) must be worked by

Scribing, Haunching, etc. — Provision for a mortice lock is shown by the double twin tenons and mortices at Figs. 1196 and 1197. At c c (Fig. 1196), the mouldings of the rail and stiles are shown mitered. An alternative method by scribing is shown at D D (Figs. 1197 and 1198). The special kind of joint between the top rail and stile is shown at Fig. 1198, where the square left from the moulding on the stile is left on to form a haunch, and a piece is mortised out above the tenon to fit over this, as shown at F (Fig. 1198). Where the bars intersect with the stiles, they should be scribed as shown at A (Fig. 1200), so as to fit over the solid mould of the stile. The intersection between a horizontal and a vertical bar is

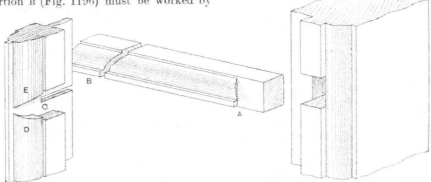

Fig. 1200.—**Bars Scribed for Fitting Together and to Stile.**

paring, or a pair of routers of the same pattern as the ovolo planes may be used with advantage. The bars should next be shown at B. The square on each side of the vertical bar is cut down to the level of the rebate, and the moulding is also cut down to the same level, thus forming a square surface from the square of the rebate to that of the moulding. Then, by cutting out a rectangular piece from the horizontal bar, as indicated at B, and then cutting a rectangular piece from the vertical bar as shown at C, the two can be pushed together. For scribing the moulding, ovolo templates are useful if the irons of the planes are carefully sharpened to the same shape. A good alternative method is to use a mitre

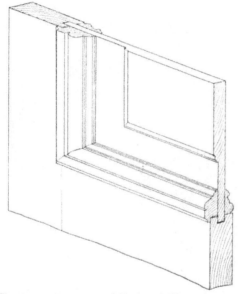

Fig. 1201.—Conventional Sectional View, showing Panel, Bolection Moulding, and Inside Moulding.

template, and first cut a mitre on the moulding, which will produce an arris ; and if this is accurately worked to with the scribing gauge, the fitting should prove satisfactory.

The Panel.—The panel is first made $\frac{1}{16}$ in. less all round than the distance between the plough grooves ; then it is gauged for the breadth and depth of the sinkings. The two sinkings across the grain should first be made either with a rebate plane or a panel plane, after first running a couple of saw kerfs across the grain. The sinking with the grain can now be made, the aim being to make all four sinkings in one plane. The moulding should first be stuck on across the

grain, and afterwards the two sides with the grain may be done. The conventional view (Fig. 1201) will make clear the construction of the panel and also of the bolection moulding mitered round the outside. The glass is fixed in with beads as shown.

Panelled Linings for Doorways.

Panelled linings for doorways, examples of which are illustrated by Figs. 1202 and 1203, are so called because they are framed—that is to say, mortised and tenoned, the panels being inserted in grooves like the panels of a door ; in fact, they are panelled to match the door that is intended to be hung to them. They are also ornamented with the same kind of moulding as the door—sometimes with moulding put in on the panel, but not quite flush with the surface of the framing, sometimes with bolection mou'ding, which fits partly on the panel and partly on the framing, the angle of the framing fitting into the rebate. More care is required in putting the latter moulding in than in the case of the former—called a sunk moulding—for, besides having the mitre to cut, the moulding has to be put on a mitre shoot and the ends planed with a trying-plane to the exact length and correct angle.

Fitting the Mouldings.—The way to ascertain the cutting length of the moulding is to take a small piece, an inch or two in length, and lay it on the panel in the same position as that in which it is to be fixed, and as near the corner of each panel as it will go. Hold a pencil against the part which rests on the framing, and draw a little line $\frac{1}{2}$ in. or so long. When each panel has been gone round in this way, the width of the rebate will be marked exactly. Cut the mitre at one end, lay the piece of moulding on, with the point to one of these marks, and mark the length $\frac{1}{8}$ in. longer than the pencil mark on the opposite side of the panel ; the spare $\frac{1}{8}$ in. will allow for planing to fit. Procure four pieces of wood, 3 in. or 4 in. long by $2\frac{1}{2}$ in. or 3 in. wide, and about the depth the panel is sunk down from the surface of the framing. Lay them flat on the panel, one at each corner, and place the four lengths of moulding on them, one at a time, as they are shot. Make them fit closely one against the other, so that when the last is inserted it will want just a slight tap of the hammer to get it down level. With the aid of a bradawl or some sharp-pointed tool, slide out the four slips of wood, each towards the centre of the panel, and drive down the moulding. Lay a flat strip of wood across the panel corner-

appearance and in other respects. A very good kind of jamb, however, is sometimes used when the doorway is in a 9-in. wall. Such jambs are called skeleton jambs. The stiles and rails are generally 3 in. by 1¼ in., but vary according to circumstances. They

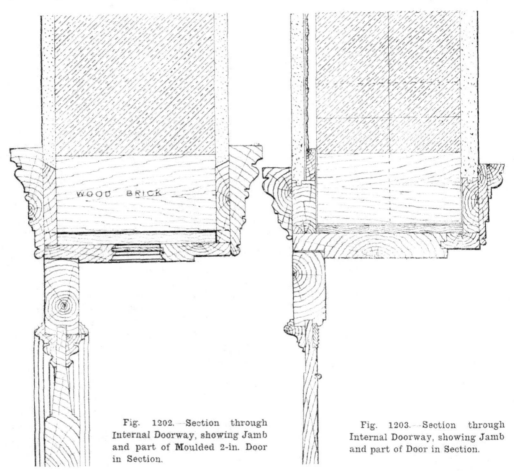

Fig. 1202.—Section through Internal Doorway, showing Jamb and part of Moulded 2-in. Door in Section.

Fig. 1203.—Section through Internal Doorway, showing Jamb and part of Door in Section.

wise, and strike this with the hammer, to prevent the moulding from being bruised. If the mitre shoot was true, every mitre will fit as closely and nicely as possible.

Skeleton Jambs.—Every doorway must have a lining of some kind to hide the rough material of the wall. A single plain board, wide enough for a 14-in. or 18-in. brick wall, would be very unsatisfactory indeed in

are planed true on one side and one edge, and mortised and tenoned, have three rails in the head (or soffit as it is called), and about four in the jambs or uprights, or perhaps five, according to height, but no panels. They are fixed in the opening, and the door is hung to them. To form the rebate, a wide strip of ½-in. stuff is nailed on over the framing, showing a rebate on the opposite edge, to correspond with the one the door shuts into. It is similar to that shown in the illustration, but in the latter the rebate is taken out of the solid material, which should be of sufficient thickness to allow of this.

Constructing Panelled Linings.—Panelled linings are more difficult to make. Fig. 1204 represents part of a horizontal section of an outer door frame, constructed of scantling, set back 4½ in. from the face of the wall in a reveal in the brickwork, and showing panelled lining. The frame is grooved on the inside face. A plain lining is used when the required width is only a few inches. One edge is rebated to form a tongue, which fits into the groove and helps to hold the lining in place. It is also nailed to wood bricks built in the sides of the opening ; the head is

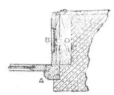

Fig. 1204.—Horizontal Section taken through Centre of Panels.

nailed to the lintel. On the edge of the lining the architrave moulding is fixed, or a framed ground is used, in which case the ground would go on first and the architrave be fixed to it. The facing is a flat frame, having two stiles and a rail ⅞ in. or ¾ in. thick at least. The rail is mortised into the stiles, and a return bead or moulding, worked on the inside edge, makes a better and bolder finish than the architrave moulding only. Exactly the same principles apply to the construction of panelled linings as in the case of ordinary doors. The first consideration in all framing is to get the stuff perfectly true on the face. Shoot one edge square and straight ; the other edge need only be jacked over. Gauge it to the required thickness if for a door, but in the case of panelled linings only one side is seen, so the back can be left rough. In setting out, put all the lines, both for the front and back edges, on one stile first, and use that for a pattern. Lay it flat on two blocks to raise it off the bench, lay the other stiles on, face to face or back to back, to get

them in pairs, and square down the lot at once, if there are not too many. If possible, reach over and square the back lines, putting the square to every line on the pattern. In wedging up, wedge the middle rail first, then the bottom, and lastly the top rail, unless there should be another between the middle and top rails, as in a six-panelled door, in which case this would be wedged before the top rail. The linings may be shot to the required width, and the rebate ploughed and planed out. The width of the rebate should equal the thickness of the door ; its depth should be ½ in. It is always best to glue or screw a block, about 1 in. thick, on the back of the lining where the hinges come. Such blocks will receive the screws and give them a better hold. Wood bricks are built in as the brickwork proceeds, to which the linings are fixed. In Fig. 1204, A signifies door frame, B panelled linings, C rough backing, D wood brick.

Front-entrance Door and Frame.

The front-entrance door, with sidelights and fanlights, illustrated by Figs. 1205 to 1207, is suitable for a villa residence, or for the entrance to a conservatory attached to such a residence, and is often executed either in red deal or in pitch pine. The frame, which is worked out of 5½-in. by 3-in. stuff, has a large ovolo moulding worked round it, the mullions and transoms being similarly moulded ; it is set on a 3-in. stone sill, in a 2¼-in. recess, in 14-in. brickwork, with 4½-in. reveals. At the angles up the jambs and over the arch are 2-in. quirked, beaded, and stopped bricks. The arch bricks are gauged, and the rise of the arch is ¾ in. to every foot of opening ; and the brickwork over the head is carried by a bressummer, a relieving arch being provided if desired. The doors and sidelights (the stiles of which are diminished) are out of 2-in. stuff, and may have an ovolo or other moulding worked on, and the top portions may be filled with plain, ornamental, or coloured glass. The bottom panels of doors are raised and bolection moulded. Linings, surmounted with architrave moulds rising from plain bases, are provided round the inside of the frame. The door is 7 ft. high by 3 ft. wide, and the sidelights are 7 ft. high by 1 ft. 6 in. wide—all

Fig. 1205.—Elevation of Entrance Door and Framing.

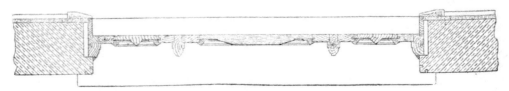

Fig. 1207. Plan of Entrance Door and Framing.

exclusive of rebates. The bottom edge of the door is rebated and throated, and *shuts against a 1-in. by* $\frac{3}{8}$*-in. wrought-iron*

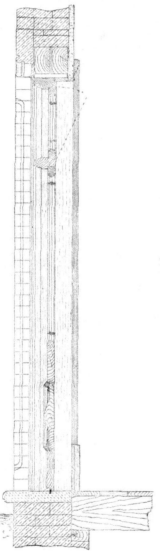

Fig. 1206.—Section through Entrance Door and Framing.

weather bar as shown in Fig. 1208, which is *sunk and cemented* $\frac{1}{2}$ *in. into the 3-in.* stone step. A similar bar is let in under

the fixed sidelights (see Fig. 1209), and is permanently grooved into the bottom rail and *bedded in red-lead. The inside face of the* bressummer is lathed for plastering. The rib extending from the head of the frame to the soffit of the arch is out of 1-in. stuff, with a scotia worked on the bottom edge. Fig. 1210 shows a part section of the door panels, which are bead flush inside ; the bevel of the *raised portion of the outside of the panel will,* of course, depend upon its width, but in no case must it rise above the line of the outside member of the bolection moulding.

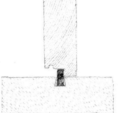 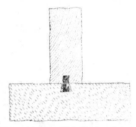

Fig. 1208.—Section through Weather Bar, showing Weathering for Bottom Rail of Door.

Fig. 1209.—Weather Joint between Bottom Rail of Side Framing and Stone Sill.

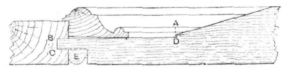

Fig. 1210.—Section through Panels and Bolection Moulding.

Fig. 1211 shows a section of the architrave moulding, which is $5\frac{1}{4}$ in. wide and $1\frac{1}{4}$ in. thick (reduced at the front edge to $\frac{5}{8}$ in.) ; plinth blocks, 12 in. deep, receive the ends

Fig. 1211.—Section of Architrave Moulding.

of the moulding, beyond which they should show $\frac{1}{2}$ in. margin at the front and ends, and $\frac{3}{16}$ in. in front of beads both ways. The rods having been prepared, the work can *be proceeded with.* The frame will be made first. The selection of the stuff is sometimes left to the workman, and, as the jambs are out of 5½ in. by 3 in., it will mean one rip down an 11-in. by 3-in. plank. In some shops this size of stuff is kept ready for use. The correct lengths for cutting will be ascertained from the drawings, to which an inch or so must be added. The planing up must be done true and parallel, and in favour of the sticking side; the backs that go next the brickwork, etc., need not be

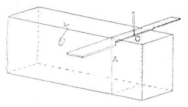

Fig. 1212.—Application of **Slip** for Setting Out Shoulders for Ovolo Edge of Posts and Muntins.

touched, except for jacking over where the mortice gauge is run down.

Frame and Linings.—Many of the processes involved in setting out and making the frame have already been explained in connection with cases previously treated. Thus,

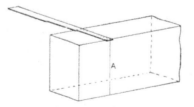

Fig. 1213.—Application of **Slip** for Setting Out Shoulders to fit to Rebates.

it will only be necessary to explain the new features. Only the sight lines, not the shoulders, may be taken from the rod. The sight lines are the distances between the intersection of the square part of the frame as indicated at A in Figs. 1212 and 1213. Then by means of a prepared slip, mark out for the moulding and rebate as follows: Now apply the slip with one edge against the sight line A (Fig. 1212); then, pricking off the breadth, as indicated at C, gives the point through which the shoulder

can be scribed, as shown at D (Fig. 1214). Then turn the stile with the rebated edge upwards, and mark for the shoulder in a similar manner to that shown in Figs. 1213 and 1215.

Setting Out Frame.—Take a thin slip of wood about 9 in. long, planed up ⅛ in. thick, and gauged off to 1⅞ in. (the extent to which the ovolo works on) for 3 in. along it; then gauge ¹⁵⁄₁₆ in. for another 2 in., which equals

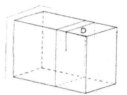

Fig. 1214.—Shoulders Set Out on Moulded Edge.

the depth the ovolo works down; then gauge ½ in. for the last 3 in., equal to the depth the rebate is worked down for the doors; these should be accurately cut away down to the gauge lines, then the slip will have the appearance shown in Fig. 1216. Lay one jamb on the bench, mark and square off

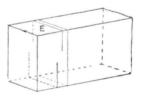

Fig. 1215.—Shoulder Set Out for Rebated Edge.

with the knife the line to cut off at bottom, mark off above this 7 ft. (the height of transom rail), then 2⅞ in. (the finished thickness of transom rail); above this mark 1 ft. 6 in., and the head is reached. Mark the other jamb and the two mullions from this, of course using the slip for the shoulder of the muntin; and at the point where the transom line crosses the mullions cut the latter in halves just midway between the shoulders. Allow for the shoulders on the jambs and mullions by using the ¹⁵⁄₁₆-in. portion of the slip (the depth the ovolo drops down) on the face edge and the ½-in. portion (the depth of the rebate) on the back edge. The jambs must be squared round in pencil at the point where the transom enters, and room must be allowed for wedging at the back. Now

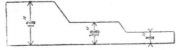

Fig. 1216.—Setting Out Slip for Moulding and Rebate for Frame.

lay the head face upon the bench (it will have been cut off to about 7 ft. 6 in. long), square a line across it 6 in. from the end (this is the allowance for horn and mortice), set off 1 ft. 6 in., then the thickness of mullion, then 3 ft., again the thickness of mullion, and finally 1 ft. 6 in., leaving the other horn and mortice. Set out the transom from this, using the slips at the ends in the same way as with the jambs; square round in pencil for the mortices and wedging room on the head; but in the case of the transom the mortices will require marking square across on the top and under side with the knife. Now, with a mortice gauge set to the flat portion of jamb, gauge both back and front faces for mortices, and all round the ends for tenons. Set gauges to the slips, and run these down the stuff, the $\frac{15}{16}$ in. down the face edge, the $1\frac{7}{8}$ in. down the

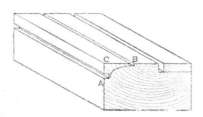

Fig. 1217.—Piece of Frame: the End Set Out for Moulding and Rebate, and Ploughed for Rebating and Moulding.

face from both edges, and the ½ in. down the back edge, these lines being for moulding and rebate.

Mortising and Tenoning Frame.—The tenons should be cut down to the shoulder

16*

lines with a half-rip saw outside the gauge lines and shoulder to within ¼ in. of saw cut. Mortices should be cut through square and true inside the gauge lines, and ½-in. wedging room cut through straight and to within about 1 in. of face.

Rebating. — Before rebating the frame, plough, with a $\frac{3}{8}$-in. iron, the back side for the inside linings, then, with the same iron set to $\frac{7}{16}$ in., plough from the back edge for the rebate to within $\frac{1}{16}$ in. of the gauge line. Chop out the intervening wood; finish down to gauge line with a rebate plane, using the ½-in. end of the slip to see that the correct depth is maintained on the back.

Moulding.—As it would probably be difficult to obtain an ovolo plane of the size re-

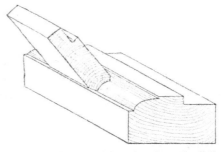

Fig. 1218.—Finishing Ovolo Mould with Hollows.

quired for this job, the moulding must be worked with hollows (unless it be machine made); and, to ensure accuracy, a zinc or cardboard template must be provided with which to mark the ends of the stuff. Plough grooves should be made as indicated at A and B (Fig. 1217); then waste c cut away with a chisel. Work as near the line as possible with a jack plane. The round of the mould can be finished with hollows, as indicated at Fig. 1218. At least two should be used, the shape of the moulding being elliptical. Of course, in most large shops the moulding and rebating would be done at the machine, and only finished by hand. Assuming that the mouldings are all worked and the top edge of the transom is weathered, the sawing down of the tenon cheeks may be finished, the cheeks knocked off, and the mitering proceeded with.

Mitering.—From the pattern of the moulding a reverse mould, about 8 in. or 9 in. long, should be worked (see Fig. 1219), and accurate mitres shot at each end. This will form the template with which to cut all mitres. Great care must be taken in this operation, the chisel not being allowed to go the least bit beyond the line cut by the

Measuring for Doors and Lights.—Before removing the frame from the bench, it will be necessary to take the measurements for constructing the doors and lights. The proper way to do this is to set out the dead heights and widths on a lath. It should be done very accurately, to avoid subsequent errors. This lath will also be found useful

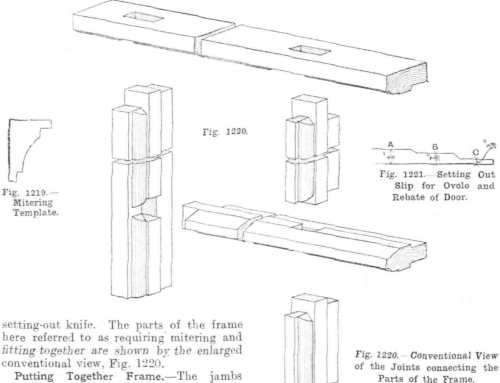

Fig. 1220.

Fig. 1219.—
Mitering
Template.

Fig. 1221.—Setting Out
Slip for Ovolo and
Rebate of Door.

Fig. 1220.—Conventional View
of the Joints connecting the
Parts of the Frame.

setting-out knife. The parts of the frame here referred to as requiring mitering and *fitting together are shown by the enlarged* conventional view, Fig. 1220.

Putting Together Frame.—The jambs should be bored for drawing; and the tenons of the mullions, where they butt-joint in the transom rail, should be bored for and held together with ¾ in. glued oak or pitch pine *dowels. The frame can now be put to*gether, pinned, and wedged up, then cleaned off, and the plough groove run along the head for the casing. Mitre the cove round the frame, square the bottoms of jambs and mullions, and up the centre of each bore a ⅝-in. hole, and drive in square galvanised *iron dowels, leaving ¾-in. projections, which* will be subsequently let into the step. The frame is now finished.

in cutting off the stuff. Door stiles and rails should be made from stuff 2 in. thick, the bars from ¾-in. stuff; and, in planing up, the best sides should, as far as possible, be selected for working the mouldings. After cutting the door stiles to the diminished size, they should be planed up, and the tried-up mark put on the back edge, as an indication of the stage to which the work has been brought.

Setting Out Doors and Lights.—The stuff for the top lights must be gauged to 2¾ in.

by 1⅞ in. ; for door and side light rails, 10⅞ in. by 1⅞ in. ; for the diminishing stiles for doors, 4⅜ in. to 2¾ in. by 1⅞ in. ; for the side lights, 4 in. to 2¾ in. by 1⅞ in. ; bars, 1⅞ in. by ⅝ in. In setting out the stiles of the door, mark off on the back edge of one of them the dead height of the door from the rod, allowing ⅛ in. for fitting. Then mark on 2¾ in. for the top rail, 10⅞ in. for the bottom rail ; and 3 ft. 2 in. from the bottom set off 10⅞ in., which will be the top edge of the lock rail ; 2 in. from the inside marks of the top and lock rails, mark ⅝-in. spaces for the transverse marginal bars. The marks must be squared round to the inside edge, and wedging room allowed on the back. Mark

rebate ; c equals the depth of the rebate for bars—namely, in this case, ³⁄₁₆ in. In setting out the vertical marginal bars, from the back edge of the pattern stile set out one bar, then, placing it on the outside of the remainder of the bars, cramp them all together, and square across both sides for mortices and tenons ; note that the shoulders should be a shade long. These may now be set on one side ; but later, when the

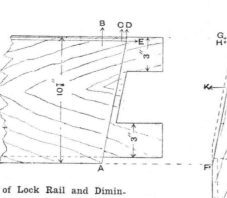

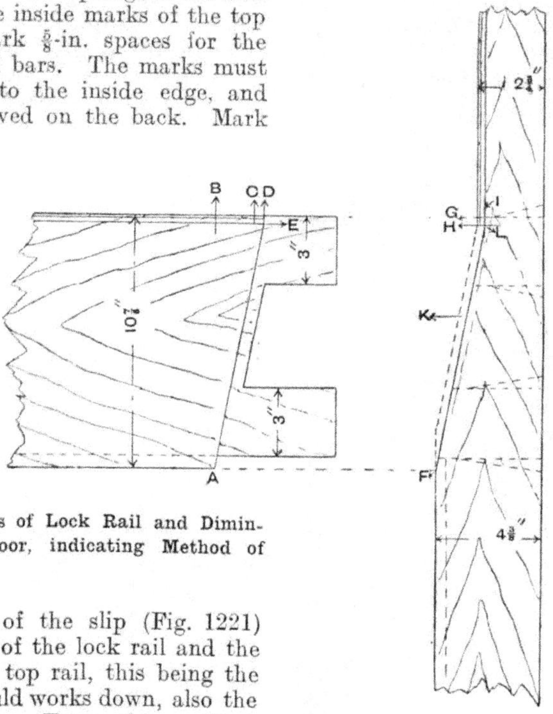

Fig. 1222.—Part Views of Lock Rail and Diminished Stile of Door, indicating Method of Setting Out.

the ⁷⁄₃₂-in. portion of the slip (Fig. 1221) inside the top edge of the lock rail and the bottom edge of the top rail, this being the depth the ovolo mould works down, also the depth of the rebate. From this pattern stile, the other door stile and the stiles of the side lights, must be set out altogether and in pairs. Previous to setting-out, it will be necessary to make the slip (Fig. 1221) for setting-out purposes. This slip, it need not be said, must agree in essential particulars with the proposed moulding and rebate. Assuming that the mould selected works ⅝ in. on and ⁷⁄₃₂ in. down, the slip, when made, will have the appearance shown, and may be explained thus : A equals the depth the ovolo works on ; B equals the depth the ovolo works down, and is also the depth of the

mortice gauge is set, run it round them as they are, and cut the shoulders with a dovetail saw, proper haunchings being left in all cases. Rails of doors and lights should be set out from the rod, the width of stiles set back, and the width of the slip at B set forward, and the shoulders squared round, the setting out of these being done as described in connection with Figs. 1198 to 1200. Cross bars may be set out from this, allowing them to be slightly longer for good joints. The setting out of the top lights is a simple matter, the height of the stiles being marked from the rod. Below is the method of setting out the lock rail, the letters corresponding to those in Fig. 1222, which is supposed to be an inside view, the opposite end to the lock ; it is therefore not double-tenoned : A is the shoulder line under side of rail ; B is the vertical line with A ; ·B to C is the difference between width of stile at top and bottom—namely 1⅝ in. ; C to D is the depth the ovolo works down ; E to A is the shoulder line. In setting out mark A on bottom of rail and B on top, set on C and D with the slip ; gauge down to E and mark E to A. In setting out the diminishing stiles, F is the point where the bottom

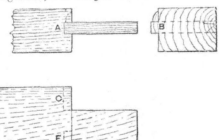

Fig. 1223.—Views showing Haunching and Scribing between Top Rail and Stile.

of the rail joins the stile, and corresponds with A ; G is the point where the top of the rail joins the stile, and corresponds with C ; H is the point where the shoulder line commences ; I is the depth the ovolo works down ; K is the surplus stuff on the edge of the stile ; L is the point of intersection of H and I ; L to F is the shoulder line. In setting out, proceed as follows : Mark F and G, mark H on with the slip, run the gauge line down I ; also from the point where H and I connect L and F. In making the joint after mortising, tenoning, moulding, and ploughing are done—at the point E on the rail, pare down square to the tenon before scribing the ovolo. After the surplus stuff K on the stile has been cut away and the shoulder cut and pared true, cut out the groove for the haunching. The ovolo will require working with gouges up to the point L on the stile, and the rebate must be pared out with the chisel up to the same point. Fig. 1206 shows a section through the door

and framing and fanlight, also the lintel and finishings inside the door frame, of which a plan is shown by Fig. 1207.

Mortising and Tenoning Door.—The setting out having been completed, the stuff must be gauged for mortices and tenons, the mortice gauge being set for a ⅝-in. chisel, and the nearest prick mark from the face being set to the slip at H. One end of the lock rail and the corresponding stile of the door must be gauged for double tenons as in the previous example, the gauge being set for a $\frac{5}{16}$-in. chisel, and $\frac{9}{16}$ in. from face for one tenon and mortice and 1¼ in. from the face for the other. The mortices being made and tenons cut down to the home line, the shoulders may be partly cut in. The stuff may now be rebated for the glass, using a sash fillister, and the ovolo worked on each portion at the same time, using No. 1 moulding plane first, and finishing off with No. 2. The bars must be worked on a sticking board, which may be made from a piece of 7-in. by 1½-in. stuff about 6 ft. long. The bars are held in position against a screw stop at one end, and by a bench knife at the other. The stiles should be ploughed for the panels, the tenon cheeks cut off, and the tenons cut to fit the mortices, proper haunchings being allowed.

Haunching, Scribing, etc.—The ends of the rails for haunching and scribing purposes are shown by Fig. 1223. These have been described and illustrated in Figs. 1196 to 1198 ; A equals the portion cut back on the end of the rails to the depth of the rebate to fit round the square left on after moulding and rebating at B on the stile ; C is a face view of the rail with dotted lines showing the direction and extent of the cutting back at A. The rails only require partly scribing through as shown at E (Fig. 1223), and a piece will be required to be taken out of the stiles to correspond with the square shoulder above these. The scribing may now be done, and particular attention must be paid to the intersection of the lock rail and stile at E and H in Fig. 1222. The scribing templates and gouges will of course be required for this process.

Panels.—The doors and lights may now be knocked together, and the size of the panels obtained. Face up the back sides of the panels; mark equal width between the stiles, and gauge the edges for the distance between the top of bottom rail and the bottom of lock rail. Set a mortice gauge to B C (Fig. 1222) and run all round. Work down to C across endwise with a saw, and with a rebate plane and side fillister with the grain. Rebate out for letting in flush beads E across endwise, and prepare and fix these beads up to the dotted lines. Square across at A and gauge the corresponding widths down both sides. Cut down from A to D with the tenon saw, and remove the waste portions. The corresponding portions with the grain should be ploughed and knocked off with mallet and chisel; afterwards clean off and smooth up with shoulder and smoothing plane. Note that the margin from B to D should be slightly bevelled towards B; this permits the bolection moulding to be planted solid in front. Of course, a thinner panel may be preferred to the bead and flush described, with a moulding planted round inside, but that is optional. When the panels are prepared, knock off a stile on one side of framing, put in the panel, and replace the stile.

Completing Front-entrance Door and Frame.—The doors and lights may now be glued, cramped, and wedged up and cleaned off, and the bolection mouldings mitred and planted in. The frame having been fixed, the bottom and top side lights must be fitted to a joint (the bottom rails of the side lights being ploughed for the water bar), but the rebates and edges should be well painted before nailing. The fanlight must be fitted so that it will swing, and it should be hung with two 3½-in. wrought-iron butts on the bottom edge, being kept in position and made to swing by using a patent quadrant fastener. In fitting the door, sufficient play must be allowed for it to open and shut easily, and it should be hung with three 4-in. wrought-iron butt hinges. A 7-in. mortice lock should be fitted, and, to facilitate the letting in of this lock, the end of the lock rail should be bored before the door is put together. To do this, bore the first hole to a depth of about 3 in., fill it up again with an easily fitting piece of round stuff (prepare a piece sufficient to make

half a dozen), and cut off long enough to stand up about ⅛ in., to allow grip with pincers; bore the second hole on the circumference of the first. Then withdraw the first core by gripping with pincers, and fill up the second hole; bore the third hole on the circumference of the second, and then remove the second core, and so on. When finished, it will be found that there is comparatively little stuff left to clear out. It is obvious that it would be impossible to bore so much out if temporary cores were not put in; moreover, the bit has a tendency to run into the adjoining hole when the holes are bored too close together. The size of bit to use in this case will be ¾ in. The bottom rail of the door must be rebated for the water bar, also throated. This bar is only inserted in exposed situations, or where there is no portico. It keeps the rain from getting under the door, and should not therefore, from the fear that it may become a stumbling block, be omitted. The linings will be prepared from the drawings, and are tongued together at the top, and fixed securely to wood pads or bricks built in the wall. The plinth blocks being fixed at the bottom, and the architrave moulding mitred at the corners and fixed round, will complete the job.

Double-margin Doors.

Double-margin doors are generally used when a door opening is wide in proportion to its height. They are framed as one piece and then hung. The arrangement of the panels varies according to the taste of the architect or the design of the building in which the door is placed. The example shown in Fig. 1224 is a door for an opening 7 ft. 3 in. high and 4 ft. 1 in. wide. The panels have been arranged to be of equal size, while the framing shows an equal margin except in the case of the bottom rail. In one method of construction the middle stile is in two separate pieces, in which case there are really two separate leaves joined to make a single door; or the middle stile can be in one piece, in which case it assumes more of the character of a muntin than of a stile. In Fig. 1225, which represents a horizontal section through the panels of Fig. 1224, the door is shown

Fig. 1224.—Double-margin Door, External Elevation.

Fig. 1225.—Horizontal Section of Double-margin Door.

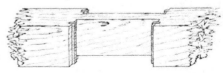

Fig. 1226.—Plan showing Centre Stile Fitted
Over Top Rail.

Fig. 1227.—View of Top Rail prepared to
receive End of Centre Stile.

with the middle stile in two pieces, but the elevation remains the same for either method. When the middle stile is constructed in two portions, each half of the

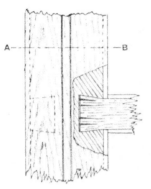

Fig. 1228.—Securing Intermediate Rails by Fox Wedging.

door is framed together, but not glued up. Then the rails are glued and wedged to each half of the middle stile and allowed to dry. The two halves of the middle stile are then shot and fitted. The ends of the tenons to the rails should be cut $\frac{1}{8}$ in. below the joining surfaces of the middle stile, so as to prevent the joint being forced

strengthen double-margin doors, iron bars are sometimes let into the top and bottom rails, and screwed to them. In this case the bars should stop short of each edge of the door, so as not to show on the edge. Another and more general method of constructing a double-margin door will be explained in a later example.

Composite Door.

This class of door is unusual in ordinary joinery work, but is occasionally employed for showing two classes of treatment in detail. This is essentially a soft pine door, with a veneered oak face, the oak face being on the room side, so as to correspond with oak furniture and fittings ; the pine, finished with either white or cream enamel, faces a corridor. The centre panel on the painted side of the door forms a notice-board for posting lecture announcements, etc., as in a school of science or college laboratory. Fig. 1231 (scale $= \frac{1}{2}$ in. to 1 ft.) represents an elevation of the corridor side (painted), and Fig. 1232 a vertical section through door and opening. Fig. 1233 shows the room side, with the oak face with round-edged framing and raised panel. Two methods of construction are illustrated by Figs. 1234 and 1235 ; the former shows the pine stiles,

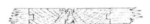

Fig. 1229.—Horizontal Section through Centre Stiles when in Two Pieces.

Fig. 1230.—Horizontal Section through Panels, Moulding, etc.

in case of shrinkage. Fig. 1226 is a plan of the top rail. When the middle stile is in one piece, the top and bottom rails are continuous, and fit over the middle stile by the use of a bridle joint, similar to that shown in Fig. 1227. To prevent the shoulders rising one above the other, cross tongues can be used as shown. The same plan of joining the top and bottom rails to the middle stile can be adopted when the stile is in two pieces. As the tenons in the middle stile cannot be wedged in the ordinary way, it is best to fox-wedge them as shown in Fig. 1228, to prevent the shoulder joints from starting (a section on the line A B is shown by Fig. 1229). Further to

etc., containing the oak sandwiched between them. The two faces of the oak are glued simultaneously, and pressed between the pine by suitable shop methods ; then, when the work is thoroughly dry, it is ripped down by hand, or sawn down the middle with a circular saw, forming two stiles, etc. When the door is framed up, glued up, and partly cleaned off, the lock stile is veneered on the edge ; the tenons being well cut back, for obvious reasons. The hinge stile is not veneered. Fig. 1235 shows an alternative way of treating the stiles, the detail showing the edge veneered first and then the face. Of course the tenons on this side will be blind, but will run through the hinge stile. All

Fig. 1233.—Room Side of Door, showing Oak Face.

Fig. 1232.—Sectional Elevation of Door and Opening.

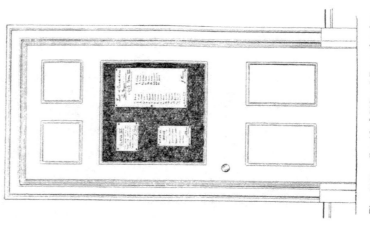

Fig. 1231.—Corridor Side of Door showing Painted Pine Face.

glue surfaces should be toothed on the veneering faces ; the sponge and hot water are freely used on the work in progress if perfect joints are required ; and the work must be done in a warm temperature.

Panels, etc., of Composite Door.—The panels are double (composite), oak and pine,

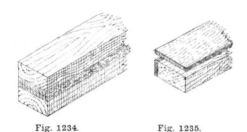

Fig. 1234.   Fig. 1235.

Figs. 1234 and 1235. Isometric Details of Methods of Jointing Oak for Door.

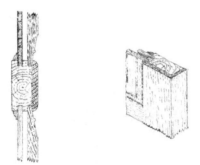

Fig. 1236.—Sectional Detail of Frieze Rail, showing Double Panels.

Fig. 1237.—Detail showing Shutting Stile Veneered on Edge.

with a composite centre panel covered with green or dark red baize, to receive the notices, etc. On the oak side the panels are raised, with a bead on the flat as shown, to be worked in the solid. Figs. 1236 and 1238 show clearly the panelling and framing. All panels must have glued pine blocks (dry) between them, as usual in double-panelled doors of thin sectioned panels. The rounded edge to the framing of the oak side will be found to accentuate the raised panels, without making the door too bold on the interior or room side. The panels on the painted side have ordinary ovolo mouldings glued and

sprigged to the framing (not to the panels). The mouldings to the baize-covered panel are neatly held in place with small brass cups and screws, to allow the easy removal of the baize when worn or damaged. Figs. 1238 and 1239 show sections of the architraves (oak and deal), the panelled jambs, plinths, blocks, etc., of the oak face. All mitres on the oak side must be well glued, screwed, and pelleted. The coke breeze lintel over the door opening (Fig. 1238) is reinforced with

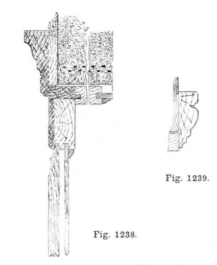

Fig. 1239.

Fig. 1238.

Figs. 1238 and 1239.—Details of Architraves and Top Rail of Door in Section.

expanded metal. The silver-grain of the oak is shown to its best advantage by arranging to have either fine-grained framing and coarse-grained panels, or the reverse. All the architraves must be dovetailed to plinth blocks. The shutting stile is prepared for a mortice lock ; and the door is hung with steel bushed brass hinges. The tenons double at every joint in the framing. The edges of the deal jambs in which the door shuts are stained oak colour. The hinge side is painted to match the exterior face.

Baize-covered Doors.

Two methods of constructing baize-covered doors will now be described. Flat surfaces on each side of the door are usual, but a flush-panel door will answer admirably if the flush side only is required to be covered. Figs. 1240 and 1241 show respectively front elevation and section, and it will be seen that in order to lighten the door, ½-in. full panels (from ⅝-in. stuff in the rough) are used, double or flush each side. To stiffen these panels, ⅞-in. thick by 6-in. wide cross-rails are tenoned into muntins

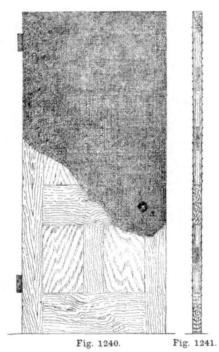

Fig. 1240.   Fig. 1241.

Fig. 1240.—Front Elevation of Double-panel Baize-covered Door.

Fig. 1241.—Vertical Section (Fig. 1240).

and side stiles respectively, as shown in Fig. 1242, which shows also the back rebating of the panels. The panels are centrally screwed to these stiffeners, thus allowing freedom for shrinkage from sides to middle. The tenons are of the usual kind for this class of door. Figs. 1243 and 1244 show an alternative method of construction, which would be cheaper if machine-worked stuff were used. The framing is mortised and tenoned, and is covered on both sides with tongued and grooved boards forming the flush faces of the door. Bracing can be used, but if the

boards are partly glued on, the door will be found to be sufficiently rigid, and free from any tendency to drop. Fig. 1241 shows a top corner, and indicates the method of construction.

Covering Doors with Baize.—The covering applied to this class of door should be of the best quality procurable, and should be obtained of such a width as to prevent waste. It is fastened on with stout tacks into plough grooves on the edges and at the sides, as shown in Fig. 1246, which represents a hori-

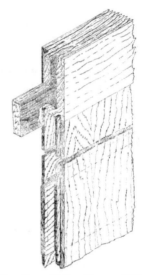

Fig. 1242.— Cross Rail acting as Stiffener to Panels.

zontal section of door stiles for a swing door, the edges of which should preferably be rounded slightly, to prevent the baize from being cut by the stretching process. It is customary to line out the panels by means of round-headed brass nails of French manufacture, and further to elaborate the baize by using green tape of a lighter or darker tint (see Fig. 1247); but these are matters of taste. The tape is neatly nailed to chalk lines sprung at the proper spacing, and should be kept moderately taut while nailing on. In the case of the double-panel door, the nails should, if possible, clear the groove. If baize-covered doors are used in entrances near the street, they are apt to harbour dust, and to be spattered with mud, and the lower part of doors in this situation should be covered with an oak board about $\frac{1}{4}$ in. thick, and of any suitable width, and painted to match.

Circular Doors.

The door (Fig. 1248) about to be described is on plan curved to a 2-ft. radius, but it

Fig. 1243.   Fig. 1244.

Fig. 1243.—Elevation showing Alternate Methods of Constructing Baize-covered Door.

Fig. 1244.—Vertical Section (Fig. 1243).

may, of course, have any sweep desired. It should always be set out from a centre, as the shoulder lines will be struck from the centre point. The panels are bead flush, with bolection moulding mitered round the front as shown in plan by Fig. 1249, also by Fig. 1250. which is an enlarged section at the top edge of the bottom rail, with the panel and moulding on the outside and the bead flush on the inside. The top space is left for glass. which is secured by means of beads screwed round the stiles and rails ; by this arrangement the

glass can be replaced when necessary without injuring the door or beads. Fig. 1251

Fig. 1245.—Detail of Top Corner (Fig. 1243).

is an enlarged section at the top edge of the middle rail with the glass and beads in position. The wood used in circular work must

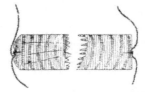

Fig. 1246.—Fastening Ends of Covering.

be thoroughly seasoned and free from defects. In beginning a job of this sort, it is necessary to make a template and set it out full size on

Fig. 1247.—Covering and Taping Baize.

a board from a centre. When this has been done, the ribs may be prepared. The plank

it is intended to use should be trued up out
of winding and gauged to the required thickness. The trammel with which the plan
has been set out can be used to line out the

Fig. 1249.—Plan of Door (Fig. 1248).

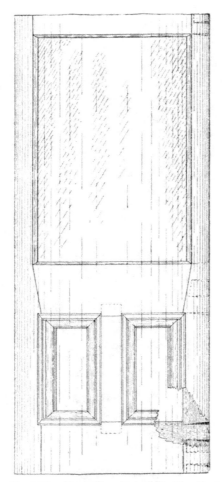

Fig. 1248.—Front Elevation of Circular Door.

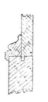

Fig. 1250.—Section
through **Top Edge**
of **Bottom Rail** (Fig.
1248).

Fig. 1251.—Section
through **Top Edge**
of **Middle Rail** (Fig.
1248).

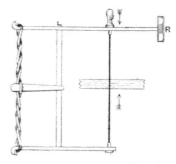

Fig. 1252.—Side Saw for Cutting Rails, etc.

ribs for the three rails ; these can be cut
out with a band-saw, or with a side-saw if the
former is not available. A side-saw (see
Fig. 1252) is much like a bow-saw, only it is
worked up and down ; L and R show respectively the position of left and right hand
when using the saw. The ribs having been
cut out and toothed ready for gluing together, it will be seen (Fig. 1253) that they are

Fig. 1253.—Building Up Rails in Ribs.

butt-jointed ; by this arrangement strength is gained, as any two pieces coming together will not be of continuous grain. Care must be taken, too, to break the joints as shown in Fig. 1253. All the butt joints may be cut before commencing gluing ; the ends may be left till the work is dry. It is necessary to warm the sides that are to be glued. All

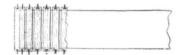

Fig. 1254.—Method of Securing Veneer for Circular Door.

these joints are secured by means of screws or nails (screws preferred), care being taken that they are not in the way of the mortices and tenons.

Panels, etc., of Circular Doors.—The panels may be got out in widths according to the width of the finished panel, which is a little over 10 in., so it may be made in three pieces and shot to a proper bevel to suit the rails, and likewise glued and set aside to dry.

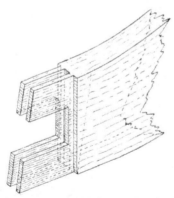

Fig. 1255.—Joint of Lock Rail of Circular Door.

The stiles call for little remark beyond the setting out of the mortices, which are indicated by dotted lines (Fig. 1248), where the rail is broken away to show the tenon in position, wedged. The rails and panels may be cleaned off to the template and taken true out of winding and prepared for covering with veneer, which must be saw-cut.

In some doors or framing this is omitted. The veneer is held in position with a caul, which is a piece of zinc large enough to cover the whole of the veneer ; pieces of wood, cut to fit the sweep, are then laid across and squeezed tightly down with thumb- or hand-screws (see Fig. 1254). Fig. 1255 is an enlarged isometric drawing of the joint at the middle or lock rail, showing $\frac{3}{16}$-in. double tenons with a $\frac{5}{8}$-in. space between to allow for a mortice lock ; these double tenons are only employed where the lock is to be fitted, the remainder of the tenons being single. All tenons must be carefully cut and fitted, so that when the door is ready to be wedged up they will be just hand-tight, otherwise the shoulders will not come well up. The panels are prepared for veneering in the same way as the rails, and are held in position with a tongue that fits into a groove, as shown by Fig. 1250. A small bead is worked on the sides of the panels, but those on the ends are mitred and planted on. The mouldings and beads are got out to suit the proper sweeps, and worked with routers, which can be obtained in sufficient variety to suit any kind of moulding.

Double-margin Door with Circular Frame and Splayed Linings.

The door and frame illustrated by Figs. 1256 to 1272 is a kind sometimes adopted for the entrances to large and important buildings in which it is more convenient to have one wide door than two folding doors, although when closed it has the appearance of two, as shown. In this class of building the door, frame, linings, architraves, etc., are frequently made of some hard wood, such as oak, mahogany, or teak, and finished by french polishing. The ordinary fixing by visible nailing, screwing, etc., is not permissible ; the holes thus formed cannot be stopped with putty as for painted work. This example will be treated so as to meet the specification and drawings demanding secret fixing as far as practicable. The leading dimensions are figured on the drawings, and the chief new features are described below.

The Rod.—For a job of this description, the rod should be clearly and fully set out,

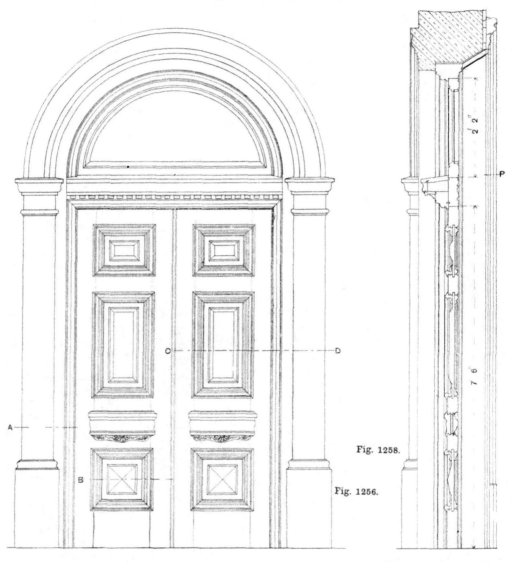

Fig. 1258.

Fig. 1256.

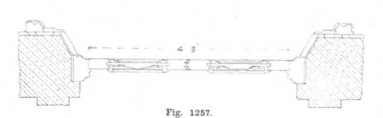

Fig. 1257.

Fig. 1256.—Front Eleva-
tion.   Fig. 1257.—Half
Horizontal   S e c t i o n
through Lower Panels,
and   Half   Horizontal
Section through Centre
Panels on Lines A B and
C D.   Fig. 1258.—Ver-
tical   Section   through
Centre of Fanlight and
Centre of Panels.

Fig. 1259.—Inside Elevation.

as an attempt to save time here may lead to costly errors. The architect's drawings should be carefully consulted, particularly the enlarged details to half-inch or larger scale, together with full-size sections of mouldings and panels, which are generally supplied by leading architects. If the building is sufficiently advanced, it is well to test between the reveals and the springing, in case any slight discrepancy has occurred in the dimensions between the masonry and the drawings. From the information thus obtained, the rod can be set out. This would show the whole of the plan or horizontal section of the woodwork as at Fig. 1257. A complete vertical section of the woodwork should be set out for the heights as at Fig. 1258. For the framing, the head and fanlight, the linings and architraves, a board should be used large enough to strike out the main lines of each of these ; and it will be found very useful for setting out these various parts.

The Frame.—The posts should be set out for the tenons of the transom. By reference to Figs. 1259 and 1260, it will be seen that the transom is built up, and that two equal - sized tenons are made, the *bottom edges of which are level with the* rebate. The mitering of the mouldings between the posts and transom is shown at Fig. 1260, and, as explained in previous examples, this must be allowed for in setting out. It should be noted that the tenon to the inner portion of the transom A *(Figs. 1259 and 1260) has the upper portion* of it haunched into the post, and does not have a tenon the whole width. The pieces to form the circular head should next be cut out. Make a template from the rod, and line out the stuff for the head " full." The head may be in two *pieces, with a joint at the crown, or in* three pieces, which is perhaps a little more economical in material, but the former no doubt involves less labour in jointing and even in moulding. The crown joint should be accurately made, and must also fit when placed on the rod. This joint may be held *together by a hammer-headed key and* tongues, or by a stout handrail screw and dowels, which is the more modern method and equally effectual. The joints between the posts and head at the springing may be held together by handrail screws and dowels, but the hammer-headed key tenon, *as illustrated at Fig. 1260, is still in favour.* When the head and posts are tried together, and the joints temporarily tightened up, the posts must be quite parallel and exactly the same distances apart, and the joints must be eased until thus correct. The soffit of the head, as it is not yet trued up, *should project over the posts a little.* This projection should be carefully scribed from the posts, and then by means of a template or radius rod the curve for the soffit can be set out. It will be found an advantage to turn the frame over and mark the other

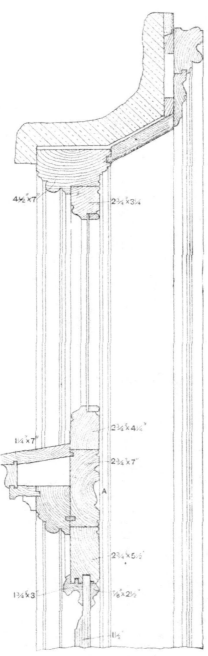

Fig. 1260.—Enlarged Detail of Upper Part of Vertical Section (Fig. 1258).

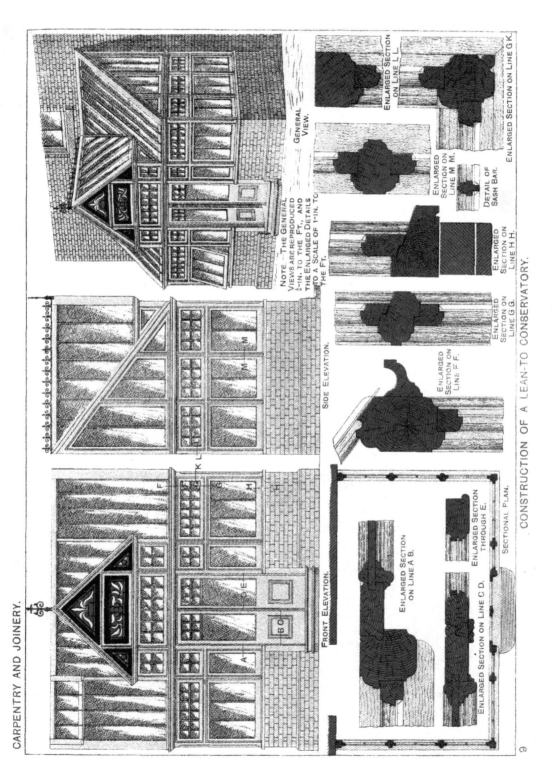

CARPENTRY AND JOINERY.

GENERAL VIEW.

NOTE.—THE GENERAL VIEWS ARE REPRODUCED ½-IN. TO THE FT., AND THE ENLARGED DETAILS TO A SCALE OF 1-IN. TO THE FT.

ENLARGED SECTION ON LINE L L.

ENLARGED SECTION ON LINE G K.

ENLARGED SECTION ON LINE M M.

DETAIL OF SASH BAR.

ENLARGED SECTION ON LINE H H.

ENLARGED SECTION ON LINE G G.

ENLARGED SECTION ON LINE F F.

SIDE ELEVATION.

FRONT ELEVATION.

ENLARGED SECTION ON LINE A B.

ENLARGED SECTION ON LINE C D.

ENLARGED SECTION THROUGH E.

SECTIONAL PLAN.

CONSTRUCTION OF A LEAN-TO CONSERVATORY.

9

side as well. The pieces can now be separated and the soffit planed to the lines by a compass plane or other similar means. The setting out, the mortising, and the tenon and rounds. For working the curved mouldings of the head, small hollows and rounds, known as thumb planes, or some other method, would be adopted, as will

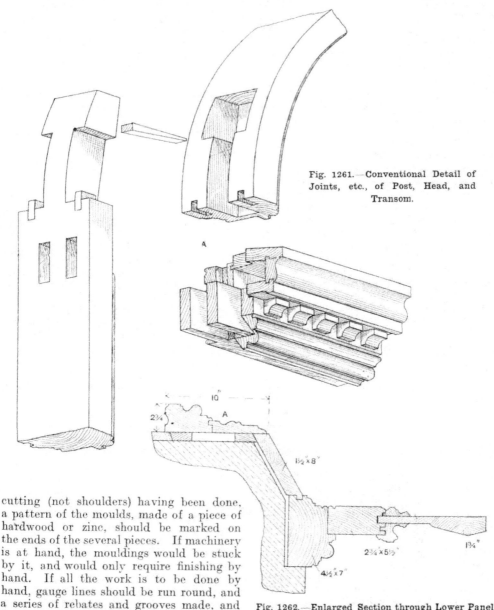

Fig. 1261.—Conventional Detail of Joints, etc., of Post, Head, and Transom.

Fig. 1262.—Enlarged Section through Lower Panel Moulding. Door Post, etc.

cutting (not shoulders) having been done, a pattern of the moulds, made of a piece of hardwood or zinc, should be marked on the ends of the several pieces. If machinery is at hand, the mouldings would be stuck by it, and would only require finishing by hand. If all the work is to be done by hand, gauge lines should be run round, and a series of rebates and grooves made, and the curved members finished with hollows be described in a later section. Sometimes these circular heads are built up of two thicknesses, the joint taking place in a line with the stop part of the rebate. One layer is formed of one more piece than the other, to "break joint," and the two layers are glued and screwed together as illustrated at Fig. 1263, the screws being inserted outside the line of the seen margin. This method is somewhat more economical in material and labour, as rebating out of the solid is avoided, and on this account it is more frequently adopted for ordinary work. Some leading architects, however, would not sanction this method of construction. The mouldings should be mitered, and the shoulders between the post and transom prepared; these should be loosened, glued, wedged and tightened up, cleaning off the flat surfaces and the junction of the mouldings. This will complete the frame, excepting the cornice portion of the outside part of the transom, which should be prepared and fitted in and, if desired, fixed; but sometimes it is more convenient to fix this temporarily only, so that it can be more easily scribed to the stonework after the frame is in position.

Splayed Linings.—The head part of these may be built up in sections or veneered and blocked. The method of preparing these will be fully described in the section dealing with the subject.

Framed Grounds.—The architraves being wide, the grounds are framed of strips of stuff 2 in. or more wide for stiles, and pieces 2 in. to 4 in. wide for rails. These are simply mortised and tenoned together, and glued and wedged in the usual way, and they would be continued round the circular head as shown at Fig. 1264. A simple way of jointing up the parts round the head is to have the rails about 4 in. wide, tenoned at each end, and the curved stiles fitting on to these with open mortices, as will be understood by reference to Fig. 1264. These would be fixed round the opening flush with the edge of the splayed linings by nailing to wood bricks or other usual means.

Architraves.—These being wide and varying much in thickness, the jamb portions are prepared in two pieces, grooved and tongued together as illustrated. The archi-

trave round the head could be prepared to fit together in two pieces, in breadth the same as those for the jambs, but of course each portion would be formed of at least two or three pieces round the semicircle, and breaking joint with the one above it. Another method would be to build up the head moulding in four thicknesses, making each thickness break joint with that below it, as illustrated at Fig. 1265, gluing together and screwing from the back. A method of connecting the circular head architrave to the vertical parts is shown at Fig. 1266. At the back a lap dovetail is made on the end of the member, fitting into a corresponding recess made in the circular one; by gluing, cramping and screwing, and leaving till dry, a good sound joint can thus be made.

Secret Fixing of Architraves.—The complete architrave must be offered up in position, the side margins accurately regulated, and, setting a pair of compasses from the arris of the head lining to the edge of the architrave less the amount of margin to be shown, scribe the bottom of the architraves to the floor. The plinth at the bottom of the outer members of the architrave not extending the whole breadth, the outer members are cut away and the plinth blocks fitted as shown at Fig. 1272, the two parts being firmly held together by gluing and screwing from the back. The system of fixing here shown is by boring a series of holes in the back of the architraves, as illustrated at Figs. 1267 to 1269. A hole is bored a little larger than the head of a screw, about $\frac{1}{2}$ in. or $\frac{5}{8}$ in. deep; then with a bit the size of the shank of the screw a second hole is bored about $\frac{3}{4}$ in. above; the wood between the two holes is mortised out, leaving a chase; then, by using a very thin chisel or other convenient tool, a **V**-shaped slot is formed on each side of this chase as indicated at A (Fig. 1267); then, taking a screw (the same size as the one to be used) turned into a piece of hardwood and allowed to project the exact required distance, insert the screw into the holes in turn, and, striking the end of the wood, drive the screw head along the **V**-shaped chase made in the backs of the architraves. Now,

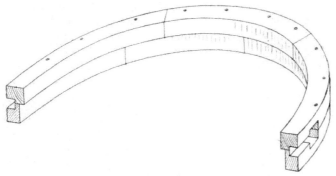

Fig. 1263.—Method of Building Up Circular Head in Two Thicknesses.

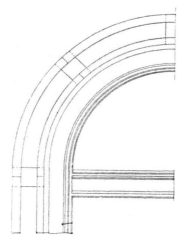

Fig. 1264.—Part Elevation of Framed Grounds.

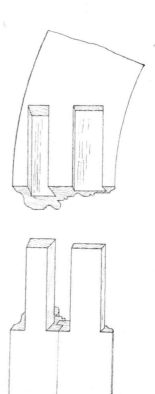

Fig. 1266.—Method of Jointing Circular Head and Vertical Architraves.

Fig. 1265.—Method of Building Up for Circular Architrave.

marking the exact positions for the centre of the screws on to the grounds (these, of course, exactly corresponding to the chases on the back of the architraves), turn screws into these so that the heads project exactly the same distance as that in the hardwood block mentioned. When the architraves are placed against the grounds and lifted about ¾ in. from the floor, the heads of the screws should sink

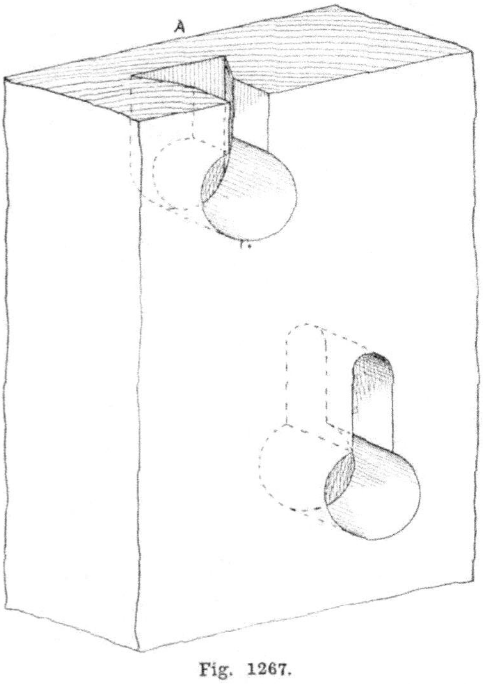

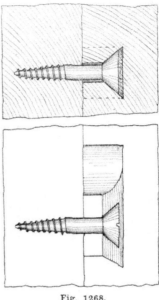

Fig. 1267.

Figs. 1267, 1268, and 1269.—Preparing Slots for Screws.

into the holes made for them in the back; then, by forcing and jarring the architraves down, the screws will firmly hold them, if the work has been carefully and accurately done. The complete architrave may now be raised and taken down, and the edge against the splayed linings rapidly glued. It is then put back and finally jarred and forced into its permanent position.

The Door.—The preparing of the framing of this will, in a general way, be similar to the cases previously treated. It is

essentially two doors. After the parts of these have been prepared and fitted together, and additional mortices made through

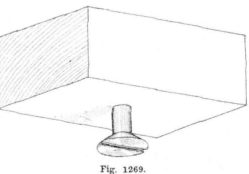

Fig. 1268.

the meeting stiles as indicated at A (Fig. 1270) to receive wedges, the two separate leaves have the joints of their meeting stiles and rails glued, cramped and wedged up without any panels or moulding being

Fig. 1269.

in. Then the two meeting stiles are shot so as accurately to fit, and both leaves are in one plane; then the meeting edges are moulded, or beaded and ploughed for tongues (see Fig. 1254). Hardwood keys or wedges are prepared, and sometimes additional gib pieces A (Fig. 1270) are also provided ; cross tongues are glued into one of the stiles ; then both the meeting edges are glued ; the wedges are also glued and quickly inserted, cramps put on, and the

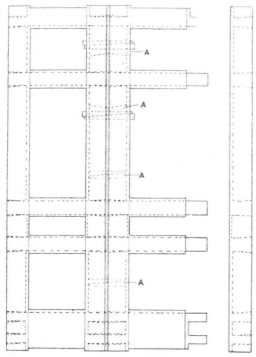

Fig. 1270.—Skeleton Framing for Double-margin Door.

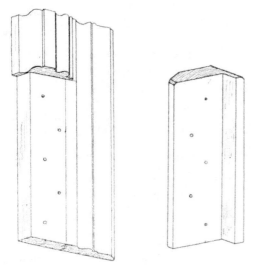

Fig. 1272.—Conventional View of Architrave and Base Block reversed at Fig. 1262.

Fig. 1271.—Conventional Sectional View of Rail, Panels, etc., at A (Fig. 1259).

wedges connecting the stiles finally driven tight. After the glue is dry, the ends of these are cut out level with the bottom of the plough grooves. The panels can then be inserted and the door wedged up. The bolection mouldings on the outside, as also those on the inside, are to be held in position without visible fixing. There-fore, these mouldings have been provided

Fig. 1273.—Outside Elevation of Doors, Frame, Fanlight, etc.

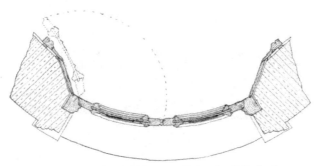

Fig. 1274.—Plan of Masonry at Level of Springing, Horizontal Section of Doors taken through Upper Panels.

with small tongues which fit into corresponding p l o u g h grooves made in the stiles and rails as illustrated. This system of construction necessitates the preparing and cleaning off (all but the final finish) of the shoulders and framework generally before wedging up, because very little of this can be done afterwards owing to the mouldings having to be fitted. Their mitres are often grooved for tongues or slip feathers, and thus have to be placed in their positions before the wedging up takes place. The conventional detail at A (Fig. 1271) will make clear the construction at the apron panel and moulding, and also that at the back. The joint at the end of the architrave with base blocks is shown at Fig. 1272.

Entrance Doors and Frame, Circular on Plan, and Circular-headed in Elevation.

Fig. 1273 shows the elevation of a pair of doors at the corner of an important stone building. The frame has a circular head, fanlight, etc. The corner is circular, forming a quadrant of 6 ft. radius, corresponding with the plan of the building. The doors and frame follow the same curve, as shown by the plan, Fig. 1274. The masonry arch is semicircular in elevation, and stilted 4 in. The outer arris of the soffit of the arch is taken as a semicircle (see A B, Fig. 1273). The reveals radiate, and therefore the soffit at the springing on each side also radiates, but it finishes level at the crown as shown at C. Thus the arris of the soffit of the arch adjacent to the head of the door frame is elliptical in elevation, as shown ; and it

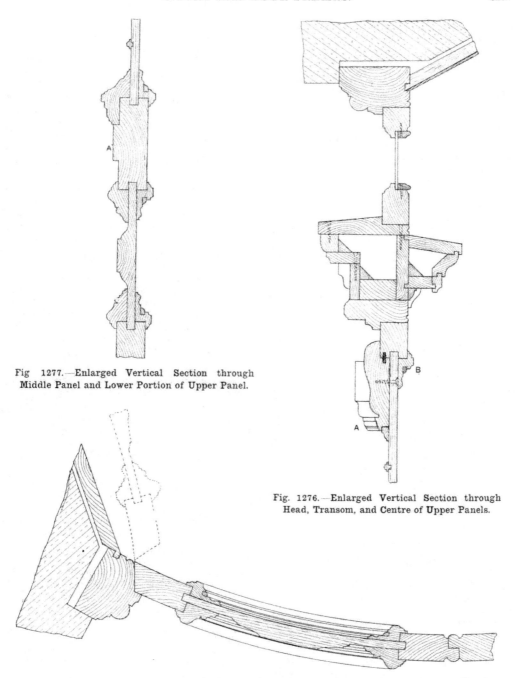

Fig 1277.—Enlarged Vertical Section through Middle Panel and Lower Portion of Upper Panel.

Fig. 1276.—Enlarged Vertical Section through Head, Transom, and Centre of Upper Panels.

Fig. 1275.—Enlarged Horizontal Section through Door, taken through Centre of Lower Panel.

is this line which must be used for working from when striking out the door frame head.

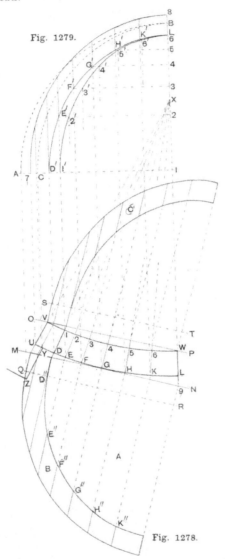

Fig. 1279.

Fig. 1278.

Figs. 1278 and 1279.—Geometrical Construction for Setting Out Face Moulds.

Setting Out for Face Moulds.—Set out the half plan of the head of the frame, as shown by z y (Fig. 1278); projecting up from z draw the dotted arc A B, for the outer arris of the soffit; project from Y and obtain the dotted curve C B, which is the inner arris of the soffit; projecting up from D obtain point D'; from B (Fig. 1279) measure down the margin for the head of the frame; then the curve D' to L is the elevation of the curve of the outer arris for the head of the frame. This line from D' to C is a little nearer, and gradually widens from the dotted line C B until 1' is reached. Take any convenient points on the curve D' to L (Fig. 1279), project down from these points to the curve in plan, and obtain a number of points in plan and elevation as shown. From these points in plan draw converging lines to the centre X, remembering these are horizontal generators. Projecting out horizontally from D', E', F', G', H', K', we obtain points 1, 2, 3, 4, 5, and 6, as shown on the centre line in elevation. Project from 1 to 1', from 2 to 2', and so on to point 6; then projecting horizontally from E', F', G', H', K', points from 1' to 6' are obtained, through which can be drawn the curve 1' to L, which represents the inner arris of the head. The outer arris of the head, as represented by the curve 7 to 8, can next be drawn in, as shown. In the plan Fig. 1278 draw the line O P through V and W, and parallel to this draw M N tangent to the curve U L; the distances between these two lines represent the thickness of the plank required. Parallel to the line M N set up the distance of the springing (in this case 4 in.), continuing the lines radiating from X from each point D to L until the Y meet M N (Fig. 1278). A number of points are obtained on M N, from which ordinates are drawn at right angles, as shown at A. Marking off the distances on each ordinate from the springing line Q R, equal to its corresponding ordinate shown in elevation (Fig. 1279), the outer face mould can be drawn as represented at B. The method for obtaining the face mould for the inside is similar, as shown at C (Fig. 1278).

Setting Out for Soffit Mould.—The half plan is re-drawn (for clearness) at Fig. 1280. At right angles to N X draw X L; on this set up distances 2, 3, 4, 5, 6, L equal to the corresponding distances at Fig. 1279.

Now with compasses set to the distance D″, E″ (Fig. 1278), and using M as centre,

with compass set to E″, F″ (A, Fig. 1278), draw the arc 8′, and cut it with the beam

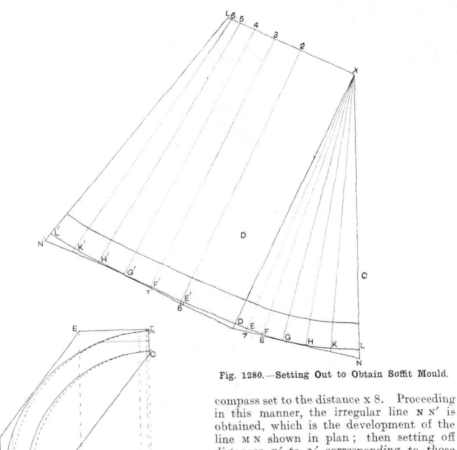

Fig. 1280.—Setting Out to Obtain Soffit Mould.

Fig. 1281.—Plan and Elevation showing Piece of Plank with Face Moulds and Bevels Applied.

compass set to the distance X 8. Proceeding in this manner, the irregular line N N′ is obtained, which is the development of the line M N shown in plan; then setting off distances E′ to L′ corresponding to those in plan from D to L, points are obtained through which the curved lines for the soffit mould can be drawn as shown.

Application of Moulds and Making Joints.—The application of the face moulds to the plank is shown at Fig. 1281, the solid curve lines representing the outer face mould applied to the face of plank. Square through the plank to the line A B, also to the line E D, which is at right angles to D F; then setting a bevel to the angle in plan X, Q, M, this can be applied to the surface E D, as shown in plan at X. The bevel can also be applied to the end A B, as indicated by the dotted

describe the arc 7′; next, with beam compasses set to X 7, cut the arc 7′, and then, lines at Y. Then by squaring over, the face mould for the inside can be applied to the back side of the plank, as indicated by the dotted curves. The next process will be to saw to these lines as near as possible, after which the joints should be made. In making these, a horizontal and

done, the joints should next be set out for handrail screws and dowels; the boring and paring for nuts should next be done, the screws and dowels inserted, and the joints drawn up tight; then the work should be tested, and any necessary easing done.

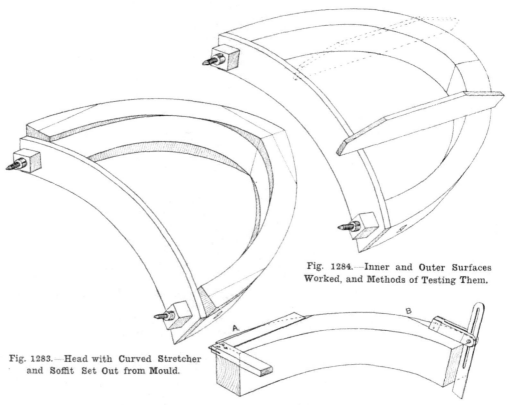

Fig. 1284.—Inner and Outer Surfaces Worked, and Methods of Testing Them.

Fig. 1283.—Head with Curved Stretcher and Soffit Set Out from Mould.

Fig. 1282.—Method of Testing Joints with Try Square and Bevel Applied to Tangent Lines.

vertical tangent line, which has been drawn on the face mould as shown, should be transferred to the face of the plank, as indicated at Fig. 1282; these lines will be found very useful for the application of the try square when testing for the face edges of the joints as indicated at A. The springing joint is also square through the plank, but the crown joint must have the bevel applied through the thickness, as indicated at B. The joints at the tops of the posts should now be planed true. This

Squaring up the Head.—Separate the posts from the head, then true up the soffit of the head, of course working exactly to the lines made by the aid of the face moulds. Prepare a piece of inch stuff the exact curve of the plan, bore each end to correspond with the joints at the springing, and fasten on as shown at Fig. 1283. Fasten this piece of board to the springing joints of the head as shown at Fig. 1283, apply the soffit mould to each piece, and mark it as shown in that figure. Then the superfluous wood on each side can be worked off to these lines, and by testing with a straightedge so that it is at right angles to the curved stretcher as indicated at Fig. 1284, the surfaces may be worked true.

Moulding.—A few leading points in the moulding are :—The patterns for moulding can be marked on each end, as it will be noted that the rebate at the springing does

transom is shown at B (Fig. 1276), as is also the general construction of the inner and outer cornices.

Setting Out Fanlight.—The method of setting out for the fanlight would be similar in almost all respects to that shown for the setting out for the head. There would be the additional advantage of being able to fit the pieces in the rebate of the frame. It should be noticed that the frame of the fanlight is made parallel on plan to allow of the insertion of the glass. This is held

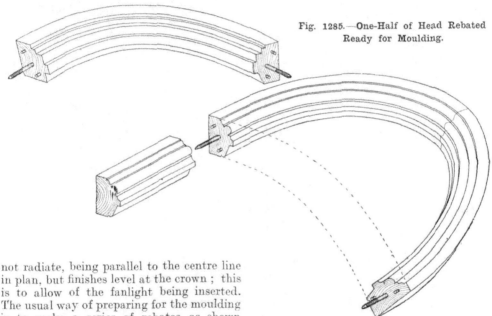

Fig. 1285.—One-Half of Head Rebated Ready for Moulding.

not radiate, being parallel to the centre line in plan, but finishes level at the crown ; this is to allow of the fanlight being inserted. The usual way of preparing for the moulding is to make a series of rebates, as shown at Fig. 1285, and then finish with small hollows and rounds, completing the moulding as shown at Fig. 1286. The mouldings of the posts are stopped near the bottom, so that the plinth finishes against the post ; a square part of this is shown in the elevation (Fig. 1273). The construction of the transom, with its cornices and method of finish to stonework, will be understood by referring to Figs. 1276 and 1277. The transom head A (Figs. 1276 and 1277) has twin tenons fitting into corresponding mortices in the posts. The intersection between the mouldings of the posts and

Fig. 1286.—Head Moulded and Completed Ready for Bolting to Posts.

in position by beads, which generally would be prepared out of the solid stuff.

Doors.—The curvature of the doors in plan being an arc of only about 2-in. rise, all the panels and rails would be prepared out of the solid. The sections of these are given in Figs. 1274 to 1277. All the horizontal mouldings are curved and worked out of the solid ; the fixing of all these is to be secret, and the method here adopted for this is a good one. The bolection

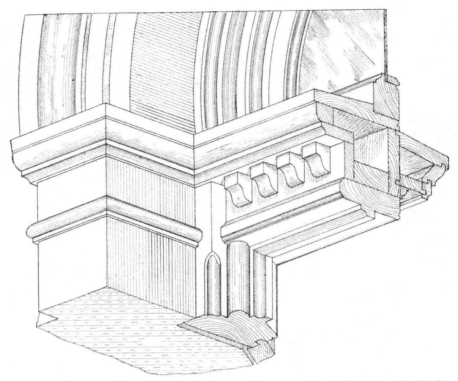

Fig. 1287.—Conventional View showing the General Construction of Transom and Cornice, and the Intersection with Stonework and Door Post.

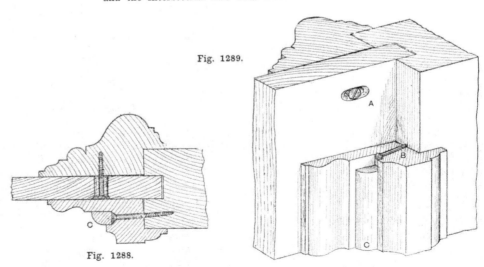

Fig. 1289.

Fig. 1288.

Figs. 1288 and 1289.—Methods of Secret Fixing of Mouldings.

mouldings on the outside are secured by screwing from the inside
of the panels as illustrated at A
(Figs. 1288 and 1289). To obviate
any chance of the panels splitting
should shrinkage occur, the holes
in the panels are made by boring
and countersinking two holes the
size of the shank of the screws, and
cutting out the material between;
thus slots are formed as shown at A
(Figs. 1288 and 1289). These slots
are made longways across the grain.
The inside mouldings are inserted
afterwards; and by preparing them
with a movable member as shown
at c (Figs. 1288 and 1289), these
can be secured by screws to the
edges of the stiles and rail, as shown,
and then the member c is glued and
inserted in the grooves made to receive it.

Apron.—The carved apron under
the moulding of the top panel is
worked on the solid of the rail as
indicated in the section at A (Fig.
1277); although a little more trouble,
this method is superior to planting
on. The small carved scroll pediments, which form a finish to the top
panels, are drawn in section at A
(Fig. 1276), where they are shown
tongued to the top rail; being elliptical on the inside allows of this
fixing being hidden by the moulding B (Fig. 1276). The inner reveals
for these frames are very often
built parallel, and sometimes a
toothing is left; the space being built
up close, or near to, and after the
frame has been placed in position;
wood blocks or other means of fixing
being built in the wall as required.
Although the foregoing description is
necessarily brief, the accompanying
illustrations are sufficiently clear to
show all details, the more ordinary
construction being similar to previous
examples.

Circle=on=Circle Work.

Figs. 1290 to 1300 illustrate an
example very similar to the last;

Fig. 1290.   Fig. 1291.

Fig. 1290.—Elevation of Circle-on-Circle Door and
Frame.

Fig. 1291.—Section of Circle-on-Circle Door Frame.

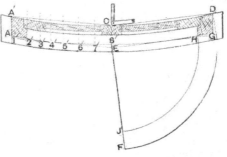

Fig. 1292.—Plan of Circle-on-Circle Door Frame.

as a rule, it would be found only in buildings of rather ordinary class. It will be seen by the plan that the jambs of the parts do not radiate; the inner arris of the soffit of the arch is a semicircle, and thus, on account of the radius of the plan not being great and the curvatures being very flat, the outer arris of the head of the frame can also be taken as a semicircle. A simple method will now be shown and explained of setting out and constructing the head with one mould only.

Circle-on-Circle Door Frame constructed with One Mould.—Fig. 1290 shows the front elevation of a 4-in. by 3-in. semicircular-headed solid door frame, with single rebated parallel jambs, oak sill, double rebated and weathered transom, a 2-in. glazed fixed fanlight, and a 2-in. four-panel door, moulded outside, with the bottom panels bead flush inside. Fig. 1291 shows a central vertical section, and Fig. 1292 the plan. In beginning a job of this description, an elevation of the head down to the transom should be set out full size on a rod, and immediately beneath it, and projected from it, the plan should also be set out. To get the thickness of the stuff required out of which to get the head, assuming that the head will be made in two pieces jointed at the centre, draw lines from the centre to the outside edges of the jambs, on the inside of the plan (see A′ C D, Fig. 1292); also draw two lines parallel with these, touching the curve on the outside, as 1′ E G, which gives the thickness of the piece of stuff required. To find its length, either go through the same process in the elevation, drawing the ends square with the tangent lines, or preferably, make the mould F G, J H squared out from the lines just drawn, using ordinates to obtain this, as in the previous example; or it can be struck out with a trammel, being a quarter of an ellipse. Cut out two pieces to this mould square from the face, and make the joints at the centre and springing the same as the end of the mould. For the horizontal cut, set a bevel as shown on the plan, and apply it on the edge of the stuff from the face. A handrail screw and a couple of cross-tongues may be used for the head joint, the nuts, of course, going in from the

top. When the joint is made, try the head, which will now have the appearance of Fig. 1293, without the lines, over its plan; its back and front faces should stand perpendicularly over the lines 1′ E G and A′ C D, and its ends completely cover the sections of the jambs (Fig. 1292). The head has now to be worked to the plan curve.

Ascertaining Plan Curves.—Divide the soffit of the head of the frame into any number of equal parts between the springing lines (as in Fig. 1290), numbering them on each side, from springing to centre; the greater the number of parts, the more accurate will be the curve. From these points drop perpendiculars into the plan, cutting the tangents or block lines of the head (see Fig. 1292), and numbering the lines to correspond with the elevation. The utility of projecting the plan from the elevation will now be apparent. Next, place the head over its plan, as shown in Fig. 1293, keeping its centre perpendicularly over the centre line in the plan; with the aid of a set square, transfer to its face the lines 1, 2, 3, 4, etc., from the like numbered points in the plan. Lines must now be drawn on the top and bottom edges from these, parallel with the joint; and to do this, take the joint bevel, and apply it to each line in succession, holding the stock level, and the inside edge of the blade to the point from which the line has to be drawn. The head now having the lines drawn as in plan and elevation must have the points marked where the curve intersects these lines. Set a pair of compasses or spring dividers to the widths 1′ A and 1′ A′ (Fig. 1292), and transfer them to the head at the springing joint on each side. Do the same throughout the series B′, transferring each width to its proper position on the top and bottom edges of the head, until all the points have been pricked off, as shown in the enlarged sketch of one side of the head (Fig. 1294). Now draw the curve through the points thus obtained, either by freehand or by the aid of a thin strip bent round the head and kept to the points. The two pieces can be worked off to the lines, keeping them straight across the face in the direction of the ordinates. They should be tested by moving a set square, held perfectly upright, carefully around the curve, and seeing whether the face fits close up to it.

Lining Out Elevation Curve.—The plan curves having been worked, the next thing is to line out the elevation curve. This is done in the manner shown by Fig. 1295. Cut in tightly between the ends of the head a stretcher as shown, and screwing it to the joints. Lay the head on the bench top, packing it up level ; then fix a small block in the middle of the stretcher, of such a height as to bring its top level with the highest point on the head. Draw a line to represent the springing line, and upon this mark the exact centre ; this will be the point

groove made in the end, thus drawing the curve as true as if it had been struck on a flat surface ; the operation should be repeated on the other side of the head, first taking the stretcher off and turning it over.

Moulding and Rebating Frame Head.— The soffit having been worked off to the lines, the head is ready for moulding and

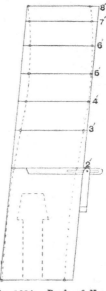

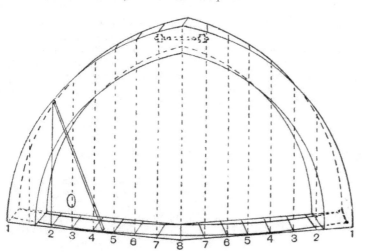

Fig. 1293.—Sketch of Head of Circle-on-Circle Door Frame before Shaping.

Fig. 1294.—Back of Head of Circle-on-Circle Door Frame, showing Method of Obtaining Points in the Curve.

from which to describe the semicircles of the elevation with the falling compass now to be described. Get a piece of light deal about 2 ft. long, ¾ in. thick, and 2 in. wide at one end, tapering to ½ in. at the other, and in it cut a slight hollow or **V** as a bed for the pencil. Mark on the edge the distances of the inside and outside of the jamb, and bore holes with a bradawl through these points, square from the bottom edge ; these will give the radii for soffit and outside lines of the head. Fix the compass as shown in Fig. 1295, and, beginning at the crown, move it steadily round, letting the pencil slip down the

rebating, as, unless the frame is going into a stone reveal, it will be unnecessary to do anything to the back of the head. The rebate should be worked first, a quirk router being used to sink a small groove, ⅛ in. deep, and 2 in. from the inside face ; run a cutting gauge, with a rounded fence, set to ½ in., round the inside face, and remove the core with a bent chisel. Finish up the rebate with a round-soled thumb rebate plane, and work a ⅜-in. bead on the rebated edge and a ½-in. bead on the outside.

Circle-on-Circle Door Frame : Completion.
The head can now be glued up at the centre joint and cleaned off, the stretcher screwed

on and this part set aside whilst the remaining portions of the frame are being worked. The transom and the oak sill will be got out to the plan moulds and worked to their respective sections after having been mortised and tenoned. In setting out the sill

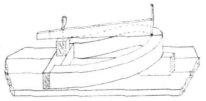

Fig. 1295.—Sketch of Head of Circle-on-Circle Door Frame Shaped to Plan Curve, with Falling Compass Describing Elevation Curve.

and transom to width, work to the sight lines of the head, making the shoulders to the quirks of the beads on it. Get out the jambs from stuff about 8 in. longer than the length between the springing and under side of the sill, to allow for the key tenon at the springing, and true up these to the

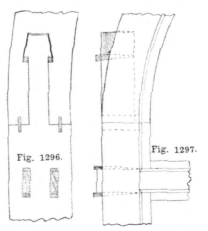

Fig. 1296.

Fig. 1297.

Figs. 1296 and 1297.—Enlarged Details of Joint at Springing of Circle-on-Circle Door Frame.

section shown at Fig. 1298. Set these out from the section, or height rod, for mortising, as shown at Figs. 1296 and 1297, *also for shoulders for key tenon as shown.* The thickness of the stem should be just under one-third of the

width of the jamb, and it will eventually be cut ¾ in. back from the face edge in order to clear the rebate in the head; reference to Fig. 1298 will make this plain. The end of

Fig. 1298.—Enlarged Section of Jamb for Circle-on-Circle Door Frame.

the key head should be tapered, to avoid weakening the head more than can be avoided, and the sides of the keys should be cut to the same bevel as the jambs are worked to. The shoulders should also be cross-tongued and stopped inside, as shown. As

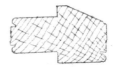

Fig. 1299.—Enlarged Section of Transom for Circle-on-Circle Door Frame.

mentioned, double tenons are used for the *transom*; this is to avoid cutting the fibres at the root of the key, and the insides of the mortices should come in line with the outside lines of the key, and in thickness they should be $\frac{1}{16}$ in. under the half of the remaining wood. Fig. 1299 is a section of the transom;

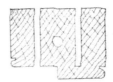

Fig. 1300.—Enlarged Section of Head at Crown of Circle-on-Circle Door Frame.

and Fig. 1300 a section of head at crown. Having set the sight lines of the jambs on the face edge of the transom, measure out and pencil lines representing the thickness of jamb, prick the mortice gauge upon these two lines—namely, the inside and outside of the jamb—and join the points by a straight line ; this will give the sides of tenons. Next gauge the jambs for the rebate and beads. These will be worked after the mortising and tenoning are completed. In wedging up, paint the tenons of the transom and enter them ; glue the keys on the jambs, cramp up, and drive the sill on dry to keep all square ; wedge up the head, then the transom, cramp the sill from the projecting ends of the keys, and wedge up, painting both wedges and tenons. If the work is true and out of winding, a straightedge applied to the sill and transom inside will touch the face of the head all round, and a set-square held to it will fit the soffit in any part. The fanlight will be managed in the same way as the head of the frame, in respect of the plan curve ; finish the fitting by scribing it into the rebate, one piece at a time, fitting the back edge in the same way, then lining round the soffit with a slip of wood the width of the margin. Joint up at the centre, and work the rebates and moulding, set the bottom rail out with square shoulders in both directions, and, when working the head, make the portion below the springing square from it. This will be mortised, the tenon being on the bottom rail ; the joint in the head will have to be glued up at the same time as the rail is wedged.

Circle=on=Circle Swing Doors with Fanlight.

Figs. 1301 and 1302 show a pair of 2½-in. circle-on-circle swing doors in a solid frame, with fanlight above, with a chord rise of 15 in. Details are given in the sectional views (Figs. 1303 to 1307). The doors are half glass, with diminished stiles and marginal bars, the upper part divided into sections with concentric radiating and inverted bars, the spandril corners filled in with sunk and moulded panels ; apron lining on middle rail ; ovolo moulded and hollowed frame, with moulded and denticulated cornice. The fanlight is fixed in the centre of the frame, with concentric and radiating bars. The head of the frame, and the fanlight, must be prepared on a cylinder. These could

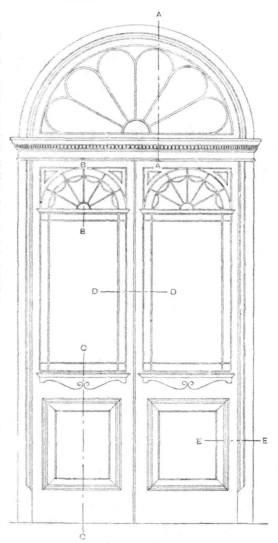

Fig. 1301.—Elevation of Circle-on-Circle Swing Doors with Fanlight.

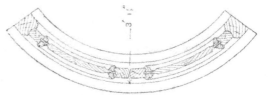

Fig. 1302.—Sectional Plan of Circle-on-Circle Swing Doors with Fanlight.

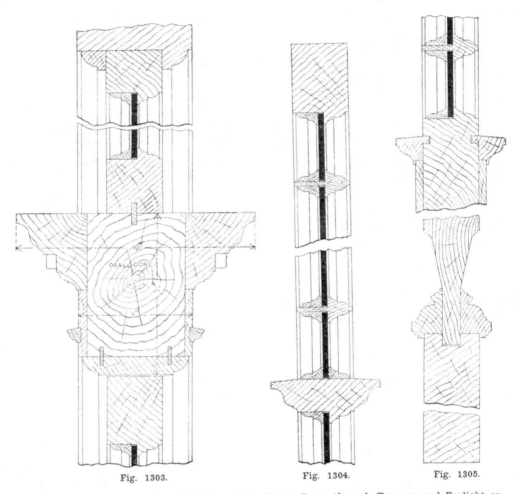

Fig. 1303.   Fig. 1304.   Fig. 1305.

Fig. 1303.—Enlarged Section of Circle-on-Circle Swing Doors through Transom and Fanlight on Line A A (Fig. 1301).   Fig. 1304.—Enlarged Section through Upper Part of Circle-on-Circle Swing Doors on Line B B (Fig. 1301).   Fig. 1305.—Enlarged Section through Lower Part of Circle-on-Circle Swing Doors on Line C C (Fig. 1301).

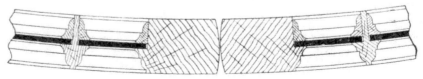

Fig. 1306.—Enlarged Section through Stiles of Circle-on-Circle Swing Doors on Line D D (Fig. 1301).

be got out by the aid of lines, but where a definite curve is required the former method is more accurate and reliable. The material will be cut as shown by the dotted lines, D E F G (Fig. 1308) showing the thickness, and A C H I (Fig. 1309) the length. When

as the upper part of the door, should be fitted on the cylinder. Joint screws should be used where practicable.

The jambs of the door frame are cut from solid material, as is also the head, but the transom is built up, as seen in Fig. 1303,

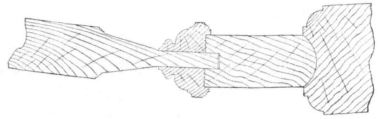

Fig. 1307.—Enlarged Section through Bottom Panel of Circle-on-Circle Swing Doors on Line E E (Fig. 1301).

fitted to the cylinder, the proper curve will be obtained by using a trammel with a rising and falling slide at the striking point,

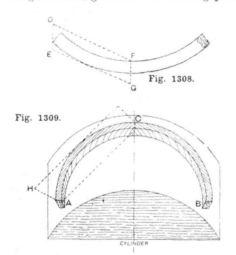

Fig. 1308.

Fig. 1309.

Figs. 1308 and 1309.—Method of Setting Out Fanlight of Circle-on-Circle Swing Doors.

similar to Fig. 1295, the head, etc., being reversed. The bars in the fanlight, as well

with a deal core and hardwood facings. The narrow frieze being prepared from $\frac{1}{2}$-in. material, it may be bent round the sweep. The neck mouldings are prepared from the solid cut to the sweep required. The cornices may be prepared as seen in section, or in separate pieces, and built up to the required size. Veneers are not used in this class of work. The door stiles are cut from solid, straight-grained material, and, needless to mention, bone dry. The rails being rather a quick sweep, they are framed to the stiles with stub tenons and haunchings, the joints being brought tight up by means of $\frac{1}{2}$-in. billiard bolts inserted from the back edge of the stiles. The nuts are let into the rails from each edge into a square mortice to fit the nut, and when in exact position the holes are plugged to prevent their moving. By this method the shoulders may be fitted perfectly with little trouble, and the trouble caused by broken, short-grained tenons is avoided. The circular holes in the edge of the stiles made by the centre-bit, and in which the heads of the bolts are embedded, are filled up with pieces of wood the same grain and colour as the stile, care being taken that the grains follow.

WINDOW SASHES AND CASEMENTS.

Sash Frame.

FOR the present purpose it will be supposed that a window frame and sashes are to be made entirely by hand, no machine-prepared material being used. The sash frame is shown by Figs. 1310 to 1312, and it will also be assumed that a rod has been set out, as explained and illustrated on p. 294, containing the two sections (Figs. 1310 and 1312). It will be noticed that the pulley stiles are drawn in line with the brick reveal, and the head in line with the head of the brick opening. When thus arranged, the size of the opening in the brickwork affords the necessary data for setting out the size of the frame.

Materials.—The first thing to do is to take off the quantities of the stuff required for the job : the lengths of the upright pieces will be taken from the vertical section, and their widths and thicknesses from the plan ; and the lengths of the horizontal pieces will be found on the plan, and their widths and thicknesses on the vertical section. Cut the stuff for the inside and outside linings $1\frac{1}{2}$ in. longer than as shown on the height rod, so that when nailed on they will run over each end. Linings are usually arranged so that two of equal width may be cut out of a 9-in. board ; the head linings should be cut $\frac{1}{2}$ in. longer than the clear length between the pulley stiles ; and the head and sill should be each 2 in. or 3 in. longer than the width of the frame over all. When fitting up, the horns should be left on, as they are sometimes handy for fixing, and, if not required, they can always be cut off. The length of the pulley stiles should be equal to the distance from the top of the head to the weathering of the sill ; and the width should be $\frac{7}{8}$ in. more than the clear width between the linings, which allows for two

$\frac{3}{8}$-in. tongues and $\frac{1}{8}$ in. for shooting. The length of the back lining should be taken from the under side of the head to the weathering of the sill, the width being $\frac{1}{4}$ in. more than from the inside of one lining to the outside of the other. The parting slips should be of similar length, and $1\frac{1}{4}$ in. by $\frac{1}{4}$ in. in section ; the parting beads, also of the same length, should be prepared out of $\frac{1}{2}$-in. by 1-in. stuff ; the two upright inside or guard beads, of the same length, should be $1\frac{1}{8}$ in. wide, and prepared out of $\frac{3}{4}$-in. by $1\frac{1}{4}$-in. stuff. The head and sill beads should be a little longer than the clear width between the pulley stiles, the sill bead to be $\frac{1}{4}$ in. wider to allow for bevelling. For the sashes, cut all the stuff the exact width, so that no labour is lost in shooting off superfluous material. Sashes require at least $\frac{1}{16}$ in. play, and if they are made to exact width, there will be still sufficient material to allow for fitting. The stiles should be $1\frac{1}{2}$ in. longer than required, and the rails $\frac{1}{2}$ in. Cut the bottom rail $\frac{3}{8}$ in. wider than the finished size, to allow for the splay. A careful workman will form the splay to fit the sill before gluing up the sash. Do not overlook the brackets or horns on the stiles of the upper sash.

Preparing the Stuff.—The side and head linings should be faced on their best sides, shot on their best edges, and gauged to thickness for $1\frac{1}{2}$ in. from their face edges ; the remaining portion of the back sides, being hidden in the casing, need not be planed, except when the lining is considerably thicker than required, then it must be either thicknessed the whole breadth or rebated out to the distance of the back of the pulley stile, otherwise the shoulder formed on the edge of the stile by the rebate for tongue would require to be splayed, and much time would be lost in

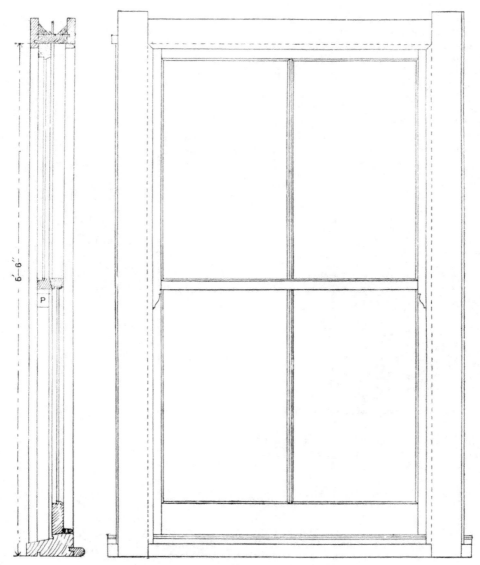

Fig. 1310.—Vertical Section
of Sash Frame.

Fig. 1311.—Outside Elevation.

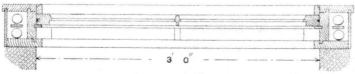

Fig. 1312.—Horizontal Section.

fitting. Put the best stuff on the inside of the frame, as nearly all the outside lining is hidden in the brickwork. The head and the pulley stiles should be planed straight and out of winding, and be gauged to width, including the tongues ; the inside edges should be gauged to thickness, the back sides being left rough. The oak sill is usually supplied of rough wedge-shape section, as shown at Fig. 1317, which facilitates working and economises material. Commence by taking the bottom or the flat side out of winding. Square the widest edge from this, and gauge it to 3 in. thick, as indicated at

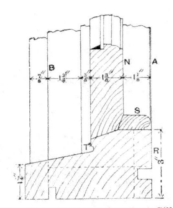

Fig. 1313.—Enlarged Detail at Sill.

A, Fig. 1318. Gauge the width ⅛ in. more than the thickness of the frame over all, in order to allow for subsequent shrinkage. Most shops keep standard patterns of sills, which should be used for marking the ends indicated at Fig. 1317 ; but if one of these is not to hand, carefully mark the section shown on the rod upon each end, gauging and ploughing from the inside to the necessary widths and depths. The best method will be to first work off the top flat square with the face (Fig. 1318), then to set the plough to the inside of the parting bead (see Fig. 1319), and run a ¼-in. groove to the depth of sinking ; gauge the outside edge 1¾ in., and work off the weathering with a badger plane. Next gauge the width of the flat, set the bevel to the sash slope, and work off to the bevel ; then work the throating T with either a No. 1

round or a proper throating plane. Plough in the two grooves for the window nosing and water bar, the former ⅜ in. by ⅜ in.,

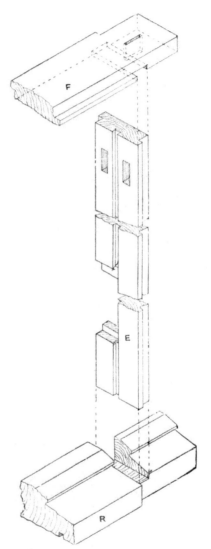

Fig. 1314.—Housings at Head and Sill.

the latter ¼ in. by ½ in., completing the planing up as shown at Fig. 1320. The position of the water bar being taken from the stone sill, it often comes in line with the outer edge of the pulley stile (see Fig. 1310). The back linings and parting slips need not be wrought. The parting beads should be stuck on the edge of a so-called

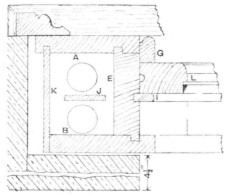

Fig. 1315.—Enlarged Detail through Pulley Stile, etc.

$\frac{1}{2}$-in. board with a $\frac{7}{16}$-in. bead plane, gauged to $\frac{7}{8}$ in. wide, and mulleted to $\frac{3}{8}$ in. tight. The guard beads can be worked on the face of a $1\frac{1}{4}$-in. board with a $\frac{5}{8}$-in. bead plane as shown at Fig. 1321, then sawn just outside

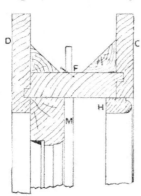

Fig. 1316.—Enlarged Detail at Head.

the quirk and planed to it, or on the edge of a $\frac{3}{4}$-in. board, as indicated at Fig. 1322, and then cut off $1\frac{1}{8}$ in. wide, and planed to the quirk for thickness.

Setting Out the Frame.—Lay the sill on the width rod with the face upwards and the face edge inwards. The sill is shown pro-

jected above the rod at Fig. 1323. Square up the inside lines or faces of the pulley

Fig. 1317.—Rough Sill with Template Applied.

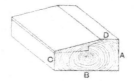

Fig. 1318.—Sill with Surfaces A B C and D Worked.

Fig. 1319.—Sill Ploughed and Set Out for Weatherings.

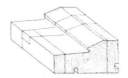

Fig. 1320.—Sill Worked, and Set Out for Housings and Linings.

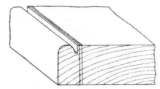

Fig. 1321.—Preparing Guard Bead on Flat of Board.

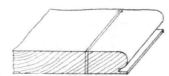

Fig. 1322.—Preparing Guard Bead on Edge of Board.

stiles, and square the lines over the top and bottom, indicated at A and B in Fig. 1323. To facilitate marking on the weathered side, make a block square or template of a piece of stuff. To do this, plane up a across the top of sill by placing the template against the shoulder lines A and E and scribing. Gauge up from the bottom of the sill for the depth of the sinking so that it will come ½ in. below the lowest point of the

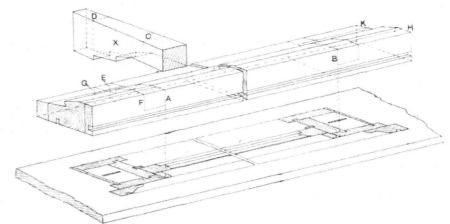

Fig. 1323.—Sill Set Out from Rod, and Application of Template A.

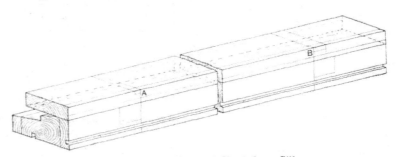

Fig. 1324.—Setting Out Head from Sill.

Fig. 1325.—End of Head Mortised for Parting Slip and Pared to receive Linings.

piece of deal 10 in. long, 5 in. wide, and thick enough to plane up, tapering from 2 in. to 1½ in. in the breadth of the sill as indicated at C and D in Fig. 1323. Next rebate one side to fit the section of the sill as shown at Fig. 1321. With this the thickness of the pulley stiles plus the wedging can be marked weathering—see the dotted line between E and G (Fig. 1323). Gauge the sinkings for the linings at each end, as shown at H and K (Fig. 1323), the outside one being set equal to the width A to B. In working, chase the cross sinking first, then rip out the lining sinkings as shown by Fig. 1314.

Setting Out Sash Frame Head.—To set out the head, place the sill on it and mark the face lines of the pulley stile, as A and B (Fig. 1324); square over the face, and gauge for the housing $\frac{3}{8}$ in., which may be made

tongues should be cut off at each end, as shown at Figs. 1324 and 1325, for the linings to bed.

Setting Out Pulley Stiles.—Set out the pulley stiles from the height rod, and mark

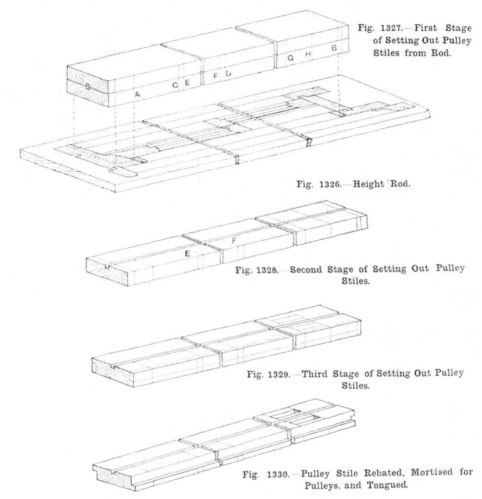

Fig. 1327.—First Stage of Setting Out Pulley Stiles from Rod.

Fig. 1326.—Height Rod.

Fig. 1328.—Second Stage of Setting Out Pulley Stiles.

Fig. 1329.—Third Stage of Setting Out Pulley Stiles.

Fig. 1330.—Pulley Stile Rebated, Mortised for Pulleys, and Tongued.

to receive a $\frac{1}{2}$-in. tongue as shown at A, or the full thickness of the pulley stile as at B (Fig. 1324). Mark off a $\frac{1}{4}$-in. by $1\frac{3}{4}$-in. mortice at each end, $\frac{1}{2}$ in. behind each pulley stile in the centre of the width for the parting slip. Gauge a $\frac{3}{8}$-in. tongue on each edge, the inside one on the back and the outside one on the face, as shown at Fig. 1324. These

off the sight lines of the head and sill. Mark off the lengths for the housings at each end, also for pockets and pulleys, as shown, run down gauge lines for the tongues, and plough the groove for the parting bead. The bead is not in the middle; its position will be found on the rod (see Fig. 1310). Next finish cutting out the pocket, which is

18

usually about 6 in. up from the sill and about 12 in. long on the inside portion of the stile ; the length will be determined by the height of the sashes. Square these

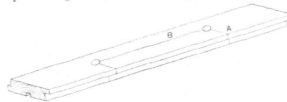

Fig. 1331.—Back of Pulley Stile Bored and Sawn for Pocket.

lines over from the inside edge to the ploughed groove E F (Fig. 1328). Mark two other lines ½ in. within them for the rebates, and square these over on the back side. In setting out the mortices for the pulleys, note must be taken of the particular kind in use. The top of the plate is usually kept 1½ in. down from the head, and the mortice is made to fit the case tightly. To ascertain the position, mark the thickness of the sash on each side of the parting groove, and gauge in the centre ; pair the pulley stiles, and strike the lines over and gauge as indicated at Figs. 1328 and 1329. The only setting out needed for the linings is the gauging for the grooves. The position of the pulley stiles, head, and back lining will be found on the width rod (see also Figs.

shown by Fig. 1311, 5 in. down and ⅝ in. on, but leave the bottom end until fitting on.

Working the Stuff.—The method of working the sill has already been explained ; it should always be prepared before setting out the frame. In preparing the head, the housing for the stiles should be worked first, then mortices cut for the parting slips, and the edges rebated for the tongues. In preparing the pulley stiles, square the ends, mortise for the pulleys (see Fig. 1330), and plough the parting groove ¼ in. by ¾ in. In cutting the pocket, if it is desired to utilise the removed material for the pocket piece, make a fine cut with a dovetail saw halfway through at the lower end, and a similar one at the top end, but undercut as shown at A in Fig. 1331, and in Fig. 1332. Turn over the stile and bore a ¾-in. hole at each end exactly where the rebate lines cross the parting groove. Stop the hole when it reaches the groove, and run the saw down on the lines halfway through, as shown by Fig. 1331. With a pad-saw cut a fine line halfway down the parting groove as near the outside as possible (see B, Fig. 1331),

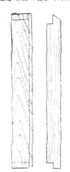

Fig. 1332.—Front and Edge Views of Pocket Piece

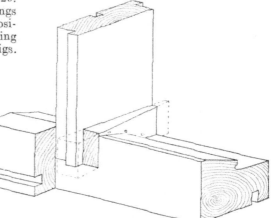

Fig. 1333.—End of Pulley Stile Wedged and Nailed in Sill.

1315 and 1316). The back lining is ploughed into the inside lining and nailed on the edge of the outside lining. The top ends of the outside linings may be cut and mitered as

but do not knock the piece out. Form rebates as on the head, and fit in the pulleys. Before screwing these in, take a shaving with a smoothing plane off the face of the stile. In preparing the linings, plough the grooves ⅜ in. by ⅜ in., as shown by the illustration, cut and mitre the outside linings in its proper housing, and wedge up. See that it beds firmly on the bottom of the housing, and drive a couple of 2½-in. nails

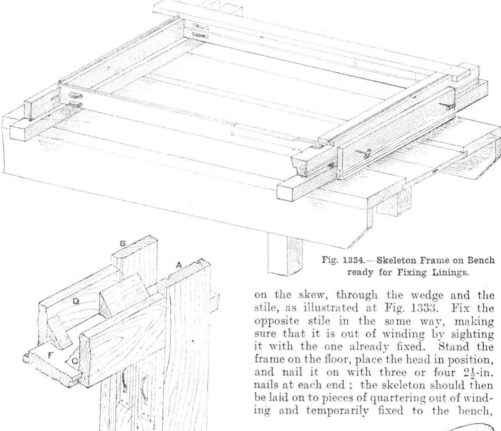

Fig. 1334.—Skeleton Frame on Bench ready for Fixing Linings.

on the skew, through the wedge and the stile, as illustrated at Fig. 1333. Fix the opposite stile in the same way, making sure that it is out of winding by sighting it with the one already fixed. Stand the frame on the floor, place the head in position, and nail it on with three or four 2½-in. nails at each end ; the skeleton should then be laid on to pieces of quartering out of winding and temporarily fixed to the bench,

Fig. 1335.—General View of Top Corner of Frame, showing Method of Fixing Head Linings.

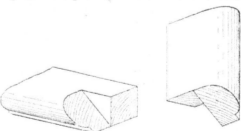

Fig. 1336.—Method of Mitering Guard Beads.

at top, and smooth the bottom ends for 12 in. up.

Fitting Together Sash Frame.—Place the sill on the bench ; put one of the pulley stiles

the inside being uppermost. Fix a piece of quartering or arras-rail on the bench end, and twist a couple of screws or gimlets through a strip nailed to the quartering into

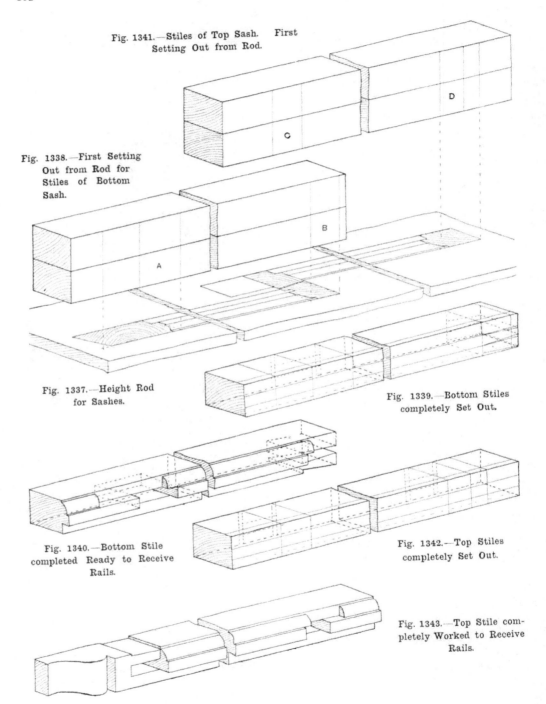

Fig. 1341.—Stiles of Top Sash. First
Setting Out from Rod.

Fig. 1338.—First Setting
Out from Rod for
Stiles of Bottom
Sash.

Fig. 1337.—Height Rod
for Sashes.

Fig. 1339.—Bottom Stiles
completely Set Out.

Fig. 1340.—Bottom Stile
completed Ready to Receive
Rails.

Fig. 1342.—Top Stiles
completely Set Out.

Fig. 1343.—Top Stile com-
pletely Worked to Receive
Rails.

the bottom of the sill as illustrated at Fig. 1334. Test the frame for squareness with a rod, pushing the skeleton in either direction until the lengths between the opposite corners are equal, then lightly drive a

$1\frac{1}{2}$-in. nails. Cut one end of the head lining square, put it in place, mark the length, and square off slightly full ; this will make a good shoulder. Keep the linings flush on the back side, and drive a nail through the

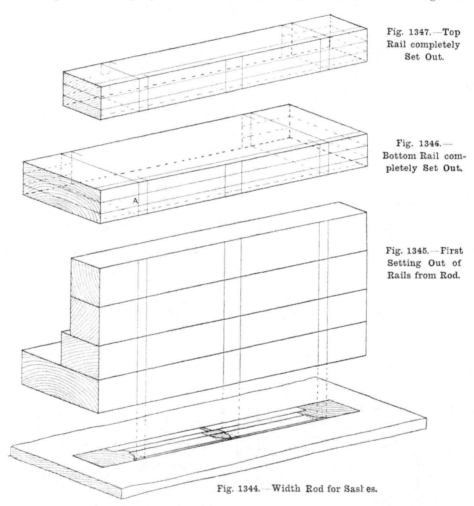

Fig. 1347.—Top Rail completely Set Out.

Fig. 1346.— Bottom Rail completely Set Out.

Fig. 1345.—First Setting Out of Rails from Rod.

Fig. 1344.—Width Rod for Sashes.

couple of nails through the back of the head into the quartering. Cut off the projecting ends of the wedges, the tongues on the head beyond the stiles, and the tongue off the side of the pocket piece. Give the lower end of the pocket piece a smart blow with the hammer from the back ; this will break it at the rebate. Next nail on the side linings with

back edge as shown by Fig. 1335. Fit in the beads tightly ; the head and sill beads should not be mitered right through, but stopped halfway, and rebated, as shown by Fig. 1336 ; drive one brad in each to keep it in place. Cut the linings off flush with the sill, and smooth up. Turn the frame over and test whether it is square—if

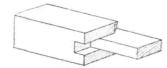

Fig. 1348.—End of Top Rail, showing Moulded Edge and Scribing.

Fig. 1349.—End of Top Rail, showing Haunching.

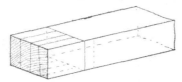

Fig. 1350.—End of Meeting Rail of Bottom Sash Set Out.

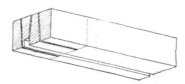

Fig. 1351.—Tenon Sawn and Ploughed.

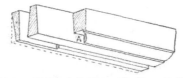

Fig. 1352.—End of Bottom Meeting Rail completed (scribed at A) for fitting into Stile.

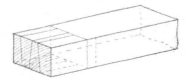

Fig. 1353.—Top Sash Meeting Rail Set Out.

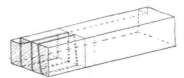

Fig. 1354.—Tenons Cut and Rebated.

not, make it so; then put in the parting slips, keeping them 1 in. short of the sill, and drive a nail through the top end, as shown at F (Fig. 1316); put on the outside linings, trimming off the sill to the level of the pulley stiles if necessary; fix the back lining with 1-in. nails; nail the head linings as before; cut off the ends and smooth up. Only ⅝ in. of the frame is seen on this side, so not much trouble need be taken, except with the sill. Stand the frame up and glue in the angle blocks on the head about 3 in. apart, care being taken to have one over each shoulder (see Fig. 1335). The frame is now finished, and may be stood aside while the sashes are being prepared.

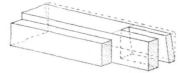

Fig. 1355.—Portion of Rail completed.

Setting Out Sashes.—Lay a stile on the rod as shown in Fig. 1337, in which the frame has been omitted for clearness, and square up the sight lines, pair the stiles, and set out mortices for top and bottom rails half the width of the stuff, or a little more (see A and D, Figs. 1338 and 1341), and the joints for the meeting rails full width, which will be dovetails instead of mortices (see B and C, Figs. 1338 and 1341). Square the line of the bracket over on the top stiles, and mark two lines on the back edge of the stiles 10 in. and 14 in. respectively from the top ends for the line groove and the knot hole. The remainder of the setting out and completion of the stiles will be understood from Figs. 1338 and 1343. Set out the bars from the stiles, allowing extra at each end for the sticking; in the case of 1½-in. sashes this would be $\frac{3}{16}$ in. Next set out the rails from the width rod (Fig. 1344), but do not overlook the sticking; place the four together as shown at Fig. 1345; square the sight lines, and continue these over the back edges of the top and bottom rails for the haunching, as shown at Figs 1346 and 1347. Set a mortice gauge to the size of the square shown, which, together with the sticking, will be ⅜ in.

Allow for the moulding $\frac{1}{2}$ in. on, and run the gauge down all the face edges, with the exception of the top meeting rail ; also set a gauge to $\frac{3}{16}$ in., and run it on both faces of all the stuff. The entire operation, from first marking off the rod to the completion of the top and bottom rails, is shown by Figs. 1345 to 1349. Mark a dovetail as shown by Fig. 1340 on the upper ends of the bottom stiles, and draw or trace the profile of the brackets on the top stiles.

Working the Sashes.—Mortise the stiles and tenon the rails, mortising the backs of these for the haunching (see Figs. 1348 and 1349) ; also cut the dovetails in the meeting rails (Figs. 1351 and 1354) ; then work the rebates with a sash fillister, and stick the moulding A (Figs. 1349 and 1352). Cut the shoulders of the tenons, and scribe them to the section of the moulding. Cut away the moulding on the stiles at the mortices to within $\frac{1}{2}$ in. of the sight line, but let the " square " remain on to form the haunching, as shown by Figs. 1340 and 1343. When scribing the bars, remember that the end of the top one which fits in the meeting rail is to be left square, as there is no moulding on this rail as shown at Fig. 1356. The moulded horns or brackets at the ends of the stiles should be worked, cramping together one or more pairs of stiles, with a piece of waste wood against the last stile to prevent the edge from breaking (see Fig. 1357). The meeting rails should be gauged to $\frac{3}{16}$ in. wider than the thickness of the stiles, and bevelled in pairs as shown by Figs. 1351, 1352, and 1355.

The bevelled sides are not to be cut at the shoulder, but must run over the faces of the stiles, and, for preference, are sunk in $\frac{1}{8}$-in. dovetailing as indicated at Figs. 1340 and 1343. Plough a $\frac{1}{2}$-in. groove in the back of each stile, down to the 10-in. line for the cord. Bore a $\frac{7}{8}$-in. hole at the

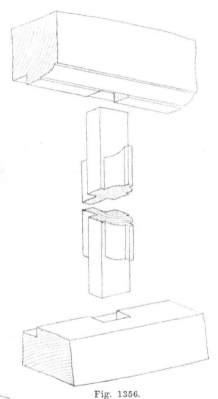

Fig. 1356.

Fig. 1356.—Joints of Bar with Top Rail and Meeting Rail.

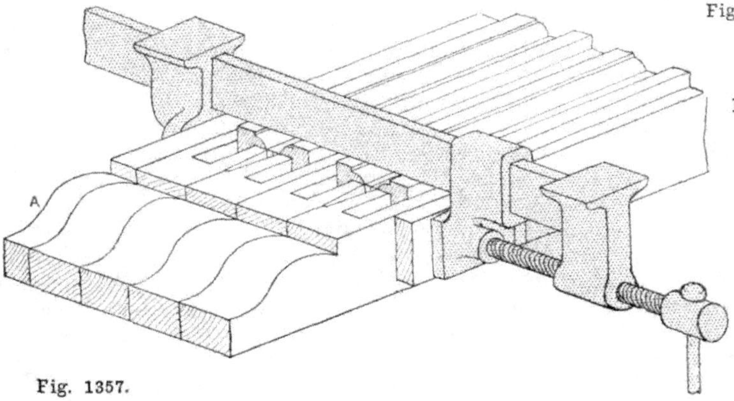

Fig. 1357.—Method of Moulding Horns of Stiles to Top Sash.

Fig. 1357.

14-in. line, $\frac{3}{4}$ in. deep, and meet it with a $\frac{3}{8}$-in. hole bored down from the bottom of the plough groove. Fit the rails into the stiles, and the bars into the rails. Take a shaving off all the inside edges and the meeting ends of the stiles, also the bottom side of the top meeting rail, and knock the sashes together preparatory to gluing them up. Lay a pair of sash cramps on the bench so that they will pinch either on or outside

sashes, cut off the horns, shoot the top rails straight, and fit the sashes into the frame. Take the side beads out, also the parting beads, cut a rod $\frac{1}{16}$ in. shorter than the clear length between the pulley stiles, and try it on the top sash; if it is too wide, reduce the stiles equally to the width of the rod. Cut the overhanging part of the meeting rail back for $\frac{5}{8}$ in. flush with the stile, try the sash in the frame, and put in the parting

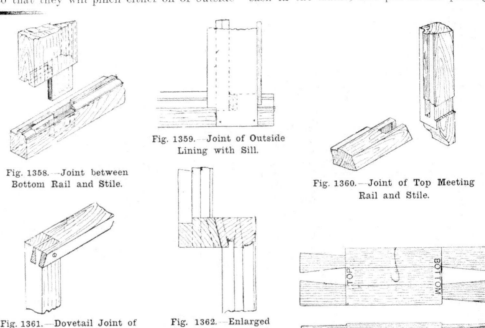

Fig. 1358.—Joint between Bottom Rail and Stile.

Fig. 1359.—Joint of Outside Lining with Sill.

Fig. 1360.—Joint of Top Meeting Rail and Stile.

Fig. 1361.—Dovetail Joint of Bottom Meeting Rail and Stile.

Fig. 1362.—Enlarged Section through Meeting Rails.

Fig. 1363.—Face of Template.

the rails; if inside they will bend the stiles. An alternative way is to lay two pieces of quartering out of winding on the bench, lay the sashes on these, and use cramps above. Knock the stiles off, glue the tenons of the bars, knock the rails on, place the ends of the tenons into the mortices, glue both mortices and tenons, and cramp and wedge up, a rod having been previously used diagonally in order to make sure that the work is quite square. Put a screw through the bottom meeting rail from the back side, and wedge the bars straight. If they are then crooked, drive the wedge most on the hollow side. When the glue is dry, clean up the

beads. The sash should run freely, with about $\frac{1}{32}$ in. clearance all round; fix it in place with a prop. Shoot the bottom sash to the width of the rod, and try it in the frame. Set a pair of compasses to the width between the top sides of the meeting rails (or a shaving less, in order to allow for shrinkage in the bottom rail), and scribe along the outside of the bottom rail from the bevel on the sill. Mark the bevel from this line on the stiles, and work the sash accordingly. Run a weather groove along the bottom edge, and bevel off the face to fit the bead. Put the sash in place, and replace the guard beads. Figs. 1358 to 1362 show details of the joints in the sashes. The letter references in Figs. 1310 to 1335 are : A, inside lining ; B, outside lining ; C and D, head lining ; E, pulley stile ; F, frame head ; G, guard bead ; H, head bead ; J, parting slip ; K, back lining ; M, parting bead ; N,

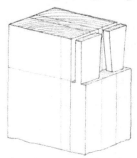

Fig. 1364.—Template applied to Stile of Bottom Sash.

sash rail ; R, sill ; S, sill bead ; T, throating. The description of the construction of the sash frame and sashes is now complete, but the next four paragraphs will discuss in detail some of the operations briefly described above, and will suggest alternative methods.

Dovetailing Meeting Rails of Sashes.

Several methods are adopted in making the joints between the meeting rails and stiles of sashes, and perhaps the most common form of joint is that of the mortice and tenon, the stiles being allowed to extend about 3 in. beyond the meeting rails, and the projecting horns being worked and moulded as brackets. This method, however, is seldom resorted to in first-class work, the joint being either mortised and tenoned in the ordinary way, or else dovetailed. In the former case—no matter how good the joint may be—as the mortice is cut with a saw in the same way as the tenon, there is always a risk of the joint becoming loosened by the frequent use of the sash ; this trouble may be entirely obviated by the adoption of the dovetail joint already illustrated and described. Assuming that the stuff for the sashes is planed up and set out, it will be necessary to prepare a template as shown in Fig. 1363 ; this should be made out of a

piece of hardwood, about 6 in. long, of the same width as the sash stuff—which may be taken at $1\frac{7}{8}$ in., and $\frac{3}{16}$ in. thick. Square each end, and, distant therefrom the thickness of the meeting rail, each end should be squared round and sunk as shown in the side view given in Fig. 1365. The mortice gauge, having been set to the work, should now be run along each face from its tried-up edge, and the dovetails marked and cut, giving them about $\frac{1}{8}$-in. bevel. One end should be legibly marked " top " and the other " bottom," and the tried-up mark should be placed on both faces and to one edge. To use the template, proceed as follows : Take the bottom pair of stiles—which should be about $\frac{1}{16}$ in. too long at the top end—lay them together on the bench with the tried-up edges outwards in each case, set to one of them, on the setting-out line of the under side of meeting rail, that portion of the template squared across and marked " bottom," so that its tried-up edge corresponds with the tried-up edge of the pulley stile, and then mark the dovetails with the setting-out knife as indicated at Fig. 1364. Serve the other stile the same by turning the template over on to it, then take the meeting rail of the bottom sash, fix it in the bench screw with its face-mark away from the bench, and set

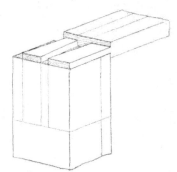

Fig. 1365.—Template applied to End of Meeting Rail of Bottom Sash.

the end of the template marked " bottom " on the end of the rail ; make their tried-up edges agree, and then mark the end of the rail as shown at Fig. 1365. If these ends are previously rubbed with chalk it will be found an advantage.

The opposite end of the same rail is marked in a similar way by turning the template over. The top stiles and meeting rails are marked in a similar manner, using the end of the template marked " top "; but the tried-up side of the template is used from the rebated side of the rail and stile instead of from the face or tried-up side. A mortice gauge set to the tops of the dovetails should now be run across the ends of the stiles, and should also be set for and run along the tops and under sides of the rails ; these lines are required to cut to, and the cutting should be done in the same manner as for ordinary

meeting rail running through for the purpose of bevelling ; while B shows a projection of the stile and meeting rail of the top sash, with the moulding cut away on the stile and the meeting rail running through on the inside for bevelling purposes. In putting the sashes together, these joints should be pinned ; and if these instructions are followed, the difficulties experienced will be few and slight.

Cutting Pockets in Sash Frames.

There are several methods of cutting pockets in sash frames, and one has been

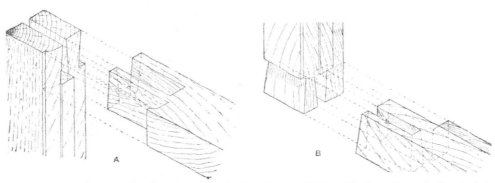

Fig. 1366.—A, Dovetailed Joint between Meeting Rail and Stile of Bottom Sash ; B, Dovetailed Joint between Meeting Rail and Stile of Top Sash.

mortising and tenoning—namely, outside the line for the tenon or dovetail, and inside for the mortice. The mortised portions on the stiles and rails should be cut out, and the shoulders should be partly cut in previous to rebating and moulding. This completed, it will be found that at the tops of the bottom stiles a small portion of the square will be left which will require chiselling off ; but on the bottom of the top stiles the whole of the mould will require removing up to the setting-out line, as well as a small portion of the square left on the rebate side. Of course, it will be noted that the shoulders on the rails require to be longer by the distance of the rebate and of the moulding. A reference to the isometrical views (A and B, Fig. 1366) will explain the processes : A showing a projection of the bottom stile and meeting rail, with the moulding left on the stile, the meeting rail scribed round it, and the back of the

described at pp. 409 and 410. An old-fashioned method is shown by Figs. 1367 to 1369, which represent the pocket as being cut in the centre of the pulley stile. This was at one time the general method, but it is seldom resorted to now, for the following reasons :—The hole, when cut in the centre of the stile, necessitated the use of a new piece of stuff to form the pocket piece, and, in varnished work, the new piece had to match in grain as nearly as possible to the pulley stile. This involved loss of time, first in selecting the piece, and afterwards in fitting it. When cords were renewed, the removal of the pocket piece broke the surface of the paint, and its outline looked ragged and unsightly when replaced, a portion of it being always exposed to view. Figs. 1370 to 1372 represent a method now generally adopted. It is superior to the old, because it takes less time, and is easier of construction. The original piece cut out is re-used to close the pocket; and, when the lines are renewed, any unsightliness occasioned by the removal of the pocket piece is concealed by the bottom sash. Let it be assumed that the frame in hand is of ordinary construction, and that, with the rest of the stuff,

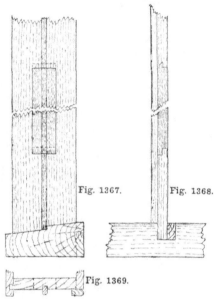

Fig. 1367.   Fig. 1368.

Fig. 1369.

Figs. 1367 to 1369.—Part Plan, Elevation, and Section of Pulley Stile, showing Sash Frame Pocket in Centre.

the pulley stiles are ready tongued, grooved, etc. Lay a couple on the bench with the tried-up marks face to face (so as to get them in pairs) and level at both ends; measure and mark across the tried-up edges, with the setting-out knife and square, a line 1½ in. from the bottom end. This is the depth that the stiles are sunk into the sill (see A, Fig. 1371); 6 in. above this, mark line B; 18 in. above it mark line C (Fig. 1370). This gives a total of 2 ft. from the top of sill to the top of pocket, and, as the weights for this class of sash seldom exceed 19 in. in length, ample depth is thus allowed for getting them in or out. Assuming that the size of the sash frame is 5 ft. 6 in. by 3 ft. 6 in., and that the sashes have a marginal bar and are glazed with 21-oz. sheet glass, the weights for the bottom sash should be

about 9 lb. each, and those for the top sash 9½ lb. each. The weights in the first case would be about 18 in. long, and in the latter, 19 in.; the thickness of each would be 1¾ in. Set the bevel to the angle shown at D D in Fig. 1371 (which, measured vertically, will rise about 1½ in.), and mark the stiles. Square these lines across both back and front (shown in dotted and full lines at E on elevation, Fig. 1370) as far as the outside of the groove of the parting bead; set a cutting gauge to this distance, and gauge the backs as deep as the cutter will allow. Fix the stiles one at a time in the bench-screw, and, with a fine tenon or carcase-saw, accurately saw down the bevelled cuts shown at D D in section, taking care not to go deeper than the width of the plough-groove. Lay them on the bench, and, with a mallet, pass a pocket chisel (resembling

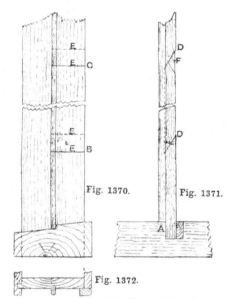

Fig. 1370.   Fig. 1371.

Fig. 1372.

Figs. 1370 to 1372.—Part Plan, Elevation, and Section of Pulley Stile, showing Sash Frame Pocket at Side.

an ordinary 1½-in. firmer chisel, but having a thin, well-tempered blade) a few times up the length of the pocket, the outside edge of the parting-bead groove forming the gauge. Do not strike the chisel too deeply, as it is not the intention to go through. This done,

give the pocket piece a tap or two with a
mallet, and it will break off. The ragged
strip of timber that formed the junction can
be pared off with the chisel on the stile and
planed off on the pocket piece. With the
bevel at the same angle as before, but reversed, mark and cut off from the top or
apex of the pocket piece the cleat F (Fig.
1371). Fix the pocket piece in its place
with a fine screw at the bottom (as shown),
say a No. 6, but do not countersink the
head, because, if the frame is painted, this
would be puttied up and would be difficult to
find when wanted. Reverse the stile, and
fix the cleat which has been cut off the
top of the pocket piece by forcing it down
into its position a trifle inside the throat of
the pocket at the top, as shown. Glue and
securely sprig it, then put it on one side to
dry. The pocket piece will be found to
stand well flush with the face of the stile, as
shown in dotted lines. Clean it off (temporarily removing screw), and run the plough
up the groove to sink it to the same depth
as that on the pulley stile. If this is not
done, the parting bead cannot be driven
home level its entire length, as it should be.

More Elaborate Method.—Figs. 1373 to
1375 represent a somewhat similar, but more
elaborate, method of constructing pockets
—suitable for better class work. The piece
is cut out and re-used in the same manner
as in the ordinary method, but the top and

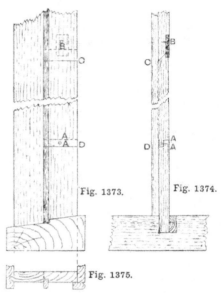

Fig. 1373.   Fig. 1374.

Fig. 1375.

Figs. 1373 to 1375.—Part Plan, Elevation, and
Section of Pulley Stile, showing Improved Method
of Cutting Pocket in Sash Frame.

bottom of the opening, etc., are dealt with
differently. The top cut is the same, but

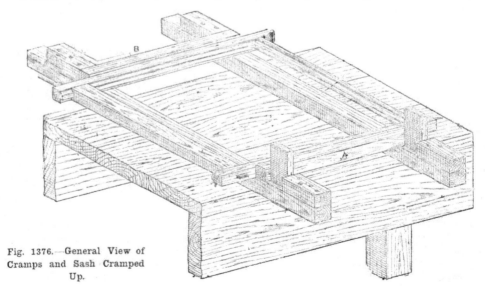

Fig. 1376.—General View of
Cramps and Sash Cramped
Up.

less acute, and, instead of a bevelled saw cut at the bottom, two transverse cuts A A (Figs. 1373 and 1374) are made, one at the back and one at the front, ½ in. apart. When the piece is knocked out, this splits vertically down the centre between the two cuts, and must be left so. At the top of the opening at back, and in the centre, let in a piece of oak B 2 in. long, 1 in. wide, and ¼ in. thick, glued and screwed in as shown. This forms a cleat against which the pocket piece is checked ; in the latter a chase is cut

fitting and fixing the hardwood stop B shown in Fig. 1374.

Wedging Up Sashes.

A method of wedging up sashes is described on p. 416. A simple improvised

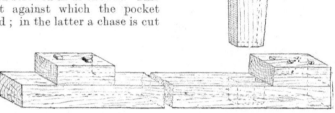

Fig. 1377.—Enlarged Detail of Cramp and Wedge for Wedging Window Sashes.

to correspond. It is important that no more should be cut out to form this chase than is absolutely necessary, so as to ensure good joints at C and D. Let the saw cuts A A be slightly pitched towards the front, as this will facilitate fitting in the pocket piece. Screw in when finished. In common work a pad-saw is used in place of the pocket chisel, the saw being started from a centre-bit hole in the corner. This method requires more care in fixing, and is less satisfactory than the use of the pocket chisel. Another method, and one that is frequently used, is to cut the top end on both sides with a pocket chisel or dovetail saw

cramp used for this purpose is illustrated in Fig. 1376. It can be made out of quartering about 3 in. by 3 in. Of course, smaller or larger sizes can be used, according to the sizes of the sashes to be wedged up. At one end a 1-in. or 1½-in. mortice must be made right through as shown at Fig. 1377, where it will be seen that the back of the mortice is vertical, and the front part splayed so as to fit the wedge. Then a cleat should be nailed or screwed on, and the end of it that comes in contact with the wedge should be splayed at the same angle as the

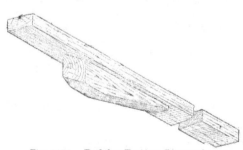

Fig. 1378.—Rod for Testing Diagonals.

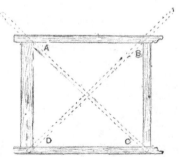

Fig. 1379.—Application of Rod for Testing Diagonals.

as shown at Fig. 1381. The joint A, formed by breaking away the pocket piece from the stile, acts as a stop, and so obviates the additional cost which would result from

mortice. An important point to keep in view is that the back of the wedge which fits against the piece that goes against the stile should all the while keep vertical while being driven down. At the other end of the piece of quartering a cleat should be nailed on. It will be found best to make the wedges of hardwood. The distance between the back of the cleat and the back of the wedge should be sufficient to allow of the insertion of a piece of stuff at each end, to fit up against each stile as shown at A and B (Fig. 1376). These pieces should be about 1 in. off at each end from the tenons and mortices. The object of these pieces is to prevent the cramps bending in the stiles. If it is desired to glue up only two or three sashes, one cramp will be sufficient. If there are several, it would be decidedly better to have two cramps, as shown at Fig.

is shown at Fig. 1378, and the method of using it is indicated by the dotted lines at Fig. 1379. Say diagonal A C is longer than that at B D, then the end of the stile at C or at A must be struck with the hammer until the diagonals are found to be equal. The next thing will be to glue the ends of the wedges. This is usually done by dipping the ends in the glue-pot. They should then be inserted in the holes which have been made for them, then the inner wedges should just be driven in hand-tight. Then the four outer wedges should be driven well home, and finally the inner wedges driven home.

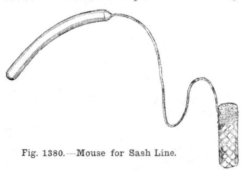

Fig. 1380.—Mouse for Sash Line.

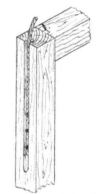

Fig. 1381.—Attaching   Fig. 1382.—Attaching
Cord to Sash by a Knot. Cord to Sash by Nails.

1376 ; because when gluing up, the four joints have to be glued at one time, and in using one cramp, whilst two joints are being wedged up, the glue is setting in the other two, whereas by using two cramps the four joints can be cramped up at once. In wedging up, the cramps must be fixed across the bench the right distance apart, and so that they are quite out of winding. The sash should be face side down across the cramps, then the stiles should be knocked down halfway off the tenons. Then the tenons and shoulders on that side should be glued. Care must be taken not to get any glue on the ovolo moulding, or to put it so near that the glue squeezes out on to the moulding. The sash should be quickly turned over, so that the face side is up, and the tenons and shoulders on that side glued. Then the joints should be driven home, and the pieces placed against the sash stiles and the wedges tightened. Next test the diagonals with a rod. The general form of this rod

Replacing Broken Sash=line.

Materials, etc.—Replacing a broken sash-line, although apparently a very easy job, is not really so. It is assumed that the disabled sash is the top one. First procure a sufficient length of sash-line ; this is sold in " knots " of twelve yards, and the proper kind is a plaited cord formed with four strands of hemp fibre, and about $\frac{3}{8}$ in. in diameter. A few town clout nails will also be wanted, and a hammer, a 1-in. or $1\frac{1}{2}$-in. chisel, pair of pincers, bradawl, punch, and a " mouse." The last is formed by rolling a small piece of sheet lead, about $2\frac{1}{2}$ in. wide, into a cylinder, in which is embedded the end of about a yard of fine cord ; and its purpose is to pass over the sheave of the pulley, taking first its " tail," and then the sash-line which is attached to it, down the inside of the frame until it reaches the pocket or hole where it can be fastened to the weight. The "mouse" is shown bent in Fig. 1380, and attached to the cord, ready for passing over the pulley; the tail in the illustration is relatively short.

Removing Beads.—First carefully remove the guard and parting beads which keep the sashes in the frame, avoiding bruising the edges of the frame with the chisel, which should be a wide one. If possible, avoid bending or breaking the nails which hold the bead in; if this operation is managed properly, the nails may be drawn out with the beads, and may be re-inserted in the original holes without driving the heads back. Insert the chisel at about the middle of the length of the side bead, and gently prise it off, working gradually towards each end, until all the nails have started, then pull firmly with one hand at the middle of the bead, so that it may be bent out in a curve towards the opposite side. Then insert the chisel between the sash and the bottom end of the bead, and cause the latter to slip past the mitre of the sill bead, when it will spring out into the hand. The end nails will probably be bent slightly in this operation, and should be straightened with the hammer on a spare piece of wood. If the sashes fit well, the opposite bead must also be removed in a similar manner; but usually there is sufficient play for the sash to be drawn out diagonally. The broken cord being in the top sash, the bottom one must be got out of the way, and with a strong assistant and a little scheming it may be possible to hold it out of the way without removing the cords; but generally one at least of these must be removed, and then the sash can be turned aside horizontally, hanging by the other.

Removing and Attaching Cords.—One method of attaching cords to sashes is shown in Fig. 1381; here the cord lies in a plough groove in the back of the stile, and its end, passing through a hole made at the bottom of the groove, comes out into a larger hole lower down, where it is tied into a knot to prevent it slipping back. To remove this fastening, all that is necessary is to slacken the line, draw out the knotted end, and untie it with the pincers, when the end can be drawn through. A commoner method of

fixing is shown in Fig. 1382; here a plough groove is made, extending halfway down the stile, and the cord is simply nailed in with clouts. To remove these, grasp the cord with the pincers as close as possible to the nail, lever it steadily out, and proceed with the others in the same way; having released the cord, with an assistant holding the sash, tie a slip-knot in the cord and let it run up to the pulley. The bottom sash having been removed, take off the parting beads. Begin at the bottom of these, driving the chisel in at the side of the bead

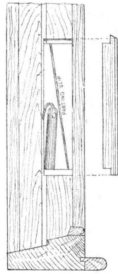

Fig. 1383.—Removing Pocket Piece of Sash.

gently, and, as it is levered down, tap the edge of the bead on each side with the hammer, when it will spring out of the groove; then carefully ease it past the shoulder on the meeting rail of the top sash, when it can be drawn out entirely: noting which side the beads come from, so as not to transpose them when replacing. Draw down the top sash and remove it from the frame, when the end of the broken cord can be extracted and the sash left hanging by the other cord for the present. The next step is to remove the pocket piece— a kind of trap-door in the pulley stile, through which the weights are inserted; it will be found either in the middle of the

pulley stile, just below the position of the
meeting rails, or between the inside edge
and the parting groove, as shown in Fig.
1314.   Insert a bradawl near the lower end
of the pocket piece, and, pulling it firmly,
tap sharply at the same time with the hammer on the face of the pulley stile, and the
piece will be released.   Next slip the hand

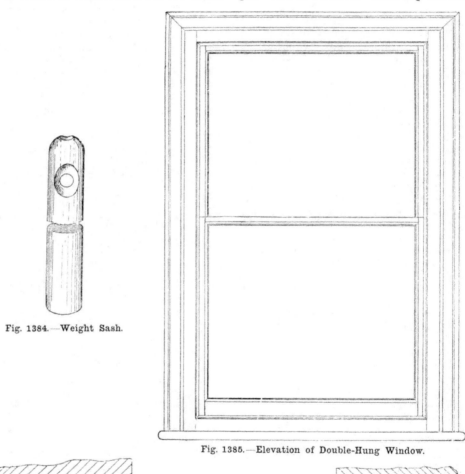

Fig. 1384.—Weight Sash.

Fig. 1385.—Elevation of Double-Hung Window.

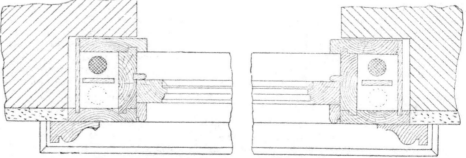

Fig. 1386.—Horizontal Section of Double-Hung Window.

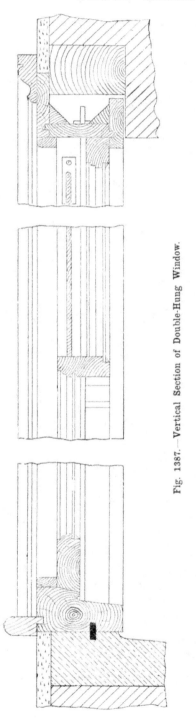

Fig. 1387.—Vertical Section of Double-Hung Window.

through the opening, pushing the loose parting slip on one side, when the disconnected weight will be found and can be drawn out.

Replacing the Sash-line.—Cut and remove the broken line from the eye, and replace the weight in the opening with its head sticking out; then stretch the cord as much as possible, cut off the end square with a sharp knife or chisel, and fasten the "tail" of the mouse to it, just as a rope-end is "whipped"; that is, by a series of half-hitches drawn tight and close to the end of the cord, as in Fig. 1380. The cord being tied securely, bend the mouse slightly, as shown, and pass it over the pulley, and, keeping the string in the middle with the fingers, allow the mouse to run down until the end of the sash-line can be passed through the face of the pulley, when the mouse may be reached with the fingers through the pocket, and the end of the line drawn down to the opening. It may next be fastened to the weight, passing the end through the hole in the head, shown in Fig. 1384, and pushing it out through the eye with a bradawl, when it can be knotted and drawn back, the knot being hammered into the eye until it is flush with the sides. Measure the distance of the end of the original cord from the top edge of the sash as shown by the nail-holes, and mark the same distance on the pulley stile from the head of the frame. Pull up the weight about 1½ in. from the bottom, and cut the cord off to the mark just made, or, in case of a knotted cord, as shown in Fig. 1381, sufficient being added to make the knot. Bring the sash up into position, and fix the cord either by knot or nails as required, taking care when replacing the weight that it goes in on the outside of the parting slip in the boxing, so that the slip lies between the two weights. Replace the pocket piece, the upper end being inserted first and then the lower knocked home, and fix the parting beads in place. To get these in, bring the top sash down to the sill, and slip the lower end of the parting bead between the overhanging end of the meeting rail and the groove; then bend it out slightly in the middle until the top end will go in its place, when the remainder may be sprung back and knocked home. The bottom sash is next

brought round into position, and the released cord pulled out straight from its slip-
knot and re-fastened in its original place.
When re-fixing the guard beads, enter the

procedure described above being in obtaining the length of the cord : in this case
measure the distance of the end of the cord
from the bottom edge of the sash, and mark

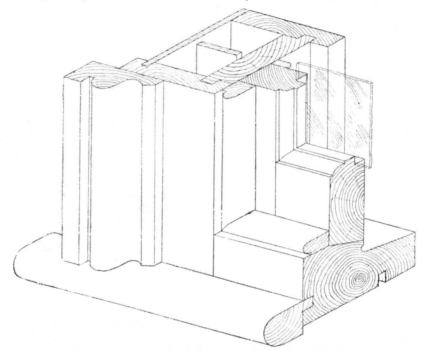

Fig. 1388.—Conventional Section of Double-Hung Window.

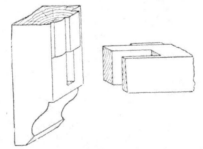

Fig. 1389.—Joint at Meeting Rail of Top Sash.

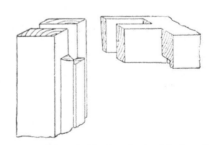

Fig. 1390.—Joint at Meeting Rail of Bottom Sash.

top end in the mitre, and bend out the
middle until the lower end will pass into its
place. Should the broken cord be in the
bottom sash, obviously only that one need
be taken out, and only the parting bead
covering the pocket on the side of the break
needs removal, the single variation in the

it on the pulley stile upwards from the sill,
and transfer the mark to the cord when the
weight is drawn nearly close up to the pulley.
It sometimes happens that cords are put in
without being stretched, and, in conse-
quence, the weights soon touch bottom, with
the result that the top sash will not keep

close up to the top and the bottom sash will not run up to its full height. The easiest way to remedy these faults is to remove the beads, push the sashes right up, take out the weights, and shorten the cords about 2 in.

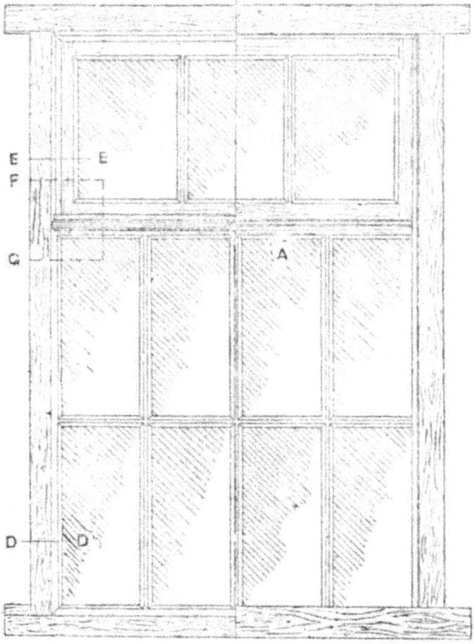

Fig. 1391.—Half Outside and Half Inside Elevations of Solid Window Frame with Movable Top Sash.

Working Drawings of Double-Hung Window.

Fig. 1385 is the front elevation, Fig. 1386 the horizontal section, and Fig. 1387 the vertical section of a double-hung window; Figs. 1386 and 1387 represent, reproduced to a reduced scale, the ordinary working drawings that are generally required. The connection of the frame with the sill and other parts is shown. Fig. 1388 is a conventional section showing the general construction of a lower corner of the frame and sash. Fig. 1389 shows the joint between the stile and the meeting rail of the top sash; it also shows the moulded horn. Finally, Fig. 1390 illustrates the joint between the stile and the meeting rail of the lower sash.

Solid Window Frame with Movable Top Sash.

The form of window shown by Fig. 1391 is used for workshops, and in other situations where it is not desirable for the lower

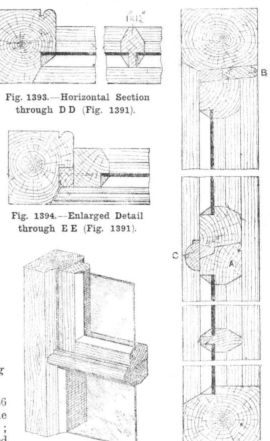

Fig. 1393.—Horizontal Section through D D (Fig. 1391).

Fig. 1394.—Enlarged Detail through E E (Fig. 1391).

Fig. 1396.—Detail of Portion indicated by F G (Fig. 1391).

Fig. 1392.—Enlarged Detail of Vertical Section (Fig. 1391).

part of the window to open. Details are shown by Figs. 1393 and 1394. The bars are framed right into the solid jambs. It is necessary, however, that the upper part of the window should open for ventilation, etc. In this instance this object is secured by having the sash hung on centres. A transom rail A (Figs. 1391, 1392, and 1396)

is tenoned into the jambs, and its lower surface is rebated and chamfered similarly to
the bars. Its upper surface is splayed and
rebated to receive the bottom rail of the sash.

follows : Jambs, head, and sill of frame,
2¾ in. by 3¼ in. ; bars, 1 in. by 1¾ in. ; sash,
stiles, and top rail, 1¾ in. by 1¼ in. ; bottom
rail of sash, 1¾ in. by 2¼ in. in its

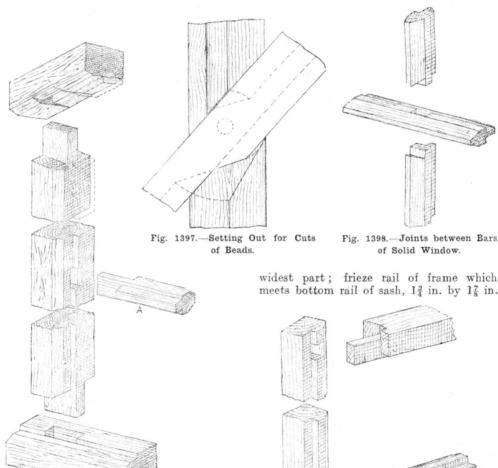

Fig. 1397.—Setting Out for Cuts
of Beads.

Fig. 1398.—Joints between Bars
of Solid Window.

widest part ; frieze rail of frame which
meets bottom rail of sash, 1¾ in. by 1⅞ in.

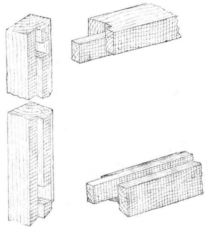

Fig. 1395. Detail of Joints at Bottom of Framings
at Transom and Head.

Fig. 1399.—Detail of Joints in Top Sash.

This is shown at Figs. 1392 and 1396.
The under surface of the head of the frame,
it will be noticed, is splayed a little. This
is to allow the bead marked B, which has
to be nailed to the top rail of the sash, to
clear the head of the frame as the sash is
opened. The leading finished sizes are as

Old=style Casement Windows.

Figs. 1401 and 1402 are respectively sections through the jamb, head, and sill of
a solid casement frame, sunk flush with the
face of a stone wall, which is duly checked

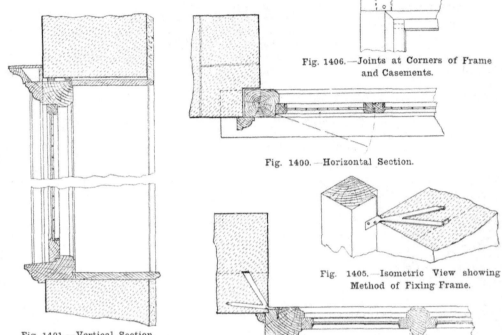

Fig. 1406.—Joints at Corners of Frame and Casements.

Fig. 1400.—Horizontal Section.

Fig. 1401.—Vertical Section.

Fig. 1405.—Isometric View showing Method of Fixing Frame.

Fig. 1402.—Horizontal Section.

Fig. 1403.—Section through Head.

in order to cover the joint, which mitres into the lower member of the cornice, as shown in Fig. 1402. A lead flashing laid

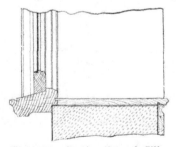

Fig. 1404.—Section through Sill.

to receive the frame. The sill is shown rebated $\frac{5}{8}$ in. deep, and should be bedded in cement, and its ends built into the wall. If the cornice is carried over the face of the wall and returned upon it, then the head of the frame can be likewise built in, which is the preferable method; otherwise the frame may be secured by wedging from the lintel, as shown in Fig. 1401. A stout moulding is planted on the face of the jambs

over the cover board of the cornice and up the face of the wall will prevent the ingress of the wet. The casements, of which there are three, open out, and are hung to the jambs, the centre leaf being fixed. The meeting stiles are rebated together. The glazing is a leaded lattice. The sections

(Figs. 1400 to 1405) show a more unusual case ; here the frame projects beyond the face of the wall. A ¾-in. check or rebate is made all round the inner edge of the frame, and the joint in this case should be made with red-lead and oil. The frame is secured by wrought-iron forked angle-ties sunk flush into the back face of the frame and built into the wall (see Fig. 1400) ; two ties on each jamb will be sufficient. The ties may either be turned up square at the ends so as to hook behind a stone, or drawn out to a pin end and sunk into a hole cut in

should be used in putting these frames to-gether, but the joints should be well painted with a thick or " round " oil paint. The wall ties should either be galvanised or painted before fixing.

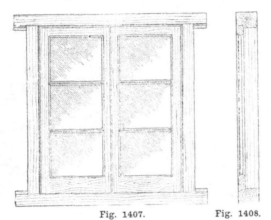

Fig. 1407.　　　　　Fig. 1408.

Fig. 1407.—Outside Elevation of Small Casement Window.

Fig. 1408.—Vertical Section of Small Casement Window.

Fig. 1409.—Horizontal Section of Small Casement Window.

the stone. The bracketing for the cornice may either be built into the wall, which is advisable if hardwood is employed for the cornice, or the stools may be secured to the wall with joint hooks. Fig. 1406 indicates the method of joining the angles of the frame and casements. The head of the frame runs over the jamb, and is cut off flush with the outside. The joint is secured with wedged tenons, which may also be pinned. In the casement, however, the stile runs through in the usual way, the rails being tenoned into the stile. No glue

Fig. 1410.—Joints of Jamb and Head, and Jamb and Sill.

Small Casement Window.

The construction of a solid frame and case-ment is shown by Figs. 1407 to 1409. The frame is made of 4-in. by 3-in. stuff, mor-tised and tenoned together at the joints as shown at Fig. 1410. The frame is also rebated ; the rebate on the sill being splayed for weathering. It will be noticed that the shoulder to the tenon on the stiles will have to be cut on the splay so as to fit the sill as shown at Fig. 1410. The ordinary method is to paint the joints and wedges when wedging up, but additional security is obtained by pinning. The stiles and head are beaded inside and out as shown. The sides and top rails of the casements are of $1\frac{3}{4}$-in. by $1\frac{3}{8}$-in. stuff (finished sizes), and are rebated for the

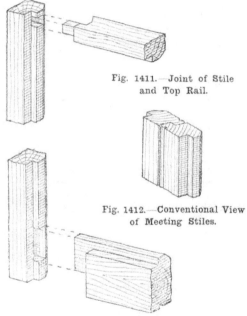

Fig. 1411.—Joint of Stile and Top Rail.

Fig. 1412.—Conventional View of Meeting Stiles.

Fig. 1413.—Joint of Bottom Rail and Stile.

glass and chamfered and mortised and tenoned together. In the joint shown at Fig. 1411, it will be seen that the shoulder on the head is scribed to fit the chamfer on the stiles. The bottom rail, $1\frac{3}{8}$ in. by $3\frac{1}{4}$ in., is also rebated and chamfered, and is mortised and tenoned to the stile as shown in Fig. 1413. The joints of these casements are glued and wedged together in the ordinary manner. The bottom rail is splayed on its under edge, and grooved to prevent moisture rising by capillary action. The meeting stiles are

rebated and beaded, the rebates being splayed as shown at Fig. 1412, so that they will open more easily.

Venetian Sash Frames.

Wide sash frames, divided into three or more lights all in the same plane, are called Venetian frames ; but if two of the lights are at an inclination to the other, the frame becomes a bay. If a frame contains two pairs of sashes in the same plane, that is, side by side, it is called a double window, or a two-light frame ; but if the frame contains two sets of sashes, not in the same plane, but behind each other, it is a double-sash frame.

Solid Mullion Venetian Sash Frame.

Venetian frames are of three varieties, each of which requires different treatment both in planning and making. The first is the solid mullion frame shown in Figs. 1414 to 1420, Fig. 1414 showing half outside elevation, and Fig. 1416 half inside elevation. This is the commonest kind, and is intended for narrow openings, where the span is not too great to be supported by a lintel and the frame. In this class it is usual to make the central sashes much wider than the side ones, and to fix these latter. Much of the information given at the opening of this chapter on the construction of a sash frame is applicable to the present case. In making a frame similar to the one shown, proceed to make a plan and vertical section, full size, on a board or rod. Beginning with the size of the opening, draw, for the width, the faces of the pulley stiles in a line with the reveals of the brickwork, or according to the architect's plans, and with these as starting points proceed to space out the mullions, linings, sashes, beads, etc., working from the specification or whatever data may be available. In setting out the rod, it is advisable to consider the practice of the shop for which the work has to be done, whether it is the custom to work to drawing, or to reputed sizes, the former being the practice in most machine shops, the latter the one in favour with hand shops ; for instance, in a machine shop the stiles for $1\frac{1}{2}$-in. sashes would leave the planing machine $1\frac{1}{2}$ in. thick exactly ; the path for them would be set out $1\frac{1}{2}$ in. wide, and the

Fig. 1414.—Half Outside Elevation of Solid Mullion Frame.

ENLARGED VERTICAL
SECTION THROUGH
FANLIGHT & PORTICO.

PART FRONT ELEVATION

ENLARGED HORIZONTAL SECTION
THROUGH DOOR AND POST.

ENLARGED VER-
TICAL SECTION
THROUGH DOOR
POST

ENLARGED VERTICAL
SECTION THROUGH
WINDOW AND
CORNICE.

ENLARGED SECTIONS OF DOOR POSTS.

DETAIL OF VERTI-
CAL SASH BAR AND
FRAME OF WINDOW.

ENLARGED HORIZONTAL SECTION THROUGH
DOOR POST, DADO, AND PILASTER.

10   DESIGN FOR A SHOP FRONT.

Fig. 1416.

Fig. 1415.

Fig. 1415.—Vertical Section of Solid Mullion Frame.
Fig. 1416.—Half Inside Elevation of Solid Mullion Frame.

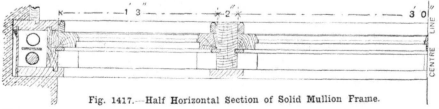

Fig. 1417.—Half Horizontal Section of Solid Mullion Frame.

subsequent cleaning off would afford the necessary clearance ; but in a hand shop 1½-in. sashes would be made from reputed 1½-in. or one-cut stuff, which, by the time in arched openings ; in this case allowance must be made for the rise, as the section shown on the rod should be a central one ; draw in the head of the frame in line with

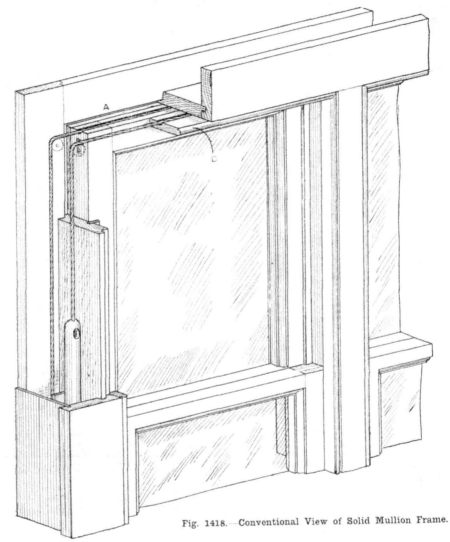

Fig. 1418.—Conventional View of Solid Mullion Frame.

it was cleaned off, would only hold 1⅜ in. bare, and the path of the sash would require drawing 1⅜ in. full. On the height rod the size of the opening is frequently taken from the top of the stone sill to the under side of the soffit of the reveal, or the springing line the under side of the arch, and the thickness of the oak sill above the sill line, and space out the remaining heights as shown in Fig. 1415. As this is a solid mullion frame with the side lights fixed, provision has to be made for carrying the cords from the central pair of sashes to the weights in the outside boxings. In the case of the top sash, this is accomplished by making a plough groove in the head of the outside side lights as shown at A in the conventional view (Fig. 1418), and in the bottom sash by taking the cord over pulleys in the mullion and pulley stile, and concealing it in the side openings by cover slips, as shown at B in the conventional view (Fig. 1418). These must be wide enough to reach from the face of the top light to the face of the frame, must be beaded on the edge to match the guard beads with which they mitre, and ploughed

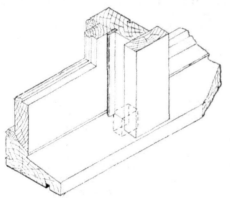

Fig. 1419.—Isometric View of Joint of Mullion and Lining with Sill.

on the rear side with a similar groove to the head of the top light. This arrangement is shown at Fig. 1418. It is assumed that the stuff is machine wrought.

Quantities for Solid Mullion Frame.— One oak sill, length 2 in. longer than out to out of frame, 3 in. by $5\frac{3}{4}$ in., wrought to section ; one head, length ditto by $1\frac{1}{8}$ in. by $4\frac{1}{4}$ in. ; two mullions, length out of head to out of sill, 2 in. by $4\frac{1}{2}$ in. ; two pulley stiles, length clear between sight lines of head and sill plus housings by 1 in. by $4\frac{1}{4}$ in. ; two back linings, length inside of head to out of sill, by $\frac{1}{4}$ in. by 5 in. ; two inside linings, $1\frac{1}{2}$ in. longer than height of frame over all, by 1 in. by $4\frac{3}{8}$ in. ; one head lining (inside), length 1 in. longer than clear between pulley stiles, by 1 in. by 3 in. ; one head lining (outside), length ditto by 1 in. by $4\frac{3}{8}$ in. ; two outside linings ditto to inside ; two mullion

linings, 1 in. longer than clear between head and sill, by 1 in. by $3\frac{1}{2}$ in. ; six parting beads, inside of head to out of sill, by $\frac{3}{8}$ in. by 1 in. ; two parting slips ditto by $\frac{1}{4}$ in. by 2 in. ; six guard beads 1 in. longer than inside of head to inside of sill by $\frac{3}{4}$ in. by $1\frac{1}{8}$ in. ; one ditto length of central opening head ; three sill guard beads, length between pulley stiles and mullions by $\frac{3}{4}$ in. by $1\frac{1}{4}$ in. bevelled to section ; two cover beads, length equal width of side lights by $\frac{3}{4}$ in. by $3\frac{1}{4}$ in. This completes the frame. Sizes for sashes will be taken from the rod in a similar manner, allowing $1\frac{1}{2}$ in. longer over all for stiles, and $\frac{3}{4}$ in. longer for rails and bars. Remember that brackets are to be worked on top

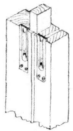

Fig. 1420.—Top of Mullion, showing Pulleys and Tenon.

sashes (see A, Fig. 1343). The above sizes are finished ones for the planing machinist. The converter will require a separate list, with an extra $\frac{1}{8}$ in. allowed for each side wrought. Thus the mullion size to him would be, length as above, by $2\frac{1}{4}$ in. by $4\frac{3}{4}$ in., and so on.

Construction of Solid Mullion Frame.— Before beginning to set out the stuff, consider how the frame is to be constructed. The pulley stiles will be housed into head and sill in the usual way, as explained at the beginning of this section. The mullions will be tenoned through head and sill as indicated in Figs. 1419 and 1420, and painted and wedged ; they will also be kept flush on the inside as shown in the half horizontal section (Fig. 1417), the outside lining being nailed on. The head linings will run right across the frame until they meet the side linings ; they are not cut between the mullions (see Fig. 1418, which shows the arrangement of the cords), as this weakens the frame. The cover beads are cut tight between pulley stile and mullion, and are held in position by the guard bead on one edge and the parting bead on the other ; remember that this cuts under the cover piece (see Fig. 1418). Pocket pieces should be cut in the centre of the pulley stiles, and not at the side, otherwise inside linings cannot be fixed for a distance of 15 in. or 18 in., which is objectionable to some architects, who consider that the pulley stiles are thus considerably weakened.

Setting Out Solid Mullion Frame.—As a precaution, run the rule over the stuff and see that the sizes are correct ; if not, note the necessary allowances to be made. Take the head, lay it on the rod, face side down, and strike over sight or face lines of pulley stiles, mullions, and parting slips. Turn it back ; mark over $\frac{1}{2}$-in. grooves for heads of pulley stiles ; set mortices for mullions $\frac{1}{4}$ in. back from the sight line, and mortices for parting slips to lines drawn ; gauge the mullion mortices from inside $\frac{3}{4}$ in. thick and one side in line with parting bead (see Figs. 1417 and 1419). Mortices for parting slips may be pencilled on in line with the parting bead (see rod, Fig. 1417). Gauge everything from the inside. Pair the sill with the head, and strike over all the sight lines with the exception of the parting slip mortices. Outside the pulley stile face lines, mark the housing, which is of the thickness of the pulley stile plus the wedging. The depth should be $\frac{1}{4}$ in. more than the lower point of the weathering ; more is unnecessary, merely weakening the sill without strengthening the stile. Square over the shoulder lines of the linings on both faces in line with the pulley stile ; run the gauge on for the sinkings ; these will be found on plan. Square over the mortices for the mullions, run the mortice gauge on the end, and transfer the lines to the sunk faces with the rule from outside. Gauge the plough grooves for the window-board and the water bar (see section, Fig. 1415).

Pulley Stiles of Solid Mullion Frame.— Lay one on the rod, and strike over the sight lines of head and sill, turn up, and mark over $1\frac{1}{4}$ in. at bottom—this amount will vary, however, with weathering—and $\frac{3}{8}$ in.

at top for housings. Set out the pocket 6 in. up from the sight line ; the length will vary as the height of frame ; usually keep them 2 in. shorter than the weights. Mark over at the top end mortices for pulleys. (It will be noted that the pulleys have to be kept close up to the head of the frame, allowing just clearance for the wheels to turn ; sometimes purpose-made pulleys are used ; where ordinary pulleys are used, file the top ends off $\frac{1}{4}$ in. above the wheel, and, when fixing, keep them $\frac{1}{8}$ in. above the shoulder of the mullion, as shown in Fig. 1420, and house them into the head, thus fixing the top ends.) Gauge the rebates, the inside one on the face, the outside one on the back, and plough groove $\frac{3}{8}$ in. for parting bead, which is as much out of centre of the stile as the guard bead overhangs the inside lining (see Figs. 1417 and 1418). With the squaring over of the shoulder of the housing on the top back side, and gauging $\frac{1}{2}$ in. full tongue on the end, the pulley stiles are finished ; the mullions may be set out from these. With the exception that no housings have to be allowed, the top ends being shouldered at the sight line, and the bottom the same inside (on outside allowing the sinking, and marking the shoulder to bevel of sill), it is a wise precaution to allow this rather full, as sills may vary. Gauge for parting grooves and tenons. To find the position for mortices for pulleys, draw lines equal to the thickness of the sashes on each side of the parting groove ; the centre of this will be the centre of the mortices—make them so that the box of the pulley fits tight. The linings require gauging for the various plough grooves, or rather, one of each kind should be sufficient for the machine, as, when once set, all will be run through exactly alike. Mark sight lines on the edge of the outside linings, and run $\frac{3}{4}$-gauge on each end for a saw cut. The pieces will be cut off when fitting up.

Cover Beads.—Draw these on the section, and gauge the groove on the back (see Fig. 1418).

Beads.—Draw on sections, and that completes the setting out of the frame. In setting out for machine work, all stuff that is moulded should have a section drawn on the face of one piece, with a reference to the number required. These sections should be drawn exactly to size, without any allowances; mark these on the length of the shoulders, etc. On the other hand, all tongues, rebates, cross-cut grooves, etc., should have allowances for fitting and cleaning off, and sections drawn accordingly; plough grooves in the direction of grain, and mortices, should be marked exactly as wanted. The setting out of sashes, having been fully described early in this section, will not be repeated here. Remember, however, that as side lights in

nailing the pulley stiles (note, avoid the pulleys); lay it on the bench, outside down, and square the frame; cramp up the mullions, paint the wedges, and wedge up. Fix the sill to a bench piece, square with a rod, and fix the head to bench top, cut off ends of wedges, level sinking of sill with pulley stile, cut away the piece of tongue on the ends of head to let lining run, up and fix on inside linings; cut the head lining tightly between these, and nail on. Keep the outside edge flush on back, clean off, and fit guard beads in; these should be rebate

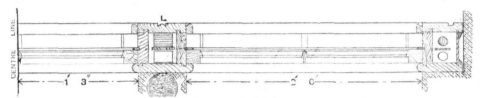

Fig. 1421.—Half Horizontal Section through Venetian Sash Frame with Double Weights.

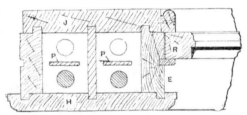

Fig. 1422.—Horizontal Section through Mullion arranged for Four Weights.

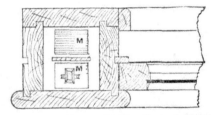

Fig. 1423.—Horizontal Section through Mullion arranged for Double Weights.

this frame have to be fixed, they should be set out rather wider than would be the case if they were hung.

Fitting Up Solid Mullion Frame.—Examine all grooves, housings, and mortices; see that they are of the required depth, and clear the wedging. Next fit the pocket pieces, running the plough groove through; fit in the pulleys, and clean off the stiles. Do the same with the mullions, and fit the parting beads; also fit, that is, mullet, all tongues. Take a shaving off the edges and bottom ends of the outside linings; these cannot be done afterwards, as the sill projects $\frac{1}{8}$ in., or should do, to allow for shrinkage. Wedge the pulley stiles in the sill out of winding with each other, drive in the mullions, and put the head on, well

mitered (see Fig. 1436). Turn over and repeat the process, first fitting in parting beads and slips, finally nailing on the back linings, and blocking the head. Rub a block on the joint.

Double Weight Venetian Sash Frame.

The second class of Venetian frames, shown in the half horizontal section (Fig. 1421), is for wider openings supported and divided by thin brick or stone mullions. In these cases the whole six sashes can be hung; but they must be about equal in size. Full details are illustrated in Figs. 1421 to 1437. If it is desired to hang all the sashes, and the size of the pier or mullion restricts the boxing to about 6 in., there would not be room for two sets of weights. A special

form of weight then used in the centre boxings is square in section, with a pulley cast
in the top end through which the cord is
passed, with its ends taken over the pulleys
and fastened to the top or bottom sashes on
each side, as shown in the sectional view,

edge of the sash stile. Fig. 1423 shows an
enlarged section of the boxed mullion ; Fig.
1425 a sectional elevation of the top end
with the outside lining off, showing weight in
position with top sash down and bottom one
up. The housings for the pulley stiles of the

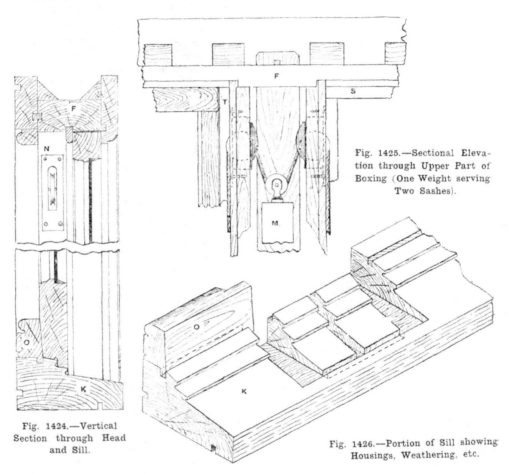

Fig. 1425.—Sectional Eleva-
tion through Upper Part of
Boxing (One Weight serving
Two Sashes).

Fig. 1424.—Vertical
Section through Head
and Sill.

Fig. 1426.—Portion of Sill showing
Housings, Weathering, etc.

Fig. 1425. Thus, for each set of three sashes,
four weights are made to answer. Of
course, in this kind the sashes must be very
similar in size, as the double weight has to
equal twice the half weight of each sash ; and
if one sash was much heavier than the other,
the lighter one would be continually pulled
up. This tendency can to some extent be
checked by inserting a piece of cork in the

mullions are stopped on the outside of the
sill as shown in Fig. 1426, and the bottom
ends of the outside linings should be tongued
on the back side into the sill. The inside
linings should run over the sill and head as
in an ordinary frame, so as to tie the frame
together, as there is no wedged mullion in
this case. The pulleys can be fixed as usual
about 1½ in. down from the top ; only one

pocket will be wanted in each mullion. Thinner pulley stiles are occasionally used in the mullions, to gain additional room.

Large Venetian Sash Frame.

The third class of Venetian frame is for the widest openings, with thick stone or brick

and sill (see Fig. 1426) to keep it in position whilst fixing the outside linings. Several variations have been introduced into this frame, which is frequently made entirely of oak. The head is made 2 in. thick, with a planted tongue on the inside to economise labour and material. The

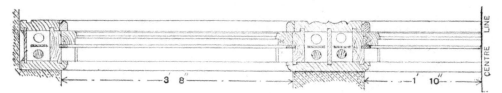

Fig. 1427.—Half Horizontal Section through Venetian Sash Frame.

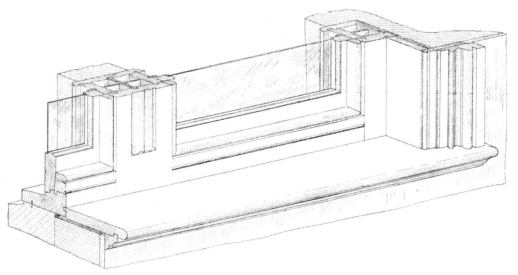

Fig. 1428.—Conventional Sectional View through Boxings and Mullions of Sash Frame.

piers and correspondingly wide mullions in the frames, in which the sashes may vary in width according to taste, and may all be hung, or part hung and part fixed. In Fig. 1427 is shown half horizontal section of a very large frame, with sashes of varying width and a wide boxed mullion to cover a brick pier, there being ample room for two sets of weights here. The conventional sectional view (Fig. 1428) will convey a general idea of the construction of this class of frame. The box is divided by a central lining, which should be housed into head

tongues are necessary because in hardwood the linings would be fixed with brads driven on the skew through the edge and hidden by the beads, which would be fixed with cups and screws. The inside linings are kept ⅛ in. back from the pulley-stile face to form a rebate for the bead and also to hide the joint. This setting back must be allowed for when housing the sill, for the shoulders on the sill abutting the linings will stand ⅛ in. in front of pulley stile (see Figs. 1424 and 1427). The parting bead is run through the head, the sides being scribed up to it. The

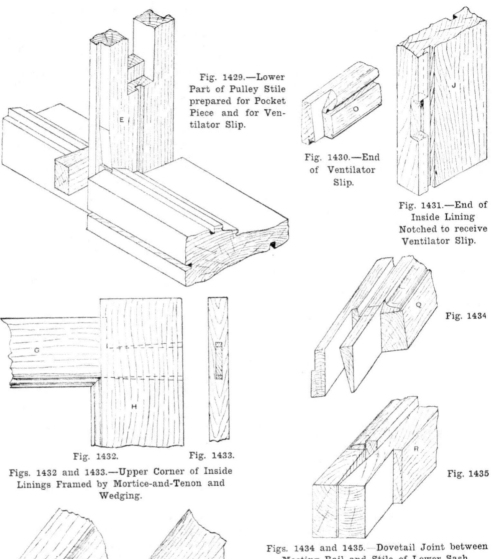

Fig. 1429.—Lower Part of Pulley Stile prepared for Pocket Piece and for Ventilator Slip.

Fig. 1430.—End of Ventilator Slip.

Fig. 1431.—End of Inside Lining Notched to receive Ventilator Slip.

Fig. 1434

Fig. 1435

Fig. 1432.

Fig. 1433.

Figs. 1432 and 1433.—Upper Corner of Inside Linings Framed by Mortice-and-Tenon and Wedging.

Figs. 1434 and 1435.—Dovetail Joint between Meeting Rail and Stile of Lower Sash.

Fig. 1436.

Fig. 1437.

Figs. 1436 and 1437.—Method of Rebating and Mitering Beads.

$2\frac{1}{2}$-in. weathered and beaded piece shown on the sill at o (Fig. 1424) is a ventilator slip. Its purpose is to allow of the window being opened 2 in. between the meeting rails for ventilation, whilst avoiding a direct draught at the bottom rail. It is weathered so that the bottom rail of the sash shall fit tightly against it when shut, but instantly release itself when lifted. It should be inserted in the frame before fixing the inside linings, being cut in between the tongues of the pulley stiles as shown in Fig. 1429. Cut a piece out of these to fit the bevel, then form a bareface tenon on the slip (see Fig. 1430) and notch the lining over it (see Fig. 1431). This secures it firmly in place. The tongue should be well painted before insertion. In this class of frame the linings are usually framed at the corners as shown in Figs. 1432 and 1433, the head linings running through and the mullion linings tenoned into them

of the ends of the meeting rail and the stile of the bottom sash, showing the best form of joint. Always leave the meeting rails rather wide, so that they can be fitted accurately when the sashes are fitted in, and thus prevent rattling. Figs. 1436 and 1437 show the rebate mitre of the beads. The reference letters in the illustrations of the Venetian sash frames (Figs. 1421 to 1426) are as follows :—E, pulley stile ; F, head ; G,

Fig. 1438.—Inside Elevation of Sash Window with Boxed Shutters.

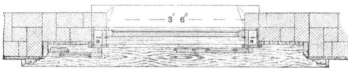

Fig. 1440.—Horizontal Section of Fig. 1438.

Fig. 1439.—Vertical Section of Fig. 1438.

and also the sill. The frames would be put together as described for the solid mullion frame, except that the joints of the outside linings would be painted, and the joints of the inside glued, before cramping up and wedging. If the centre lining of the boxed mullions is crooked, it may be kept in position for getting the face lining on by cutting little struts tightly between it and the backs of the pulley stiles, which are knocked away by the weights when they go in and removed through the pocket. Figs. 1434 and 1435 are perspective sketches

head lining ; H, outside lining ; J, inside lining ; K, sill ; L, mullion ; M, weight ; N, pulley ; O, ventilator slip ; P, parting slip ; Q, meeting rail ; R, sash stile ; S, guard bead ; T, parting bead.

Sash Windows with Boxed Shutters in Brick-and-a-Half Wall.

The sash window shown in Figs. 1438 to 1440 has boxed shutters, and is built in a brick-and-a-half wall. Details are illustrated by Figs. 1441 to 1449. Figs. 1441 and 1442 are conventional views of parts of the outside, which is built in Flemish bond on the face, with old English backing. A camber gauged arch is shown, having a straight extrados and cambered intrados, the depth of the arch at the springing being 12 in. Fig. 1443, which is a sectional view of the window as seen from the inside,

various parts being prepared to the sizes indicated in the enlarged detail (Fig. 1448). The sill is ploughed and connected to the stone sill by a 1-in. by $\frac{1}{4}$-in. galvanised iron bar. The wooden sill is also prepared to receive the tongue of the window-board as shown at Fig. 1447. The head lining and

Fig. 1441.—Conventional Sectional View of Top Corner of Window.

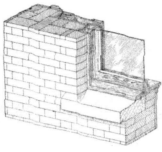

Fig. 1442.—Sectional View of Bottom Corner of Window showing Sill, etc.

Fig. 1443.—Sectional View of Window from Inside.

shows the top corners of the sash, shutters, and architraves. Conventional views showing the details of the brickwork inside are given at Figs. 1445 and 1446; $4\frac{1}{2}$-in. reveals are provided for the sash frame, and there is a $4\frac{1}{2}$-in. recess for the shutters. A wooden lintel is shown, on which a core is formed, and on

the upper ends of the inside lining are grooved to receive the tongue of the soffit lining. Both this lining and the window-board have to be cut round the frame as shown at Fig. 1447. The window-board is

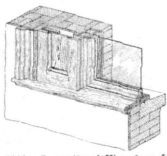

Fig. 1444.—Conventional View from Inside of Lower Corner of Window, Shutters, etc.

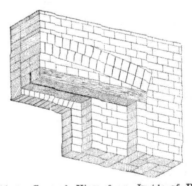

Fig. 1445.—General View from Inside of Upper Part of Opening

this a two-ring relieving arch is built. The stone sill is shown 6 in. by 11 in., with level stools at each end for brick jambs. It is tool-sunk and weather-throated, and grooved for the metal weather-bar as shown at Fig. 1449. The cased sash frame with double-hung sashes is of the ordinary character, the

ploughed to receive a small moulding underneath, as shown at Figs. 1438 and 1439. It is also prepared with a nosing, and returned at the ends as shown. The vertical inside linings are grooved to receive the tongue of the fillet A, to which the shutters are hung. Linings B (Fig. 1448) are provided with rebates on the outer edge for the beads of the shutters to stop against. These linings are tongued into the soffit linings and into the window-board (see Fig. 1447), and are also slightly bevelled to form a key for plastering. A fillet c (Fig. 1448), with a bead stuck on its edge, is fixed to the back edge of the vertical linings of the sash frame, and is

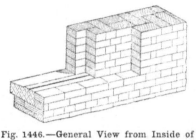

Fig. 1446.—General View from Inside of Lower Part of Opening.

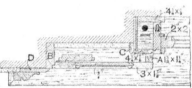

Fig. 1448.—Enlarged Detail of Horizontal Section.

Fig. 1449.—Detail of Stone Sill.

slightly splayed to form a key for the plastering. As will be seen, there is one framed and panelled shutter on each side, the panels being bead-flush on the inside, and with mouldings planted on to the face side. Each shutter is hung with 3-in. wrought-iron butts to the fillets previously mentioned. The shutters are made with a flap, which (being narrow) is not framed, but is formed of a piece of board and clamped at each end to prevent warping. The shutter and flap are connected by 2½-in. back-flap hinges. A shutter bar of an ordinary form

is shown at Fig. 1438. On the inside, the opening is finished with 5-in. by ¾-in. facing grounds D (Fig. 1448), the outer edge being ovolo-moulded and the back edge splayed to receive plastering. On these facing grounds 4½-in. by 1½-in. architrave mouldings are fixed as shown.

French Casements to Open Inwards.

Figs. 1450 to 1457 show a pair of French casements hung to a solid frame, with transom, fanlight, splayed linings, etc.

Frame and Linings. — It will not be necessary to describe in detail the construc-

Fig. 1447.—Joints in Linings, etc.

tion of these, as it is similar to work that has already received attention. General views of the joints in the frame are shown by Figs. 1456 and 1457. A part of the horizontal section is shown on a larger scale by Fig. 1453.

Casements and Fanlight.—The construction of the casements and fanlight is identical with that involved in sash work already treated; it is therefore only necessary now to enumerate the special features of this example. The casements are constructed to open inwards, and when therefore they occupy exposed situations, arrangements must be made for excluding wet and draught. The sill, made either of oak or

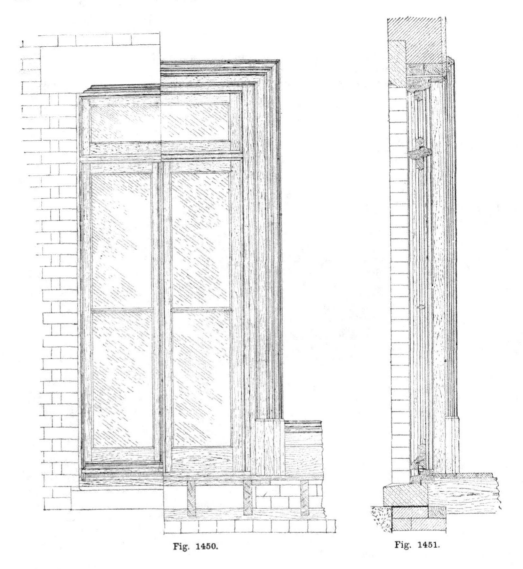

Fig. 1450.

Fig. 1451.

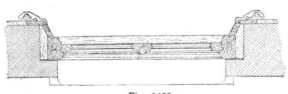

Fig. 1452.

Fig. 1450.—Half Outside and Half Inside
Elevation of French Casements
to Open Inwards.

Fig. 1451.—Vertical Section of
Fig. 1450.

Fig. 1452.—Horizontal Section. of
Fig. 1450,

teak, is double-sunk and splayed to receive
a special water bar as shown.   This is probably one of the best methods of excluding
wet.   The water bar is hinged, and, when
the casements are closed, is held up against
the moulded weatherboard (as shown at
Fig. 1454) by the striking plate screwed to
the bottom rail of the casements.   The under
side of the sill is throated for weathering,
and ploughed for a metal water bar, which is

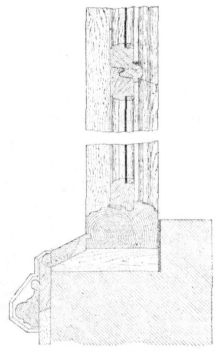

Fig. 1453.—Enlarged Detail of Horizontal Section
(Fig. 1452).

inserted to prevent water finding its way
between the wooden sill and the stone sill.
The wooden sill is to be also ploughed on
the inside to receive the floorboards.   A
section through the sill and water-bar when
the casements are open is presented by Fig.
1455.   The frame is ovolo-moulded outside,
and lamb's-tongue moulded inside.   The
jambs are moulded inside and out, rebated
with hollow sinking to receive round projection of stile of casement, and ploughed on
the inside to receive splay linings, as shown

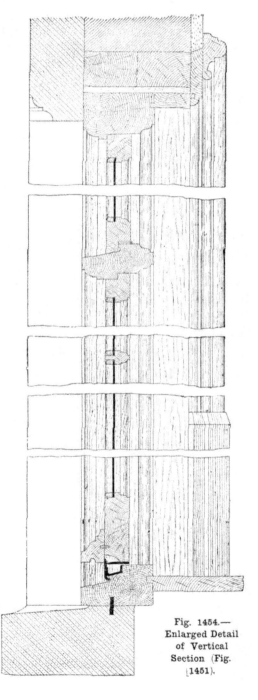

Fig. 1454.—
Enlarged Detail
of Vertical
Section (Fig.
1451).

at Figs. 1452 and 1453. The transom is moulded inside and out, and rebated to receive the head of the casement ; and on the upper side is sunk, splayed, and throated to receive the bottom rail of the fanlight (see Figs. 1451 and 1454). The head of the frame is moulded, rebated for the head of the fanlight, and ploughed to receive the head of the splay linings, as shown at Figs. 1451 and 1454. The casements

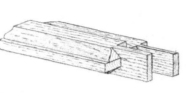

Fig. 1455.—Section through Sill and Water Bar (Casements Open).

are ovolo-moulded and hung folding, the meeting stiles having a hooked joint with moulded fillet on the outside. This fillet may be worked on the solid as shown in the illustrations, but it is frequently nailed on. The glass is shown fixed in with beads from the outside. The bottom rail is prepared for the metal water bar, and a moulded weatherboard is fixed to it (see Figs. 1451 and 1454). The fanlight is hung to the

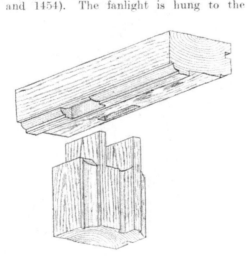

Fig. 1456.—Joint between Jamb and Head.

transom to open inwards, as shown. The ground is ploughed to receive the tongue of the linings. It will be seen that this ground also forms a facing, thus representing part

of the architrave, which is stopped at the bottom by a plinth as shown.

Elliptical=headed Window with Casements and Fanlight in Solid Frame.

The case illustrated by Figs. 1458 to 1465 shows a solid frame with a transom and an

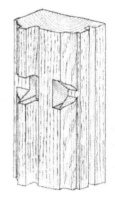

Fig. 1457.—Joint between Transom and Jamb.

elliptical head, a pair of casements opening inwards, and a fanlight which, being hinged to the transom, also opens inwards if desired. Casements opening inwards are less frequently adopted than those opening outwards. Objections to the former are that in exposed situations it is comparatively difficult to make them weather-tight ; while, if they are not kept securely fastened, they are apt to be blown open by a sudden gust of wind, when more or less serious damage may be done. They also interfere with the window hangings, furniture, ornaments, etc., which are often placed near windows. In spite of their obvious disadvantages, however, inwardly opening casements are sometimes adopted ; hence it has been deemed desirable to treat of a typical example here.

The Frame.—The principal points in the construction of the frame are as follows. As will be seen from Figs. 1461 and 1463, the outside edge of the moulding has a large ovolo moulding worked on, while the inner edge is finished with an ogee. The jambs have tenons wedged into mortices in the oak sill, as illustrated in previous cases. The oak sill is rebated, throated, splayed, and weathered on the under edge, and

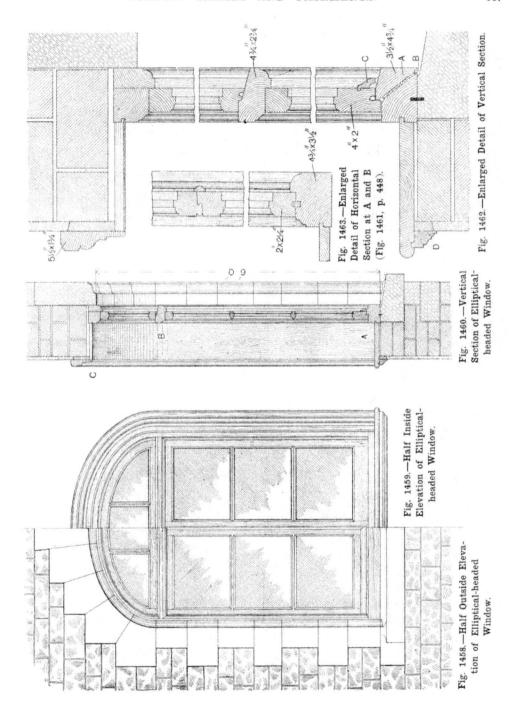

Fig. 1463.—Enlarged Detail of Horizontal Section at A and B (Fig. 1461, p. 448).

Fig. 1462.—Enlarged Detail of Vertical Section.

Fig. 1460.—Vertical Section of Elliptical-headed Window.

Fig. 1459.—Half Inside Elevation of Elliptical-headed Window.

Fig. 1458.—Half Outside Elevation of Elliptical-headed Window.

ploughed to receive the tongue of the window board and the metal water bar as illustrated in section at A (Fig. 1462). Any moisture finding its way under the bottom rail of

(Fig. 1462). The head is constructed in two thicknesses, each layer breaking joint as indicated at A B C (Fig. 1464). The jambs are cut to receive each thickness of the head

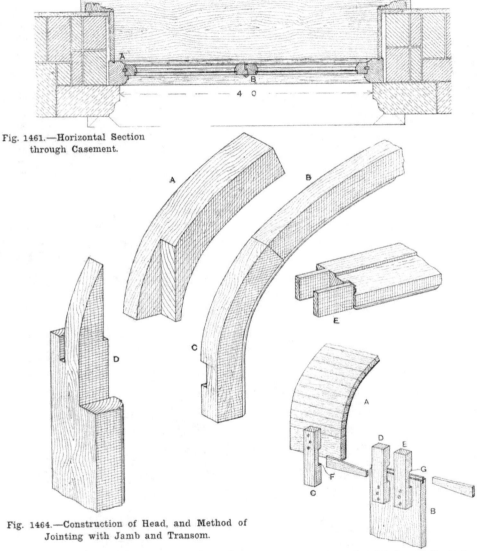

Fig. 1461.—Horizontal Section through Casement.

Fig. 1464.—Construction of Head, and Method of Jointing with Jamb and Transom.

Fig. 1465.—Method of Jointing Soffit Lining and Jamb Lining.

the casements would drip into the throating of the sill, and for carrying off this moisture three or four holes should be bored from the throating to the weathering of the outside bottom edge of the sill, as shown at B

as shown at D (Fig. 1464). Mortices are prepared as indicated at C and D to receive the tenons of the transom shown at E (Fig. 1464). When the several pieces have been fitted together satisfactorily, the pieces forming the head are glued and screwed together, the screws being inserted from the outside layer, behind the portion that will be hidden by the reveal of the arch.

Casements and Fanlight. — The casements are ovolo-moulded inside, and rebated on the outside for glass. The hanging stiles are ploughed with a round-edged plough iron, so as to fit over a weather-bead as shown at A (Fig. 1463). The meeting stiles are rebated, and have a hooked joint, and the edges are splayed to facilitate opening. The bottom rails of the casements are rebated, and have a moulded weatherboard fixed on as shown at C (Fig. 1464). A metal water bar is not shown, but the arrangement illustrated, in which the bar is formed entirely in the wood, will be found at once simple and quite effective. The principal point to notice in the construction of the fanlight is that the head is formed of two or three pieces jointed and fixed together with handrail screws or hammer-headed keys.

Grounds, Linings, and Architraves. — The grounds are of the usual form, and are fixed to the brickwork in the usual manner. They are clearly shown in the sections. The linings are square. In the case of painted work, the head pieces might be either formed of pieces jointed together and sawn out by a bandsaw to form the soffit, or else saw-kerfed. The latter method, however, is usually unsatisfactory. Undoubtedly the better method is to have a veneer wide enough and long enough to bend over a cylinder prepared for this purpose, and to fit and glue on staves at the back. A portion of the lining prepared in this manner is shown at A (Fig. 1465). The whole process of preparing the soffit lining on this principle will be dealt with in a later section. The soffit lining and jamb lining are connected by grooves and a tongue, as illustrated at Fig. 1465. The head having been jointed with the jamb, the key-block C is between the blocks D and E ; and then by gluing the joint and inserting a pair of wedges, all is held firmly together. The wedges in the blocks should be cut so that

the wedges press against F in the block C, and against the upper edges G of the blocks D and E. The architrave is made in two or three pieces, and prepared as explained in examples previously described. The window board has a tongue which fits into a groove in the frame as shown. The front edge has a rounded nosing, and is finished off with a moulding underneath as shown at D (Fig. 1462).

French Casements with Boxing Shutters to a Segmental-headed Opening.

Fig. 1469 shows the half elevation of a casement opening with segmental head, fitted with a transom head with casements opening outwards. Figs. 1466 to 1468 respectively represent the elevation, plan, and section, which show the inside of the frame, casements, fanlight, boxing shutters, architraves, etc., complete. The main dimensions are figured on the illustrations.

The Frame. — Mortice and tenon joints are used between the head and jambs, and also between the jambs and the transom. The intersection of the mouldings is formed by mitering. The jambs are ploughed with a rounded iron in the rebate, to receive a weathering bead, which is worked on the solid of the hanging stiles of the casements, as illustrated. The sill should be of oak or teak, but the material for the other parts may be of deal, pitch-pine, oak, mahogany, or teak, according to requirement or class of building. The general construction of the frame is shown in the illustrations, and reference to similar examples previously given will probably be found sufficient to render further description superfluous.

Casements. — In this example, as in others, only the parts being immediately dealt with are shown by the view representing the rod, the complete setting out of the rod, showing the frame, etc., being considered superfluous for the purpose in hand. Fig. 1478 represents the height rod for the casements. Projected up from this is represented a stile (Fig. 1479), set out ready for mortising, rebating, and moulding. The stiles would be, of course, set out in pairs, the method of procedure being very

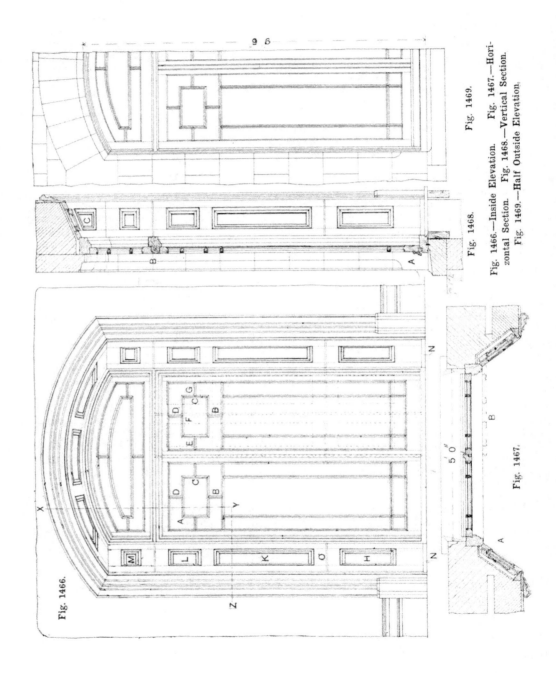

Fig. 1469.

Fig. 1468.

Fig. 1467.

Fig. 1466.

Fig. 1466.—Inside Elevation.　Fig. 1467.—Horizontal Section.　Fig. 1468.—Vertical Section.　Fig. 1469.—Half Outside Elevation.

similar to that explained early in this section when dealing with sashes. The lower portion of the stile is mortice-rebated, so that they can be rebated and moulded in reasonable lengths. Fig. 1483 shows a portion of the width rod for casements.

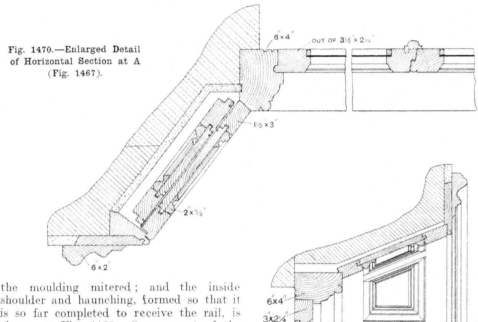

Fig. 1470.—Enlarged Detail of Horizontal Section at A (Fig. 1467).

the moulding mitered; and the inside shoulder and haunching, formed so that it is so far completed to receive the rail, is shown at Fig. 1482. On account of the hooked joint between the meeting stiles, the top and bottom rails should have double tenons in breadth fitting to these stiles. When only single tenons are used, the short end-grain of the tenon is liable to break away when being formed into the hooked joint. For the hanging stiles, single tenons would be best, because of the projection. Fig. 1480 represents a vertical bar set out for tenons and mortised. Fig. 1481 shows one strip of stuff set out for parts of the light B, C and D (Fig. 1466). When gauging for the mortices and tenons for the intersection of the bars, it should be noted that the mortices and tenons are thinner where the bars intersect than where they join the stiles and rails. This is shown at B and E (Fig. 1473), and also where the bars are set out in Figs. 1480 to 1486, and 1491. By placing four strips together and setting them out, and then cramping them together as shown at Fig. 1491, the shoulders may be entered with a dovetail saw as indicated,

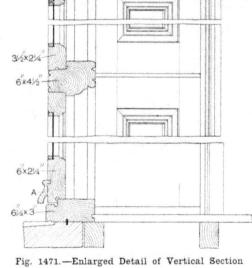

Fig. 1471.—Enlarged Detail of Vertical Section at A, B, and C (Fig. 1468)

Fig. 1472.—Conventional Sectional View of Inside Top Left-hand Corner

Projected above it, at Fig. 1484 (which shows the top rail completely set out), a top cross-bar is shown set out for mortices and tenons (Fig. 1485). A horizontal bar as at F (Fig. 1466) is shown set out at Fig. 1486. The ends A and B are afterwards dovetailed and mitered. Four of these bars will be required, and they should be set out together. The method of setting out and entering for the shoulders and tenons for the four bars forming the marginal square in the top of the

Fig. 1489 shows a piece of the horizontal similarly prepared, with dovetail socket cut ready for completion as shown at Fig. 1477. Where the tongues of the bars mitre together, they may be strengthened by cutting a chase as shown at A and B (Figs. 1475 to 1477), and gluing in a small hardwood key.

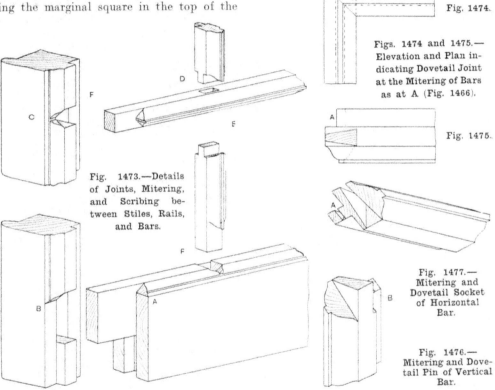

Fig. 1474.

Figs. 1474 and 1475.—Elevation and Plan indicating Dovetail Joint at the Mitering of Bars as at A (Fig. 1466).

Fig. 1475.

Fig. 1473.—Details of Joints, Mitering, and Scribing between Stiles, Rails, and Bars.

Fig. 1477.—Mitering and Dovetail Socket of Horizontal Bar.

Fig. 1476.—Mitering and Dovetail Pin of Vertical Bar.

casements is shown at Figs. 1488 and 1489. To make a good job of these four angle-joints of the marginal squares (one angle of which is lettered A, Fig. 1466), they should be dovetailed and mitered as shown by the enlarged elevation and plan (Figs. 1474 and 1475), but the joint will be more clearly understood on reference to Figs. 1476 and 1477. Fig. 1488 shows the setting out, the cutting of the dovetail pin, and the entering of the shoulders of a bar ready for moulding (shown completed at Fig. 1476).

Rebating and Moulding Sash Bars on the Sticking Board.—A short length of a suitable form of sticking board for the rebating and moulding of the bars is shown at Fig. 1490. It is made of a board 6 in. to 9 in. wide, and of any suitable length, the base being dovetail-keyed on the under side as indicated at D and E (Fig. 1490) to prevent warping. A rebate is made equal in depth to the rebate of the bar, so that the tongue may properly bed as shown at G whilst the opposite side is being rebated.

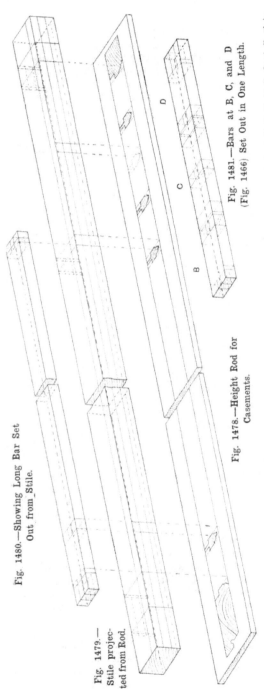

Fig. 1480.—Showing Long Bar Set Out from Stile.

Fig. 1479.—Stile projected from Rod.

Fig. 1478.—Height Rod for Casements.

Fig. 1481.—Bars at B, C, and D (Fig. 1466) Set Out in One Length.

At F is shown a piece of bar in position for the first rebate. The strip C is ploughed so as exactly to fit the tongue of the bar whilst the sticking of the moulding is being done, as indicated at H and K. The rebate and plough grooves must vary according to the size of the bars being worked; hence several sticking boards, each for a different size, will be found in most workshops. A stout screw is generally used for a stop, and the back end is held by a bench knife being driven into the cheek and board. In this example mitering is shown at the intersection of the mouldings of the stiles and rails (see A B, Fig. 1473), and where the bars intersect with each other scribing is illustrated, as at D and E (Fig. 1473).

Fanlight.—The only point that calls for special attention in this is that both the head and the curved marginal bar may be got out in one piece.

Fitting the Casements.—The meeting stiles are rebated, splayed, and hooked together as shown by the section (Fig. 1470). When this joint is found to be satisfactory, the bead on the inside to break the joint is worked, and the moulded weathering ploughed for and fitted in. The top and bottom of the casements have next to be fitted to the head and sill, the bottom rail being rebated and throated as shown. The casements are now placed together, with their meeting stiles fitting, and the width between the rebates of the frame accurately marked off on the hanging stiles at the top and bottom. The hanging stiles are next rebated, sufficient being left on to form the projecting weathering beads, which are rounded as shown. Then the stiles are applied to the frame, and the necessary easing is done, so as to produce a good fit. Between the meeting stiles, the hanging stiles, and the frame a sufficient joint must be provided to allow for painting or polishing, and for easy opening and closing, without any binding. The outside of the bottom rails are ploughed (the plough groove extending to the edge of the stiles) to receive a moulded weatherboard, as shown a tA (Fig. 1471). This should be secured with screws, the joining parts being first painted. Each casement is hung with three 4-in. wrought-iron or brass butt hinges. The

most suitable forms of fastening are espagno-
lette bolts.

Boxing Shutters.—The arrangement of

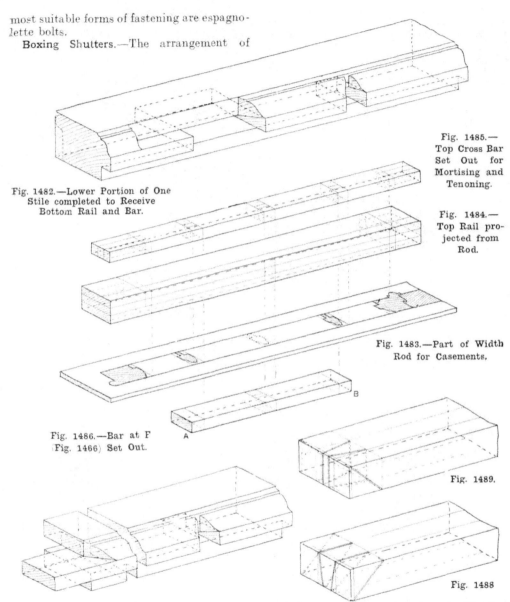

Fig. 1485.—
Top Cross Bar
Set Out for
Mortising and
Tenoning.

Fig. 1482.—Lower Portion of One
Stile completed to Receive
Bottom Rail and Bar.

Fig. 1484.—
Top Rail pro-
jected from
Rod.

Fig. 1483.—Part of Width
Rod for Casements.

Fig. 1486.—Bar at F
(Fig. 1466) Set Out.

Fig. 1489.

Fig. 1488

Fig. 1487.—Part of Top Rail Mortised.

Figs. 1488 and 1489.—Ends of Bars for Marginal
Square.

the shutters when in their boxings is shown
by Fig. 1467. At B in Fig. 1466, the right-
hand half is indicated by dotted lines as
closed. The wall being hollow, and thus
thick, allows a sufficient recess for the

shutters to be formed of four leaves. In the
case of thinner walls, six or even eight leaves
might be used. The shutters may be made
to open and close in one length, which would

include the panels H, K, and L, or may open and close in two sections. The lower part,

panels K and L, the bottom edges of the leaves being rebated at O on the back so as

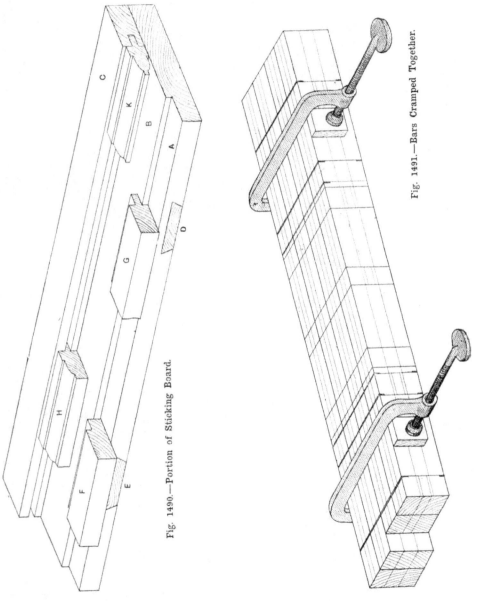

Fig. 1490.—Portion of Sticking Board.

Fig. 1491.—Bars Cramped Together.

containing the panel H, is closed first, the upper edge having a rebate and bead as shown at O. The upper part contains the

to fit against the lower part. Generally the fanlight is left free, and thus the framing at M is dummy. The bottom ends of the

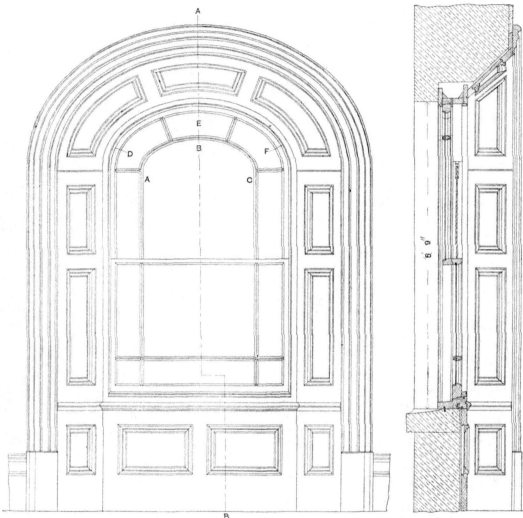

Fig. 1492.

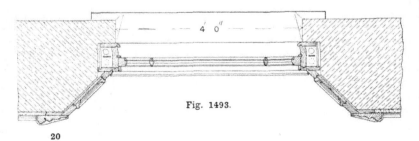

Fig. 1493.

Fig. 1494.

Fig. 1492.—Inside Elevation of Elliptical-headed Sash Window.

Fig. 1493.—Horizontal Section.

Fig. 1494.—Vertical Section on A B (Fig. 1492).

20

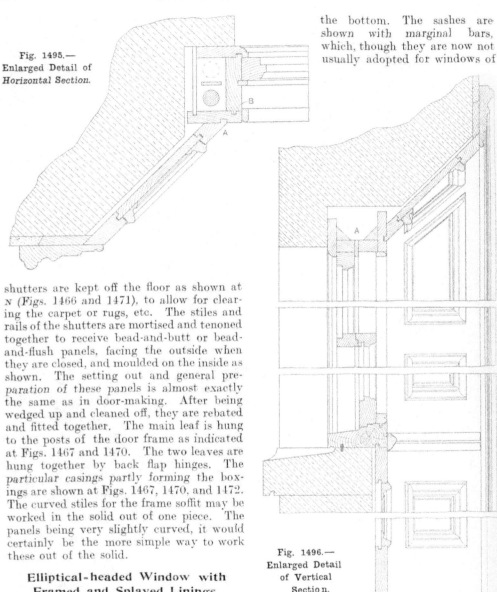

Fig. 1495.—
Enlarged Detail of
Horizontal Section.

the bottom. The sashes are
shown with marginal bars,
which, though they are now not
usually adopted for windows of

shutters are kept off the floor as shown at
N (Figs. 1466 and 1471), to allow for clearing the carpet or rugs, etc. The stiles and
rails of the shutters are mortised and tenoned
together to receive bead-and-butt or bead-
and-flush panels, facing the outside when
they are closed, and moulded on the inside as
shown. The setting out and general preparation of these panels is almost exactly
the same as in door-making. After being
wedged up and cleaned off, they are rebated
and fitted together. The main leaf is hung
to the posts of the door frame as indicated
at Figs. 1467 and 1470. The two leaves are
hung together by back flap hinges. The
particular casings partly forming the boxings are shown at Figs. 1467, 1470, and 1472.
The curved stiles for the frame soffit may be
worked in the solid out of one piece. The
panels being very slightly curved, it would
certainly be the more simple way to work
these out of the solid.

Elliptical-headed Window with Framed and Splayed Linings.

Figs. 1492 to 1494 represent the inside
elevation, plan, and vertical section of an
elliptical-headed window with double-hung
sashes, cased frame, framed panels, splayed
linings, soffit, and window back with elbows.
The inside of the opening is finished with
architrave mouldings, with plinth blocks at

Fig. 1496.—
Enlarged Detail
of Vertical
Section.

this class, may be conveniently introduced
here with the object of illustrating and explaining the method of bending curved sash
bars. The preparing of the stuff, setting out,
mortising, tenoning, and other processes in

the making of the frame and sashes, being generally identical with cases previously treated, it is here only necessary to describe the new features.

Construction of Elliptical Head of Sash Frame.—This may be made in two or three pieces, cut out of the solid, jointed a little above the springing, and fastened, as shown at Fig. 1498, with screws. The crown joints or radial joints (as the case may be) are fastened by dowelling or handrail screws. In

side, and planed true with a compass plane. The pieces to form the parting bead should be similarly treated. The head just above the springing is secured to the pulley stiles with screws as indicated at E (Fig. 1498). It will be seen that the head does not finish at the springing, but at a sufficient distance above to allow of $\frac{1}{2}$ in. projection, as at F (Figs. 1497 and 1498). This is to allow the stiles of the sashes to butt against and prevent the bottom sash becoming jammed.

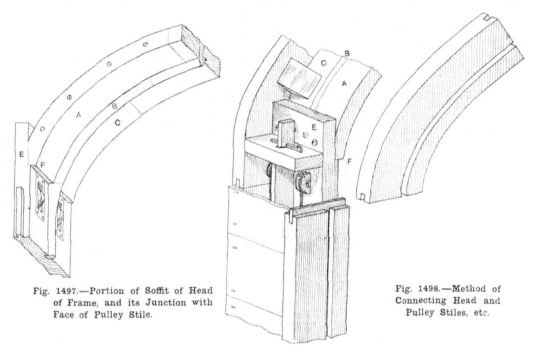

Fig. 1497.—Portion of Soffit of Head of Frame, and its Junction with Face of Pulley Stile.

Fig. 1498.—Method of Connecting Head and Pulley Stiles, etc.

this method a plough groove is worked out of the solid to receive the parting bead, which also is worked out of the solid. Another method, which is equally good, is to form the head in three laminations, as represented in section at A (Fig. 1496), and also by A, B, and C in the conventional views (Figs. 1497 and 1498), where it will be seen that the thickness on the inside of the parting bead may be made of three pieces round the curve, and the portion of the parting bead also of three pieces, while the outer portion is made of four pieces. These pieces should be accurately sawn out on the soffit

Elliptical-headed Linings. — These are got out of the solid in two or three pieces, which are jointed together and connected to each other, and to the straight linings, by cross tongues (see Fig. 1498). The joints are, of course, glued, and the linings are nailed on in the usual manner. The head linings are glued and blocked as represented at Fig. 1498. The inside head linings may be ploughed to receive the tongue of the soffit, so as to correspond with the straight inside linings.

Elliptical Head of Sash.—This is made of three pieces, the joints occurring as at

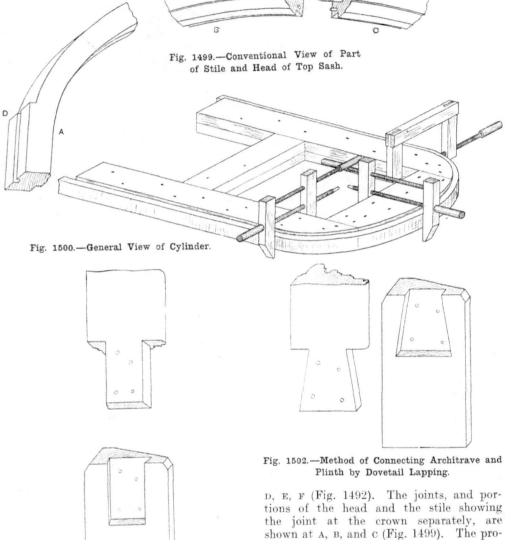

Fig. 1499.—Conventional View of Part
of Stile and Head of Top Sash.

Fig. 1500.—General View of Cylinder.

Fig. 1501.—Connecting Architrave and Plinth
with Slip Dovetail Tenon.

Fig. 1502.—Method of Connecting Architrave and
Plinth by Dovetail Lapping.

D, E, F (Fig. 1492). The joints, and portions of the head and the stile showing the joint at the crown separately, are shown at A, B, and C (Fig. 1499). The projecting shoulder D (Fig. 1499) is for butting against the stop of the head, shown at Fig. 1498.

The Arch or Cot Bar.—This may be formed of two pieces and worked out of the solid, joints occurring at the crown and a little below the springing; but a more satisfactory job results when the bar is made in one continuous piece to meet with the two marginal bars A and C (Fig. 1492). A rib or cylinder round which to bend the bar must be specially made. A suitable form for this purpose is illustrated at Fig. 1500, where it is shown constructed of two thicknesses. The strip of wood for the bar should be obtained as straight-grained as possible, and should either be steamed or be soaked in boiling water, steaming yielding the better results. Then, by means of hand screws, the strip of wood should be gradually bent round and fastened to the cylinder as illus-

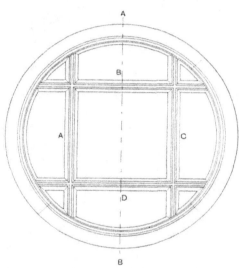

Fig. 1503.—Elevation of Circular Bull's-eye Frame with Central Sash.

trated. A piece of hoop-iron bent round with it on the outside will be found useful in preventing fibres of the wood from bursting out. When the bars are thick, it is a good plan to form them of two thicknesses, gluing them together as they are bent round the cylinder. It is best to let the strips remain on the cylinder a few days, so that they may become thoroughly set to shape; and, on taking off, they should be kept to their shape by means of a couple of stretchers. They can then be rebated and moulded. It is not necessary to weaken the cot bar by mortising for tenons of radial bars, as these latter need only be scribed to fit the cot bar, and then each one secured by a fine

long screw inserted through the cot bar. It is more satisfactory if these bent bars are of straight-grained oak, ash, or other hardwood that is pliable. Two methods of connecting the architraves with the plinths by dovetailing are illustrated at Figs. 1501 and 1502. After the joints are made satisfactory, the parts will be glued and screwed together.

Circular Bull's-Eye Frame with Square Centre Sash Hung on Centres.

Figs. 1503 and 1504 show, respectively, the elevation and the vertical section of

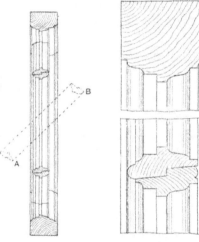

Fig. 1504.—Vertical Section through A B (Fig. 1503).

Fig. 1505.—Enlarged Detail at B (Fig. 1503).

a circular bull's-eye frame which is moulded inside and out. The elevation is divided by four stout bars, the bars forming the centre square being in two parts, rebated together so as to form a square sash, which is hung on centres as illustrated in the section (Fig. 1504), the construction being shown more clearly in the conventional sectional view at Fig. 1506. The frame is constructed of four pieces, as indicated at Fig. 1503. The joints may be held together with handrail screws and dowels, or with hammer-headed keys and tongues. The bars are moulded and scribed to intersect with the mouldings of the frame, and connected by mortice-and-tenon joints. The inside of the bars A, B, C, and D are not moulded, but are rebated and slightly splayed, so as to facilitate the opening and closing of the sash (see A B in section, Fig. 1504, and see also Fig. 1505). The sash is made of four pieces, which are rebated and splayed, and fit the bars just mentioned, which are moulded so that when

tical section at Fig. 1508. The point calling for special attention is the setting out of the cutting of the beads so as to allow of these cuts properly clearing as the sash is

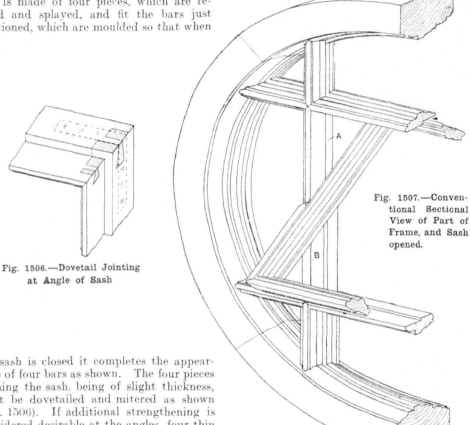

Fig. 1506.—Dovetail Jointing at Angle of Sash

Fig. 1507.—Conventional Sectional View of Part of Frame, and Sash opened.

the sash is closed it completes the appearance of four bars as shown. The four pieces forming the sash, being of slight thickness, must be dovetailed and mitered as shown (Fig. 1506). If additional strengthening is considered desirable at the angles, four thin brass brackets may be let in flush, screwed on as shown. The rebates of the vertical bars to meet those on the stiles of the sash are on the inside of the upper portion as at A, and on the outside of the lower half as at B (Fig. 1507). The hanging of the sashes on pivots is similar to that explained and illustrated in the next example.

Sash Hung on Pivots.

A solid frame with weathered, throated, and sunk sill, and sash hung on pivots, is shown as closed, and also as opened, by ver-

opened or closed. When setting out the rod, have at least a portion of the sash opened to the full extent required, the beads being included, as indicated at A B (Fig. 1508). Where the outer edges of the beads on the sash when opened intersect with the lines of the beads fixed to the frame as at C and D, between these two points draw the line C D; draw C F, and C E at right angles to C D, and then C D and D F are the lines of the cuts for the beads fixed to the frame. With centre G and radii G and D, determine the points H and K. Again with G as centre, and E and F as radii, determine the points L and M. Then clearly H L and K M are the splays for the beads fixed to the sash ; and when the sash is closed this would meet C E and F D respectively. When the pivot is fixed on the frame, and the slotted plate on the stile of the sash, a small chase has to be made in each stile, as indicated by the dotted lines from G to M. When the pivots are screwed on to the stiles, then the chases have to be made in the frame. In order that the head of the sash shall not bind as it is opened, the head should be prepared a little out of the square, as illustrated.

Circle-on-Circle Sashes and Frames.

Fig. 1510 represents a horizontal section (looking up) showing the soffit of arch, sash,

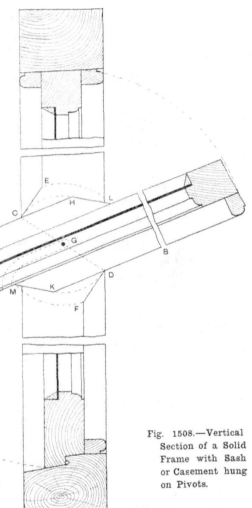

Fig. 1508.—Vertical Section of a Solid Frame with Sash or Casement hung on Pivots.

frame, linings, etc., and Fig. 1511 represents a part outside and part inside elevation of an upper portion of an opening, with a cased frame and double-hung sashes, for a window which is semicircular in elevation and circular on plan. This is commonly known as circle-on-circle work. Sometimes frames of this description are made with the faces of their pulley stiles radiating as shown by the lines A and B (Fig. 1509). When this method is adopted, the sashes cannot be inserted into their positions from the inside. The projecting portion of the

lower half of each outside lining must be movable, so that the sashes may be placed in position from the outside. The sashes are therefore troublesome to hang, and

the renewal of sash lines is difficult. The most common method of constructing these sashes is illustrated at Figs. 1510 and 1511, which show the geometrical setting out for obtaining moulds, bevels, etc., for the head of the frame and the head of the sash. The outer arris of the soffit of the arch is a semicircle (A B, Fig. 1511), and therefore the inner arris becomes elliptical, as shown by C B (Fig. 1511). On X Y (Fig. 1513) set up the elevation of the curve C' B', the same as C B (Fig. 1511). This is the line

between the head and the pulley stile. Mark off any convenient points on C', B', as M', N', O', P' (Fig. 1513). Project these down to the plan (Fig. 1514), giving the

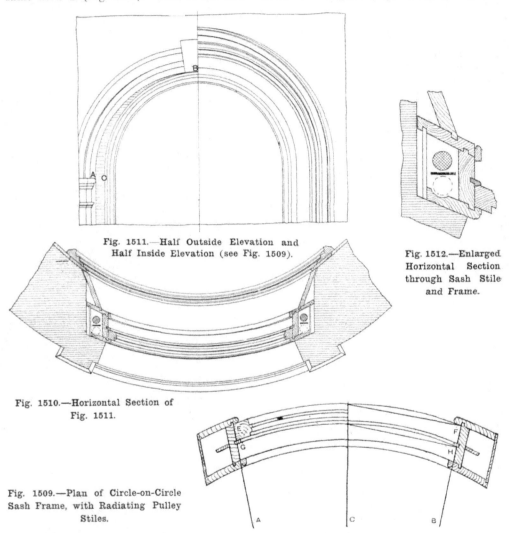

Fig. 1511.—Half Outside Elevation and Half Inside Elevation (see Fig. 1509).

Fig. 1512.—Enlarged Horizontal Section through Sash Stile and Frame.

Fig. 1510.—Horizontal Section of Fig. 1511.

Fig. 1509.—Plan of Circle-on-Circle Sash Frame, with Radiating Pulley Stiles.

elevation of the frame, and also the top edge of the sash. Projecting down, draw the half plan of soffit shown by Fig. 1514. The thickness of the head may be drawn in as shown by the lines D', E', F', G', H'. The breadth of the sash may also be drawn in, as indicated by K' L'; also the connection

plans of the generators of the soffit C, M, N, O, P, B. The soffit mould can now be obtained. At right angles to the projector Q C', draw a line Q S (Fig. 1515), and along it mark off Q, 5, 6. 7, 8—respectively C', M', N', O', P', B' (Fig. 1513). Then, projecting at right angles from the points on the line Q s (Fig. 1515), and then from M, N, O, P, and B (Fig. 1514), parallel to Q S, the points M, N, O, P, and B (Fig. 1515) are obtained. Through these intersections the curve may be drawn in as shown. The soffit mould may be completed by obtaining points and drawing the curve Q T, and also for the head of sash by U V.

Face Moulds for Head of Frame.—For the face mould for the head of the frame, join the points Q R in plan ; and where the plans of the generators M, N, O, P, and B cross this line, as in points 1, 2, 3, 4, 5, draw ordinates at right angles to Q R, making these the same lengths as the ordinates M', N', O', P', and B' in elevation (Fig. 1513). Then the curve may be drawn in.

Face Moulds for Head of Sash.—It is here assumed that the stiles of the upper sash will continue part of the way into the head, and then there will be a joint as shown at W (Fig. 1513). There will be then a crown joint as represented at L'. By projecting down from W, to the tangent line drawn from C, and also from Q, and

Fig. 1515.

**Figs. 1513 to 1515.—
Geometrical Setting Out.**

drawing ordinates at right angles to this tangent line and making them equal in length to the corresponding ones in elevation (Fig. 1513), the face mould from the springing to the joint W can be drawn as shown at A (Fig. 1514). From the face mould shown at B (Fig. 1514), draw the tangent line 10. Then where the plans of the generators meet this line, project up ordinates, and then the face mould B can be obtained as previously explained, the joints being shown at C W and B'.

Preparing the Head.—Probably the most satisfactory way of making the head of the sash frame is to prepare it in two pieces out of the solid, with a joint at the crown, held together by one or two handrail screws, and fitting into laps made into the pulley stiles as illustrated at Fig. 1517. At Fig. 1516, half the plan of the soffit is represented by dotted curved lines, and the rhomboid A, B, C, D enclosing it

20 *

shows the thickness of the plank required. Projected above are shown the surface of the plank and the application of the face mould. By cutting square through the plank and making the surface L M horizontal, the bevel can be applied as represented at B C in plan. By cutting square through and level at N F, the bevel can be applied, and then the face mould can be used for marking the other surfaces of the plank

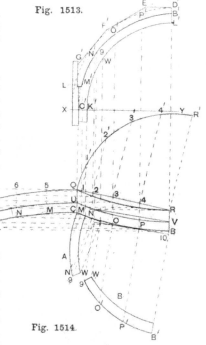

Fig. 1513.

Fig. 1514.

as shown by the dotted lines F. Assuming that the two pieces to form the head have been sawn out, and the soffit planed true to the face-mould lines, the joints made, and the head fitted and screwed to the pulley stiles as shown at Fig. 1517, the soffit mould can be applied and the soffit lined out as illustrated. In order to keep the pulley stiles equidistant from each other until the linings are fixed, the curved stretcher should be temporarily fixed to the head of the stiles as represented at Fig. 1517. If the sill has been prepared to the proper curve when the superfluous wood is planed off the head, a long straightedge should fit on the sill, stretcher, and head, in all positions such that it is parallel to the pulley stiles.

Inside and Outside Linings.—In preparing these, it will involve a little more

Fig. 1516.—Application of Bevel and Face Mould for Head of Frame.

Fig. 1517.—Head of Frame Jointed and Fixed to Pulley Stiles, and Soffit Mould Applied.

Fig. 1518.—View from Outside of a Portion of Head of Frame.

to the lines. Then the pieces should be accurately jointed, tongued, glued, and blocked in position, and the surface of the inside lining smoothed off so as to fit a straightedge applied to the sill and the lining, as explained in connection with Fig. 1517. For the outside lining, only a little more than the seen margin need be cleaned off, if it is to fit against brickwork; but in the case of fitting against masonry, a fair amount of accuracy would be necessary in cleaning off. The edge of the outside lining must be worked so as to project the proper distance from the soffit, and then moulded, if moulding is shown and specified; the edge of the inside lining, of course, finishing flush. The parting

labour, but will make a much more satisfactory job, if the joint, instead of being at the springing, is made some little distance above it, as illustrated at Fig. 1518. This will keep the frame more rigid, and the pulley stiles parallel to each other. So as not to have the grain of the other parts of the head linings too short, it will be necessary for them to be in two pieces. It will be found the simplest plan to have the stuff thick enough to work the inside surface to fit the head, and from this to gauge full for the outside, and rough off

bead will have to be in three or four pieces, and is cut off a thin board to the proper curve, and may be bent sideways into the plough groove, being then glued in position.

Preparing the Head of Sash.—By reference to the plan Fig. 1509 and the enlarged detail Fig. 1512, it will be seen that the stiles are not square in section, but these of course would be worked to a bevel set to the angle U, C, N (Fig. 1514). By reference to Fig. 1513, it will be seen that part of the bead and a straight stile are formed in one; and the face mould for the curved part is shown at A (Fig. 1514). The face mould is applied to each side of the plank in a similar manner to that explained for the head of the frame (Fig. 1516). It will be seen that a portion of the head w to 1' (Fig. 1513) is shown in the plan (Fig. 1514) as enclosed by a parallelogram, which shows the thickness of the stuff required. The face mould is applied to each side of the stuff by using a bevel set to the angle shown at B (Fig. 1514). The next process is to true up the top edges, apply the falling mould plane nearly to the lines, and then fit each piece to the frame in its proper position, when the joints should be made and bolted together. The meeting rail is of course of the curve shown in plan, but in other respects it is dovetailed to the stiles as explained for previous examples. If all fits in square, the next thing will be to mould and rebate the head and stiles. As the section of the moulding varies round the head, it should be formed by small rebate planes and hollows of compass pattern.

Guard Beads.—Up the sides, these are of the ordinary pattern, the edge of the sash being planed off so that the joint between it and the guard bead is square to the pulley stile, as will be seen by reference to Fig. 1512. The guard bead to the soffit may be in sections worked out of the solid, which will be found the easier method ; but, alternatively, this could be bent in a square state, and moulded afterwards ; but this method of course would be more expensive, and would involve the construction of a special cylinder on which to bend it. Fig. 1518 is a conventional view of a portion of the head of the frame from the outside. It will be seen that the back of the splay lining is represented as formed of a veneer with staves jointed and glued at the back. The development and construction of this will be treated of in a subsequent section. Of course there are other ways of constructing the frame, such as building it up in three thicknesses, the middle thickness forming the parting bead ; but as the head would practically be formed of two separate parts extra, twice the amount of setting out would be necessary, and therefore, though a little stuff might be saved, a greater expenditure of time would be involved. In superior polished work, the head would be constructed of a veneer and staves at the back. In this case, of course a special cylinder would have to be made on which to bend and block the veneer.

MOULDINGS: WORKING AND SETTING THEM OUT.

Introduction.—A moulding is a curved surface whose section is continuous. The essentials of a good moulding, apart from suitability of design, are, that its surfaces shall be perfectly regular and smooth, its edges sharp, and its curves flowing and

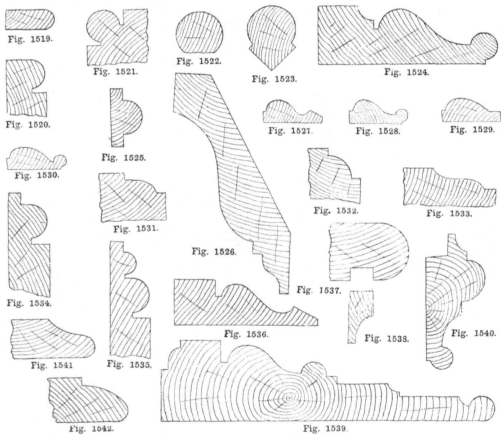

Fig. 1519. Fig. 1521. Fig. 1522. Fig. 1523. Fig. 1524. Fig. 1520. Fig. 1525. Fig. 1527. Fig. 1528. Fig. 1529. Fig. 1530. Fig. 1531. Fig. 1526. Fig. 1532. Fig. 1533. Fig. 1534. Fig. 1537. Fig. 1536. Fig. 1538. Fig. 1540. Fig. 1541 Fig. 1535. Fig. 1542. Fig. 1539.

Fig. 1519.—Parting Bead. Fig. 1520.—Quirked Bead. Fig. 1521.—Angle or Returned Bead (when of Large Section, generally known as a Staff Bead). Fig. 1522.—Section of Bead or Round frequently Fixed as a Staff Bead. Fig. 1523.—Staff or Angle Bead. Fig. 1524.—Cavetto Quirked Ogee and Bead Architrave. Fig. 1525.—Astragal and Fillets. Fig. 1526.—Cyma Recta, or Reverse Ogee, with Fillet and Ogee Cornice Moulding. Fig. 1527.—Quirked Grecian Ogee Panel Moulding. Fig. 1528.—Quirked Ogee and Bead Moulding. Fig. 1529.—Quirked Ovolo and Fillet. Fig. 1530.—Quirked Ovolo and Bead. Fig. 1531.—Ordinary or Grecian Ovolo. Fig. 1532.—Roman Ovolo. Fig. 1533.—Lamb's Tongue. Fig. 1534.—Torus. Fig. 1535.—Double Torus. Fig. 1536.—Grecian Ogee Base or Architrave Moulding. Fig. 1537.—Nosing. Fig. 1538.—Scotia. Fig. 1539.—Double Face Architrave. Fig. 1540.—Bolection. Figs. 1541 and 1542.—"Thumb."

without perceptible break. In the hand manufacture of mouldings, the above requirements are obtained by the employment of planes and routers of suitable contour, and by the careful cleaning off of the surfaces with glasspaper, wound round rubbers of wood or cork. Hand manufacture has now given place largely to machine work, the vertical spindle moulding machine producing mouldings cheaply and well.

Varieties of Common Mouldings.

Common mouldings used in joinery include those shown on p. 468. It may be said that bead is a general term applied to small mouldings of circular section. Quirked bead is a small moulding semicircular in

in Figs. 1543 to 1545. Bead planes are made in sets of nine, from $\frac{1}{8}$ in. to 1 in., increasing in size by $\frac{1}{16}$ in. up to $\frac{1}{2}$ in., and thereafter by $\frac{1}{8}$ in. The better makes of these tools have the working faces in boxwood, the body of the plane being in beech, and the smaller sizes are " slipped "—that is, one side of the plane is rebated, and a loose slip fitted in and secured with screws, the object being to enable the tool to be worked down into a rebate or over a projecting surface, such as the bead on the edge of the bottom moulding shown in Fig. 1546. Figs. 69 and 70 on p. 15 are end views of the stocks of a pair of hollows and rounds, with sections of the mouldings produced by them. These are made in sets of nine pairs, their curves

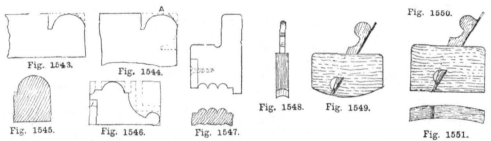

Figs. 1543 to 1546.—Beads.   Fig. 1547.—Reeding Plane.   Figs. 1548 and 1549.—Thumb Hollow.
Figs. 1550 and 1551.—Curved Thumb Rebate Plane.

section, stuck level with a surface, and separated from it by a groove known as a quirk (see Fig. 1520). Return bead and staff bead have quirks on adjacent surfaces, generally meeting at right angles (see Figs. 1521 and 1523), but occasionally at obtuse and acute angles. The terms return bead and staff bead generally include all beads that are at an angle; but beads of small section are often distinguished as return beads, and those of larger section as staff beads. In the North of England the inside $\frac{5}{8}$-in. bead to sash frames is frequently called a staff bead, whereas in London and the South of England it is generally known as an inside bead or guard bead.

Planes for Straight Moulding.

Figs. 66 to 68 on p. 14 show a bead plane and its cutter or " iron," a typical example of a tool for producing such beads as are shown

being portions of circles whose radii increase by $\frac{1}{8}$ in. Figs. 71 and 72 on p. 15 are sash moulding planes, named respectively ovolo and lamb's-tongue; such planes are made in pairs, one to follow the other, and in three sizes, namely, $\frac{1}{2}$ in., $\frac{5}{8}$ in., and $\frac{3}{4}$ in., these being the distances the moulding works on the edge of the stuff, the respective depths being $\frac{3}{16}$ in., $\frac{1}{4}$ in., and $\frac{5}{16}$ in. Fig. 73, p. 15, is a cabinet-maker's ogee moulding plane, which may be had in four sizes, from $\frac{3}{8}$ in. to $1\frac{1}{2}$ in. These, in common with most moulding planes, require holding at an angle of about $20°$ with the side of the stuff being worked, the line shown on the fore end of the plane being kept vertical, or in line with the surface, against which the fence of the plane works. Beads, on the contrary, are always held upright when being worked. Fig. 1547 is a reeding plane, which may be made for two, three, and five reeds, the one illustrated being a three-reed plane. They are usually slipped similarly to the beads, and for the same reasons, and in some the fence is movable so that the margin may be varied. Fluting planes, the converse of these, are also made, but are not much used, as the work can be done equally well with rounds. The foregoing are the principal planes for working straight mouldings, but there are

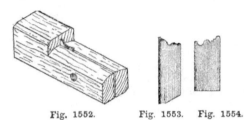

Fig. 1552.   Fig. 1553.   Fig. 1554.

Figs. 1552 to 1554.—Scratch Tool and Scratches.

in addition a few special moulding planes used by joiners, such as shop-front lamb's-tongue, stair-nosing, and scotia planes; these, however, are not likely to be of general service.

Tools for Shaped Mouldings.

Thumb Hollows and Rounds.—Figs. 1548 and 1549 show respectively an end and a side view of a thumb hollow for working rounds on curved work, and a complete set can be had matching the set of ordinary hollows, and also rounds; however, with a little ingenuity in their use, three or four can be made to work almost every imaginable curve. They are generally made of some very hard wood, such as box, ebony, and lignum vitæ, and are from 2 in. to 4 in. long, 2 in. deep, and from $\frac{3}{8}$ in. to $\frac{5}{8}$ in. thick. The irons may be purchased as blanks from the tool-dealers, or may be easily made from a piece of sheet steel.

Curved Thumb Rebate Plane.—Figs. 1550 and 1551 are side and plan views of a curved thumb rebate plane; these planes are made to various sweeps and thicknesses, and are very useful in working rebates and squares on curved work. These rebates are also made with circular soles, as in Fig. 1549. The thumb planes above mentioned would be indispensable for making work of double curvature, a good example of which is

given at Figs. 1283 and 1286, pp. 394 and 395.

Scratch Tools.—Scratch tools (Fig. 1552) are much used by wood-workers and joiners for working small mouldings on sweep work, and occasionally on straight work also. They consist of small pieces of hardwood sunk at one end, to form a fence, and with a slot or saw-kerf to receive the cutter or scratch, which is gripped tightly in the slot by means of a wood screw turned into the end of the stock. The scratch is a piece of thin hard steel filed up to an exact reverse of the moulding it is desired to produce. It is then rubbed square across with an oilstone slip to remove the file marks, and finally the edge is turned with a bradawl or scraper — that is, the bradawl is rubbed very firmly along the edge until a sharp burr is produced on each side; this burr forms the cutting edge, and is, of course, soon worn away, especially on hard wood. The proper formation of this burr is the secret of producing good work with the scratch. Figs. 1553 and 1554 are views of two steel scratches, the one shown in the stock in Fig. 1552 being a sash ovolo. In using the scratch, the cutter should not be projected the full depth at first, but after some of the surplus wood has been removed it is pushed out to its proper projection, the

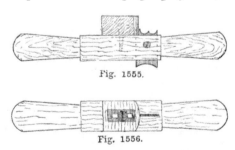

Fig. 1555.

Fig. 1556.

Figs. 1555 and 1556.—Scratch Tools.

correct amount being shown by a file mark on the face. As much as possible of the surplus wood should be removed with gouge, chisel, or plane, as may be most convenient, and the scratch is then rubbed backwards and forwards a few inches at a time until it is down, a final rub continuously through the length being given with a sharp scratch. Figs. 1555 and 1556 show respectively a side and an under-edge view of a scratch tool as supplied by tool-dealers. This is about 8 in. long, and is made of beech, with turned handles and a boxwood movable fence, one face being square, the other round ; into the fence a slotted brass plate is let, through which a screw slides, giving about 1 in. adjustment, and more can be obtained by re-entering the screw. The fence is also slotted slightly to pass over the cutter, which prevents side movement when working, the cutter being secured by a set screw

Fig. 1557.—Quirk Routers.

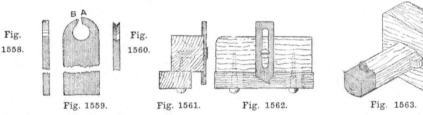

Fig. 1558.   Fig. 1560.

Fig. 1559.   Fig. 1561.   Fig. 1562.

Figs. 1558 to 1560.—Cutter.   Figs. 1561 and 1562.—Quirk Router.

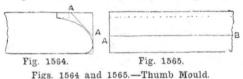

Fig. 1563.

Fig. 1563.—Cutting Gauge.

at the side. This tool is useful on larger mouldings, and is shown with a two-reed scratch in it.

Quirk Routers.—Quirk routers (Fig. 1557) are made of malleable iron, and have adjustable fences of various shapes, with a pair of clips and thumbscrew for holding the cutter. Three views of the latter, to a larger scale, are given in Figs. 1558 to 1560. This tool is used primarily for cutting quirks for beads and other mouldings, but is also very handy for sinking grooves in any curved work where a plough would not be available ; it is used similarly to the scratch tool. The cutter (Fig. 1559) is made of sheet steel, from $\frac{1}{16}$ in. to $\frac{3}{16}$ in. thick, the ends forming parts of two concentric circles, one slightly larger than the other. The larger one A, filed to a **V** section, as shown in Fig. 1560, forms the gauge or side cutter of the proposed groove, and must always be on the front or forward side of the tool, but the side B is filed square to a chisel edge, and

this part forms a cutter for removing the core. The circular hole inside the cutting edges is to receive the dust or shaving removed in the stroke, and the tool requires repeated lifting from the cut to clear this throat. Figs. 1561 and 1562 illustrate a home-made quirk router that will in many cases prove quite as serviceable as the more elaborate shop-made article. A block of hardwood, about 2 in. long by $\frac{3}{4}$ in. square, is grooved across one face (see Figs. 1561 and 1562) to receive the cutter, which should fit tightly. This is a piece of saw blade, $\frac{1}{16}$ in. thick, $1\frac{3}{8}$ in. long, and $\frac{3}{8}$ in. wide, with a $\frac{1}{8}$-in. slot in it ; a stout $\frac{1}{2}$-in. wood screw passing through this slot secures the cutter at any height desired. A movable fence, made of a slip of hardwood $\frac{3}{4}$ in. by $\frac{3}{8}$ in.

Fig. 1564.   Fig. 1565.

Figs. 1564 and 1565.—Thumb Mould.

by $2\frac{1}{2}$ in., is secured to the stock with a couple of round-headed screws, which pass through two small slots in the fence. This proves a very efficient tool for inlaying strings in marqueterie work.

Cutting Gauge.—This tool is illustrated at Fig. 1563, and is used for cutting the edges of the various members of a moulding, more especially when working them across the grain, the object being to obtain a clean-cut edge. The small movable cutter can be adjusted to various depths and taken out for sharpening on the oilstone. When using it, the flat face of the cutter must always be turned towards the side on which the clean edge is desired.

Working Thumb Mould.

To work a thumb mould (Figs. 1564 and 1565) on the edge of a table or sideboard top, first square up the edges to the required size, then with the cutting gauge set to the breadth of the moulding, gauge a line on the top all round. Next set a plough to a shaving less than the gauge mark and to the exact depth of the sinking required, as shown by the dotted lines in

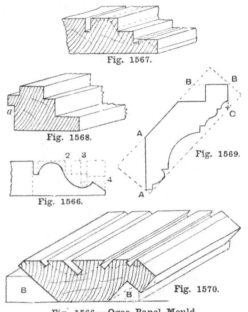

Fig. 1566.—Ogee Panel Mould.
Fig. 1567.—Working Ogee Panel Mould.
Fig. 1568.—Working Bolection Mould.
Fig. 1569.—Cornice Mould.
Fig. 1570.—Working Cornice Mould.

Fig. 1565, and work this on the two side edges of the top. Then fix a straightedge to the gauge line across the end by means of two handscrews or clips, and run in a tenon saw to the requisite depth. Again run in the saw; then the plough may be worked across the ends to remove the core to the proper depth. Chamfer off with the jack plane as indicated by A B, Fig. 1565, and, selecting two hollows of suitable size, work the larger one down first. Then turn the top up on edge and finish off the curve

with the smaller one. The side moulds should be worked first, and a mitre line drawn on each end with a mitre template, the end moulds being then worked down to this line, as the shape of the mould cannot well be marked on the moulded edges.

Working Ogee Panel Moulding.

In Fig. 1566 the dotted outline represents the end of the board on the edge of which the moulding is to be stuck, and the full line is the shape of the moulding which should be marked on each end by means of a pattern template. For this the ends of the board should be planed and rubbed over with chalk, so that the pencil line shall show clearer. Having shot the edge of the board straight and square, gauge the edge to the required thickness from the face, and try up the back. Set the plough with No. 1 iron and plough the grooves as indicated at 1, 2, 3, and 4 (Fig. 1566) to the distance on of the sticking as shown at 1. The core between grooves 2 and 3 can be removed with a chisel and levelled with a rebate plane, when the board will have the appearance shown in Fig. 1567. Next work off the salient angles of the core in a series of chamfers with the rebate plane as indicated by the dotted lines, and with a suitable round work out the hollow part of the moulding. A reverse of the template, cut out of card or thin wood, is useful for testing the depth. Begin the round with a suitable hollow. If the hollow will not work on the off side quite down to the bottom of the quirk, a snipe-bill plane must be employed to finish. This is specially for working in the quirks of mouldings.

Working Bolection Moulding.

To work a bolection moulding (Fig. 1546, p. 469), a piece out of which the mould is to be produced should be planed up accurately to size all round first and grooves run in as indicated by the dotted lines till it assumes the shape of Fig. 1568. Work the return bead on the front edge, removing the side slip for that purpose. Then the upper round or astragal should be worked, and two thin marking slips prepared exactly the width of the two side flats or "fillets." These are to be drawn along the sides of the astragal as a guide to the marking knife to cut in the edges of the hollows, which are thereafter worked with suitable rounds.

Working Cornice Moulding.

Fig. 1569 is a section of a cornice mould set up in the position it would occupy when fixed, the dotted outline indicating the necessary size of the piece required to produce it; this piece should be planed true

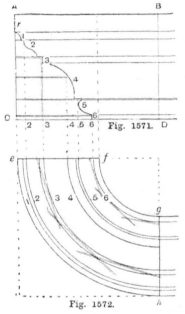

Fig. 1571.—Elevation of Circular Cornice.
Fig. 1572.—Plan of Circular Cornice.

on the back or worse side, and the edges shot to the width as shown. Then the outline having been marked on each end as shown by the full lines, gauge lines should be run on from the points A, B, and C, and the piece planed off to these bevels, which should be at right angles or square to each other. Then the rebate for the cover board should be formed by ploughing a groove, and other grooves are then to be run in to form the fillets as shown in Fig. 1570. The piece is here shown supported by two strips B nailed to the bench; it will,

of course, require turning round when being ploughed from the edge A (Fig. 1569). The core is removed with gouges, and preliminary rebates are formed as guides for the depth of the stickings, as described previously for the ogee moulding.

Working Circular Cornice Moulding.

Figs. 1571 and 1572 are the elevation and inverted plan of a circular corner to the cornice shown in section at Fig. 1569. The rectangle enclosed by the lines A B and C D (Fig. 1571) shows the size of the block required. Two templates will be necessary, one to the plan curves e h and f g (Fig. 1572),

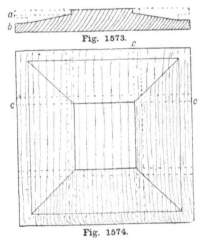

Fig. 1573.—Section of Raised Panel.
Fig. 1574.—Plan of Raised Panel.

and one as shown in Fig. 1569. The joints e f and g h are made first, square with each other and the top surface; then the plan template is applied at the top and bottom and the outlines are marked, the front and back edges of the piece being worked off with chisel and gouges, or cut with a bow-saw and finished with a spokeshave. The section template is next applied on each joint, the edge B C (Fig. 1569) being kept flush with the front, and the outline marked; pencil lines are then gauged round from the front on the bottom surface of the block from all the members of the moulding, as indicated by the lines numbered 1 to 6 in Fig. 1572, the numbered points in Fig. 1571 corresponding. The block is now fixed on the bench, and saw-cuts are run in tangent to the various curves, and are met by corresponding cuts from the face of the block, as indicated by the double lines. These are the commencements of a series of rebates to be finished with chisels and rebate planes, and they must be carefully worked one after the other, commencing with the member marked 6 and keeping the margins equal all round. After the rebates are finished, the moulds are worked with thumb hollows and rounds. To prevent the edges at the joints breaking out, the outlines should be

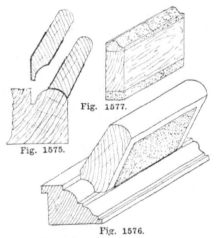

Fig. 1577.

Fig. 1575.

Fig. 1576.

Fig. 1575.—Rubbers for Beads.
Fig. 1576.—Holding Glasspaper on Rubber.
Fig. 1577.—Cork-faced Rubber.

carefully cut in for about ¼ in. with suitable gouges or chisels, from the outside towards the middle, and in working the planes take care not to go below this cut portion, the cleaning off being left until the piece is glue-jointed and dowelled to the straight portions of the moulding.

Working Raised Panel.

Figs. 1573 and 1574 are the section and plan of a sunk, raised, and fielded panel. It is prepared by trying up the back and gauging to thickness, and then cutting and squaring off to the finished size, as indicated by the outline in Fig. 1573. Next the sinking a is gauged all round the edges from the

face, and also the thickness of the tongue b, which is gauged from the back. The lines c c (Fig. 1574) are then gauged in with the cutting gauge, working from the edges of the panel. A plough groove is run along the sides, close to the gauge lines, to the required depth of the sinking, and similar grooves are cut across the grain with the tenon saw, as indicated by the dotted lines in Fig. 1574, the core being removed with a small chisel and the plough run across to regulate the depth. The panel is now turned up on edge in the bench-screw, back outwards, and a plough groove b (Fig. 1573) run around the edges to a depth slightly under the amount by which the tongue will enter the groove in the framing. Next work out the rebate a all round, and with a shoulder plane or a rebate plane set very fine and laid flat, go carefully around the sides of the sinking, working off clean and square to the gauge lines. Then chip away the core between the top and bottom of the chamfer, and true up with the shoulder plane. The cross or end grain should be worked first, and a mitre line drawn at each angle, when the sides can be worked off to these lines, which will produce correct mitres.

Cleaning Up Mouldings.

Mouldings are cleaned up by rubbing the surface with glasspaper of different degrees of fineness, applied to the various members of the moulding by means of suitably shaped rubbers of soft wood or cork; the latter is the better material, as it does not get heated so readily with the friction. This heating melts the glue with which the glass is attached to the paper, and causes it to come to the surface, where it adheres to the dust produced in rubbing, quickly clogging the paper and rendering it useless. The rubber must be made to fit the curves of the moulding exactly, and in the case of a large moulding the rubbers may fit different parts of the surface; generally, however, quirks and flats are best rubbed with separate rubbers, as in Fig. 1575, although, if care is taken, the flats on a small moulding may be cleaned up with the rubber used for the curved parts, as shown in Fig. 1576. The flat parts of the rubber should, however, pass well beyond the arrises of the fillets, to avoid the danger of the glasspaper rising on the edge of the rubber and taking off the sharp edge of the moulding. The paper should not be folded double on the rubber, and must be pressed very tightly round the latter; it should also not be allowed to overhang the ends of the rubber, but should be kept short. Wood rubbers are shaped with hollows and rounds, and are usually about 4 in. long by $2\frac{1}{2}$ in. wide. Cork rubbers, being generally made from odd pieces of cork (bottle corks), are of various lengths; they are sometimes glued in $\frac{1}{4}$-in. slabs to wood blocks and shaped

ing raises the grain and results in a much smoother finish.

Fixing and Fitting Mouldings.

Fig. 1578 is a sketch of a small door showing the method of inserting a planted moulding; the shorter pieces are mitered and shot to the exact length between the stiles, and the longer pieces are shot rather full in length, so that they can be forced in without bruising the ends when the other pieces are removed. Having fitted each piece separately, try each mitre to see that the mouldings meet correctly, which they should

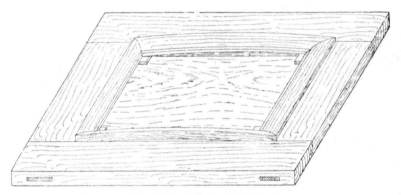

Fig. 1578.—Fitting Mouldings.

as required after drying as shown in Fig. 1577. A sharp gouge and a rat-tail file are useful tools for roughly shaping cork; then bed a piece of fine glasspaper round the moulding, and rub the rubber over this, which will fin'sh it accurately to shape. Oakey's cabinet glasspaper should be used, and is made in several degrees, the most useful being : M2, for the first cut in large soft wood mouldings; F2, for a similar purpose in hard wood; $1\frac{1}{2}$, a good general paper and suitable as a first cut for small mouldings, and a second cut for large deal mouldings; I, a fine paper to finish hardwood mouldings; and O, or flour paper, an extremely fine cut, only used when a very high finish is desired. In cleaning up hardwood mouldings, after the first papering up is finished, slightly damp the moulding with hot water, and when this has dried, rub down with the finer paper; this damp-

do, in all but the last mitre, if the lengths have been cut off consecutively. The last mitre, joining up the opposite ends of the length, will probably vary slightly in section, and should be trimmed before being planted in. All being ready, cut four little blocks and place one at each corner, as shown; then insert the end pieces of moulding, resting them on the blocks. Next spring in the side pieces by pressing the end down with the left elbow, pulling the middle of the length up into an arch with the left hand, and pressing the other end down into place with the right hand. When the end is entered, let go the middle, and the piece will spring in straight, bringing the mitre up tight.

Setting Out Mouldings.

Diminishing Moulding.—The following geometrical method will be found both simple and effective, both for diminishing and enlarging mouldings. To find the section of a moulding smaller than that represented at A, E, D, C (Fig. 1579), draw from A a horizontal line A G, at any convenient distance, produce the line E A to any convenient point H; from C draw the perpendicular C F, and then take any suitable points in the profile of the moulding as 1, 2, 3, and 4; from these project horizontals to meet C F in 5, 6, 7, and 8. Then projecting down from points 2, 3,

H F in f, draw the horizontal f, a, which cuts the radiating lines in b, e, g, and h; then from f' mark off corresponding distances as shown by a', b', e', g', h', and raise projectors as shown; then from the point where the radiating lines from G cut d, f', in 5', 6', 7', 8', project horizontally, and thus show points in the profile of the moulding as 1', 2', 3', 4'.

Enlarging Mouldings.—To enlarge the moulding to a breadth equal to G M and to a thickness of H N, on the vertical line

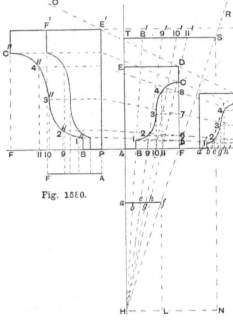

Figs. 1579 and 1580.— Diminishing and Enlarging Mouldings.

Fig. 1579.

Fig. 1580.

G M mark off the breadth, project a horizontal to cut the radiator G O in E'; draw the vertical line E' P, then produce the other radiators to meet the line as shown. Then project horizontally from H, and on H N mark off the given thickness, cutting the radiating line H R in S; draw the horizontal S T; produce the other radiators to meet this line as shown. From P (Fig. 1580) mark off distances B, 9, 10, 11, and F, to correspond with those at S T (Fig. 1579); projectors raised from these points will determine points 1″ to c″ in the curve, which can then be drawn as shown. By P, E', F' is shown the profile of a moulding which is much broader than but of the same thickness as the original moulding A, E, D, the setting out of which will be clearly understood without need of further explanation.

and 4, we obtain 9, 10, and 11. Now from H draw radiating lines passing through points B, 9, 10, 11, and F, also from G draw radiating lines passing through points 5, 6, 7, 8, and D. At G set up a perpendicular, and then make G K equal to the breadth of the moulding required. From K draw a horizontal line cutting G D in d, then from d draw the perpendicular d f'. Next from H draw a horizontal line making H L equal to the thickness required, then from L draw a line parallel to H A, so that it cuts

Raking Mouldings Round External Angle.—Fig. 1581 illustrates the process of obtaining the sections of raking mouldings round the external angle of a wall which is square. In this case the returned pieces are horizontal. The raking moulding A B C being given, it is required to find the profiles of the mouldings mitering into it. Take any number of points in the section given, and project two lines from each, one to the back of the section A C, and another parallel to A C to a line P L drawn from A at right angles to A C ; A′ C′ and A″ C″ are the elevations of the faces of the returning walls. Take any point P in C′ A′ and C″ A″ produced, and draw P L at right angles to it, and on that side on which is the moulding.

required sections. To avoid confusion of lines the points taken should be numbered, the lines drawn through them, and the points in the projection lines numbered in the same way. The above example also represents the moulding which occurs in the pediment.

Raking Mouldings on Internal Angle.—Fig. 1582 illustrates the various sections of a moulding running round the interior angles of a wall. It also shows the different forms of sash-bar moulding required for a certain kind of lantern light, and the shape of the

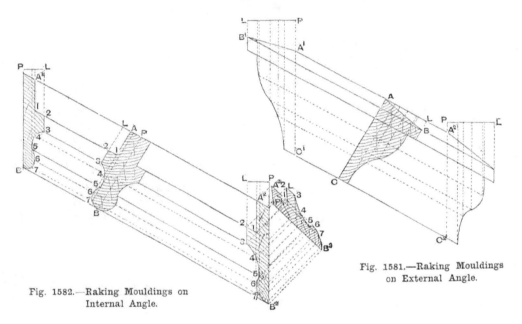

Fig. 1581.—Raking Mouldings
on External Angle.

Fig. 1582.—Raking Mouldings on
Internal Angle.

P L might be called the projection line. It is required to transfer the points on it to the other lines P L. To do this take a strip of paper, and, placing the edge against the projection line, make marks on it over the points. The strip can now be placed against the new lines and the several points marked off from it, taking care that P is over A and L above B. Project lines from the points in the projection lines, parallel to the backs of the mouldings A′ C′ and A″ C″. Draw also lines, to meet these, through the points taken on the given profile, and parallel to the lines A′ C″ and A′ C″. The intersections of these lines are points on the

angle bar in a shop front. A B is the section given or known. Points 1, 2, 3, etc., are taken in the profile, the points in the curved portions being taken sufficiently close to admit of a freehand curve being drawn through the resulting points. The process is the same as for the previous example. A′ B′ and A″ B″ are the sections of the mouldings mitering into the raking one, which are horizontal. When the mouldings on both faces of the wall are on the rake or at an angle, one of the methods illustrated at the right of the figure is adopted. In one case the points on the profile are projected to the back line of the section, and lines are drawn

from these at the angle to which the moulding is to be fixed. A new back line is then
drawn across these lines and at right angles
to them. A line is now drawn from one of
the front points, say A″, at the new rake of
the moulding. By drawing the projection

responding lines drawn through the projection line.

**Raking Moulding intersecting an Obtuse
Angle with a Horizontal.**—A B, C (Fig.
1583) is the plan of the faces of two walls
intersecting at B at an obtuse angle, A B

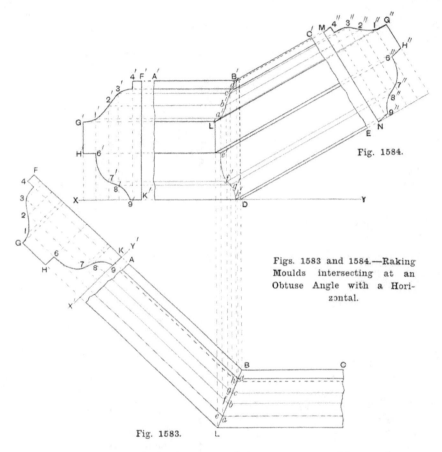

Fig. 1584.

Figs. 1583 and 1584.—Raking
Moulds intersecting at an
Obtuse Angle with a Hori-
zontal.

Fig. 1583.

line P L, and lines through the points in it,
the point A‴ is found. Join this to the end
of the new back line, and draw lines parallel
to it from the other points in the line. The
intersections of these lines, with those
drawn through the line P L, are points on
the new section. The other method, which is
the shorter when the elevation of the mitre
(A″ B″) has been drawn, is by drawing lines
from the points on the profile, and at the
rake of the moulding to intersect the cor-

being a level moulding and B C a raking
moulding. Parallel to B C draw X Y,
projecting up from B the back of the mitre,
the point D (Fig. 1583) is obtained ; draw
the inclination of the moulding as shown
by D E. Draw X′ Y′ at right angles to
A B (Fig. 1584), and draw the section of
the level moulding as shown by K, F, G, H, 9,
and from convenient points as shown by
1, 2, 3, and 4 project down and draw the
plan of the moulding and also the mitre

B L as shown; repeat the section of the horizontal moulding on X Y as shown by K', F', G', H'. Now consider the points 1 to 9 and 1' to 9' as the elevations of horizontal lines on the moulded surfaces, which have been projected from the sections and are repeated on plan. These lines in plan intersect at the mitre line in both the horizontal and raking moulds. Now projecting up from the intersections in the plan of the mitre and horizontally from corresponding points G', 1', 2', 3', 4', in the elevation, points L', a', c', d' are obtained. Then from these points draw lines as shown at the proper rake to cut

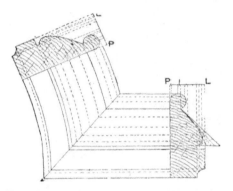

Fig. 1585.—Section of Ramp Moulding to have Straight Mitre.

the line M N, which is at right angles to B' C'; make the ordinates from G'', 1'', 2'', 3'', 4'' the same lengths from M N as the corresponding ordinates from F' K', thus obtaining the section of the upper portion of the raking moulding so that it shall intersect with the horizontal moulding in a vertical mitre. Points in the profile of the lower part of the section of the raking mould are shown at H'', 6'', 7'', 8'', 9''; these are obtained in an exactly similar manner to the upper portion, and no difficulty will be found in following the working illustrated. Each imaginary line in the upper moulded surfaces is represented by one straight line in plan in every case but one; in plan one line is made to represent an imaginary line directly under it on the lower moulded surface.

Mitering Mouldings.—When two pieces of moulding are joined together by mitering, the mitre is straight when the pieces are straight, or of the same curvature. To facilitate the operation of mitering, the mitre is often made straight, and the contour of one of the pieces of moulding is varied to suit it. Fig. 1585 illustrates the method. The profile of the straight piece is given; it is necessary to find that of the curved one. Take points on the given section, and project them to a projection line P L and to the mitre line. Draw a line across the curved moulding, radiating to the centre from which the curve is struck (or normal to the

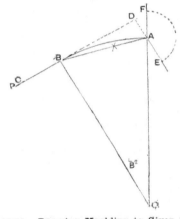

Fig. 1586.—Ramping Moulding to Given Point.

curve if it is not circular). Continue the lines drawn to the mitre line to this line, and project lines thence at right angles. Draw the projection line P L as shown, and obtain the intersections on the points on the required profile.

Ramping Moulding to Given Point.—Fig. 1586 illustrates the problem of ramping a moulding to a given point, such as occurs when a dado rail is continued from the level up a staircase, or from a staircase to a landing. C B D represents the line of the moulding. It is required to ramp it to the point A. Draw a line F A O vertical. The centre of the curve of the ramped portion will be in this line. It will also be in a line drawn from a point in C B D at right angles to it; the point in C B D being the place at which the straight portion meets the curved. Through A draw a line D E at right angles to C D. Take points F and E in A F and A E equidistant from A. With the compasses set to a distance about equal to the line joining F E, describe arcs with F and E as centres. Join A to the intersection of these arcs, and produce it to meet C D in B. From

Appliance for Marking Mouldings for Mitering and Scribing.

At Fig. 1588 is illustrated an appliance designed to do away with the difficulties commonly met with in marking mouldings for the purpose of returning the ends in the solid, or for scribing. Properly manipulated,

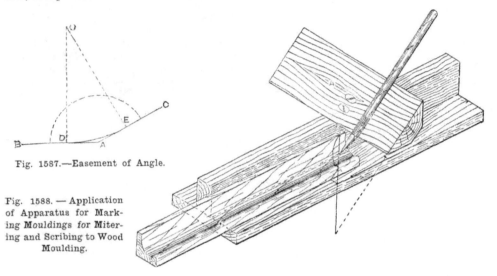

Fig. 1587.—Easement of Angle.

Fig. 1588. — Application of Apparatus for Marking Mouldings for Mitering and Scribing to Wood Moulding.

B draw B O at right angles to C D. Then O is the centre of the curve joining B to A, and to which the line C B is tangential. [Any difficulties met with in solving similar practical problems should be submitted to the Editor of "Building World" for solution in that journal.]

Easement of Angle.—In Fig. 1587 the method of easing an angle is shown. Take points D and E in A B and A C equidistant from A, and draw lines from these points at right angles to the lines in which they are. These intersect at O, which is the centre, while O D or O E is the radius of the curve.

it will give any kind of scribe or mitre that may be required. The pencil has a flat side planed throughout its length, so as to bed fairly on the plane, and is used as indicated in Fig. 1588, which shows the application of the apparatus : the moulding is held in position by the left hand, any marks that have been made on it being adjusted to the point of the pencil, which is then worked all round the surface of the moulding, care being taken that it always lies flat upon the plane. The curved line shows the path traced round the surface of the moulding by the point of the pencil.

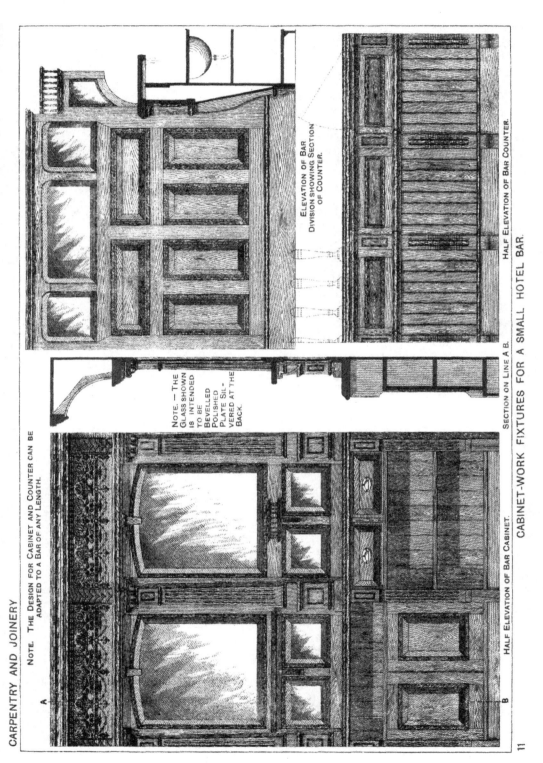

CARPENTRY AND JOINERY

NOTE. THE DESIGN FOR CABINET AND COUNTER CAN BE
ADAPTED TO A BAR OF ANY LENGTH.

NOTE. — THE
GLASS SHOWN
IS INTENDED
TO BE
BEVELLED
POLISHED
PLATE SIL-
VERED AT THE
BACK.

ELEVATION OF BAR
DIVISION SHOWING SECTION
OF COUNTER.

HALF ELEVATION OF BAR COUNTER.

SECTION ON LINE A B.

HALF ELEVATION OF BAR CABINET.

CABINET-WORK FIXTURES FOR A SMALL HOTEL BAR.

11

SKIRTINGS, DADOS, PANELWORK, LININGS, ETC.

Lining Material.

ALL pine woods, besides many hardwoods, are used in the production of lining. The hardwoods generally employed are mahogany, walnut, maple, oak, etc. Linings may be plain **V**-jointed, double **V**-jointed, beaded, or reeded. The **V** bead or reed should always

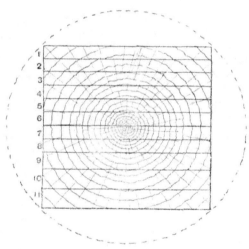

Fig. 1589.—Method in which Log is Cut into Boards.

be formed on the face, or best-dressed side. Linings vary in thickness from ⅜ in. to ¾ in., ⅝ in. being the size mostly used. First quality Bjorneborg pine (red and white) furnishes good material for lining. Small, firm knots are not objected to. Pitchpine should be free from knots and sap; likewise first quality yellow pine. The latter is most expensive, on account of its nature, appearance, and adaptability. Half an inch is allowed for the tongue on all pine woods; ⅜ in. on hardwood. A beautiful grain in pitchpine

lining is an important feature which should always be considered, but which cannot be obtained if the stuff is cut athwart the grain. The best figure is always obtained from the outside of sawn or hewn logs, the nearer the alburnum of the wood the better. Fig. 1589 shows graphically how this occurs. Let it be assumed that the dotted outer ring was the original size of the tree in the round. The square represents the size to which it was reduced for exportation. Between the two outer circular lines the wood is sappy; between the second outer line and the heart it is resinous. The parallel lines represent saw-cuts. All the planks up to No. 8 are 3½ in. thick; No. 9, 4½ in.; Nos. 10 and 11, 5 in. As pitchpine lining is generally 3½ in. broad in the rough, it may be assumed that this log was planked out by the log-frame for lining purposes. Plank No. 1 would yield about forty ⅝-in. boards, but not one would show a good figure. The result of No. 2 would be almost similar. With No. 3 a good figure would appear, especially on the outer edges. Nos. 4 and 5 would produce a finer figure still. Nos. 6, 7, and 8 would be beautiful across their entire breadths. No. 9 should be cut the opposite way from the planks that precede it; if cut the same way, considerable loss would result, and no figure could be obtained. Nos. 10 and 11 should also be cut as No. 9. In the butt of a good tree there should be no knots outside of 4-in. radius from the heart, but the heart of any tree (whether butt or top) cannot be had free of knots. Yellow lining is commonly ⅝ in. by 3 in.; four boards, at ⅝ in. full, are got from 3-in. deals. All yellow wood should be dried in a kiln before being wrought by the machine. Thin saws only should be used for breaking it out; saws thicker than No. 15 gauge are not suitable. With No. 16 gauge, four pieces at 1 1/16 can

21

be got from a 3-in. deal ; $\frac{1}{32}$ in. is sufficient for machine dressing and working. Narrow lining is preferable to that which is broad or double **V**-jointed ; however well seasoned the broad stuff may be, shrinkage is more or less perceptible in a few months after it is placed in position. Hardwood lining cannot be manufactured with the same ease as pine-wood lining ; with hardwoods more caution must be observed, and more time taken in the sawing and machine working.

Fixing Wooden Plugs for Grounds.

Plugs are pieces of wood or metal, or of wood encased in metal, that are inserted in the joints of brickwork or stonework, or are driven into holes bored into brickwork or stonework, for the purpose of affording a

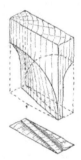

Fig. 1590.—Twisted Plug.

holding-place for those fixtures that are erected by carpenters and joiners. The twisted plug shown by Fig. 1590 is made from dry straight-grained deal, the opposite corners of which are cut off as illustrated ; the thin or entering edge of the plug is left of equal thickness along its width, with its edges parallel.

Securing Skirting in Cottage Work.

For securing skirting in ordinary cottage work, joints about 3 ft. apart are made in the brickwork with a plugging chisel. The plugs should be of the shape shown in Fig. 1590, each fitted in its own joint, and driven in up to the shoulder (Fig. 1591). A chalk line is then stretched from the first plug to the last, the line being held sufficiently far from the surface of the bricks to allow for the thickness of the plaster. The

plugs are marked at this point, and the superfluous wood is cut off. When the wall has been plastered, the end of the plug should be not quite level with the surface of the plaster. Fig. 1592 shows the common way of fixing the skirting.

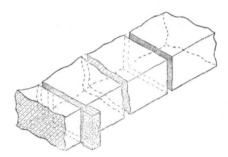

Fig. 1591.—Plug Fitted to Joint.

Securing Skirting in First-class Work.

In buildings of a more pretentious character, and superintended by a clerk of works, a higher class of work is expected. Suppose an order has been given to fix two sets of grounds for a 9-in. skirting, the walls of the room to be perfectly square and plumb when plastered. When the room is on the ground floor, the floor-level, if it has not been marked, must be obtained. It may sometimes be got from an adjoining room, where the floor-level has been determined by the floor joist running over one of the cellars. The room must now be squared by the outside wall, working through the window

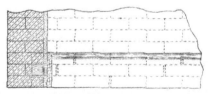

Fig. 1592.—Fixing Skirting in Common Work.

opening (Figs. 1593 to 1595), the inside walls being plumbed from floor to ceiling, trying the window side first to see if it inclines inwards or outwards at the top or at the bottom. Then the joints to receive the plugs (beginning with the window side) must be cleared, making the joints 18 in. apart along the wall, and placing the bottom ground or row of plugs about 1 in. or 1½ in. from the floor, and the top ground ¼ in. or ⅜ in. below the top edge of the skirting. When all the plugs have been driven in on the window side, fasten the chalk line across the room to the bottom row of plugs. Then place a rule across the window opening on

rule; B, parallel rule; C, plugs; D, chalk line; E, equal thicknesses. Plugs, to which blocks of wood (sometimes called "soldiers") must be nailed, should also be placed in the joints between the top and bottom rows; these plugs must be about 2 ft. apart (F F, Fig. 1596). The grounds should be so fixed that the walls may be perfectly straight, plumb and square when plastered; and the

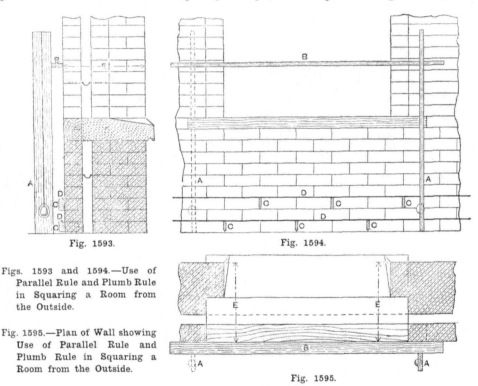

Fig. 1593.

Fig. 1594.

Figs. 1593 and 1594.—Use of Parallel Rule and Plumb Rule in Squaring a Room from the Outside.

Fig. 1595.—Plan of Wall showing Use of Parallel Rule and Plumb Rule in Squaring a Room from the Outside.

Fig. 1595.

the inside, keeping it parallel with the outside of the wall (Figs. 1593 to 1595), and drop a plumb line from the rule as a guide to set the chalk line, which should be parallel with the rule. The chalk line should be kept back the thickness of the grounds, so that the latter may finish flush with the face of the plaster. If the face of the wall is not quite straight or flat (it is sometimes rounded a little), the face of the ground must be ⅞ in. from that part of the surface that protrudes most. The reference letters in Figs. 1593 to 1595 are as follows : A, plumb

least amount of plaster on any part of the walls must be of the specified thickness (say ⅞ in.). Figs. 1596 to 1599 show a method of preparing the grounds and fixing skirting sometimes adopted in first-class work. If the wall on the window side slopes towards the inside of the wall, say, ½ in. at the top, the faces of the grounds must be fixed ⅞ in. from the most protruding point at the bottom, and there will then be 1⅜ in. of plaster at the top. When the chalk line is parallel with the outside face of the wall, the line may be run along the top row of plugs, keeping it perfectly plumb to the bottom line, *using for this purpose* a small plumb rule about 18 in. long by 2½ in. by ½ in. All the

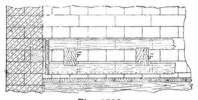

Fig. 1596.

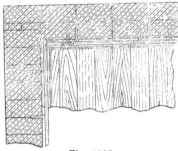

Fig. 1597.

squaring the other sides, taking first the side to the right and plumbing the wall from floor to ceiling as before, making due allow-

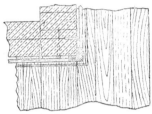

Fig. 1598.

Fig. 1599.

Figs. 1596 to 1599.—Method of Fixing Skirting in First-class Work.

plugs may now be marked and cut off, and the window side will be ready to receive the grounds. Before, however, any grounds are

ance in setting the ground for any roundness or other irregularity on the face or at the top or bottom of the wall. To square

Fig. 1600.—Method of Preparing for and Fixing Double-faced Skirting

fixed, the other three sides of the room must be made perfectly true. The plugs already fixed can be used as a base to work from in

this side from the plugs on the window side, a square with arms about 8 ft. long will be required, one edge being placed against the face of the sawn plugs, and the chalk line set parallel to the other edge. This side may now be plugged, and the chalk line run along the plugs as before, after which the other two sides of the room can be done. When the plugs for the skirting ground have been cut off, the window and door openings must be plugged to receive the grounds for the architraves. The joint grounds must be plumb and parallel, and the heads must be square, level, and $\frac{1}{4}$ in. or $\frac{3}{8}$ in. from the outside edge of the architrave. The joints must be lined out to the grounds behind the window linings and the door casings. All the internal angles of the grounds must be grooved and tongued, and the external angles mitered. When the grounds have been fixed, the blocks (F F, Fig. 1596) may be fixed between them, one being placed on each side of the external and internal angles to form a solid angle. The blocks should be let into the grounds. When the grounds are fixed, a nail must be driven into the wall on each side of the corners of the room near the ceiling, plumb with the face of the grounds. The nails must not be more than 9 ft. apart, and on each nail a pat of plaster must be laid, to which the plasterer will afterwards have to work. At Fig. 1600 is illustrated the method of fixing double-faced skirtings,

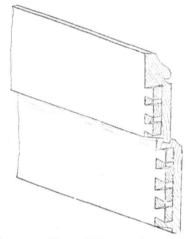

Fig. 1601.—Piece of Skirting for Internal Angle Mitered and Dovetailed.

which are usually in two parts as shown. The lower member is prepared with a tongue for fitting into a groove made in the flooring; its upper edge is ploughed to receive a tongue of the upper member. The two parts are usually of nearly equal thickness, and as the lower one projects this necessitates preparing and fixing blocks as shown at A; these fit to the back of the skirting. The grounds c are as previously described. Where the skirtings meet at the internal angles, they are usually grooved, tongued, and scribed, as illustrated. The skirting at D is purposely drawn off from E the better to illustrate the construction. Usually the external angles are simply mitered, and for painted work are nailed together; where

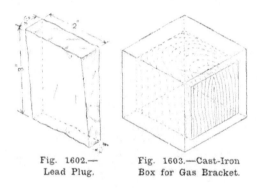

Fig. 1602.—   Fig. 1603.—Cast-Iron
Lead Plug.   Box for Gas Bracket.

nailing is not permissible, secret dovetailing gives the best result. This is illustrated at Fig. 1601, which shows the mitre and sockets cut to receive the other pieces.

Lead Plugs and Iron-cased Plugs.

All plugs and woodwork should be kept at least 9 in. from fireplaces or flues, and if a plug is required within the prescribed area, it must be made of lead (see Fig. 1602). If a gas bracket is to be fixed on the chimney breast or opposite a flue, the wood block must be tightly encased in a $\frac{1}{2}$-in. cast-iron box or $4\frac{1}{2}$-in. cube (outside dimensions), the grain of the wood being vertical (see Fig. 1603). If these boxes are to be fixed after the walls are up, they must be fastened with iron wedges, then made tight all round with good mortar. All wood that is to be plastered over, such as plugs, is kept back $\frac{1}{4}$ in. from the face of the plaster.

Round Wooden Plugs.

The round wooden plug, to contain which a hole is drilled in brick or stone, is often found to be useful, but the larger the plug the more in proportion will be the shrinkage in drying. The plug, therefore, should bear some relation to the size of the screw. A ½-in. or ⅝-in. plug is quite large enough for

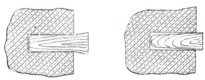

Fig. 1604.—Correct Form   Fig. 1605.—Incorrect
of Round Plug.   Form of Round Plug.

a No. 12 or 14 screw. *Fig. 1604 shows the* best form of round plug; Fig. 1605 is a bad form, being too tapering, but it is often used.

Dado Sham Framing.

Figs. 1606 and 1607 show a method of constructing a cheap and effective dado to a room. The framing is a sham, all its members being mounted on the face of the panelling. It is, of course, neither so strong

nor so durable as a framed dado would be, but, given good fixing, will prove serviceable for many years. An inspection of the detail drawings (Figs. 1608 to 1611) will make it

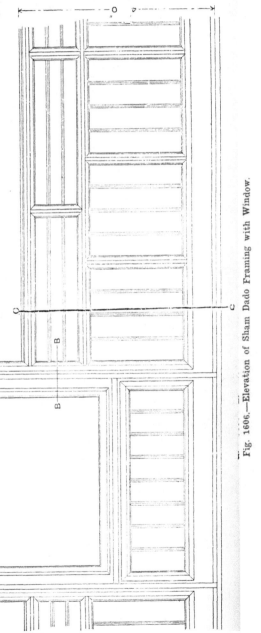

Fig. 1606.—Elevation of Sham Dado Framing with Window.

clear how the effect of panelled framing is
obtained. The linings to the door and
window openings (one of the latter being
shown in Figs. 1606 and 1607) are carried
over the framed grounds bounding the openings, and they project ⅛ in. in front of the
architraves. The mouldings of the surbase,
plinth, and dado rail are replicas of the outer
members of the architrave, up to the level of
the first flat, and the various members are
mitered into the architrave and finish flush
with the flat D (Fig. 1609). The architrave
and dado rail ground are rebated to receive
the panelling as shown at E (Figs. 1608 and
1612), the grounds finishing ¾ in. thick.
These grounds must be fixed first, plumb,
square, and level; the top edge of the
dado ground from ¼ in. to ⅜ in. below the
line of the dado, which in this case is 4 ft.
high. The grounds are fixed to wood plugs
driven into the joints of the brickwork about
every 2 ft. One of these plugs is shown in
dotted outline in Fig. 1609. The plastering
should be finished flush with the grounds,
and after it is set the fixing of the dado can
be proceeded with. Assuming all the mouldings and the battens for the panels to have
been prepared, proceed to fix the door and
window linings as shown in Fig. 1609. These
must be made to project over the face of
the grounds 1¾ in., the edges being plumb
and out of winding. Nail them to the grounds
and cross backings as shown. Next set out
on one or more rods, as may be required, the
net sizes between the linings, and between
the faces of the grounds, on the ends of the
room where no linings occur. These dimensions must be taken very exactly. Take a
piece of the architrave mould, and with its
back edge on the lines representing the
linings set off its width. Then space out
the muntins as shown in the plan (Fig. 1607),
in each case using a piece of the requisite
moulding for a gauge by which to obtain
the widths. Immediately beyond the first
section, and working from the same boundary
lines, set out a section of the lower panelling,
upon which the lower muntins are drawn
as they occur. No moulding need be drawn,
only the sight lines of the muntins. One
such rod should be prepared for each bay, and
from these the lengths of the plinth, surbase,
and dado rails may be obtained, the sight

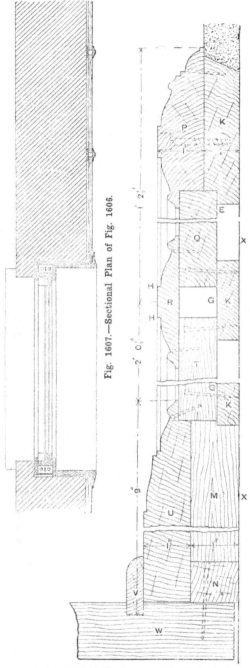

Fig. 1607.—Sectional Plan of Fig. 1606.

Fig. 1608.—Detail Section on C C (Fig. 1606).

lines of the respective architraves and muntins being squared across their edges with the marking knife. From these points the mitres may be cut with the aid of a mitre template. A similar height rod, showing the exact heights of the three horizontal members of the framing, should be prepared. Before setting this out, the state of the floor should be ascertained. If it is much out of level, or irregular, sufficient allowance must be made for scribing the plinth level, and it would be preferable to allow an extra $\frac{1}{2}$ in. on the width of the plinth for scribing, so that its finished height

adopted. In fixing the grounds, etc., take the height at the centres of the grounds G G as shown on the rod, and plug the walls to receive them. The best way to do this is to drive plugs at the required height from the floor at each end of a bay, and resting a plumb-rule in the rebate of the top ground E (Fig. 1608), plumb it upright, and with a piece of the thin ground as a gauge drive in the plug until the former will just clear the edge of the rule. Having done this at each end, drive a nail into each plug, and stretch a chalk-line between. The remaining plugs are then driven or cut off flush with the

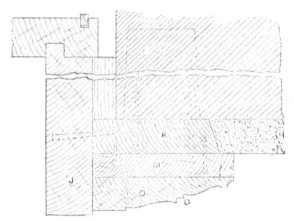

Fig. 1609.—Detail at A A (Fig. 1606).

Fig. 1610.—Detail of Panel.

should remain as given—namely, 9 in. As there is a skirting fillet, an extra fit for the lower edge of the plinth is not necessary.

Fixing the Sham Dado.—If this dado is to be prepared at a shop some distance from the building where it is to be fixed, the prepared rods should be sent on there, and the joiners would fit together the parts as shown, marking with a chisel all the mitres on the back for identification, and cutting off the battens to size as shown for the panels. In this case it would be safer not to cut the mitres in the architraves until they had been fitted round the linings; and the horizontal rails should be cut off $\frac{1}{32}$ in. full at each end, so that they should fit close when sprung in place. If, however, the entire work were prepared on the job, then a slightly different procedure in relation to the fitting would be

line. Next the floor fillets are nailed in place as shown in Fig. 1608, the exact position being ascertained from the rod. The thickness of the plinths is carefully noted, and these are fixed straight by the aid either of a spring line or of a long straightedge. After these are fixed, the plinth backings, pieces of 1 in. by $1\frac{1}{2}$ in., are fixed to plugs about every 3 ft., and finished flush with the floor fillet. Upright grounds are then fixed behind each muntin, as shown by the length rod. Assuming that the work is to be fitted and fixed on the job, cut the plinths in tight between the linings, fitting the piece between the window linings first, and cutting in a temporary stretcher between the door linings.

Scribe the wall angles of the plinth, as shown in Fig. 1611. Having fitted the plinth all round, next fit the architraves round the openings, cutting the uprights first. Having fitted these, mark them, and then take down. Fix the lower sets of panels as shown in Fig. 1608 ; if the tongues fit tightly, they will only need a brad here and there. They should have been previously all cut to size, as shown on the rod, with plenty of clearance between their edges. The upper panels may now be fixed. These will only require bradding on the lower edges, with an occasional brad in the upper edge to keep them from falling over. Remove the plinth first, however,

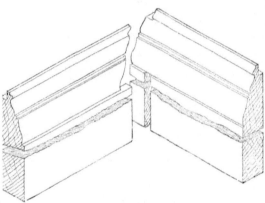

Fig. 1611.—Isometric Detail of Plinth.

pencilling a line along its face on the floor, and, taking it to the bench, set out the mitres for the muntins from the rod, as described previously. All the fitting up can now be done on the bench, as the external sizes are given in each direction by the architraves and plinths respectively. To get the lengths of the muntins, after fitting accurately an architrave with the first and second rails, turn it up on edge, and square over the mitre or sight line, on one top and one bottom muntin, and use these as length rods for setting out the remainder. A template applied to the sight lines will give the length of shoulder where it cuts the edge of the moulding (see H, Fig. 1608). The first fitted architrave will also be a guide by which to set out the remainder, care being taken to keep the bottom ends flush, which will

21*

ensure the dado being level all round the room. It should be noted that in fitting the dado rail the top member does not come into the mitre, but is stopped square against the architrave, a beaded backing piece being fixed to the framed ground to take this, as shown in Fig. 1609. All being fitted together, the preliminary to fixing is to run all round the room the groove for the skirting shown in Fig. 1608. This would be done with a grooving plane working against a thin strip equal in width to the thickness of the plinth, and resting against the floor fillet. This groove should just take the pencil line previously made out, so that the plinth is nipped tightly between the skirting and fillet. Proceed to fix the plinth as shown in Fig. 1608, bradding the top edge

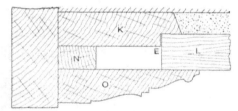

Fig. 1612.—Detail at B B (Fig. 1606).

under the ogee, where the holes will be out of sight. Do not nail the lower edge ; drive the skirting in tightly, preferably gluing into the groove, but not to the plinth. Next fix the architraves to the ground and backings, and spring the rails between them, bradding the surbase mould as shown in Fig. 1608. Also insert the lower muntins, forcing the surbase moulding down tightly on to them. Next fix the top muntins, and finally the dado rail. The latter, having to keep the whole in position, is better fixed with screws, which in painted work would be turned in flush and puttied over, and in polished work should be sunk into holes (as shown in Fig. 1608), which are afterwards filled in with turned pellets of similar wood. Figs. 1606 and 1607 are reproduced to a scale of $\frac{1}{4}$ in. to 1 ft., and Figs. 1608 to 1610 half full size. The following is a list of letter references not explained in the text : J, jamb-lining ; K, ground ; L, panelling ; M, backing ; N, fillet ; O, architrave ; P, dado

rail ; Q, top panel ; R, surbase mould ; T,
bottom panel ; U, plinth mould ; V, skirting ; W, flooring ; X, wall line.

Oak Panelwork.

In the framing up of oak panelwork,
the essential requirement for sound, true
work is a proper regard to the joints. The
material is used so thin that the panelling

it touches the grounds, and will wedge the
top and bottom here and there, in order to
bring the work true and straight both ways.
A common fault, causing a great deal of
trouble, is the neglect to thickness down the
panels before final insertion in the framing.
This prevents the framing from touching
the grounds, and therefore the rough plaster
filling or screeding must be hacked or

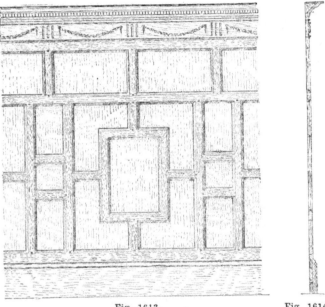

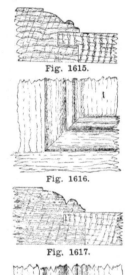

Fig. 1615.

Fig. 1616.

Fig. 1617.

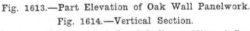

Fig. 1613.   Fig. 1614.

Fig. 1613.—Part Elevation of Oak Wall Panelwork.
Fig. 1614.—Vertical Section.
Figs. 1615 and 1616.—Details of Ordinary Mitered Corner.
Figs. 1617 and 1618.—Details of Mason-Mitered Corner.

Fig. 1618.

should be stiffened with stout canvas glued
to the back, which is sometimes primed with
red-lead paint to afford protection from
damp when the work is placed in a new
building. It is usual to frame up the work
with stuff varying in thickness from $\frac{7}{8}$ in.
to $1\frac{1}{4}$ in., finished sizes. In some shops it
is not considered necessary to thickness the
framing, which is only prepared on the face
and two edges ; in which case a little more
trouble is given to the fixer on the building,
who, if his grounds are straight and true, will
traverse the back of the panelwork where

scraped away to allow of the panels going
back without firring out the grounds. Complicated methods of framing require the
use of draw-bore pins and oak dowel pins
when finally gluing up the framing on the
benches. A point that must be emphasised
is that the tenon should be kept in the centre
of the thickness of the framing, because
under the pulling influence of the cramp
the stile or rail always turns to the weaker
shoulder ; and when such work has a
shoulder $\frac{1}{2}$ in. deep at the front or moulded
side, and another shoulder $\frac{5}{16}$ in. or $\frac{1}{4}$ in.

deep at the back, extra labour is necessary in order to bring the face side to a true surface, while the extra planing may injure the moulding on the edges, whereas a slight extra thickness of material would obviate all the trouble and risk. The panelwork shown in Fig. 1613 is framed with 1¼-in. stuff, got to thickness and widths as shown in Fig. 1614.

Fig. 1619.   Fig. 1620.

Fig. 1619.—Detail of Cornice Frieze, Main Framing. and Skirting.

Fig. 1620.—Vertical Section through Fig. 1619.

The centre framing is mitered together at the corners, which are further strengthened by the insertion of a cross-tongue joint, while the side rails and top and bottom muntins are tenoned and pinned like all the other tenons in the framing. The moulding stuck on the edges (or in the solid) is not mitered in the same way as for ordinary work (see Figs. 1615 and 1616); but the mitered corners are worked as shown in Figs. 1617

and 1618, thus forming butt joints with mason-mitered corners to all moulded edges. These corners are worked by the joiner on the bench after the panelling is glued up and cleaned off. The bottom rail is tongued into the skirting as shown in Figs. 1619 and 1620, and the top rail meets the festooned frieze board under the small necking mould as shown, the frieze board being tongued to the dentilled cornice also. This cornice is double-dentilled, one row of dentils being cut farther back than the other, as shown in Figs. 1619 and 1620. As usual with built-up cornices, this section can be worked on the four-cutter moulder or on a spindle machine; the dentilling, however, is best cut by hand. A cover-board lies at the back of the cornice, which is back-rebated to receive the front edge of the cover-board. The three flutes over the top muntins have rounded-out top ends, and finish at the bottom on a splay; whilst the festoons are preferably cut out of the solid, but are generally planted on unless otherwise specified. In fixing this class of work, which is, as a rule, screwed up, all fixing screws should be hidden, or the holes should be bored to take "corks" a little larger than the screw head, and the "corks" should be cut from wood closely matching that in which the hole is bored. The framing must be fixed as true and upright as possible (especially at external corners where mitered vertical joints occur) and well scraped and cleaned down after the fixing is done. The illustrations are reproduced to the following scales :— Figs. 1613 and 1614, ¾ in. to 1 ft. ; Figs. 1615 to 1618, half full size ; Figs. 1619 and 1620, 1½ in. to 1 ft.

Fixing Hardwood Dado.

Fig. 1621 shows an elevation of a panelled dado with moulded skirting and capping. The dotted lines at c and D show the fixing fillets. Fig. 1622 shows a section of the dado. At A and B are grounds fixed to wood plugs or coke-breeze bricks built into the wall. The upper ground is wide enough to take 2 in. of the top rail of the dado and 2 in. of capping, making 4 in. in all. It is splayed on the top edge to receive and to form a key for the plaster. On ground A at c are dovetailed fillets (see Figs. 1623 and 1627), fixed at convenient intervals of about 3 ft. apart. The mortices shown in Fig. 1624 are cut into the top edge of the dado rail and the under edge of the capping; these should be set out on the bench and cut in before sending on to the job. The fillets c being

In fixing the dado, it should be placed against the grounds in its exact position, with the dovetailed fillet projecting above the upper edge. Two screws should then be driven

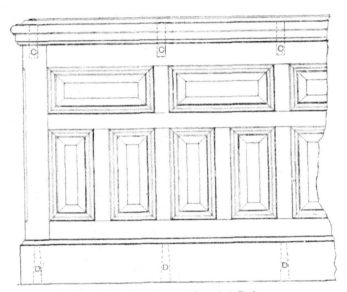

Fig. 1621.—Elevation of Panelled Dado.

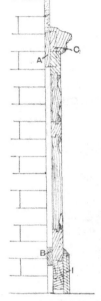

Fig. 1622.—Vertical Section.

prepared and fitted into the mortices and the capping fitted on, the fillets D (Fig. 1625) are cut, tapered in shape, and dovetailed as before. Skirting is prepared with a moulding tongued into the upper edge. Before

into each piece, which will securely fix the whole; the capping can then be gently dropped on in position. It will be advisable

Fig. 1623.—Fixing Ground and Fillets.

gluing in the moulding permanently, the skirting should be grooved on the rear side as shown in Fig. 1626, and fitted on to the fillets D. The skirting is prepared in this manner to enable the grooves to be cut easily and clean. As the piece forming the skirting slides on to these fillets, it tightens itself, and so a good secure fixing is obtained.

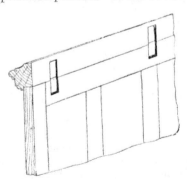

Fig. 1624.—Back View of Upper Part of Dado.

to use a little glue on the fillet and in the mortices in each case, also along the top edge between the dado and capping. It will be seen by referring to Fig. 1625 that the bottom rail of the dado is built up partly of

Fig. 1625.—Showing Fixing Fillets for Skirting.

Fig. 1626.—Back of Skirting with Grooves for Fixing.

hardwood and deal, E being deal. Fig. 1628 is an enlarged detail of rail and skirting.

Fig. 1628.—Detail Section of Lower Part of Dado.

Fig. 1627.—Conventional View, showing Method of Fixing Upper Part of Dado and Moulding.

Geometrical Head Linings to Door and Window Openings.

In the case illustrated by Figs. 1458 to 1465 (pp. 447 and 448), an elliptical-headed casement window was shown, finished with square jamb linings, and the soffit prepared by veneering and blocking. A

$1\frac{1}{2}$ in. by 2 in., having been jointed, is nailed firmly to the edges of the ribs, and then planed off true to the elliptical curve. The board to form the veneer is from $\frac{1}{8}$ in. to $\frac{1}{4}$ in. thick, and wide enough

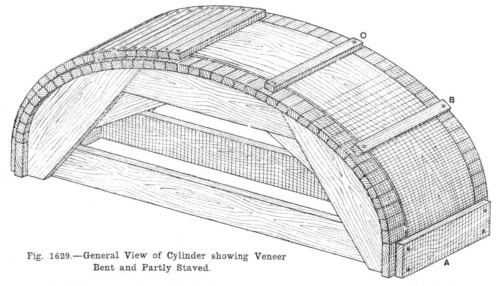

Fig. 1629.—General View of Cylinder showing Veneer Bent and Partly Staved.

to obviate joining. It is gradually bent over the cylinder by first fixing a broad stave just below the springing A (Fig.

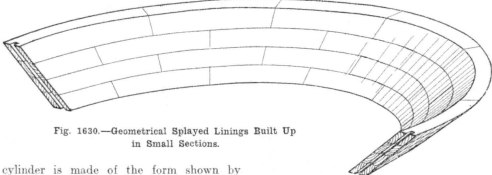

Fig. 1630.—Geometrical Splayed Linings Built Up in Small Sections.

cylinder is made of the form shown by Fig. 1629, which is of similar construction to a centre. Two ribs must be prepared, each rib consisting of two thicknesses nailed together and cut accurately to the curve required, the thickness of the lagging being, of course, deducted. The lagging, which may consist of strips about

1629). Then, with a piece of wood, it is gradually pressed and worked forward a distance of about 12 in. or 18 in. A stave is then screwed down temporarily as indicated at B, and is followed by a second at C, and so on until the veneer is bent round to fit the surface of the cylinder. For ordinary-sized openings, thin veneers can usually be bent round dry ; for narrow openings, it is necessary to steam the veneer, or to soak it with boiling water, and similar preparation is nearly always necessary for thick veneers. The staves are, of course, a little out of the square, and are sufficiently long to project beyond

Fig. 1631.

Fig. 1632.

Fig. 1633.

Figs. 1631 to 1633.—Method of Obtaining Shape of Veneer for Geometrical Splayed Linings.

the veneer as illustrated. At each end a hole is bored. Each stave is fitted to the preceding one, and glued to it and to the veneer. A screw is then inserted in each of the holes. These screws hold to the lagging of the cylinder. If the veneer is of pine, mahogany, oak, or other hardwood, and is thick enough to allow of its being faced up, it is advantageous to do this before bending ; but where it is neces-

sary to use a thin veneer, the finishing is done afterwards, by means of scraping, etc.

Geometrical Splayed Linings, Built Up in Small Sections.

Fig. 1630 illustrates the manner of building up a curved splayed soffit lining for the soffit of the splayed linings illustrated by Figs. 1256 to 1259 (p. 382). By this system, pieces are cut out through the thickness of plank, which may be from $1\frac{1}{2}$ in. to 2 in. thick, a mould being made for the proper curve at each joint. It will be seen that this curve increases from the inner to the outer edge of the lining, because of the latter being conical. By marking out on each side of the plank, the conical cuts can be at once made through the plank, much labour being thus saved. If a bandsaw with a tilting table is available, pieces can at once be cut accurately to the proper sweep. The butt joints of the ring forming the inner edge are carefully fitted over a board. These are cross-tongued and glued. The next layer is fitted and jointed on this, the joints being grooved and cross-tongued. Each piece is then bored for the insertion of at least two screws, to connect with the lower ring. On the joints being found satisfactory, the screws are released, and the joint between the rings and the butt joints, with their grooves and tongues, are glued, and the screws re-inserted. This process is repeated for each layer. Then the soffit is worked true, and the ends trued and grooved for connecting with the jambs at the springing. This latter form of joint was illustrated by Fig. 1465. This method is not generally adopted for hardwood work, owing to the variation of grain distinctly showing when polished, but it makes a good sound job for painted work.

Geometrical Splayed Linings Veneered for Varnished or Polished Work.

The example illustrated by Figs. 1631 to 1634 is for the head of a similar door or window opening to that treated of in the previous section, except that it is assumed here that it is desired to show the proper side grain of the material, which, of course, necessitates the preparing of a conical - shaped cylinder, and the shaping of a veneer, also the staving, bending, and gluing the back in a nearly similar manner to that explained for the example illustrated by Fig. 1629 (p. 494). It is therefore only necessary to illustrate and explain the method for geometrically setting out the veneer. Figs. 1631 and 1632 show respectively the half elevation and half plan of the conical-shaped veneer. Continue the splay of the lining (shown by

a veneer wide enough to obviate jointing. In the case of some timbers--oak, for instance—it is sometimes possible to get a piece of compass timber, the object being to show as far as possible the continuous grain.

Splayed and Panelled Soffit Lining for Elliptical-headed Opening.

The preparation of the elliptical splayed and panelled soffit lining for the window illustrated at Figs. 1492 to 1494 (p. 457)

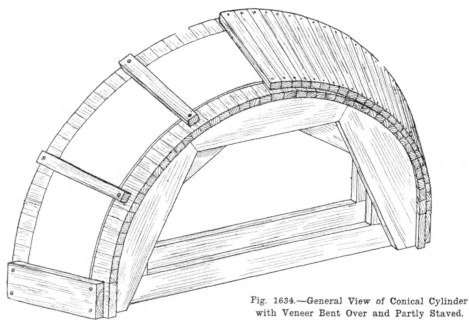

Fig. 1634.—General View of Conical Cylinder with Veneer Bent Over and Partly Staved.

I C, Fig. 1632) till it cuts the centre line in D. Then A I D is the plan of quarter of the imaginary conical surface. Adopting the well-known principle of the development of the cone, with centre D and radius I describe the arc I 9, also the arc C 10; divide the quadrant (Fig. 1631) into any number of convenient parts, as shown by I' to 9', then mark off distances equal to these along the arc I to 9 (Fig. 1633). Then C 1, 9 10 represent one half the development of the veneer. The other half is, of course, exactly the same shape. Where possible, it is advisable to obtain

may now be described. The lining being elliptical, the development requires rather more elaboration than it would if it were semicircular. The method of construction consists in building up in three thicknesses on a cylinder. One thickness is equal to the amount to which the stiles and rails project beyond the faces of the panels as shown at Fig. 1639; the strips forming the stiles being of course cut out to shape. The method of setting out for these shapes will be described in due course. The pieces for the sham stiles are bent over the prepared cylinder as shown at Fig. 1639, and are held in position by staves, which are placed at intervals as indicated at A B C (Fig. 1639). Pieces of similar thickness are then accurately fitted in between the stiles, so as to form sham rails as shown at D and E (Fig. 1639). Then a thin veneer is bent over. This veneer need not be all in one piece, but can be and rails. Then, on the back of the veneer, the staves are jointed to each other, and glued and screwed down as indicated on the right-hand half of Fig. 1639. When doing this kind of work, many joiners consider it advantageous to glue a layer of coarse canvas on the back of the staving, in order to give additional strength.

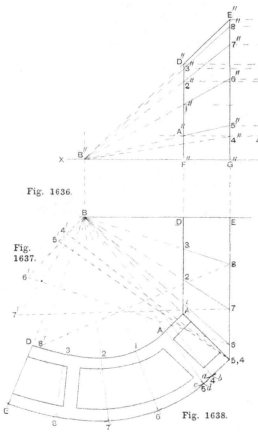

Fig. 1635.

Fig. 1636.

Fig. 1637.

Fig. 1638.

Figs. 1635 to 1638.—Geometrical Setting Out for Development of Elliptical Conical Soffit Lining.

Construction of the Cylinder.—As will be seen by Fig. 1639, the cylinder is a frustum of an elliptical cone. Two ribs are made, one for the smaller curve shown by F' D', while the larger rib is made to the curve G' E' (Fig. 1635). The edges of these ribs are bevelled to receive the lagging, which must be nailed on as previously explained, and as illustrated at Fig. 1639, which shows the cylinder as being somewhat wider than the lagging, to allow the projecting staving to be secured temporarily to the lagging with screws.

Geometrical Construction for Soffit.—Begin by drawing the springing line A' C', and then draw the elliptical curve A' D', which is the line of intersection between the soffit and the head of the sash frame. Continue the springing line to the left, then at any convenient point A'' erect a perpendicular. Project horizontally from D' to cut this line at D'', and thus A'' D'' is the side elevation of the curve A' D'. Now project down and draw the plan of this curve as shown by A D. Through D draw the line B E parallel to A'' C. Then B F is the direction of the plan of the axis

formed of three pieces, the joints being made along the middle of the two sham rails, one of which is shown at D (Fig. 1639). An advantage in having three separate pieces is that then the straight grain may be so arranged as to run tangential to the centre of the panel, giving a much superior finish when the work is stained or polished. These veneers are glued and bent down to the backs of the sham stiles of the cone. From A set off the plan of the angle of the jamb lining as shown by A 5. Continue this line till it cuts B E. Then B is the plan of the vertex. Project up as indicated by B B''. At D'' set off the angle of the crown of the soffit (which is here shown as having the same angle as in the plan). This will give the line E'' B'', and B'' is the side elevation of the vertex. Then through B'' draw the horizontal line X Y, and, continuing the minor axis D' C to cut X Y, obtain the elevation

5'', 6'', 7'', 8'', E''. From these points project horizontally to meet the elevation of the generators, obtaining points 5', 6', 7', 8', E', as shown. Through these points the outer curve can be drawn. From the points 4'' in the springing line, draw the elevation and side elevation of an additional generator. From 4' to E' straight piece has been added, so as to work from the level of the vertex. This expedient will simplify the working for obtaining the development.

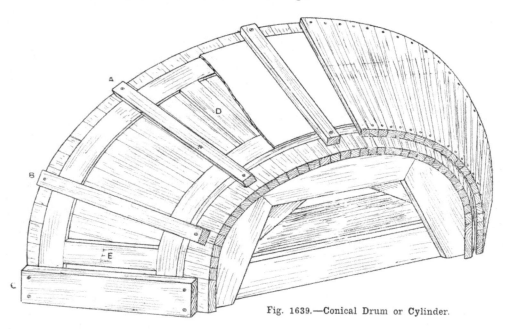

Fig. 1639.—Conical Drum or Cylinder.

of the vertex B. Dividing A' D' into any convenient number of equal parts, as here shown, through the points A', 1', 2', 3', draw lines radiating from B. It will be an advantage to continue these lines indefinitely. We now have the elevation of the generators of the conical surface. Project horizontally from 1', 2', 3', thus obtaining points 1'', 2'', 3'', 4'' (Fig. 1636). From B'' draw radiating lines through these points, thus obtaining the side elevation of the generators. Complete the half-plan of the lining as shown by D A 5 E, project up, and draw G'' E''; then the generators will cut this line in points

Obtaining the Development of the Soffit.— In dealing with an elliptical cone, the generators gradually increase from the minor to the major axes of the section. Therefore, before the development can be set out, it is necessary to obtain the true length of each generator. Stated in geometrical terms, the problem is, " Given the plan and elevation of an oblique line, determine its true length." At right angles to the generator 5 B (Fig. 1637), draw B 5', making it equal to G'' 5'' (Fig. 1636). Join 5' to 5. This gives the true length of the generator. The others, shown by 6', 7', 8' (Fig. 1637), are obtained in a similar manner. Now, with compasses set to the length 5 4′, using B as a centre, describe an arc as shown by *a*, and with compasses set to G″ 4″ (Fig. 1636), using 5 as centre, draw the arc *b*. Where the arcs intersect at 4, join to B. With compass set to the true length 5 5′,

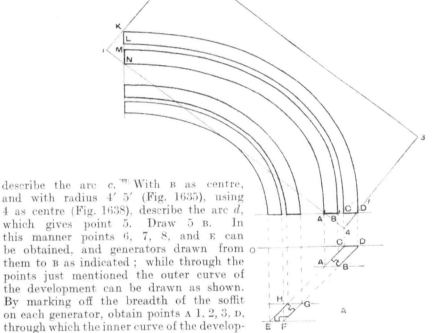

describe the arc *c*. With B as centre, and with radius 4′ 5′ (Fig. 1635), using 4 as centre (Fig. 1638), describe the arc *d*, which gives point 5. Draw 5 B. In this manner points 6, 7, 8, and E can be obtained, and generators drawn from them to B as indicated ; while through the points just mentioned the outer curve of the development can be drawn as shown. By marking off the breadth of the soffit on each generator, obtain points A 1, 2, 3, D, through which the inner curve of the development can be drawn. This completes the outline of the development for one-half the soffit. The stiles and muntins can next be set out as shown. From this development moulds can be made for the stiles, and the shape of the veneers for the panels can be ascertained.

Obtaining Development Direct from Cylinder.—For obtaining the development, a method which, while not quite scientific, is nevertheless practical, and has been largely used by joiners in the past, is to make a cylinder as true as possible, and then to set out the soffit on the laggings ; then, by applying cardboard or

stiff paper to the laggings, to ascertain the slope of the development, in this way obtaining moulds for marking and cutting out the stuff which was to be bent on the cylinder.

Preparing Soffit out of the Solid.—The principal points involved in setting out and making splayed and panelled soffit lining by cutting and working up the several pieces out of the solid without any bending are as follows. At A (Fig. 1640)

Fig. 1640.—Setting Out Edge Moulds.

are shown in plan the ends of the stiles of the soffit at the springing. The parallel lines A B, C D show the thickness of the plank required. Projecting up from C D and A B, draw the curves D′, K, C′, L, B′, M, A′, N, which represent the face moulds C′, D′, L, K, for application to one side of plank, A′, N, M, B′, the face mould for the other side of plank. Then, setting a bevel to the angle O, C, A, and applying it to a plank which has been cut for the joints as shown by A B, the rectangle 1, 2, 3, 4 indicates a piece of plank. By cutting

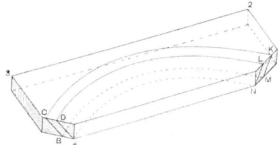

Fig. 1641.—Conventional View of Plank Set Out from Moulds ready for Sawing.

along the lines A', D', the plank is made to assume the form shown by Fig. 1641. Now apply and mark the shape of the inner face mould on the under side of the plank as indicated by A B M N (Fig. 1641). Then the outer face mould can be applied to the plank and marked as indicated by the curves C K and D L (Fig. 1641). The piece for the stile can now be sawn out as represented at Fig. 1642. A band saw with a tilting table is advantageous for doing this kind of sawing. The inner edge is worked square to the soffit of the stile as represented by E F (Fig. 1643). The stiles for the part of soffit adjacent to the frame are set out and prepared in a similar

Fig. 1643.—Stile with Inner Edge worked Square to Soffit.

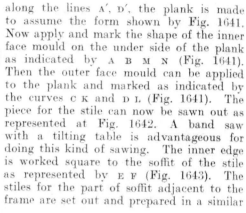

Fig. 1642.—Stile Cut Out.

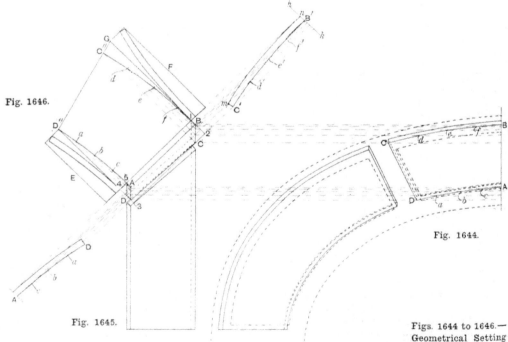

Fig. 1646.

Fig. 1645.

Fig. 1644.

Figs. 1644 to 1646.— Geometrical Setting Out for Head Panel.

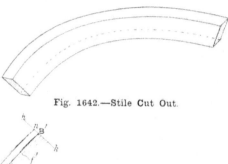

manner. The curved stiles, having been
so far prepared, can be set out, and mortices
made for the rails, which are then ploughed
for the panels. The joints at the crown
are most satisfactorily connected with
hand-rail screws.

Setting Out for Panels.—A half-elevation
of the soffit is indicated at Fig. 1644, the
framing being shown by dotted lines and
the panels by solid lines. Having set out
the half elevation of the panels as shown,
project across and make a sectional elevation as shown at Fig. 1645. As each of
the moulds for the panels is obtained by
an identical method, attention may be
confined to the necessary working for the
top panel, and setting out for one half will
be sufficient. Points A, B, C, D (Fig. 1644)
have been projected across, giving new
projections of these points at Fig. 1645,
and the thickness of the panel is represented
by A B, 1 5. A 3, C B, it should be noted,
represents the curved face of the panel.
Inclose this sectional elevation of the panel
by a rectangle as represented by 1, 2, 3, and
4. This shows the thickness and the breadth
of the plank required for the panel. Fixing
on any convenient points in C B, as d, e, f
(Fig. 1644), project horizontally to C B
(Fig. 1645). Then from the points just
obtained, project at right angles to 1 5
(Fig. 1645). Make each of these projectors
the same length from 1 5 as they are
from A, B in Fig. 1645, and thus obtain
points d, e, f, and c'' (Fig. 1646), through
which the curve can be drawn as shown.
Similarly projecting horizontally from D a,
b c, A to A D (Fig. 1645), the curve passing
through points 5 c b, a D'' (Fig. 1646)
is obtained. Joining c'' to D'' gives the
shape of the mould to apply to the face
of the plank. For the moulds to apply to
the edge of the plank, proceed as follows :—
The points c, d, e, f, B are projected to
B C (Fig. 1645). Projecting from these
points parallel to A B, and drawing a line
at any convenient distance at right angles
as h k, then measuring from this line
and making each projector the same length
as the corresponding one from A B, points
c', d', e', f', B' are obtained, giving
points in the curve, which can be drawn
through them as shown. The second curve

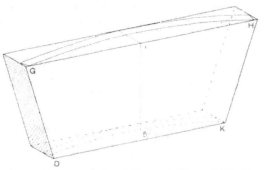

Fig. 1647.—Conventional View of Piece of Plank
with Edge Moulds Applied.

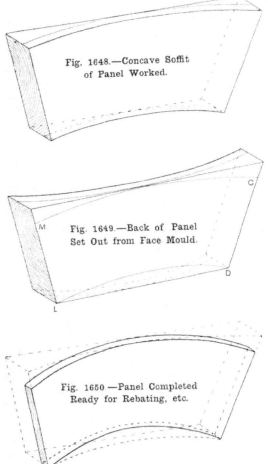

Fig. 1648.—Concave Soffit
of Panel Worked.

Fig. 1649.—Back of Panel
Set Out from Face Mould.

Fig. 1650 —Panel Completed
Ready for Rebating, etc.

m n can be drawn parallel, representing the thickness of the finished panel. The

ally the same shape as at 4 D'' G 1 (Fig. 1646), and thus D G, H K (Fig. 1647) represents the piece of plank for the top panel cut to its first shape 1, 5, being the centre line. On the top edge is shown the applica-

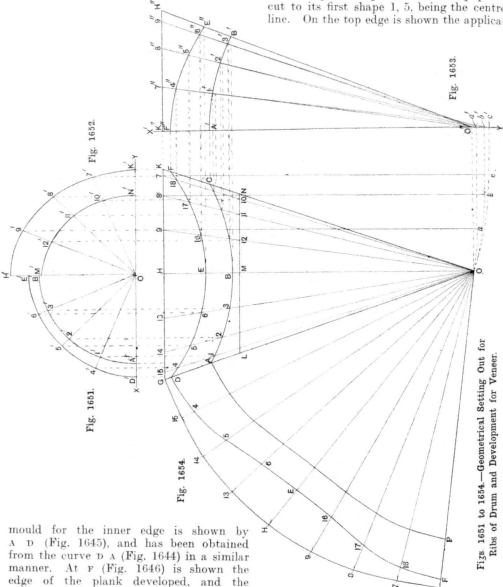

Figs. 1651 to 1654.—Geometrical Setting Out for Ribs of Drum and Development for Veneer.

mould for the inner edge is shown by A D (Fig. 1645), and has been obtained from the curve D A (Fig. 1644) in a similar manner. At F (Fig. 1646) is shown the edge of the plank developed, and the outer edge mould applied. The application of the inner edge mould is shown on the developed edge of plank at E. At D G 1 5 (Fig. 1647) is shown convention-

tion of the outer edge mould, while the application of the inner edge mould is indicated by dotted lines. Fig. 1648 represents the concave face side of the panel worked, and at Fig. 1649 the face mould C D, L M, applied to the back of the panel. The inner and outer edges are worked to these curved lines, and then the panel is gauged and worked to thickness. This completed state is fully shown by Fig. 1650.

Veneered Splayed Lining to Opening with Circular Head and Segmental on Plan.

At Figs. 1510 and 1511 was shown a circle-on-circle window opening with splayed linings. Here will be described and illustrated methods of setting out and constructing the conical soffit lining, which is constructed of a veneer staved or blocked at the back. Draw the plan of the face of the lining as shown by D, E, F, C, B, A; continue D A, F C, and E B, giving point O, this being the plan of the vertex of the conical surface. Projecting up from A and B, draw the curve a' b', which, as explained in connection with Fig. 1511, will probably be a quarter of an ellipse. Fix on any convenient points in A' B', as 1', 2', 3', and draw radiating lines to 0', which are the elevation of generators. Projecting down from 1', 2', 3', on A B, we obtain point 1, 2, and 3. Draw through these points lines radiating to 0, which are the plans of the generators. It is now necessary to draw a side elevation. Make X' Y' parallel to B O. Project up from A, 1, 2, 3, and B, thus obtaining points A', 1', 2', 3', and B', through which the curve can be drawn, representing the inner edge of the lining. Project from 0, obtaining point 0''. From this draw the elevation of generators, passing through points 1', 2', 3', 4', B'. As the frustum of a cone on which to block and bend the veneer will have to be constructed, the plans and elevations for the curves of the ribs may next be drawn. At 2 in. or 3 in. away from D and F, draw G K at right angles to H O. Also at a couple of inches from B draw L N. Then G L N K will represent the outline plan of the frustum of a cone (commonly called the cylinder) required. Continue the generators to G K, giving points K, 7, 8, 9, H. Obtain the elevation

of these points on the side elevation (Fig. 1653) as shown by K'', 7'', 8'', 9'', H''. From these points in plan project up to the elevation, and then make 7' from X Y the same distance as 7'' from X' Y', and 8' the same distance from X Y as 8'' is from X' Y'. Points 9' H' are obtained in the same manner; then the curve drawn through these points as shown from K' to H' is the shape of the outer rib, including the thickness of the lagging. The curve in elevation from N' to M' is obtained by projecting from points in M N, where the plans of the generators cut it, and from these points project up to the elevations of the generators as shown. This curve is for the smaller rib, including the thickness of the lagging.

Development for Veneer.—The curve from K' to H' is not the quarter of a circle, and as it is a section at right angles to the axis of the conical surface, the generator O H will be the longest, the others gradually decreasing to O K. It is therefore necessary to obtain the true lengths of each generator. From 9 draw a line parallel to the axis, then, with 9 as centre, and O as radius, obtain point a. Project up to X' Y', giving point a'. Join to 9''. Then this line a' 9'' is the true length of the generator 0 9. Points b b', c c', and thus the true lengths of the other generators, have been obtained in the same way as 0 9. With compass set to distance K' 7' (Fig. 1652), and G as centre, describe the arc 15. With beam compass set to 7'' c' (Fig. 1653), using O as centre, cut the arc 15 (Fig. 1654). Again, with radius equal to 7', 8', and 15 as centre, describe arc 14; and then with 8'' b' (Fig. 1653), using O as centre, cut arc 14. The remainder of the development of the conical surface is completed by the same procedure. The development of the line G K having been obtained, now by measuring the distances 7'', 4'', 8'', 5'' (which are obtained on the true lengths of the generators), from 15 to 4 and 14 to 5 respectively (Fig. 1654), we obtain points in the curve. As all the other distances are ascertained by the same expedient, there is no difficulty in completing the shape of the veneer as shown by the irregular curved lines F D and P A (Fig. 1654). The

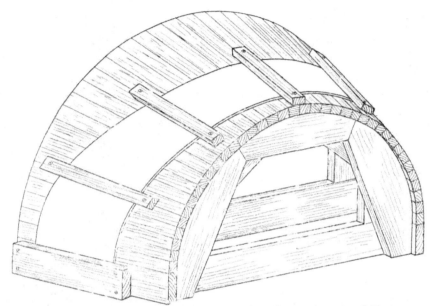

Fig. 1655.—Conventional View of Drum with Veneer Shaped and Bent.

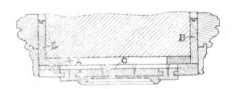

Fig. 1656.—Horizontal Section of Jamb.

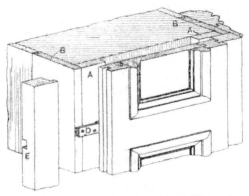

Fig. 1657.—Conventional Sectional View of Portion of Face of Framed and Panelled Jamb.

conical drum (Fig. 1655) shows the veneer cut to shape and bent over ready for staving. The method of doing this having been previously dealt with, it is not necessary to recapitulate.

Fixing Hardwood Door=casings, etc.

A general view of a door and casings, etc., in hardwood is represented at Fig. 1658, and the details showing the method of secret fixing are given by Figs. 1656 to 1659. In each figure, A represents the framed ground which is fixed to masonry in one of the usual ways. The side grounds marked B are fixed to the edges of the framed grounds as indicated. Fillets c, with each end cut to the form of a dovetail tenon D, are prepared and screwed to the back of the rail of the framed and panelled jamb as shown at c (Figs. 1656 and 1659). The jambs are then fixed in position by screwing the dovetail ends to the grounds as shown at D (Figs. 1657 and 1658). The portions of the stiles forming the rebates, having been prepared with slots as shown at E (Fig. 1657) to fit the ends of the fillets as shown at D, can be glued and fixed into their positions. The archi-

Fig. 1658.— General View of Door.

traves are fixed by preparing hardwood strips with dovetail edges as represented

Fig. 1659.—Conventional Sectional View of Back of Jamb.

the architrave as shown at G and H (Fig. 1660), so as accurately to fit the dovetail slips. The inner member of the architrave is fixed in position first, and the outer member H afterwards. Thin glue should be

Fig. 1660.—Back of Architrave, showing Dovetailed Slots.

Fig. 1661.—Panelled Wainscoting: End Wall showing Fireplace and Doorway.

Fig. 1662.—Enlarged Part Section on Line C D (Fig. 1661).

at F (Fig. 1659). Corresponding slots are made in the back of the two parts forming

brushed on the dovetailed slips and groove, and along the tongue edge of the inner section.

Fig. 1661 shows the treatment of an end wall in which a fireplace and doorway occur. Fig. 1666 shows a side wall with a three-light recessed window in it. Fig. 1663 gives an enlarged section on line A B (Fig. 1661), showing details of the over-door A, archi-

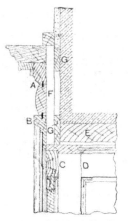

Fig. 1663.—Enlarged Section on Line A B (Fig. 1661).

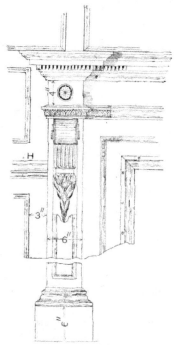

Fig. 1664.—Enlarged Elevation of Left-hand Jamb of Wood Mantel.

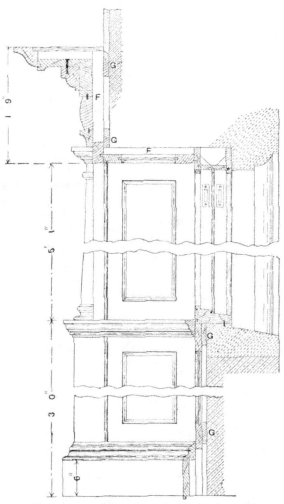

Fig. 1665.—Enlarged Part Section on Line E F (Fig. 1666).

Panelled Wainscoting.

The wainscoting here shown could be executed either in hardwood polished, or in yellow pine painted and finished white.

trave B, the top portion of the five-panel door C, and the framed jamb linings D, lintel E, framing F, and grounds G. The top member of the over-door intersects with the top member of the wainscoting. Fig. 1664 represents a part enlarged elevation of the left-hand jamb of the wood mantel, also portions of the wainscoting. The surbase moulding H is stopped by the wood mantel, but the skirting intersects with the plinth of

linings, and the pulley frames. Fig. 1667 represents the plan of the stone mullion, showing boxings for the sash weights. For greater convenience in making and fixing, it would be better to make the framing above the surbase moulding independent of

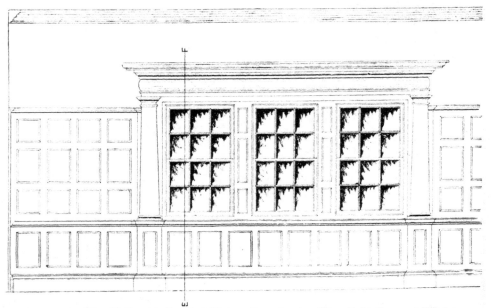

Fig. 1666.—Side Wall, with Panelled Wainscoting, showing Three-light Recessed Window.

the jambs. Fig. 1662 shows an enlarged part section on line C D (Fig. 1661) of the mantelpiece and the wainscoting over it. A brass frame is screwed on to the mantel,

Fig. 1667.—Plan of Stone Mullion showing Boxings.

breaking the joint between the wood mantel and the firebrick lining to the dog grate. Fig. 1665 represents an enlarged part section on line E F (Fig. 1666), showing details of the cornice, the pilasters, the panelled

that below, the framing above and below being connected by a rebated joint, which is covered by the surbase moulding. Figs. 1661 and 1666 are reproduced to the scale of $\frac{1}{4}$ in. to 1 ft., and Figs. 1663 to 1665, and Fig. 1667, to the scale of $\frac{3}{4}$ in. to 1 ft. For Fig. 1665, which represents an enlarged part section on line E F (Fig. 1666), see p. 507.

Wall Panelling and Enriched Cornice for Billiard-room or Dining-room.

Figs. 1668 to 1685 show the preparation and fitting of a dado and wainscoting suitable for a billiard-room, dining-room, hall, or other similar apartment in a first-class villa, or in a town mansion or country residence. Various hardwoods are used for such work, but probably oak is most popular. Fig. 1668 shows the side of the room in which the doorway and its fitments are situated; Fig. 1669 the side of the room containing the fireplace. The dressing round the lower portion of the fireplace usually consists of marble or other masonry, up with pilaster jambs and soffit (see the perspective sketch, Fig 1671). The mouldings are worked on the solid of all the stiles and rails. The general principles of set-

Fig. 1668.—Elevation of Door and Portion of Panelling.

while the frieze directly over the fireplace, the cornice shelf, and the overmantel with pediment, etc., are constructed of wood. The woodwork against the external wall, and the inside elevation of the opening (illustrated at Fig. 1670), are shown fitted with casements and frame having an elliptical fanlight, the opening being fitted ting out rods, preparing the stuff, setting out for mortising and tenoning, mitering, and so forth, having been already dealt with in other sections of this book, it is unnecessary to repeat such details, and therefore the accompanying illustrations have been prepared with the object of showing more particularly the fitting and

Fig. 1669.—Elevation of Mantelpiece, and Panelling Adjacent to Internal Angle of Room.

connecting of one part to another, and the general fixing of the entire work. It is assumed that provision for fixing has been made by building-in wood bricks, or by some other method in general use.

A block B (see Fig. 1674), stub-tenoned and notched for the rebated ground and plinth, or for the skirting moulding, is fixed to the wall and floor. The skirting is fixed by being tongued into the floor, and dovetail-

Fig. 1670.—Elevation showing Treatment of Casement Opening.

Fig. 1671.—Pilaster, Linings, Soffit, Cove of Ceiling, etc.

Fig. 1673.—Vertical Section through
Pilaster, Mouldings, and Cornice.

Fig. 1672.—Enlarged Detail of Pilaster, Cornice,
Frieze, etc.

22*

grooved so as to slide on to the hardwood dovetail slip shown at D. The skirting moulding is ploughed on the under edge P to receive the top edge of the plinth, or of the skirting. The rebated ground H is next screwed to the back of the plinth mould as indicated at K. Then, the mould and the ground being placed in position, the bottom edge of the latter is held by the short stub-tenons in the tops of the blocks, and is nailed on through the rebated portion into the wall ground, firmly fixing the plinth. The bottom edge of the dado is rebated, leaving a barefaced tongue in

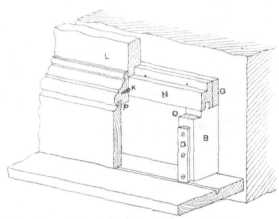

Fig. 1674.—Method of Fixing Plinth and Bottom Edge of Framing.

the front, and can then be dropped into position as shown at L (Fig. 1674). The top edge of the dado frame is sufficiently wide to allow of its being screwed to the wall ground G as indicated at N (Fig. 1675), these screws being afterwards hidden by the dado moulding. The ground M having been screwed to the dado moulding at N, these are placed in position with the tongue of the ground fitting into the top of the dado as shown. The ground M is then fixed to the wall ground G by nailing through the rebate as shown at R. The dado mould and the ground R have a groove formed between them, in which the barefaced tongue on the bottom edge of the upper framing O fits as shown. The top end of the upper framing is rebated to receive the bottom edge of the curved facia D, and is screwed

to the ground G (Fig. 1676); the fixing being hidden by the astragal mould C, which is fixed either with screws and slots, or preferably by dowelling to the top of the upper framing. The ground E having been screwed to the back of the curved facia, the upper part of the ground is nailed into the ground G. The ground E,

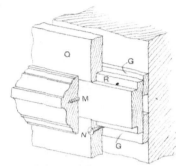

Fig. 1675.—Method of Fixing Top Edge of Lower Framing, Dado Moulding, and Bottom Edge of Upper Framing.

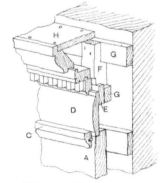

Fig. 1676.—Method of Fixing Cornice to the Top of the Framing.

projecting above the ground G, forms a rebate at the back, allowing the bottom ends of the blocks F to fit into the rebate. These blocks are secured to the back of the mouldings forming the cornice, and then the top ends of the blocks are notched and nailed as indicated. Figs. 1677 and 1678 show the method of fixing the astragal moulding, curved facia, and top cornice moulding. A wall-ground F, which has been rebated on the top edge and splayed on the bottom, is fixed as shown. To this the astragal moulding should be fixed with screws and slots or dowels and screws, inserted from the upper part of the round. These are out of sight. The bottom edge of the curved facia fits into a rebate made in the ground G,

Fig. 1673 a vertical section taken through the centre, to indicate the method of building up and fixing. Fig. 1679 shows a horizontal section through the bottom panel of the pilaster; Fig. 1680, a horizontal section taken through the pilasters and upper panelling. These views show in section how the pilasters are mitered, jointed, and tongued at the angles, and rebated to fit against and between the stiles of the

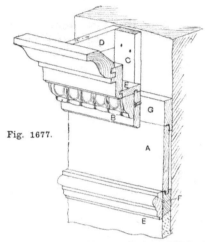

Fig. 1677.

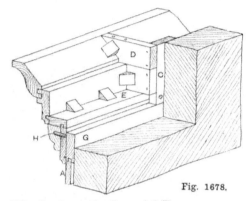

Fig. 1678.

Figs. 1677 and 1678.—Method of Fixing Main Cornice under Cove of Ceiling.

Fig. 1680.—Enlarged Horizontal Section through Fluted Pilaster and Panelling.

and is fixed to it by screws. The cornice is fixed by screwing the member B to the ground G, and then vertical blocks C are screwed to the member E. Bracketing pieces D are screwed to the top moulding

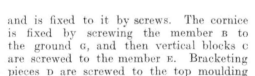

Fig. 1679.—Enlarged Horizontal Section through Lower Panel of Pilaster.

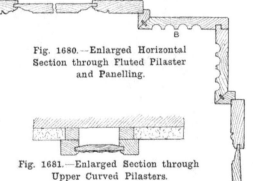

Fig. 1681.—Enlarged Section through Upper Curved Pilasters.

as indicated, and should also be glued and blocked; then the bracketing pieces are nailed to the vertical block C, and these in their turn can be nailed to the wall. The carved egg-and-tongue moulding is fixed by means of gluing and dowelling as indicated at H (Fig. 1678). Figs. 1672 and 1673 show enlarged details of a pilaster; Fig. 1672 representing an elevation, and

framing. At B (Fig. 1680) is shown the method of connecting the pilasters intersecting at an internal angle. Fig. 1681 is a section through the top of the pilasters that have carved panels. Fig. 1682 is a conventional detail view showing the plinth of the dado grooved, and the floor also grooved, as represented at A, B, and C, to receive the plinth forming the base of the pilaster. Fig. 1683 is a conventional view showing principally the back of one of these bases. In the same figure, tongues are shown for fitting into A, B, and C (Fig. 1682), with the moulding scribed to fit the mouldings of the plinth. Various expedients for fixing these bases, by screwing fillets to the inside of the base and screwing these fillets to the floor, will suggest themselves. Then a rebated fillet is fixed to the inside of the moulding, a portion of which is shown at A (Fig. 1683) and in section at A (Fig. 1673). The bottom ends of the lower pilasters are made to fit ac-

The method just described for the fixing of the lower pilaster is adopted for the upper pilasters, the bottom end of each being connected as indicated at B, C, D, and E (Fig. 1673). Fig. 1685 is a conventional sectional view of the bottom left-hand corner of the door, and of the base of the adjacent pilaster, etc.

Framed and Panelled Linings with Boxing Shutters to a Doorway.

Fig. 1686 is the half outside and half inside elevation of a circular-headed door

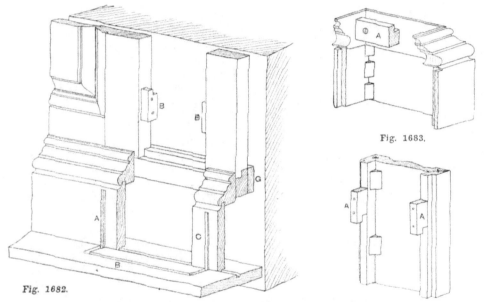

Fig. 1682.

Fig. 1683.

Figs. 1682 and 1683.—Methods of Fixing Base of Pilaster.

Fig. 1684.—Back of Pilaster with Fixing Buttons Screwed On.

curately behind the base moulding, and are prepared with a barefaced tongue so as to fit into the rebate of the fillet A. The pilasters are also held to the edges of the framing by buttons, two of these (A A, Fig. 1684) being shown screwed on to the back of the returned edges of the pilasters; and buttons are screwed on to the edges of the framing as shown at B B (Fig. 1682). When the pilaster is placed against the framing, and slid down into its base, the buttons A and B clip together, and thus firmly hold the pilaster to the framing.

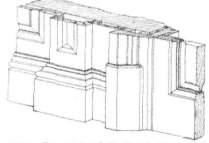

Fig. 1685.—Conventional Sectional View showing Bottom Left Corner of Door Base of Architraves, Pilaster, etc.

Fig. 1686.

Fig. 1688.

Fig. 1687.

Fig. 1686. — Half Inside and Half Outside Elevation of a Door with Marginal Lights, Panelled and Splayed Linings, and Boxing Shutters.

Fig. 1687.—Horizontal Section.

Fig. 1688.—Vertical Section.

opening. The inside of the opening is finished with framed and panelled splayed bottom lining A (Fig. 1686); the intermediate portion B is framed and panelled

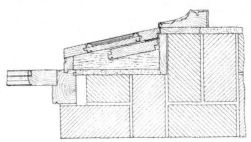

Fig. 1689.—Enlarged Detail of Horizontal Section at A (Fig. 1687).

with boxing shutters 1¼ in. thick. As will be seen, these are provided for covering the glass portion of the door, thus taking the place of a lifting shutter, being more

Fig. 1690.—Conventional View of Brickwork at Upper Part of Opening.

convenient. The head lining has a rail C (Fig. 1686) following the curved head of the door frame, but the outer edges of these linings are square. The curved rail

of the head lining follows the splay all round, and hence its outer surface is conical, but very flat, and thus not necessitating any complicated geometrical setting out. It should be cut to the circular form out of a board about ¾ in. thicker than the other stiles of the framing, having a joint

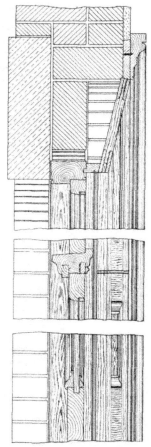

Fig. 1691.—Enlarged Detail of Vertical Section.

at the crown as shown. Then the bevel should be marked on each end; then by running a gauge on each edge from the lines on the ends, the conical surface can be formed by planing down to these gauge lines. The joint at the crown may be formed by grooving and tonguing, by halving together, or by making the muntin of a piece of thicker material, so that a projecting portion of it can run down behind the stile, and be screwed to each half. Another way of making the curved rail would be to form it of two thicknesses glued and screwed together, the front portion being in two parts and the back in three. In either case it would, of course, be ploughed to receive the panels, and at the springing it would be stub-mortised to receive the tenons of the rails. The stiles and top rail meeting on the splay would have to be mitered together as illustrated at D (Fig. 1686); the most satisfactory method of jointing here probably would be by an open mortice and tenon. From the illustrations it will be clear that the face side of the panel is a twisted surface, its square

edges being in a vertical plane and its circular edge starting conically; but for the example illustrated it will be found that stuff about $1\frac{3}{4}$ in. thick would be sufficient. This would require jointing up, the grain running in the direction as shown at Fig. 1686; then, with the bevel set to the splay of the linings applied at the springing and crown end of the panels, and running a gauge line round the curved edge, the face side of the panel could be worked to shape. To readers who have perused preceding descriptions, the general construction of the door, shutters, etc., will be clear. Fig. 1690 is a conventional view showing part of the brickwork of the arch and the reveal prepared to receive linings and shutters.

PARTITIONS AND SCREENS.

Setting Out Panelled and Moulded Framed Partition.

For the panelled and moulded framing of which an elevation is shown by Fig. 1692,

a rod 13 ft. long by 11 in. wide, and proceed to set out the plans of Fig. 1692, seen on rod 2 (Fig. 1693), the stiles and muntins being 4½ in. wide. First square across the rod two lines 12 ft. apart. Draw four lines parallel

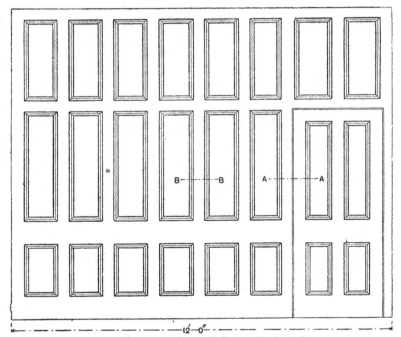

Fig. 1692.—Elevation of Panelled and Moulded Framing.

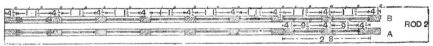

Fig. 1693.—Sectional Plans on Width Rod of Panelled and Moulded Framing.

one piece of deal framing is required, 12 ft. by 10 ft. by 2 in., *three panels in height*, moulded both sides, with a door at one end, with rebated and beaded joints. First take

with each other, 1⅞ in. apart, to represent the *thickness of the framing*. From the right-hand end mark off 4¼ in. for the outer stile B (Fig. 1693). Form the rebate with mould

Fig. 1694.—Enlarged Section on Line A A (Fig. 1692).

Fig. 1695. Enlarged Section on Line B B (Fig. 1692).

Fig. 1696.—Vertical Sections on Height Rod of Panelled and Moulded Framing.

(Fig. 1694). When a mortice lock is to be used, the rebate must be formed out of the centre to enable a half-rebated lock to be let into the door; but sufficient strength must be left for the bead to form the stop. Set off from the rebate 2 ft. 8 in., the width of the door, and fill in 4¼ in. for the stiles and 4¼ in. the centre for the muntin, leaving 9⅝ in. sight for the panel. Fill in the stile of the framing next the door, and also the left-hand outer stile, with moulds (Fig. 1694). Now divide the space between into six panels as shown. Set these spaces out accurately, and fill in the muntin with the mould (Fig. 1695). Fill in the lines representing the panels, also the moulding on both sides. To show what is required, it is sufficient to fill in one panel with the section of the moulding. A (Fig. 1693) may now be set out from B, as the plan for each is the same, except for the two panels over the door. The hanging and shutting stiles carry the 4¼-in. line to the top rail. Turn the rod over to the other side, and set out the sections as shown, and to figured dimensions. C (Fig. 1696) is a section through the framing; D is a section where the door occurs. Fig. 1697 represents the quantity board. In a big shop, the foreman joiner or setter out takes off the material, books it, and hands it to the chalk-line foreman, who marks out from the board all the material required. The board is then handed to the machinist, and finally to the material clerk, to be entered in the prime cost account. First take off the

Job No. 30.
1" Piece Deal Painted Framing. 12'0" × 10'0"
2" thick. moulded both sides.

	No.	Ft. In.	Ft. In.	In.	Deal.
Framing Stiles	3	10 3		4½	2
,, Rails	2	12 2		9	,,
,, ,,	2	9 0	9		,,
,, Muntin	6	3 0		4½	,,
,, ,,	5	3 10½			,,
,, ,,	5	2 0			,,
,, Panels	6	2 8½	1 2	½	,,
,, ,,	2	-	1 2½		,,
,, ,,	6	3 8	1 2		,,
,, ,,	6	1 9			,,
Door Stiles	2	6 10		4½	2
,, Rails	2	2 9	9		,,
,, ,,	1			4½	,,
,, Muntins	1	3 7			,,
,, ,,	1	2 0			,,
,, Panels	2	1 9	11	½	,,
,, ,,	2	3 8½			,,
Mouldings	2	7 9	1½	⅞	,,
,,	6	7 7			,,
,,	6	9 3			,,
,,	2	8 2			,,
,,	6	5 8			,,
,,	2	5 2			,,

Remarks.
This Partition is to be fitted together, dry-wedged, cleaned off and primed, taken to pieces and put together on the job.

Fig. 1697.—Quantity Board.

framing, stiles, rails, and muntins, then the panels, the door stiles, etc., in the same order. The whole of the mouldings may be booked as one item if taken from a stock pattern,

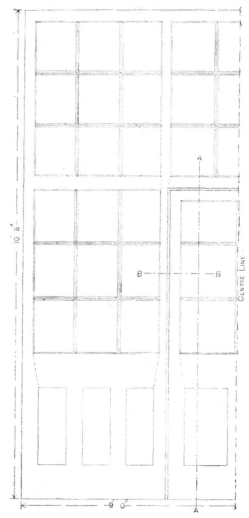

Fig. 1698.—Half Elevation of Glazed Partition.

but if the moulding is to be prepared to a special section it is best ordered in single lengths, sufficient to mould one panel. Odd lengths of stuff can be used up in this way without waste. Any remarks necessary should be added in the space reserved for the

purpose at the bottom of the board as shown (Fig. 1697).

Setting Out Glazed Partition.

For the glazed partition shown in half elevation by Fig. 1698, one piece of deal framing is required, 9 ft. by 10 ft. 8 in. by 2 in., three panels in height; the lower panels being square, the two upper panels divided into six squares, each with moulded bars, etc., for glass; a door to match to be formed in the centre with rebated and beaded joints, and prepared for a 6-in. mortice lock (half rebated). Take a rod 12 ft. long by 11 in. wide, and set out the plan and sections of Fig. 1698, which shows the elevation of the glazed partition. E (Fig. 1699) represents the plan below the transom rail, and shows the square framing and the position of the bars and diminish to stiles. The dotted lines on the rails in the elevation show the diminish. F (Fig. 1699) is a plan above the transom rail, and is set out from E, the lines being perpendicular from the middle rail upwards. The hanging and shutting stiles are carried up the diminished width above the transom, forming a muntin, and the transom rail is cut as shown in Fig. 1698. Determine the position of the door, and set out as before; the diminish to stiles is set off from the inside edge in each case. Note that the openings for the glass are equal in width, which should be obtained as follows: First set out with moulds, illustrated by Figs. 1694 and 1695 (p. 521), the position of the muntins and stiles. Place the bars exactly in the centre of the muntins. The panel width works out at 7 in. sight; 2 in. on each side to the centre of the muntins makes 11 in.; ½ in. deducted on each side for the half thickness of the bar leaves 10 in. sight. Now, 10 in. is wanted for the squares on each side. The stiles must therefore be diminished 1 in., and the required 10 in. is obtained. This will therefore be the width of all the openings above the transom rail; and the side framings below the openings in the door work out at 11 in. G and H (Fig. 1700) are sections through the side framings and where the door occurs. Fig. 1694 is an enlarged detail of the door on the line A A (Fig. 1692), and Fig. 1695 is an enlarged detail on the line B B (Fig. 1692).

Set these sections out in the same manner as before. Make a quantity board (Fig.

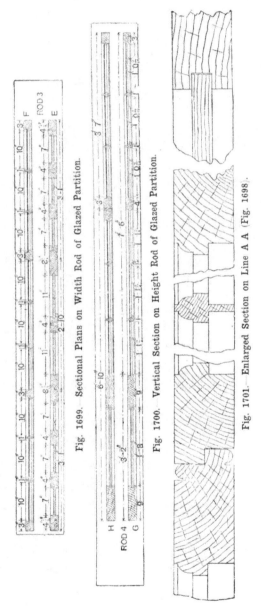

Fig. 1699. Sectional Plans on Width Rod of Glazed Partition.

Fig. 1700. Vertical Section on Height Rod of Glazed Partition.

Fig. 1701.—Enlarged Section on Line A A (Fig. 1698).

1703), as described in the previous paragraph, and book the material on it as shown:

Hall Screen with Door.

Fig. 1704 shows an elevation of a hall screen, the approximate size of which may be taken as 10 ft. high by 8 ft. wide. Fig. 1705 shows a vertical section, and it will be

Fig. 1702.—Enlarged Section on Line B B (Fig. 1698).

observed that the thickness of the two outer jambs above the transom c is reduced 1 in. on the rear side D; while the three muntins E F G (Fig. 1704) are worked to the same thickness (their finished sizes being 3 in. by 3 in.), and are wrought quite square. They

Fig. 1703.—Quantity Board.

are simply rebated on the rear side to receive the leaded lights and bead, and are tenoned into the transom c and curved rail H (Fig. 1705), the top and rear sides of which are also rebated in the same way as for the muntins. Fig. 1706 gives a horizontal section, showing the outer jamb A and door jamb B, which are worked from 4-in. by 4-in. and 4-in. by 3-in. stuff respectively. The

and muntins. On both the front and the rear sides the top rail has sunk and moulded spandrils, as shown at J (Fig. 1704), and this rail, when fixed to its proper position, sets

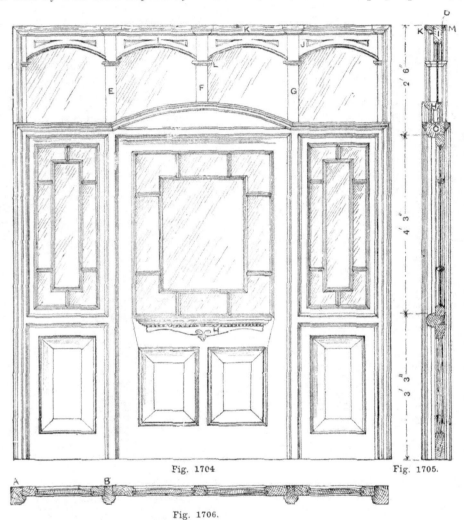

Fig. 1704.   Fig. 1705.

Fig. 1706.

Fig. 1704.—Elevation of Hall Screen.   Fig. 1705.—Vertical Section of Hall Screen.
Fig. 1706.—Horizontal Section of Hall Screen.

top rails I (Fig. 1705) are tongued $\frac{3}{4}$ in. into the jambs and muntins, and are cut, wrought, shaped, grooved, and ovolo-moulded from an 11-in. by 2-in. sound plank, and framed flush with the rear sides of the outer jambs

back from the face of the screen 1 in., which allows for the moulding K to be mitered and returned to jambs and muntins as shown, as is also the neck moulding L, which should be grooved $\frac{1}{4}$ in. into the jambs and muntins, the whole piece of framing being afterwards strengthened on the rear side by the top moulding M. which is glued and planted on (Fig. 1705). All the framing below the transom C is rebated and beaded to receive the door, sidelights, and under-side framing, all of which is 2 in. thick, and is finished flush to the rear side of the framing, the front side of which is ovolo-moulded. The sidelights and the top portion of the door are also rebated for glass, and ovolo-moulded, and are constructed as shown in the elevation. All the bars, except those framed into the upper edge of the middle rail of the door, should be mortised and tenoned through both stiles and rails, and glued and

round-headed screws. The screen may be executed in good yellow deal or pine twice sized and twice varnished with good copal. If it is made in either oak or mahogany it may be oiled or French-polished. The door should be hung with one and a half pairs of 4-in. brass butts, with steel washers, and fitted with a good 6-in. mortice lock. Good bold brass handles and finger-plates should be chosen, and the top portion of the screen above the transom should be fitted with leaded lights glazed with tinted glass ; tinted and white Muranese glass, bedded between strips of chamois leather, being used for the door and for the sidelights. Fig. 1709 is an enlarged section of the architrave.

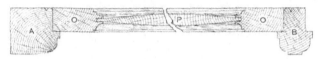

Fig. 1707.—Enlarged Horizontal Section of Framing of Hall Screen.

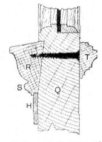

Fig. 1708.—Enlarged Section of Bar of Hall Screen.

Fig. 1709.—Enlarged Section of Architrave of Hall Screen.

Fig. 1710.—Enlarged Section through Top Portion of Middle Rail to Door.

wedged in the usual manner. The ends should be simply stumped into the central mitered bars, and screwed through the rebates. The lower panels are moulded and raised as shown, A indicating the outer jamb, B the door jamb, O O the stiles of the framing, P the panel (Fig. 1707). The bottom portion of the door is in every particular identical with the corresponding part of the side-framing, and has diminished stiles (see Fig. 1698). Fig. 1710 shows an enlarged section of the upper portion of the middle rail of the door ; Q indicating the rail ; R moulding with returned ends, sunk into and planted on to rail, with dentils cut in the bead S, and fixed through the rail with screws ; T moulding with returned ends, glued and planted on to cover screws ; H the shaped apron (see also elevation, Fig. 1704). Fig. 1708 is an enlarged section of the bar to the sidelights and the top portion of the door. The moulded fillets U should be fixed with brass

Corridor Screen and Door.

The screen and door shown in Figs. 1711 and 1712 is suitable for a public office or building or for a private dwelling. If it is used for a dwelling, stained and leaded glass of good design should be inserted ; but for a public building, the top sashes should be filled with quarter-plate polished glass, while the doorlight and sidelights will be of embossed glass, with a suitable design or lettering advertising the business that is carried on. Fig. 1711 shows the front view. The height is 9 ft. 8 in., the width 6 ft. The door jambs and wall jambs are 3½ in. by 5 in., rebated for the door and sidelights, grooved for the raised panels, and ovolo-moulded on the front edges, and beaded on the back edges. The transom is 7 in. by 4 in., sunk double moulded, and rebated for the top and lower sashes and door. The jambs are stub-tenoned to the transom,

Fig. 1711.

Fig. 1712.

Fig. 1711.—Front Elevation of Corridor Screen and Door.
Fig. 1712.—Vertical Section of Corridor Screen and Door.

and also connected by short rails, which are tenoned and scribed to them, forming divisions for the side panels and lights. The semicircular frame is of the same section as the jambs, and is double moulded, rebated, beaded, and tenoned to the transom (see Fig. 1713), and it is connected to the top rail with a short mullion, which is mortised,

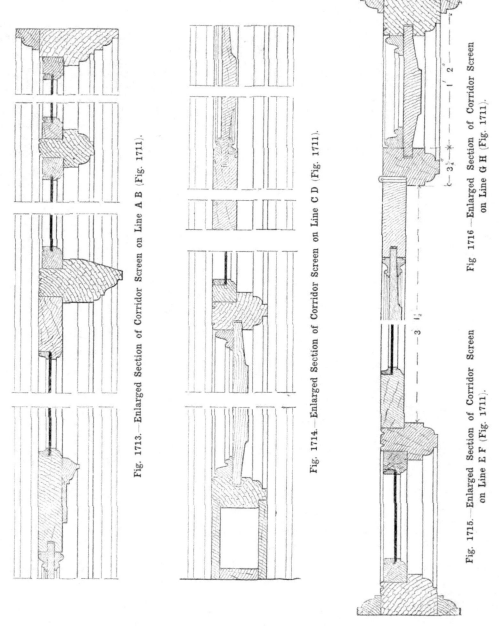

Fig. 1713.—Enlarged Section of Corridor Screen on Line A B (Fig. 1711).

Fig. 1714.—Enlarged Section of Corridor Screen on Line C D (Fig. 1711).

Fig. 1715.—Enlarged Section of Corridor Screen on Line E F (Fig. 1711).

Fig 1716.—Enlarged Section of Corridor Screen on Line G H (Fig. 1711).

tenoned, and scribed to both members. The jambs above the transom are stub-tenoned to the latter, and mortised to receive the top rail. The latter is secured to a joist above, while the wall jambs are fixed to wood or breeze bricks in the usual way. The two side and three top frames are 1¾ in. by 2 in., moulded and rebated, and fitted with loose beads. The door is 6 ft. 6 in. high by 3 ft. 3 in. wide. It has diminished stiles, which are 7 in. and 5 in. wide respectively at top and bottom, by 2 in. thick. The upper rail is 5 in. wide. The top panel of the door is moulded, rebated, and fitted with shifting beads, fixed with brass cups and screws. The lock and bottom rails are each 9 in. wide. grooved to receive the raised and moulded panel, and mortised, tenoned, and wedged to the stiles. The shoulders

apron is fixed to the door, and capped with a moulded rail with return ends, as shown. Fig. 1713 shows a section taken on A B, while Fig. 1714 represents a vertical section taken through the side panel and short rails, showing the method of construction at the lower end of the panel at D (Fig. 1711). Figs. 1715 and 1716 respectively represent horizontal sections on E F and G H, showing in enlarged detail the sashes, panels, etc. The door is hinged on three 4-in. brass butts, fitted with pull handles, brass mortice lock, and finger-plates   This class of door is

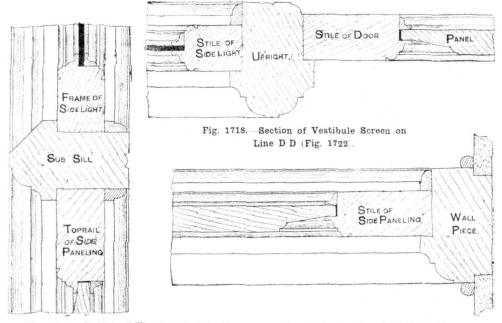

Fig. 1718.—Section of Vestibule Screen on
Line D D (Fig. 1722).

Fig. 1717.—Section of Vestibule Screen on
Line C C (Fig. 1722).

Fig. 1719.—Section of Vestibule Screen on
Line E E (Fig. 1722).

between the middle rails and stiles are diminished as illustrated, and in the case of the lock edge of the door the middle rail and stile are connected by twin double mortices and tenons with a solid haunching (as illustrated and explained in connection with Fig. 1196, p. 360), so as to make provision for a mortice lock, which is the kind that would be likely to be provided and fixed. A suitable moulding is fixed round the panel, and a shaped

often hung with a patent single or double spring hinge, the former closing the door by its spring action after opening, while the latter allows of the door being opened both inside and outside, the action of the hinge closing it automatically. The material may be pitchpine or red deal, sized and varnished, or painted and grained oak; or, if the work is executed in hardwood, as teak, mahogany, or oak, it may be finished in oil or French-polished.

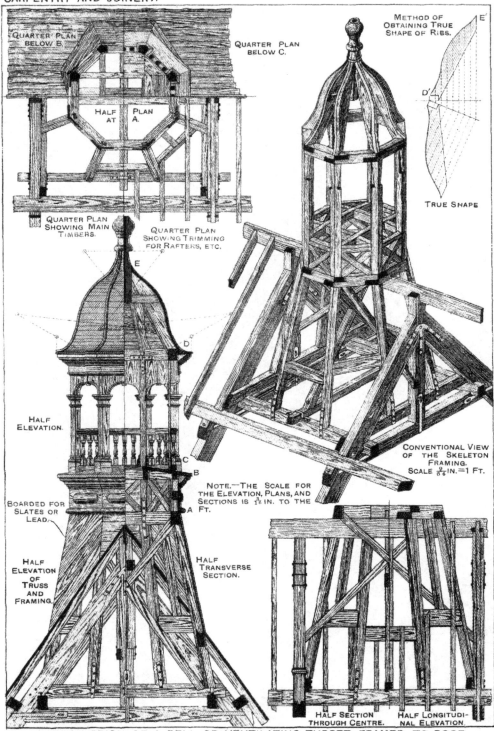

QUARTER PLAN BELOW B.

QUARTER PLAN BELOW C.

METHOD OF OBTAINING TRUE SHAPE OF RIBS.

E'

D'

TRUE SHAPE.

HALF PLAN AT A.

QUARTER PLAN SHOWING MAIN TIMBERS.

QUARTER PLAN SHOWING TRIMMING FOR RAFTERS, ETC.

E

D

HALF ELEVATION.

C

B

A

NOTE.—THE SCALE FOR THE ELEVATION, PLANS, AND SECTIONS IS $\frac{1}{16}$ IN. TO THE FT.

CONVENTIONAL VIEW OF THE SKELETON FRAMING. SCALE $\frac{3}{64}$ IN.=1 FT.

BOARDED FOR SLATES OR LEAD.

HALF ELEVATION OF TRUSS AND FRAMING.

HALF TRANSVERSE SECTION.

HALF SECTION THROUGH CENTRE.

HALF LONGITUDI-NAL ELEVATION.

12   CONSTRUCTION OF A BELL OR VENTILATING TURRET FRAMED TO ROOF.

Vestibule Screen.

The screen shown by Figs. 1717 to 1724 is easily adaptable to any opening of reasonable width and height. The panels being equally spaced out, they can be reduced or extended in width at pleasure, the only difference being that the two doors will necessarily become one only in the case of any excessive reduction taking place in the width. The height and width shown in the details are 10 ft. 4 in. (or, adding a 7-in. joist and 1-in. flooring board, 11 ft.) and 8 ft. respectively. The height, if so required, could be reduced either by omitting the rough beam and cornice or by shortening the fanlight. The tendency of vestibule and entrance-hall screens being towards excessive height, a subsill is introduced with the object of rendering the height less conspicuous; and if a dado is constructed, it should line up as much as possible with this subsill. The screen will look well if executed in oak or mahogany; but if deal is the material selected, it should be painted rather than stained and varnished. The framework is of 5-in. by 3½-in. section, with the exception of the transom, which is 5 in. by 4 in., and dentilled; but as these dentils are rather expensive to cut, the transom may, if cost is a consideration, be run straight through. The doors, fanlights, and sidelights are of 2-in. stuff, rebated for glass, and moulded. The circular sinkings at the heads of the fanlights are ⅜ in. deep, chamfered ¼ in. down and on. The doorway is 3 ft. 9 in. wide, and fitted with a pair of doors, which are less unwieldy than a single door, although, for ordinary purposes, the one door affords sufficient passage room. The wall upright may be of a lighter section than the other timber (the illustration shows it 5 in. by 3 in.), but this is not absolutely essential; it can either be of the same size as the other timbers, and fixed directly against the wall to wood bricks, or be of lighter section, and fixed to grounds; the latter method obviates the necessity of grooving for plaster. The wood panels are raised, 1⅛ in. thick at the centre, and reduced to ½ in. at the edges, the stiles and rails being rebated to receive them; they are secured in the rebates by a moulding mitered round on the inside. The upper panels and fanlight may be glazed with leaded lights, or plain or bevelled plate glass, the latter being the most suitable, owing to the rather severe character of the design. To

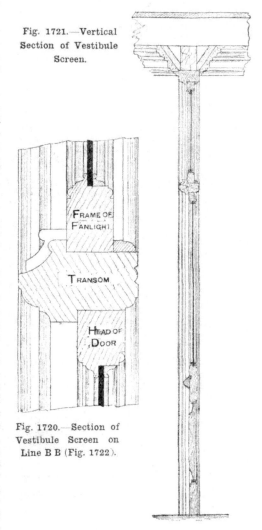

Fig. 1721.—Vertical Section of Vestibule Screen.

Fig. 1720.—Section of Vestibule Screen on Line B B (Fig. 1722).

form the cornice, rough brackets are fixed at intervals to receive the plaster, a suitable key being formed on the head of the screen by the bead, as shown in the detail. For sections on the lines indicated by lettering in the usual way on the elevation shown by Fig. 1722, see Figs. 1717 to 1721.

Fig. 1722.—Elevation of Vestibule Screen.

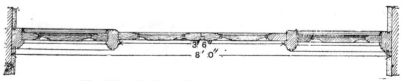

Fig. 1723.—Sectional Plan of Vestibule Screen.

Vestibule Framing and Swing Doors for the Main Entrance of a Large Building.

A more important case of vestibule framing with swing doors than either of those preceding is here illustrated by Figs. 1725 to 1748. The framing is prepared and fitted to an opening with an elliptical stone arch springing from imposts as illustrated.

with the specification clause to the effect that "The material and workmanship must be the very best of their respective kinds." It is almost superfluous to state that joinery of this character would be made of hardwood. It is necessary, in jobs of this description, for the setter-out to give some consideration to the arrangement of the fitting together of various parts where carving is introduced. The carving is not

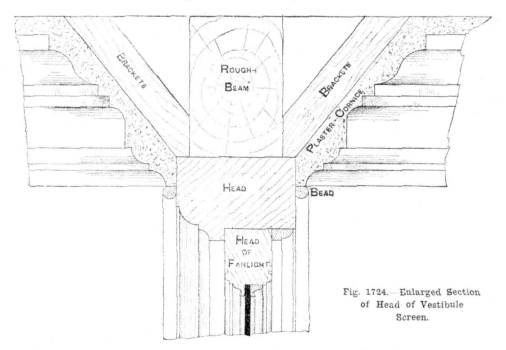

Fig. 1724.—Enlarged Section of Head of Vestibule Screen.

The whole is designed for the treatment of a vestibule, the walls of which are of masonry, there being a plinth or skirting at the bottom, and a surbase, or dado moulding; the next portion of the wall finishing with an intermediate cornice and fascia, and above this being a frieze and a cornice connected with the ceiling (but not shown). It will be seen that the main horizontal members of the wood framing are of similar section to those of the masonry, with which some of them intersect. The methods of construction which will be described and illustrated are amongst the best adopted for first-class work, so as to comply fully

of course, introduced as specimens of joinery, but at the same time it is part of the treatment of the whole design, forming the ornament of panels, friezes, etc. It falls to the task of the joiner to plough and tongue and fit together the various parts, and, when these operations are found satisfactory, to hand over to the carver the pieces he has to deal with, which can be afterwards fixed in their respective positions in the piece of framing. It must here be noted that pieces of wood carved and "stuck on" are not here illustrated, nor would they be tolerated for a job of this description. Of course sometimes the carver

Fig. 1725.—Perspective View of a Portion of Vestibule Framing seen from Outside.

Fig. 1726.—Outside Elevation of Vestibule Framing with Swing Doors.

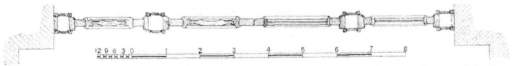

Fig. 1727.—Half Horizontal Section through Lower Pilasters and Panels, and Half Horizontal Section through Upper Pilasters and Glass Panels.

would not commence any of his work until the joiner had finished, so far as bench work was concerned.

Setting Out.—Usually, for important work, architects supply enlarged detail drawings in the form of elevations and sections ; and in addition, where there are mouldings and similar members of special design, full-sized sections are supplied ; while full-size sketches for carvings are commonly provided.

From the specification and the drawings the rods would be set out. Naturally it would be well for the actual opening to be tested for measurement, in case of any discrepancy having crept in. One rod *should be set out, giving a horizontal section* at a level through the bottom panels of the doors and lower pilasters ; another section should be given at a level of the upper cornice of pediment, the tympanum, and the horizontal cornice, showing the relation of this with the pediment, etc. Whole elevations, or at least halves, drawn full size, and in connection with the vertical sections, *should be set out for the pediments at the* tops of doors, dado moulds and apron panels, quadrant corners, etc. On these should be indicated, by coloured pencils or other

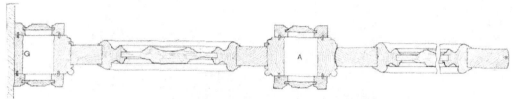

Fig. 1728.—Enlarged Horizontal Section through Lower Pilasters and Panels.

pilasters and glass panels of doors, similar to section Fig. 1727. It would be very convenient to have a third horizontal section taken through the frieze. Probably the best plan would be to have the rod sufficiently wide, so that these horizontal sections could be side by side, and thus *their relation to each other be apparent at* a glance. The following vertical sections would be necessary. The rod for the vertical section should show a section taken through the centre of the pilaster to the top of the cornice as Fig. 1730. Another section should be taken through the centre of one of the *swing doors and head of frame as repre-*sented at Fig. 1731. The construction of the means, the methods of connecting the pediments, mouldings, carved panels, and such parts, to the stiles and rails ; also the sizes and number of tenons connecting the stiles and rails.

The Frame.—As will be seen, this consists of a head and six jambs, each $2\frac{1}{2}$ in. thick. There being pilasters on each side, the central jambs are distant from each other to the extent of 4 in., as represented at A (Figs. 1728 and 1729). Beyond the jamb at each end is a ground G ; these are fixed to the masonry. The central jambs are connected and stiffened to each other by blocks screwed between, as represented at A, B, and C (Figs. 1730 and 1735). The

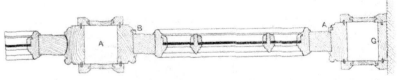

Fig. 1729.—Enlarged Horizontal Section through Upper Pilasters and Glass Panels.

fixed doors being similar in many respects to that of the swing doors, it would only be necessary to show the section taken through *the marginal bars. This of course might be* done adjacent to the section taken through the swing doors. Another vertical transverse section is taken through the crown of the soffit of the arch, giving sections of the head of the frame, the fanlight, the jambs at each end are also similarly connected to the ground G. The central jambs are hollowed and moulded for the hanging stiles of the swing doors. The other jambs are rebated for the fixed doors, and also ploughed to receive a tongued slip A (Fig. 1729), which forms a member of the moulding, and keeps the fixed doors in position. The angles of the jambs are

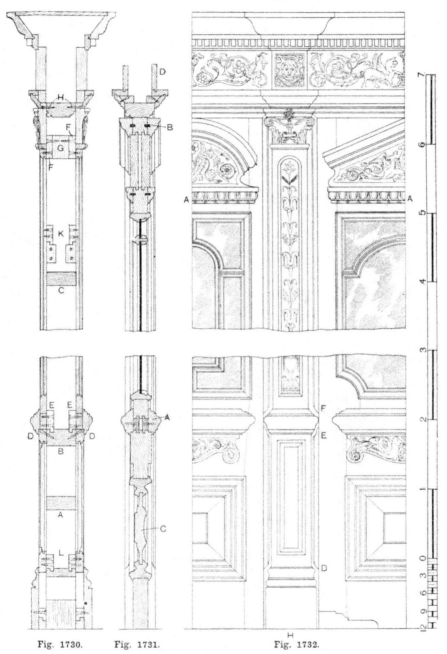

Fig. 1730.   Fig. 1731.   Fig. 1732.

Fig. 1730.—Enlarged Vertical Section through Centre of Pilasters (Fig. 1725).
Fig. 1731.—Vertical Section through Centre of Swing Door (Fig. 1725).
Fig. 1732.—Enlarged Detail of a Portion of Swing Door, Pilaster, Cornice, and Fixed Door.

moulded and mitered so as to intersect with similar mouldings stuck on the head, to which the jambs are connected by mortice and tenon joints as illustrated at A (Fig. 1733). By reference to Fig. 1726, and D, E,

clearly shown in the illustrations. A deal fillet F is fixed to the head of the frame (Fig. 1733); then to this fillet, and the head of the frame, the moulding E is fixed by the insertion of screws through the fillet to the

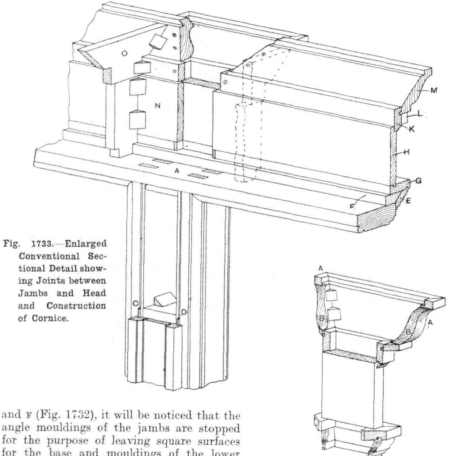

Fig. 1733.—Enlarged Conventional Sectional Detail showing Joints between Jambs and Head and Construction of Cornice.

Fig. 1734.—Conventional View of Portion of Cornice which Breaks Forward over Pilaster.

and F (Fig. 1732), it will be noticed that the angle mouldings of the jambs are stopped for the purpose of leaving square surfaces for the base and mouldings of the lower part of the pilasters to butt against. The jambs are ploughed to receive tongues, which are inserted in corresponding plough grooves made in the back of the stiles of the pilasters, as represented in section at Figs. 1728 and 1729, and C, D (Fig. 1733). The bottoms of the jambs are fixed to the stone step or floor by the insertion of two copper dowels in each.

Cornice.—The different pieces of this, and the methods of connecting them, are

back of the moulding. The members G, H, K, L, M, can be screwed together, and the cornice on each side and the fascia connected together by cutting cradling pieces between, and fixing these by a few screws, and gluing blocks to the cradling pieces and backs of the members, as represented at N and O (Fig. 1733). In this way the cornice could be built up of three main sections, and thus be easily placed in position. Fig. 1734 represents principally a back view of one of the portions of the cornice which breaks out over the pilaster. The external angles A are shown as mitered, glued, and blocked on the inside. The intersections of the mouldings at the internal angles are shown as scribed, not mitered, although of course either method may be adopted.

Pilasters.—The lower pilasters have stiles and rails with mouldings stuck on the solid, and are connected together by stubbed mortices and tenons. The panels are sunk and worked with a moulded splay as represented in the sections. The stiles are sufficiently long to run down to the floor so as to allow of the moulded plinth with its mitered returns being glued to them. The plinth is further secured to them by the insertion of a few screws from the backs. The mitres of the plinths, if not dovetailed, should be grooved and tongued, and of course glued. The lower pilaster is fixed in position by screwing on buttons at the back which have rebated ends, the tongue portions fitting into rebates or grooved blocks fixed between the jambs, as represented at L (Figs. 1730 and 1735). The heads of the lower pilasters are fixed to a block between the jambs by a couple of screws being inserted obliquely, as indicated at D (Fig. 1730). The mouldings at the heads of the lower pilasters have thicknessing pieces fixed at the back as shown at Fig. 1730, and to these buttons E are screwed the tongues which fit into grooves made in the block B (Figs. 1730 and 1735). It will be noticed that these buttons are allowed to project above the thicknessing pieces so as to receive the ends of the upper pilasters as shown at E (Figs. 1730 and 1735). The top ends of the upper pilasters have blocks screwed to the back of them as indicated, and these blocks in their turn are screwed to the block G, which is fixed between the jambs. The carved capital, with its neck moulding, is connected to the top of the pilaster by a couple of dowels; then the top of the capital is secured to the head of the frame by screws as indicated at H (Fig. 1730). To prevent any vacancy occurring between the pilasters and the edges of the

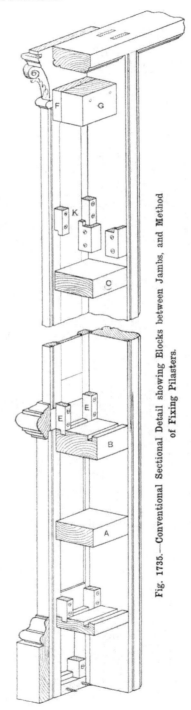

Fig. 1735.—Conventional Sectional Detail showing Blocks between Jambs, and Method of Fixing Pilasters.

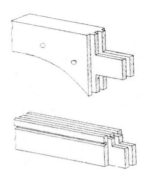

Fig. 1736.—Joints between Stile, Top Rail, and Frieze Rail of Swing Door.

Fig. 1740.—Joints between Top Rail and Frieze Rail of Fixed Door.

Fig. 1737.—Tympanum Panel Prepared Ready for Carving.

Fig. 1741.—Tympanum Panel Prepared for Carving.

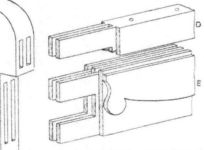

Fig. 1738.—Joints between Stile and Middle Rails, with Apron Prepared for Carving.

Fig. 1742.—Joints at Middle of Door.

Fig. 1739.—Joint between Bottom Rail and Stile.

Fig. 1743.—Joint at Bottom of Door.

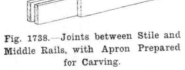

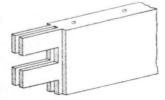

1736), whereas the groove at B is for receiving a tongue that is fixed into the back of the pediment moulding (see B, Fig. 1736). At C the quadrant corner is shown formed on

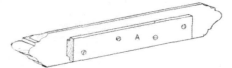

Fig. 1745.—Dado Mould with Fillet Screwed On at Back.

the solid stile, and not inserted as would be the case in more ordinary work. The rail D is shown having tenons passing right through the stile; this would undoubtedly

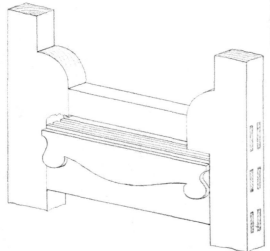

Fig. 1744.—Rails Ploughed Ready to Receive Fillet of Dado Mould.

jambs, these are held securely by means of buttons, as shown at K (Figs. 1730 and 1735).

Doors.—In the construction of these, twin mortices and tenons have been adopted. At Fig. 1736 is shown the upper part of the stile of the swing door mortised and haunched, with the top rail and frieze rail tenoned, haunched, and ploughed to receive the tympanum panel that is shown at Fig. 1737, and is represented as rebated and tongued ready for carving. The plough grooves A in the frieze rail are for the purpose of receiving the tongue, which is inserted in a similar plough groove made in the back of the egg-and-tongue moulding (see A, Fig.

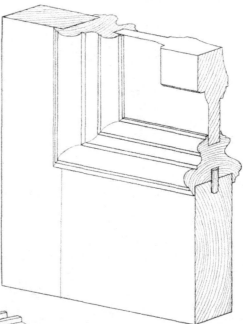

Fig. 1746.—Conventional Sectional Detail of Framed Mouldings and Panel, showing Moulding Saddled Over Stile and Rail.

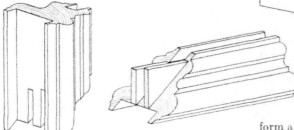

Fig. 1747.—Method of Framing Mouldings by Slot Mortice and Tenon Joint.

form a better job than stubbing them in. The rail E, it will be seen, is of thicker material, so as to allow of the apron (which is to be carved) being worked on the solid as represented. Both these rails are ploughed on their inner edges for the purpose of receiving rebated fillets, the tongues of which are held in these grooves—the object being to fix the dado moulds. These fillets are screwed and glued to the back of the dado moulds, as represented at A (Figs. 1731 and 1744), and then the framing, having been put together, is placed in position.

Panels.—These are raised, with a small scotia worked on the edge of the raised part, the margin of the panel being worked to the form of a flat ogee. The inner side of the panels has a flat margin, with a bolder moulding worked on the raised part, and the

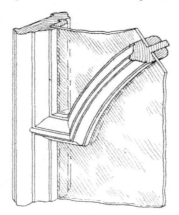

Fig. 1748.—Conventional Detail showing a Portion of Saddle Moulding and Curved Bar for Glazing.

face is sunk as shown in section at c (Fig. 1731). The bolection mouldings round the lower panels are solid through, so as to saddle over the edges of the stiles as represented at Fig. 1746 The inner edges are ploughed to receive the panels as shown (Figs. 1746 and 1747). The angles of the mouldings are mitered and framed by slot mortices and tenons, as represented at Fig. 1747. These mouldings would have the panels inserted and the joints glued before being placed between the stiles and rails. The framed mouldings are firmly secured to the stiles and rails by means of dowels, as indicated at Fig. 1746, and the holes for these are represented in the stiles and rails at Figs. 1736 to 1743.

The mouldings round the glazed part of the doors may be framed separately and inserted similarly to what has been described for the lower panels; or each piece may be fixed by gluing and dowelling, and then the mitres fitted together and

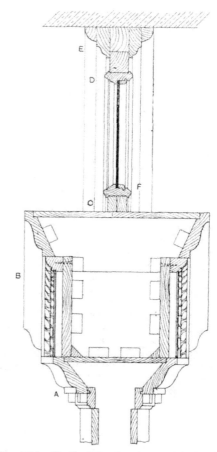

Fig. 1749.—Vertical Section through Horizontal Cornice, Centre of Pediment, and Fanlight.

treated almost in the same manner as a case of solid moulded work, but probably the former would be found in the end the most satisfactory manner. At Figs. 1740 to 1743 are shown the joints between the stiles and rails of the fixed doors; the general principles of construction are the same as has been set forth for the swing doors, but of course varying in details as illustrated. The circular-headed marginal bars in the fixed doors may be cut and worked out of the solid, in which case they would of course have to be made of two pieces and butt-jointed, at the crown or near it, and springing, to the vertical bars. A neater and stronger job would be produced by steaming and bending, in which case a cylinder would have to be made and the method adopted as explained in connection with Fig. 1500, page 460. As the bars are rather stout, it would be found necessary to joint up three thicknesses over the cylinder; the thicknesses should be so arranged that the joints would not fall in the curved parts of the mouldings, but in the fillets or square parts.

Pediment and Fanlight.—At A (Fig. 1749) is shown the continuation of the section D (Fig. 1731), which is a section through the horizontal cornice; above this is shown the section through the centre of the main curved pediment and tympanum. The general construction is shown at Fig. 1749, where it will be seen that the several parts are screwed together as far as practicable, and further strengthened by cradling pieces being inserted and glued and blocked as illustrated. The top is boarded as shown by c, and then a curved rail prepared in two thicknesses is fitted on as shown in section at c, this forming the bottom rail of the fanlight. The top rail of the fanlight and head of frame fitting to the soffit of the stone arch are represented in section at E, where it is shown the whole is built up of five thicknesses. The mouldings D and F are worked out of the solid in convenient lengths. The curved bars being of flat curvature, it will probably be found the most convenient method to prepare these out of the solid rather than to bend material for them. These are mitered and tongued to the short straight bars, and then these latter are connected to the radial bars by any of the usual methods. It is not necessary to enter fully into every detail of the construction, as the illustrations have been carefully prepared so as to make clear all the most essential particulars; while the general principles have been sufficiently expounded in other sections of the book.

Hanging Swing Doors.

In hanging swing doors of the description here shown, it is essential to success that the doors shall be prepared accurately out of winding, and that the frame shall be fixed out of winding. If these precautions are carefully observed, no difficulty will be

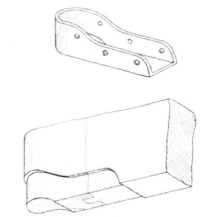

Fig. 1750.—Shoe, and Heel of Door Prepared to Receive It.

Fig. 1751.—General View of Spring Hinge, with Shoe Removed.

experienced in getting the doors to close and meet in line. Only general methods can be given for fixing the springs, of which there are many different kinds on the market, each requiring more or less special treatment. Printed instructions are generally sent out with each hinge by the maker, and these of course should be followed as closely as possible. The spring hinges are usually enclosed in a cast-iron box which has projecting flanges (see Fig. 1750), and to these the brass plate is secured with screws. The box is sunk so that the brass plate is flush with the surface of the floor.

In the instance of a wood floor, two or three floor-boards are taken up, and trimming pieces are fixed between the joists, and also the firring pieces that may be necessary at the side (as indicated at Fig. 1753), to support the flanges or lugs of the box. The box is next placed in the hole, and its exact position found in the following manner. The shoe when at rest (A, Fig. 1754) must have its sides quite parallel to the plane of

Fig. 1752.—Hole Cut in Masonry Floor to Receive Box of Spring Hinge.

Fig. 1753.—Trimming and Firring Piece to Joists, etc., to Receive Hinge.

When the floor is of stone or concrete, the position and size of the hole for the box are marked out by the joiner, and then the hole is cut by a mason as represented at Fig. 1752.

the jambs, and its centre line opposite the centre of the hollow of the jamb as indicated. The shoes should next be turned into the positions shown at B and C, so as to clear the jambs when the door is opened at right angles, as indicated at B and C (Fig. 1754). The flanges or lugs should next be screwed to the trimming pieces, then the flooring made good ; after which the brass plate can be applied, the flooring being marked round from it and then paved, while finally

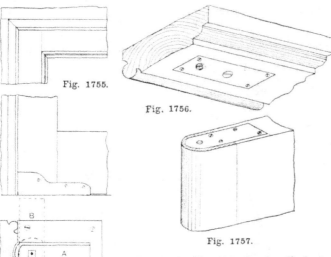

Fig. 1755.

Fig. 1756.

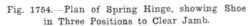

B

A

C

Fig. 1754.—Plan of Spring Hinge, showing Shoe in Three Positions to Clear Jamb.

Fig. 1757.

Fig. 1755.—Elevation showing Heel of Door and Side of Shoe.

Fig. 1756.—Pivot let into Head of Frame.

Fig. 1757.—Pivot Plate let into Door.

Fig. 1756. The exact position of the pivot must be obtained from the centre of the movement of the shoe. In the top of the stud (on which the shoe is fixed) a punch mark can usually be seen, which indicates the axial centre. The distance of this centre marked from the hollow of the frame gives the exact distance of the centre of the pivot from the hollow of the jamb. Care must also be taken to keep the pivot in the centre of the thickness of the door and the hollow in the post. The forms of the posts vary

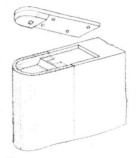

Fig. 1758.—Head of Door Recessed to Receive Pivot Plate.

the brass plate is dropped in position and fastened to the iron flange with screws, as shown at Fig. 1751. Where the floor is of stone or concrete, the iron box must be temporarily fixed with a few small hardwood wedges, of course in the exact position it will have to occupy, as explained above. The box is finally secured by filling in the spaces between the stone with cement. The next operation is to fit the shoe on the bottom of the door as represented at Fig. 1750. This will require to be very accurately done, especially in fitting the shaped sides of the shoe. The shape may be obtained by applying the side of the shoes to the side of the door. The pivot and plate can next be let in to the head of the frame as shown at

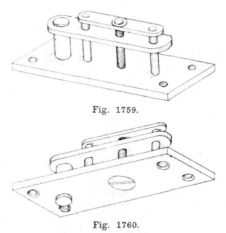

Fig. 1759.

Fig. 1760.

Figs. 1759 and 1760.—Mode of Action of a Pivot Plate.

according to the different makers, but are mostly constructed on the principle that the head of a fixed screw, when turned with a screw-driver, acts on a plate to which the pivot is fixed, which may be raised or lowered according to the direction in which the screw-head is turned. Two views of a very good form of pivot are shown at Figs. 1759 and 1760, from which the action is easily inferred. A plate to receive the pivot has to be let into the head of the stiles and top rails as shown at Figs. 1757 and 1758.

a very useful arrangement for doors through which there is very much traffic.

Sliding and Folding Partitions.

Figs. 1761 to 1772 illustrate a useful form of folding and sliding partition, which has been largely used for dividing large rooms, and for separating spaces under galleries from the main part of halls, and in other similar positions. The general arrangement is not so up-to-date as a number of forms that are the subjects of current patents.

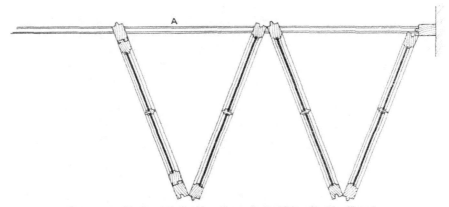

Fig. 1761.—Horizontal Section through Partition Partly Closed.

Fig. 1762.—Enlarged Section through B B (Fig. 1763).

The exact position of the plate is determined by the centre of the hole to receive the pivot, as previously explained. Some improved forms of pivots have an adjustable side screw, by which the doors may be adjusted sideways, so as to correct any slight fault in case of winding of the doors or frame. Another form of fitting provides an arrangement for throwing the door slightly forward or backward as may be required. To some forms of springs adjustable arrangements are provided by which, on taking up the brass plate, the mechanism can be got at ; and, on turning adjustable screws or studs, the shoe is regulated in such a manner that the doors may always be kept in one plane. This is

The special feature in all these inventions lies not in the general construction of the joiners' work, but in the special mechanism introduced to produce lightness and accuracy in movement, so that large openings may be provided with movable folding partitions, which can be expeditiously folded up. The partition here illustrated is intended for an opening of moderate dimensions, and for that service is found to work satisfactorily. The general arrangement is as follows :— At least one piece of framing is so constructed as to form a door as represented at Fig. 1763, where it will be seen that a rebated frame is made to receive the door, which is also rebated to fit the frame as shown in the section (A and B. Fig. 1764). This frame is

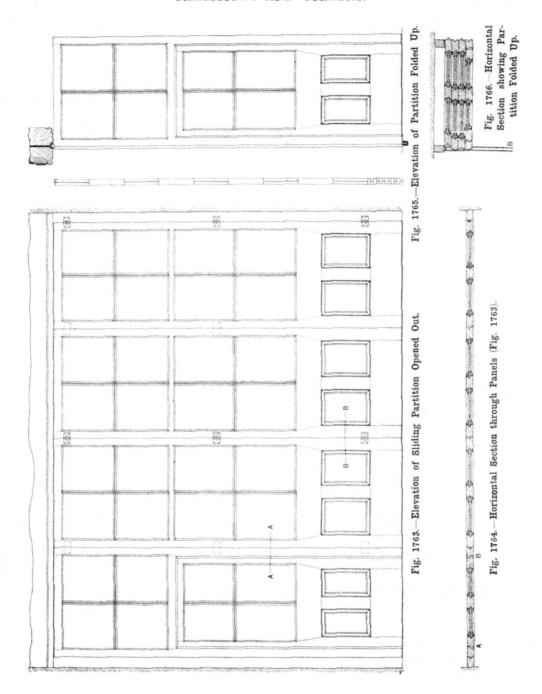

Fig. 1765.—Elevation of Partition Folded Up.

Fig. 1766.—Horizontal Section showing Partition Folded Up.

Fig. 1763.—Elevation of Sliding Partition Opened Out.

Fig. 1764.—Horizontal Section through Panels (Fig. 1763).

held together at the head and transom in the usual way, by mortice-and-tenon joints. The other pieces of the framing are of the ordinary door-like construction, with diminished stiles having wood panels below and mouldings inserted ; the upper part of each frame being prepared for glass, having bars with movable beads which are fastened by screws. A wall post or stile is fixed to the

Fig. 1770.—Top Bracket Plate and Roller.

Fig. 1771.—Bottom Bracket Plate and Caster.

Fig. 1769.

Fig. 1767.

Fig. 1767.—Transverse Section through Channel-iron let into Floor, and End View of Caster and Bracket Plate.

Fig. 1769.—Section through Channel-iron let into Soffit of Beam, and End View of Bracket Plate and Roller.

wall at each end, and either the door end of the partitioning, or the opposite end, may be attached to the wall stile with large brass back flap hinges as shown. Each piece of framing is connected to that next to it also with

Fig. 1768.—Longitudinal Section through Channel-iron, and Side Elevation of Caster and Bracket Plate.

back flap hinges, but these hinges have to be fixed on each side alternately, as illustrated at Fig. 1763, where the hinges on the opposite side are indicated by dotted lines. Each piece of framing has one stile with a tongue formed on the solid, and the other side with a groove as shown at Figs. 1761 and 1762. To insure the tongues entering the grooves,

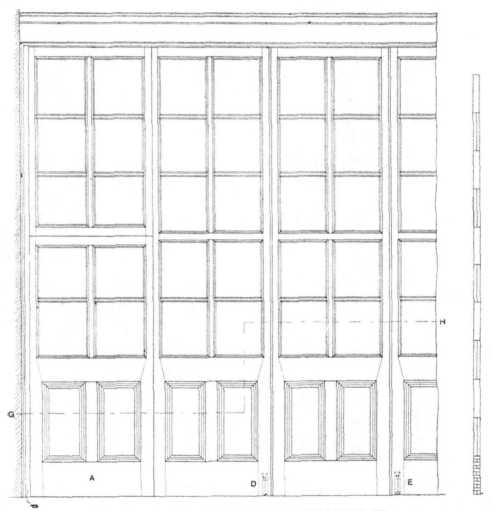

Fig. 1772.—Elevation of Part of Sliding and Folding Partition.

Fig. 1773.—Horizontal Section on Line G H (Fig. 1772).

the former are made tapering in section, as shown by the enlarged detail, Fig. 1762.

At Fig. 1761, a plan is given showing the partitioning partly closed. Figs. 1765 and

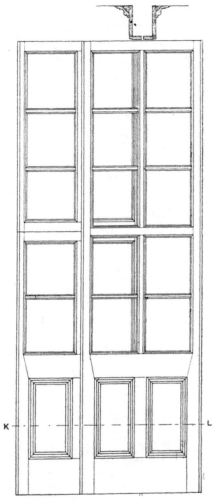

Fig. 1774.—Elevation of Door and Pieces of Framing when Folded Back to Wall.

Fig. 1775.—Horizontal Section through K L (Fig. 1774).

1766 are respectively elevation and plan. Let into the floor is an iron channel which runs the whole length of the partition. A portion of this is shown in the plan (A, Fig. 1761; and B, Fig. 1766). An enlarged transverse section of the channel is given at Fig. 1767, and at Fig. 1769 a portion of the longitudinal section through the channel-iron is given. The views will make clear the forms of bracket plates to which the roller casters are attached. The bracket plates are let into the stile and bottom edges of each piece of framing as shown. The body of the caster can revolve on pivot P (Fig. 1768), which passes through the bottom of the bracket plate. A conventional view of one of the bracket plates and casters is given at Fig. 1771. At Fig. 1769 is given a cross section through the channel-iron let into a flitch beam to which it is fixed by screws. Of course in the case of an iron girder the channel-iron would be bolted to it. An end elevation of the bracket iron and roller is also shown at Fig. 1769. Fig. 1770 is a conventional view of the top bracket plate and roller complete.

Sliding and Folding Partition.

A sliding and folding partition of modern construction, more particularly as regards the mechanism (at the head and feet of the framing) to promote ease and speed of movement, is illustrated by Figs. 1772 to 1777. This form of sliding partition has been largely adopted in schools, where, during the time of ordinary instruction, it divides a large room or hall into classrooms, and can be quickly folded up when the larger space is required for mass assemblage. The principal points to note in this example are as follows:—Where a doorway is provided as shown at A (Fig. 1772), the door, with a piece of glazed framing immediately above it, is often hung to a wall stile as shown at B (Figs. 1772 and 1773). The usual arrangement admits of the door and the glazed frame above it folding back against a wall as shown in elevation and section at Figs. 1774 and 1775, where it will be noticed that the door and framing above project about half their width beyond the other pieces of framing. The other parts of the partition are formed of pieces of doorlike framework, which are centrally hung at the top to a carriage and wheel arrangement, the wheels or pulleys having hollow rims, and running on parallel rails, which are supported by malleable-iron boxlike hangers, fixed to the flange of a rolled-iron girder by means of bolts. These hangers are fixed at intervals which vary according to the weight of the partitioning to be supported. The bolt which passes through the carriage has, at the lower end, a flange which is let in and bolted to the top rail of the framing as shown in section at Fig. 1776. The upper end of the bolt passes through the carriage, and has a collar on it so as completely to support the piece of framing, and at the same time allow the latter to be turned round as desired. Then immediately above this collar is a cog pinion, which works in a toothed rack, as shown in section at Fig. 1776. The bottom of the framing is also supplied with a plate, which is let into the bottom rail (having a projecting stud on which a pinion can revolve), and is in contact with a rack; while a boxlike channel similar to that represented at Fig. 1777 is let into the floor as shown. The object of the rack and pinion is to produce an equal movement at the top and bottom, without binding, which would sometimes occur if the pieces of framing get out of the vertical. Where a door is provided, as here illustrated, the pieces of framing are connected by groove-and-tongue joints. One advantage of this centre-hung system is that, by connecting one piece of framing to the adjacent pieces by rebated joints, it can be used as a door as indicated at c (Fig. 1773), thus allowing two spaces a little less than half the width of a piece of the framing, which is sufficient for persons to pass through. When this method is adopted, it is usual to provide the two adjacent pieces with a flush bolt, as indicated by the dotted lines at D and E (Fig. 1772). The piece of swing framing is provided with a brass mortice latch, with flush drop handles. There are many other systems of constructing sliding and folding partitions, many of them protected by letters patent, but the examples that have been given suffice to explain the general principles involved.

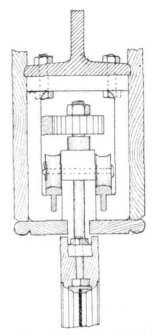

Fig. 1776.—Enlarged Transverse Section showing Mechanism for Hanging Framing.

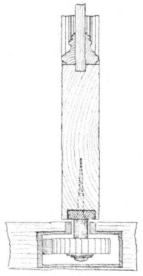

Fig. 1777.—Transverse Section through Box Channel, showing Mechanism at Bottom of Framing.

BEVELS : FINDING AND SETTING OUT.

Introduction.—Writers on this subject usually assume a considerable amount of geometrical knowledge on the part of the reader, and give "short-cut" methods without explaining the geometrical principles that are involved. The principles and methods employed in each particular case may possibly have been adopted after much geometrical study and experimenting on the part of the writer, who, however, is apt to forget that the lack of such preliminary grounding may place his readers at a serious disadvantage. A student, following the methods thus superficially described, may obtain correct results for similar cases, with little or no geometrical reasoning on his part. Hence, when he is called upon to deal with more complicated cases, he finds himself in difficulties, as he is unable to adapt and apply principles that he has never really mastered ; because they have only been described to him, not explained or expounded. Everyone who desires to become proficient in setting out for bevels, so as to be able to deal promptly, decisively, and effectively with the various cases that arise in practice, must first possess a clear idea of the principles involved, and of their application to varying circumstances. In other words, it is necessary for him to study a few of the fundamental principles of solid and descriptive geometry. This preparation is best obtained in a class under a competent teacher ; but where class-work is out of the question, the necessary knowledge may be obtained from a good text-book on geometry. Such study does not involve any very great expenditure of time ; and the principles, once mastered, are not easily forgotten. It is beyond the scope of the present work to deal fully with the subject of solid and descriptive geometry ; but several geometrical problems, forming

the foundation of some of the more general methods of setting out bevels, will be here explained and illustrated. The direct application of geometrical principles to typical

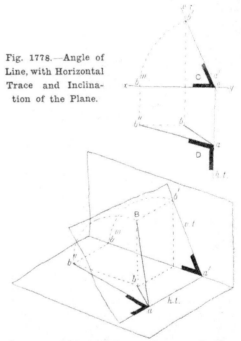

Fig. 1778.—Angle of Line, with Horizontal Trace and Inclination of the Plane.

Fig. 1779.—Pictorial View of Geometrical Working of Fig. 1778.

examples of joinery will afford a useful combination of theory and practice. Bevels, and some of the geometrical principles involved in setting out as applied to roofing, are illustrated and explained in the section on Roofs, etc., beginning at p. 167, and it will be found an advantage to study, in connection with the present section, the descriptions and illustrations there given.

Bevels for circle-on-circle work, etc., have also been shown, and the methods of setting them out geometrically explained, in connection with the special subject to which they belong.

Geometrical Problems Practically Applied.—Fig. 1778 shows a very useful geometrical problem—namely, " Given an inclined plane and the plan of a line in that plane, find the angle the line makes with the horizontal trace." Briefly, the working is : At right angles to the *h.t.* (horizontal trace), draw

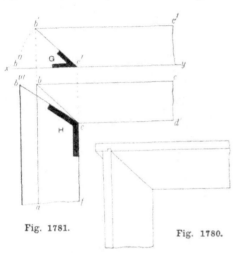

Fig. 1781.

Fig. 1780.

Fig. 1780.—Plan of One Corner of Linings or Trough.

Fig. 1781.—Method of Obtaining Bevels for Case shown at Fig. 1780.

x y intersecting it in *o*. Draw the *v.t.* (vertical trace) at the given angle as shown by the bevel c. Let *a b* be the plan of the line, then *a' b'* will be its elevation. With *o* as centre and *b'* as radius, draw the arc *b'''*. Projecting from this, parallel to *h.t.*, and projecting out from *b* parallel to *x y*, point *b''* is obtained. Joining this to *a* gives the angle the line makes with *h.t.* as shown by the bevel D. Evidently the oblique plane with the line has been rotated into the horizontal plane. The complete working is shown pictorially at Fig. 1779.

Bevels for Trough or Linings.—Now apply the foregoing problem to obtaining

the bevels for the face sides of splay linings and troughs, etc. A corner of such an example is shown in plan at Fig. 1780. Consider the inner surfaces as geometrical inclined planes, and draw a portion of these surfaces in plan *a b c d e* and *f* (Fig. 1781). Produce *f e* to *e'*. Consider it the inner surface of the inclined plane *a b e f*. Through *e'* draw *x y* at right angles to *f e'*, and set up the angle of the inclined plane and

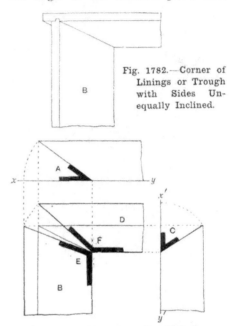

Fig. 1782.—Corner of Linings or Trough with Sides Unequally Inclined.

Fig. 1783.—Geometrical Construction for Obtaining Bevels for Fig. 1782.

height of the trough as shown by *b' c'*. Then the bevel shown at G will be that required for the edge of the stuff that is to fit the edge of the frame or bottom of the trough. Now consider *b e* and *b' e'* as the plan and elevation of the given line on the inclined plane. Using *e'* as centre, *b'* as radius, point *b'''* is obtained as previously explained. Joining this to *e* gives us the bevel for application to the side of the stuff for the oblique cut of the end as shown by the bevel H. At Fig. 1782 is shown in plan an object in which the sides are unequally inclined ; and at Fig. 1783 is shown the geometrical working. Two elevations are given, namely, one at $x\,y$ and the second at $x'\,y'$, each $x\,y$ being at right angles to a horizontal trace. The bevel at A is for application to the edge of the piece of stuff shown in plan at B, the bevel at c being for that of the piece of stuff at D, and the bevel E F for application to the surface of the stuff.

Bevels for Hexagonal Hopper.—Fig. 1784 shows the part plan of a hexagonal hopper

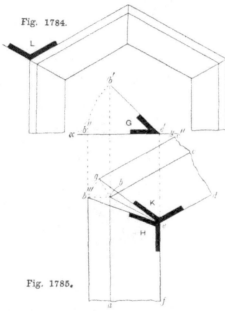

Fig. 1784.

Fig. 1785.

Fig. 1784.—Plan of Part of Hexagonal Object with Inclined Sides and Mitered Angles.

Fig. 1785.—Geometrical Construction to Obtain Bevels for Fig. 1784.

or similar article, the sides of which are inclined and mitered together as shown, the edges being in parallel planes. The method of obtaining the bevels about to be explained is also applicable to special mitered linings, inclined mitered fascia boards, and similar work. Set out the plan of two adjacent surfaces as shown at Fig. 1785; then, proceeding as before, take the representation of the bottom arris of one shown by $e\,f$. Produce it to any point e', and make it the horizontal trace of an inclined plane. Draw $x\,y$, and,

proceeding as previously described, obtain the bevel H If the sides are equally inclined as shown, the bevel at K will be the same as that at H. The bevel for the edge of the stuff is indicated by G. If the edges of the stuff are bevelled so as to be in parallel planes, the bevel to be applied to the edge of the stuff for mitering will be as that shown at L (Fig. 1784).

Bevels for Edges Square to Inclined Surfaces.—Taking examples similar to those already dealt with, but assuming that it is necessary to obtain the bevel for the

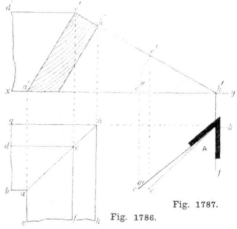

Fig. 1787.

Fig. 1786.

Fig. 1786.—Mitered Angles, and Top Edges Square.

Fig. 1787.—Obtaining Bevel for Square Edge of Fig. 1786.

mitre, which has to be applied to an edge that is at right angles to the surface, the proceedings will be as explained below. Figs. 1786 and 1787 show the part plan and sectional elevation of one corner of an object mitered at the angles as shown. Through a' draw $x\,y$, produce $e'\,h'$ (Fig. 1786), which gives $e'\,h'$ (Fig. 1787), the vertical trace. At right angles to $x\,y$, of course parallel to $a\,c$ (Fig. 1786), draw $e\,h$ parallel to $a\,e\,h$ (Fig. 1786), project up to the vertical trace, giving point e'. With h' as centre, and e' as radius, draw the arc $e'\,e''$. Projecting down and horizontally from e gives point e'''. Joining this to h gives the bevel required for the application to the edge for mitering the stuff. It is now obvious that the edge of the stuff forms an inclined plane, and the mitre $e\,h$ the plan and elevation of a line in that plane. By rotating the inclined plane and line into the horizontal plane as previously explained,

reference to Fig. 1791, which shows pictorially the principle of working, the construction will be plainly evident.

Bevels for Mitered Angles of Triangular Hopper.—At Fig. 1792 are shown the plan and elevation of a triangular hopper. Let it be required to find the angle between the two surfaces, and the bevel for mitering the

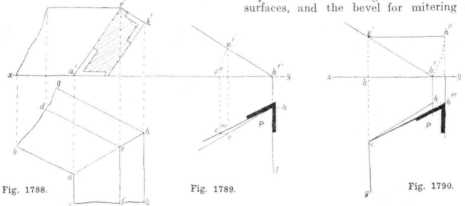

| Fig. 1788. | Fig. 1789. | Fig. 1790. |

Fig. 1788.—Sides meeting at an Obtuse Angle and having Square Edges.
Figs. 1789 and 1790.—Alternative Methods for Obtaining Bevel for Mitre on Square Edge.

the bevel is ascertained. Fig. 1788 shows the part plan and part sectional elevation of a corner of an object of which the sides form an obtuse angle and are mitered together. The line of the bottom arris $a\,c$ being produced is considered as the horizontal trace of a plane. Then, at any convenient position, draw a line $l\,h'$ parallel to $c\,a'$, and work the problem as shown at Fig. 1789 ; the working being identical with that explained in connection with Figs. 1786 and 1787. Sometimes it is very convenient to imagine the inclined plane as being brought into a horizontal position by rotating it about a level line above the horizontal plane. It should be carefully noted that this line should be taken parallel to the line originally fixed on as the horizontal trace. The method of working is shown at Fig. 1790, where it will be seen that $e'\,h'$ is the inclination of the stuff (and the vertical trace), $h\,h'$ the horizontal trace, $q\,k$ the level line of rotation. Then imagine the line and plane to move into the position shown by elevation $e'\,h''$. Projecting down from h'' and horizontally from h gives point h''' and the bevel required as shown at P. By

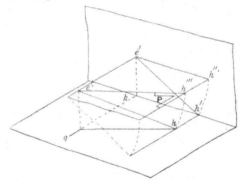

Fig. 1791.—Pictorial View of Geometrical Working shown at Fig. 1790.

edges of the stuff. Draw $a\,b$ (Fig. 1793) parallel to the intersection $a\,b$ (Fig. 1792). Then at Fig. 1793 complete the outline plan of the corner by drawing $a\,d$, $a\,e$. Draw $x\,y$ parallel to $a\,b$. Project up from a, and mark off a' from $x\,y$ equal to d' from $x\,y$ (Fig. 1792). Then draw the outline of the top edge as shown by $a'\,c'$ (Fig. 1793). From any convenient point in $a'\,b'$ set out a line at right angles meeting $a'\,c'$ in c'. Now imagine this line $c'\,f$ to represent a

24*

plane at right angles to $a'b'$; projecting down from c', it will be clear that this plane will cut the top arrises of the surfaces as shown in plan at d and e, and the portion of

Now draw b' b parallel to the lines in the plan, as shown, and c b parallel to $a'b'$; then join b to a. It should be noted that this work is identical with that explained

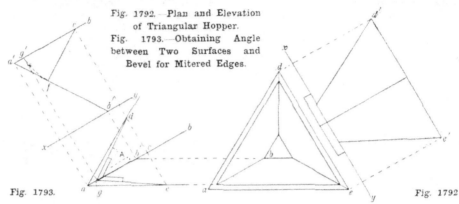

Fig. 1792.—Plan and Elevation of Triangular Hopper.
Fig. 1793.—Obtaining Angle between Two Surfaces and Bevel for Mitered Edges.

Fig. 1793.   Fig. 1792

the plane fitting between the two surfaces from d and e will be triangular in shape. An edge view of this triangular portion of plane is represented by $c'f$. Now imagine this triangular plane rotated about the line $d e$ until it is in a horizontal position. The apex shown at f in elevation would move to the point g', and thus projecting down from g' to the intersection line $a b$ obtains point g; then joining g to d and to e gives the angle between the surfaces as shown by the bevel. Half the angle as indicated by the bevel dotted at A will give the bevel for application to the mitered edge. The geometrical problem here introduced is thus stated: "Given two inclined planes and their intersection, determine the angle between them." This is known as the dihedral angle.

Setting Out Mitre Lines on Mouldings.

Assume that two pieces of cornice moulding are to be joined at right angles; that is, an angle of 90°. Let the section of the moulding be as shown in Fig. 1794. Draw the plan of the mouldings and mitre as at Fig. 1795. Then set a bevel to the mitre line c d. This will be the bevel to apply to the top edge, as indicated by the line c d (Fig. 1796). For the bevel for the sloping back, through the angle at A' (Fig. 1794) draw A' B'. With A' as centre and c' as radius, draw the arc c' B'.

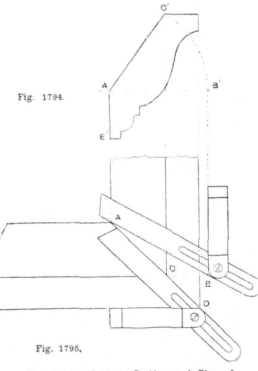

Fig. 1794.

Fig. 1795.

Figs. 1794 and 1795.—Section and Plan of Portion of Mitered Cornice, and Geometrical Construction for Mitre.

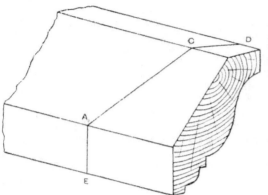

Fig. 1796.—Mitered Lines Drawn on Moulding.

angle. If there are several mitres to be made, and all meet at the same angle, a simpler plan is to construct a mitre box which will hold the moulding to the exact angle, as shown at Fig. 1797, and the mitres can be cut in the usual manner.

Setting Out Mitre Lines.

When setting out a mitre block for mouldings meeting at right angles as shown at A (Fig. 1798), it is only necessary to draw a square on the top block as shown at A B C D (Fig. 1799), and then the diagonal A C is the

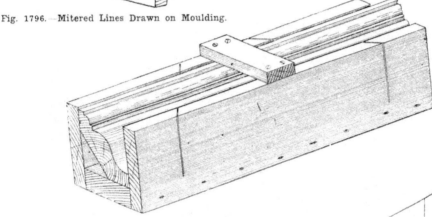

Fig. 1797.—Moulding in Mitre Box Ready for Cutting.

in connection with Fig. 1781. Set the bevel as indicated, and apply it to the sloping back of the moulding and mark it. This will give a line as indicated by

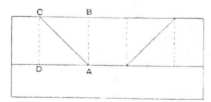

Fig. 1799.—Setting Out for the Cuts on a Square Mitre Block.

A C (Fig. 1796). As A′ E′ is a vertical sur-face, the line A E indicated at Fig. 1796 can be drawn square. This principle can be applied for mouldings meeting at any

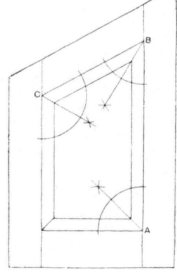

Fig. 1798.—Setting Out the Angles for Obtuse and Acute Angles of Panel Mouldings.

mitre line. When the mouldings meet at an obtuse or acute angle, as B or C (Fig. 1798), the better plan is to set out the mitre on a piece of board, as at Fig. 1800. Smooth up a board and shoot the edge, then gauge a line about $\frac{1}{2}$ in. (say) away from the edge and set out the required angle, as indicated at F H G ; now bisect this angle, and then H K is the mitre line. A bevel should now be set to the mitre line (see Fig. 1800), and then applied to the mitre block, as illustrated at Fig. 1801.

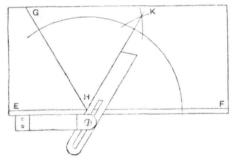

Fig. 1800.—Setting Out Angles for Mitres.

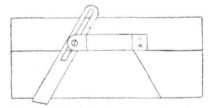

Fig. 1801.—Setting Out Block for Obtuse Mitres.

Mitre of Moulding on Oblique Corner.

To draw the elevation and plan of a moulding mitre round an oblique angle of a wall, and, inclined on both walls (Fig. 1802), draw a' e, the elevation of the corner of the wall. From a' draw a' d' at the slant of the moulding on one wall. Draw d' f at right angles to d' a'. On d' f draw the section of the moulding. Draw a d the plan of the face of wall, and b f parallel to it at a distance equal to the thickness of the moulding. From a draw a line at an angle equal to the

angle of the wall. Draw a line parallel to it, corresponding to b f, intersecting it at b. Join a b, which is the plan of the mitre line. Fig. 1802 shows the elevation and plan. The elevation of the point b' is, of course, found by projecting up from b to the line b' c', drawn through the section of the moulding parallel to a' d'. From a' draw a' c' at right angles to b' c' ; a' b' c' is the elevation of the right-angled triangle. b c, being parallel to the vertical plane, is seen in true length in the elevation. A c is equal to the width of the upper surface of the moulding, and is seen in the section ; with these lines the

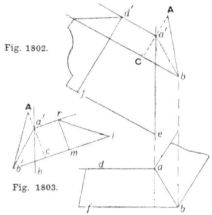

Fig. 1802.

Fig. 1803.

Fig. 1802.—Elevation and Plan of Moulding to an Oblique Angle.

Fig. 1803.—Elevation of Return Mould.

triangle can be drawn. The other bevel for the cut will obviously be the angle d' a' e. To obtain the bevels for the return piece, draw the elevation shown in Fig. 1803, where a' h is the corner of the wall, a' r the slope of the moulding. a' b' will be equal to a' b' (Fig. 1802), and can be drawn by projecting lines across. Draw b' l parallel to a' r. Draw the section of the moulding r m l, making m l equal to the thickness of the front moulding, and r m at right angles to a' r. From a' draw a' c at right angles to b l. Produce c a' to A, make c A equal to r l, join A b'. A b' c is the angle required.

INDEX.

(Illustrated subjects are denoted by asterisks.)

INDEX

Printed by
Cassell and Company, Limited, La Belle Sauvage,
London, E C

PRACTICAL VOLUMES

ON

Arts, Crafts, and Industries.

PAUL N. HASLUCK.
Photo: H. Moyse, Putney.

By PAUL N. HASLUCK,

Editor of "Work," "Building World," etc.

IN this Catalogue will be found a list of books for men with hobbies, for mechanics, and for all interested in handicrafts. The books are eminently practical and up-to-date, and are fully and instructively illustrated.

AUTOMOBILE, THE : A Practical Treatise on the Construction of Modern Motor Cars, Steam, Petrol, Electric, and Petrol Electric. 3 Vols., 12s. 6d. each. (To Subscribers only.) 1,275 pages, super royal 8vo (6½ × 9 in.). With 1,260 illustrations and 12 plates.

BAMBOO WORK. Cloth, 1s. net. 160 pages, f'cap 8vo (4¼ × 6¾ in.). With 177 engravings and diagrams.

BASKET WORK OF ALL KINDS. Cloth, 1s. net. 160 pages, f'cap 8vo (4¼ × 6¾ in.). With 189 engravings and diagrams.

BEEHIVES AND BEEKEEPERS' APPLIANCES. Cloth, 1s. net. 160 pages, f'cap 8vo (4¼ × 6¾ in.). With 155 engravings and diagrams.

BENT IRON WORK : Including Elementary Art Metal Work. Cloth, 1s. net. 160 pages, f'cap 8vo (4¼×6¾ in.). With 269 engravings and diagrams.

BOOKBINDING. Cloth, 1s. net. 160 pages, f'cap 8vo (4¼ × 6¾ in.). With 125 engravings and diagrams.

BOOK - KEEPING FOR BUILDERS, DOUBLE ENTRY. 6d. net. 96 pages, f'cap 8vo (4¼ × 6¾ in.).

BOOK - KEEPING FOR BUILDERS, SINGLE ENTRY. 6d. net. 96 pages, f'cap 8vo (4¼ × 6¾ in.).

BOOT AND SHOE PATTERN CUTTING AND CLICKING. Cloth, 2s. 160 pages, crown 8vo (5 × 7½ in.). With 235 illustrations.

BOOTMAKING AND MENDING : Including Repairing, Lasting, and Finishing. Cloth, 1s. net. 160 pages, f'cap 8vo (4¼×6¾ in.). With 179 engravings and diagrams.

BRICKS AND BRICKMAKING. 6d. net. 96 pages, f'cap 8vo (4¼×6¾ in.). With 90 illustrations.

BRICKWORK, PRACTICAL. Cloth, 2s. net. 160 pages, crown 8vo (5 × 7½ in.). With 368 illustrations.

BUILDERS' WORK IN ITS LEGAL ASPECTS. 6d. net. 96 pages, f'cap 8vo (4¼×6¾ in.).

BUILDING CONSTRUCTION. By Prof. HENRY ADAMS. 7s. 6d. net. 568 double-column pages, royal 8vo (6¼ × 8¾ in.). With 2,284 illustrations and 12 coloured plates.

CASSELL & COMPANY, LIMITED, *La Belle Sauvage, London, E.C.*

PRACT. VOLS., 8.06

Practical Volumes on Arts, Crafts, and Industries.

BUILDING STONES. 6d net 96 pages, f'cap 8vo (4¼ × 6¾ in) With 39 illustrations

CLAY MODELLING AND PLASTER CASTING. Cloth, 1s net 160 pages, f'cap 8vo (4¼ × 6¾ in) With 153 engravings and diagrams

COFFIN MAKING AND UNDERTAKING. 6d net 96 pages, f'cap 8vo (4¼ × 6¾ in) With 107 illustrations

CYCLE BUILDING AND REPAIRING. Cloth, 1s net 160 pages, f'cap 8vo (4¼ × 6¾ in) With 190 engravings and diagrams

CYCLOPÆDIA OF MECHANICS. 4 Vols, each containing 384 double-column pages, super royal 8vo (7 × 10¼ in) With about 1,200 illustrations 7s 6d each

DECORATIVE DESIGNS OF ALL AGES FOR ALL PURPOSES. Cloth, 1s net 160 pages, f'cap 8vo (4¼ × 6¾ in) With 277 engravings and diagrams

DRAUGHTSMEN'S WORK, PRACTICAL. Cloth, 2s net 160 pages, crown 8vo (5 × 7½ in) With 226 illustrations

DWELLINGS, CHEAP. 1s net 64 pages, crown 4to (10½ × 7½ in) Over 200 Plans, Sections, Elevations, Details, and Specifications of Dwellings actually built at costs ranging from £75 to £300 each and upwards

DYEING TEXTILES, APPLICATION OF COLOURING MATTERS TO. Cloth, 2s net 160 pages, crown 8vo (5 × 7½ in) With illustrations

DYNAMOS AND ELECTRIC MOTORS. Cloth, 1s net 160 pages, f'cap 8vo (4¼ × 6¾ in) With 142 engravings and diagrams

ELECTRIC BELLS, HOW TO MAKE AND FIT THEM. Cloth, 1s net 160 pages, f'cap 8vo (4¼ × 6¾ in) With 162 engravings and diagrams.

ELECTRO-PLATING. Cloth, 1s net. 160 pages, f'cap 8vo (4¼ × 6¾ in) With 77 engravings and diagrams

ENGRAVING METALS. Cloth, 1s. net. 160 pages, f'cap 8vo (4¼ × 6¾ ins). With 117 engravings and diagrams.

ESTIMATING FOR BUILDERS' WORK. 6d net 96 pages, f'cap 8vo (4¼ × 6¾ in) With 7 illustrations

GAS FITTING, PRACTICAL, INCLUDING GAS MANUFACTURE Cloth, 2s net 160 pages, crown 8vo (5 × 7½ in) With 120 illustrations

GLASS WORKING BY HEAT AND ABRASION. Cloth, 1s net 160 pages, f'cap 8vo (4¼ × 6¾ in) With 300 engravings and diagrams

GLASS WRITING, EMBOSSING, AND FASCIA WORK Cloth, 1s net 160 pages, f'cap 8vo (4¼ × 6¾ in) With 129 engravings and diagrams

GRAINING AND MARBLING, PRACTICAL. Cloth, 2s net 160 pages, crown 8vo (5 × 7½ in) With 76 illustrations

HANDRAILING, PRACTICAL. Cloth, 2s. net 160 pages, crown 8vo (5 × 7½ in) With 144 illustrations

HARNESS MAKING. Cloth, 1s net 160 pages, f'cap 8vo (4¼ × 6¾ in.) With 197 engravings and diagrams

HOISTING MACHINERY, BUILDERS' 6d net 96 pages, f'cap 8vo (4¼ × 6¾ in) With 83 illustrations

HOUSE DECORATION. Comprising Whitewashing, Paperhanging, Painting, etc Cloth, 1s net 160 pages, f'cap 8vo (4¼ × 6¾ in) With 79 engravings and diagrams

IRON: ITS SOURCES, PROPERTIES, AND MANUFACTURE. Cloth, 2s net 160 pages, crown 8vo (5 × 7½ ins) With illustrations

IRON, STEEL, AND FIREPROOF CONSTRUCTION. Cloth, 2s. net 160 pages, crown 8vo (5 × 7½ in) With 211 illustrations

LATHE CONSTRUCTION. 24 Coloured Plates (13 × 18 in), drawn to scale. 6s the set, or 4d each Plate.

CASSELL & COMPANY, Limited, *La Belle Sauvage, London, E C*

Practical Volumes on Arts, Crafts, and Industries.

LEATHER WORKING. Cloth, 1s net 160 pages, f'cap 8vo ($4\frac{1}{4} \times 6\frac{3}{4}$ in) With 152 engravings and diagrams

MEASURING BUILDERS' WORK. 6d net. 96 pages, f'cap 8vo ($4\frac{1}{4} \times 6\frac{3}{4}$ ins). With 24 illustrations

METAL PLATE WORK, PRACTICAL. Cloth, 2s each 160 pages, crown 8vo ($5 \times 7\frac{1}{2}$ in) With 247 illustrations

METALWORKING: A Book of Tools, Materials, and Processes for the Handyman 9s 760 double-column pages, royal 8vo ($6\frac{1}{4} \times 8\frac{3}{4}$ in) With 2,206 illustrations

MICROSCOPES AND ACCESSORIES. Cloth, 1s net 160 pages, f'cap 8vo ($4\frac{1}{4} \times 6\frac{3}{4}$ in) With 140 engravings and diagrams

MODEL BOATS, BUILDING. Cloth, 1s net 160 pages, f'cap 8vo ($4\frac{1}{4} \times 6\frac{3}{4}$ in). With 168 engravings and diagrams

MORDANTS, METHODS, AND MACHINERY USED IN DYEING Cloth, 2s net 160 pages, crown 8vo ($5 \times 7\frac{1}{2}$ ins) With illustrations

MOTOR CYCLE CONSTRUCTION. Cloth, 1s net 160 pages, f'cap 8vo ($4\frac{1}{4} \times 6\frac{3}{4}$ in) With engravings and diagrams.

OPTICAL LANTERNS AND ACCESSORIES. Cloth, 1s net 160 pages, f'cap 8vo ($4\frac{1}{4} \times 6\frac{3}{4}$ in) With 147 engravings and diagrams

PAINTERS' OILS, COLOURS, AND VARNISHES. Cloth, 2s net 160 pages, crown 8vo ($5 \times 7\frac{1}{2}$ in) With 51 illustrations

PAINTERS' WORK, PRACTICAL. Cloth, 2s net 160 pages, crown 8vo ($5 \times 7\frac{1}{2}$ in) With 45 illustrations.

PAPERHANGERS' WORK. 6d net 96 pages, f'cap 8vo ($4\frac{1}{4} \times 6\frac{3}{4}$ in) With 38 illustrations.

PATTERN MAKING, PRACTICAL. Cloth, 2s net 160 pages, crown 8vo ($5 \times 7\frac{1}{2}$ in) With 300 illustrations.

PHOTOGRAPHIC CAMERAS AND ACCESSORIES. Cloth, 1s net. 160 pages, f'cap 8vo ($4\frac{1}{4} \times 6\frac{3}{4}$ in). With 241 engravings and diagrams

PHOTOGRAPHIC CHEMISTRY. Cloth, 1s net 160 pages, f'cap 8vo ($4\frac{1}{4} \times 6\frac{3}{4}$ in) With 31 engravings and diagrams

PHOTOGRAPHY. Cloth, 1s net 160 pages, f'cap 8vo ($4\frac{1}{4} \times 6\frac{3}{4}$ in) With 76 engravings and illustrations

PHOTOGRAPHY, THE BOOK OF Practical, Theoretic, and Applied 10s 6d 744 double-column pages, royal 8vo ($6\frac{1}{4} \times 8\frac{3}{4}$ in) With 935 illustrations, and 48 full-page plates

PIANOS: THEIR CONSTRUCTION, TUNING, AND REPAIR. Cloth, 1s net 160 pages, f'cap 8vo ($4\frac{1}{4} \times 6\frac{3}{4}$ in) With 74 engravings and diagrams

PICTURES, MOUNTING AND FRAMING. Cloth, 1s net 160 pages, f'cap 8vo ($4\frac{1}{4} \times 6\frac{3}{4}$ in) With 240 engravings and diagrams

PLASTERERS' WORK. 6d. net 96 pages, f'cap 8vo ($4\frac{1}{4} \times 6\frac{3}{4}$ in) With 53 illustrations

PLUMBERS' WORK, PRACTICAL. Cloth, 2s. net 160 pages, crown 8vo ($5 \times 7\frac{1}{2}$ in) With 298 illustrations

QUANTITIES, BUILDERS'. 6d net 96 pages, f'cap 8vo ($4\frac{1}{4} \times 6\frac{3}{4}$ in) With 59 illustrations.

RANGE AND STOVE FIXING, AND OVEN BUILDING. 6d net 96 pages, f'cap 8vo ($4\frac{1}{4} \times 6\frac{3}{4}$ in) With 61 illustrations

ROAD AND FOOTPATH CONSTRUCTION. 6d net 96 pages, f'cap 8vo ($4\frac{1}{4} \times 6\frac{3}{4}$ in) With 38 illustrations

ROPES AND CORDAGE, KNOTTING AND SPLICING. Cloth, 1s net 160 pages, f'cap 8vo ($4\frac{1}{4} \times 6\frac{3}{4}$ in) With 208 engravings and diagrams

SADDLERY. Cloth, 1s net 160 pages, f'cap 8vo ($4\frac{1}{4} \times 6\frac{3}{4}$ in) With 99 engravings and diagrams

CASSELL & COMPANY, Limited, *La Belle Sauvage, London, E.C*

Practical Volumes on Arts, Crafts, and Industries.

SANITARY CONSTRUCTION IN BUILDING, Cloth, 2s. net. 160 pages, crown 8vo (5 × 7½ in.). With illustrations.

SANITARY CONVENIENCES AND DRAINAGE. Cloth, 2s. net. 160 pages, crown 8vo (5 × 7½ in.). With illustrations.

SEWING MACHINES: THEIR CONSTRUCTION, ADJUSTMENT, AND REPAIR. Cloth, 1s. net. 160 pages, f'cap 8vo (4¼ × 6¾ in.). With 177 engravings and diagrams.

SIGNS, TICKETS, AND POSTERS, HOW TO WRITE. Cloth, 1s. net. 160 pages, f'cap 8vo (4¼ × 6¾ in.). With 170 engravings and diagrams.

SMITHS' WORK. Cloth, 1s. net. 160 pages, f'cap 8vo (4¼ × 6¾ in.). With 211 engravings and diagrams.

STAIRCASE JOINERY, PRACTICAL. Cloth, 2s. net. 160 pages, crown 8vo (5 × 7½ in.). With 215 illustrations.

TAILORING: HOW TO MAKE AND MEND TROUSERS, VESTS AND COATS. Cloth, 1s. net. 160 pages, f'cap 8vo (4¼ × 6¾ in.). With 184 engravings and diagrams.

TAXIDERMY: SKINNING, MOUNTING, AND STUFFING BIRDS, MAMMALS AND FISH. Cloth, 1s. net. 160 pages, f'cap 8vo (4¼ × 6¾ in.). With 108 engravings and diagrams.

TELESCOPE MAKING. Cloth, 1s. net. 160 pages, f'cap 8vo (4¼ × 6¾ in.). With 218 engravings and diagrams.

TERRA - COTTA WORK: MODELLING, MOULDING, AND FIRING. Cloth, 1s. net. 160 pages, f'cap 8vo (4¼ × 6¾ in.). With 245 engravings and diagrams.

TEXTILE FABRICS AND THEIR PREPARATION FOR DYEING. Cloth, 2s. net. 160 pages, crown 8vo (5 × 7½ in.). With 43 illustrations.

TIMBER. 6d. net. 96 pages, f'cap 8vo (4¼ × 6¾ in.). With 56 illustrations.

UPHOLSTERY. Cloth, 1s. net. 160 pages, f'cap 8vo (4¼ × 6¾ in.). With 162 engravings and diagrams.

VIOLINS AND OTHER STRING INSTRUMENTS. Cloth, 1s. net. 160 pages, f'cap 8vo (4¼ × 6¾ in.). With 177 engravings and diagrams.

WOOD FINISHING. Cloth, 1s. net. 128 pages, f'cap 8vo (4¼ × 6¾ in.). With 12 engravings and diagrams.

WOODWORKING: A BOOK OF TOOLS, MATERIALS, AND PROCESSES FOR THE HANDYMAN. 9s. 760 double-column pages, royal 8vo (6¼ × 8¾ ins.). With 2,545 illustrations.

Weekly Journals Edited by PAUL N. HASLUCK.

Weekly, 1d.; Monthly, 6d.

Weekly, 1d.; Monthly, 6d.

" Remarkably well written ; the illustrations are well executed ; all the items are first-class, and the only wonder is that such a paper can be given to the public for a penny. It is crammed full of information, and is fully illustrated."—*Sun.*

Half-yearly Volume, 4s. 6d.

" It is a curious reflection, but soundly true, that there is not a person of ordinary average intelligence and strength who could not learn from 'WORK' . . . how in a short time to make a living."—*Saturday Review.*

Half-yearly Volume, 4s. 6d.

CASSELL & COMPANY, LIMITED, *La Belle Sauvage*, London, E.C.

CPSIA information can be obtained
at www.ICGtesting.com
Printed in the USA
BVHW062054231019
561882BV00004B/62/P